Urban Forestry: Planning and Management

Urban Forestry: Planning and Management

Edited by Malcolm Fisher

SYRAWOOD PUBLISHING HOUSE
New York

Published by Syrawood Publishing House,
750 Third Avenue, 9th Floor,
New York, NY 10017, USA
www.syrawoodpublishinghouse.com

Urban Forestry: Planning and Management
Edited by Malcolm Fisher

International Standard Book Number: 978-1-68286-679-5 (Hardback)

Cataloging-in-Publication Data

Urban forestry : planning and management / edited by Malcolm Fisher.
p. cm.
Includes bibliographical references and index.
ISBN 978-1-68286-679-5
1. Urban forestry. 2. Forests and forestry. 3. Forest management. I. Fisher, Malcolm.
SB436 .U73 2019
635.977--dc23

Printed in China.

TABLE OF CONTENTS

PREFACE

Urban forestry is concerned with the growth, management and sustenance of varied species of trees in urban settings. Planning and management of individual trees and parks in cities and other urban spaces fall under this area of study. Such green spaces are crucial in metropolitan areas for a number of purposes including refining air quality, maintaining ecological balance, improvement of microclimate, etc. This book aims to present concepts and principles related to planning and management of urban forestry in a comprehensive manner while also bringing modern techniques and models into focus. Some of the diverse topics covered in this book address the varied branches that fall under this category. It studies, analyzes and upholds the pillars of urban forestry and its utmost significance in modern times. This book includes contributions of experts and scientists which will provide innovative insights into this field.

The information contained in this book is the result of intensive hard work done by researchers in this field. All due efforts have been made to make this book serve as a complete guiding source for students and researchers. The topics in this book have been comprehensively explained to help readers understand the growing trends in the field.

I would like to thank the entire group of writers who made sincere efforts in this book and my family who supported me in my efforts of working on this book. I take this opportunity to thank all those who have been a guiding force throughout my life.

Editor

Financial Feasibility of Ghg Reduction. The Case of Baturité Mountain – Brazil – 2013

Ana Milena Plata Fajardo* and Romano Timofeiczyk Junior

Federal University of Paraná, Brazil

Abstract

Forest carbon sequestration is a mechanism to remove greenhouse gases. Trees, through the process of photosynthesis, absorb carbon dioxide – CO_2 from the atmosphere and store it as biomass. The objective of this study was to quantify and assess the financial viability of the generation of carbon credits in the Baturité Mountain, an area of 7,000 ha of tropical montane sub humid forest in Ceará state of Brazil. GHG reduction was estimated by non-destructive methods (based on forest inventory estimates). Economic criteria used to evaluate the project were the Net Present Value (NPV) and Internal Rate of Return (IRR). The results showed that, as a result of project activitiesin the Baturité MountainGHG emissions were reduced annually by 903,120 Tones of CO_2. Based on prices and costs in 2013, forestry projects for carbon sequestrationarenon-viable if traded under the Clean Development Mechanism. The project is financially viable with returns under the New Zealand Emission Trading Scheme (IRR=28%) and in the Voluntary Carbon Standard (VCS, IRR=27%). This study provides methodological guidelines for economic evaluation of carbon capture projects. In the right environments and with the right financial incentives, keeping standing forests intact can provide benefits for the environment and higher economic benefits compared with other extractive uses.

Keywords: Carbon credit; Tropical montane subhumid; Transaction cost; Cash flow

Introduction

The Agriculture, Forestry and Other Land Use (AFOLU) sector is responsible for up to one third of global Greenhouses Gases (GHG) emissions [1]. Moreover, this sector is increasingly vulnerable to climate changes and hence requires adaptation measures to maintain food and fibre production and enhance future potential for carbon storage and sequestration. However, it is not immediately apparent which options deliver the most economically efficient reductions in GHG. In that regard, it is necessary identify low-cost options that generate significant co-benefits in the form of improved agricultural production systems, resilience, biodiversity conservation or other ecosystem services [2].

Under the Kyoto Protocol seven types of projects are considerate as a AFOLU: Afforestation, Reforestation and Revegetation (ARR), Agricultural Land Management (ALM), Improved Forest Management (IFM), Reduced Emissions from Deforestation and Degradation (REDD), Avoided Conversion of Grasslands and Shrublands (ACoGS), Wetlands Restoration and Conservation (WRC) [3].

The Baturité Mountain located in north-central Ceará State in Brazil, has diversified soils and it is covered by subtropical montane forests, semideciduous forest and lower altitude deciduous forest (BRASIL, 2004) [4]. Within this area an ARR project was implemented that was consistent with the requirements for CDM A/R (2011) project [5], with reforestation referring to activities that create a forest on land that was non forest for at least 10 years.

The VCS (2014) defines eligible ARR projects as "those that increase carbon sequestration and/or reduce GHG emissions by establishing, increasing or restoring vegetative cover (forest or non-forest) through the planting, sowing or human-assisted natural regeneration of woody vegetation.

These projects aim to create a financial value for the carbon stored in forests, offering incentives for developing countries to reduce net emissions from forested lands and invest in low-carbon paths to sustainable development. However, while there is a wide range of technical solutions, project developers must consider the overall financial feasibility of the project opportunity at each specific site. Financial viability is a critical element of project success and net discounted carbon credit revenues (taking into account transaction costs) must be sufficient to cover implementation and/or opportunity costs.

Measuring and estimating forest carbon stocks and changes in carbon stocks is an important first step in implementing a carbon sequestration project. This requires transparent and verifiable methods, and appropriate monitoring systems for carbon stocks. The second step is to evaluate the financial feasibility of the project, the study use two decision criteria from Cost – Benefits Analysis: Net Present Value (NPV) and Internal Rate of Return (IRR). These are not a requirement of the Standards for implementing CDM or other voluntary GHG emission reduction programs, but it is relevant to provide measures that help project developers assess the financial viability of a project. The objective of this study was to quantify the potential increased stock of CO_2 that might be achieved through ARR project and to evaluate the economic performance of the project under three scenarios for the sale of carbon credits.

Methods

Area

The study area encompassed the rural land of the APA of Baturité,

***Corresponding authors:** Fajardo AMP, Ph.D, Federal University of Paraná, Brazil, E-mail: aplatafa@gmail.com

a Brazilian region consisting of eight municipalities in the Ceará State (Figure 1). The predominant soils classes in the APA of Baturité are Oxisol and Utisol. According to Lima [6], the Oxisol soils in the APA of Baturité are usually deep or very deep soils (averaging over 1.50 m), varying with relief.

The Baturité Mountain represents an area of great biological importance and is priority for biodiversity conservation in Brazil [4]. In the region, the Baturité Mountain constitutes a geographic exception in comparison with surroundings emiarid neighbouring municipalities. It has high precipitation (annual average 1500 mm) and constitutes one of the most important sources of water for the metropolitan region of Fortaleza [7].

Such climatic and geomorphological characteristics has enabled the evolution of a complex vegetation cover, with the general attributes of tropical rainforest that can be divided into two large units of land cover: Tropical montane subhumid, which occupies the highest altitudes, and Tropical montane dry, which predominates in areas between 400 and 600 m altitude [8].

Forest inventory

Forest inventory is the systematic collection of data and forest information for assessment or analysis. The aim of the statistical forest inventory is to provide comprehensive information about the state and dynamics of forests for strategic and management planning. Accurate estimates of carbon in forests are crucial for forest carbon management, carbon credit trading, [9].

Traditional forest inventories provide information on stand volumes, but not on biomass or carbon stock [10]. Thus, the available volume estimates had to be converted in to biomass and carbon budget estimates. Data from these inventories can be converted to biomassand therefore to the carbon in one of two ways, depending upon the level of detail reported [11].

The forest inventory of the area was divided into four strata, with differing forest characteristics [7], (Table 1). The assessed sites ranged in altitude from 600 to 940 masl. The forests were similar in average height and ranged in density from 735 to 1741 stems/ha, in basal area from 14.1 to 31.2 m^2/ha and from 10.5 to 13.4 cm in average DBH.

Biomass quantification

Forest biomass can be estimated using direct and indirect methods [12]. Direct methods involve cutting a sample of trees and their components and drying and weighing each component. Indirect methods can use published allometric relationships to convert individual tree measurements from forest inventory data, to aboveground biomass or satellite images and stand level estimates of biomass, usually integrated in a geographic information system (GIS).

The net GHG reductions associated with a reforestation project can be estimated using either, or a combination of these methods. To assess project viability, proponents need to provide estimates of the values of future sequestration that are not available before the start of the Crediting Period and commencement of monitoring activities. Project proponents are generally required to adopt a conservative approach in making these future estimates.

Direct estimations of biomass based on measurements of forest

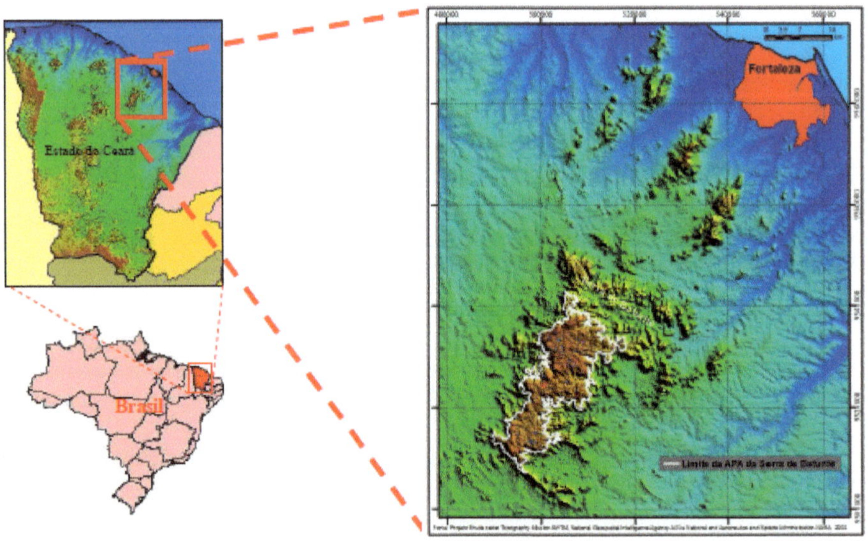

Source: Rodrigues (2011)

Figure 1: Study Area. Baturite Mountain.

Forest sites	Elevation (MASL)	Sample plot area (ha)	Density (stems/ha)	Basal Area m^3/ha	Average DBH (cm)	Average Height (m)
Arvoredo	935	0.66	1215	14,1	10,5	9
Lagoa	940	0.46	1741	31,2	12.9	8.3
Sinimbu	695	0.5	1597	25,5	11.8	9
Taveiras	600	1.08	735	15,5	13.4	8

Source: Oliveira (2005)

Table 1: Structural Parameters of Forest at Serra de Baturité.

inventory by destructive harvesting are verifiable and more efficient than indirect estimates [13]. On the other hand, develop allometric relationships are time and cost consuming because they requires the destructive harvesting of a large number of trees. Stratifying forest types or ecological zones and using generalized allometric relationships is more feasible in the tropics because the diameter breast height explains over 95% of the variation in carbon stock above ground of the rainforest even in very different regions (Brown 2002).

Remote sensing techniques have been widely used in forestry studies, since it is possible to estimate biophysical parameters such as biomass, carbon and timber volume, based on the spectral properties of the components of the vegetation [14]. Imagery from satellites such as Lands at has been used to estimate forest carbon stocks, by developing statistical relationships between ground measurements and vegetation indices observed by satellite [15]. However, the method was not used for this study because the tropical montane sub humid forest is highly diverse in physiognomy and composition, complicating the determination of biomass.

Thus, for this study biomass estimates were based on forest inventories. Immediately after determining the parameters, the equation of Brown et al. [16] was used as the most robust (R2=0.97). This equation was also used by Watzlawick [17] and Fernandes et al. [18] for estimating biomass in the Atlantic Forest:

$$Y = exp\left[-3{,}1141 + 0{,}9719 * \ln(dbh^2 * h)\right] \quad (1)$$

Where

Y: Biomass

dbh=Diameter Breast Height

h=Height

Carbon stocks were estimated assuming that carbon makes up 50% of the dry biomass [18]. To convert carbon in to CO_2, the mass of carbon was multiplied by the ratio of the molecular weight of carbon dioxide to the atomic weight of carbon (44/12 or 3.67). GHG emissions and removals in the project will result from re growth of the residual forest and the effects of silvi cultural interventions such fencing, tree clearing, site preparation (mowing, slash burning, soil cultivation or drainage), tree planting (tillage, planting, fertilizer application) and tending (for example, pest and weed control). These were incorporated into baseline estimates.

Baseline

In broad terms, the baseline for a CDM or other carbon sequestration project activity is the scenario that reasonably represents the anthropogenic changes in carbon stocks in pools and emissions of GHGs by sources that would occur in the absence of the proposed project activity. A baseline shall cover both significant carbon stock changes in all relevant pools and significant emissions by sources of all GHGs that would occur within the project boundary [19].

In order to estimate the carbon stock component, the VCS AFOLU documents mandate the use of the 2006 IPCC Guidelines for National Greenhouse Gas Inventories and this was recommended for the project design document [20]. However there is no one accepted approach for determining the baseline scenario. Different GHG programs and different GHG accounting standards specify various means for determining the baseline scenario in a standardized and impartial manner. In general, though, there are two basic kinds of baseline determination approaches, the project-specific approach and the performance standard. The performance standard is usually developed by a GHG sequestration or reduction program to simplify the accounting requirements for project proponents and to reduce the potential for subjective selection of a baseline scenario. A performance standard typically establishes a set rate of baseline emissions by analyzing the emission rates from all the potential baseline candidates [3].

A number of GHG programs have adopted a form of performance standard for reforestation projects. Under these programs, once the eligibility requirements have been met, a zero baseline is adopted. A zero baseline is adopted because in most cases the carbon stocks on non-forest land areas are assumed to not change, meaning that there are zero GHG emissions or removals from these lands.

The project specific approach for determination of the baseline scenario is more commonly applied by a GHG emissions reduction programs. These can range from simply assuming continuation of the existing land use to undertaking detailed evaluation of a range of land use alternatives. The project – specific approach assesses what would have occurred in the absence of the project, based on the specific economic and other circumstances of the individual project.

Quantifying emission and removals in tropical sub humid forests in the baturité mountain

The estimation of net GHG removals through reforestation involved stratification of the project area (Table 1) and calculation of mean annual increase in aboveground biomass carbon stock, converted to carbon dioxide equivalent.

Cost benefit analysis

Costs: Total production costs of the project include transaction costs, project implementation costs (for year 1), project maintenance costs (after the 2 year), verification and certification costs and opportunity costs [21]. Verification and certification of sequestered carbon are required to prove to investors that the estimated levels of carbon have been sequestered and can be traded. Costs were determined from project literature: Baalman and White [3], Conservation International Peru [22] and were consistent with the recommendations of Cifloreslas Brasil [2].

Transaction costs: Dudek and Wiener [23] divided transaction costs of emissions projects into six categories: (1) search (2) negotiation, (3) approval, (4) monitoring, (5) enforcement, and (6) insurance. For this study, the transaction costs of carbon projects were grouped into 2 categories based on the general AIJ guidelines/CDM project cycle model. This incorporates the categories defined by Dudek and Wiener, along with an additional two cost groupings; design and validation and registration and certification of the Emission Reductions (tCERs).

Design and validation costs: These include the cost of monitoring techniques and verification protocols, baseline and project scenario measurements and feasibility studies. The Project Design Document (PDD) is the central component in the design and validation cost. The procedures including in the PDD are: the baseline methodology and assessment of additionally, the selection of crediting period, a monitoring plan, the assessing environmental impacts and the stakeholder comments [24].

According to Social Carbon [25], the cost of developing a PDD (Estimate of baseline and with-project deforestation rates and Measurement of carbon stocks) is in average US 70.000 for the Latin

American context, and the cost of validation the PDD is in average US \$20.000.

Registration and certification cost: The registry provides credibility to sell tCERs. The registry allows project developers to initiate the project registration process, upload documents and request issuance. The majority of VCS projects are registry in APX, therefore it is used as a proxy of the cost. To 2014 the cost of registry in APX is US \$0.05 by tCERs. The certification is the total fees charged by standards body (VCS, NZ ETS or CDM etc.). The cost of certification used in the study is US \$ 0.03 by tCERs.

Implementation costs: These are incurred in the first year. CI Florestas [26] recommended consideration of the cost of land acquisition and the cost of activities to reduce deforestation: Site assessment and planning, fencing, tree clearing: mowing, slash burning, Soil flooding or drainage, Tree planting: tillage, planting, Fertilizer application, fighting cutting ants. Operation cost: Personnel, Buildings, vehicles and equipment.

Maintenance cost: After the second year, the standing forest requires the Maintenance activities to reduce deforestation, the operating cost and the monitoring (Figure 2).

Revenues

A number of international standard and guidance documents exist that have built on the guiding rules of the Kyoto Protocol but that are designed to be scheme neutral and can be applied to any GHG reduction program. Numerous programs that include land base project activities have been developed especially in Australia, the United States, Canada and the New Zealand. There are many similarities between these different GHG programs, but they vary in the price of tCERs sold in the program.

Three different prices were used for analysis based on the payment for forest management CERs at the end of 2013: the Clean Development Mechanism (CDM, at a price of \$ 1.1); the New Zealand Emission Trading Scheme (NZ ETS, with a value of \$ 7.90),and the Voluntary Carbon Standard (previously called the Verified Carbon Standard, US \$ 7.6).

The VCS market is the largest market in transaction volume of carbon credits [27]. After the category of renewable energy, forest management was the category with the highest number of projects transacted in the year 2013, with 29 million tCO_2. The New Zealand market was chosen because the forestry sector was the first category to be created (since 2008), due to the great importance of the forestry sector has the obligations to reduce emissions of greenhouse gases [28]. The CDM was the first institution to be created after the Kyoto Protocol, and to be regulated internationally.

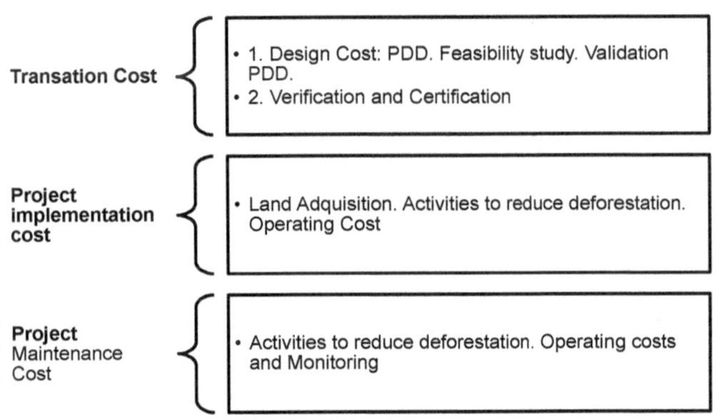

Figure 2: Disaggregation of total cost.

The carbon market does not have a specific form of payment and various contract structures are allowed Peters [27]. The most recognized form of payment is the Spot Transaction. This type of trading is performed after verification of the claims. Therefore, the temporary Certified Emission Reductions (tCERs) may expire at the end of the commitment period and must be verified every five years. After verification, a tCER can either be re-issued (if the sequestered carbon remains intact). This means that the designer of carbon project should ensure that the benefits realized by the project are sustainable with allowances for risks in the event of fire, drought and change in government policy to ensure the sustainability of the project [27].

For the case study, it was considered the sale of ex ante credits which means that after PDD auditing, the contract is signed and the project is allowed to start. Each 5 years the Emission Reduction are verified and the investor get the revenues. The project expires at the end of the registered credit period (30 years for sustainable forest management) [29].

Cost Benefit Analysis is used to support decisions about whether to proceed with a particular project or investment or not [30] using the presentation of benefits and costs in a common framework. In this study, net present value (NPV) and relative profitability (Internal Rate of Return, IRR) were used to assess different projects.

Cash flow is the combined estimates of costs and revenues resources over time. The difference between costs and revenues is the net cash flow [31]. This can also be used to compare different projects. In this study we prepared a cash flow containing the key costs and revenue. The planning horizon of the project was 30 years, the minimum time for the proposed Forest Management activity to maintain sequestration for under the recommendations of the IPCC [13].

Forestry projects involve long term costs and benefits, often incurred and realized over decades. To bring these into a standardized accounting framework, a discount rate is used to discount future costs and revenues. The discount rate traditionally used in forestry projects typically varies between 6-12% per year [6]. In this study a real discount rate of 10% per year was used, as recommended by the Center for Integrated Studies on Environment and Climate Change of the Ministry of Environment in the document published in MMA [32], which deals with the Proposal Revised Eligibility Criteria and Indicators for the Evaluation of Projects in the CDM market in Brazil.

An alternative way to present the information is in terms of annuities. An annuity is simply a constant annual value which, when discounted and summed, produces the net present value.

Internal rate of return

The net present value rule requires the use of some predetermined social discount rate rule to discount future benefits and costs (formula 3). An alternative rule is to calculate the discount rate which would give the project a *NPV* of zero and then to compare this "solution rate" with the pre-determined social discount rate. In other words, the benefit and cost streams are presented in the next equation form:

$$NPV = \sum_{j=0}^{n} Rj(1+i)^{-j} - \sum_{j=0}^{n} Cj(1+i)^{-j} \quad (1)$$

$$PVA = \frac{VPL*i}{\left[1-(1+i)^{-n}\right]} \quad (2)$$

$$\sum_{j=0}^{n} R_j(1+IRR)^{-j} = \sum_{j=0}^{n} C_j(1+IRR)^{-j} \quad (3)$$

Where R is the revenues, C the cost, j the planned horizon of the project and *i* is the rate of discount that solves the equation. Once i is determined, the rule for accept-reject and for ranking of options is to adopt any project which has an internal rate of return (IRR) in excess of the predetermined social discount rate. As with the NPV rule, then, it remains essential to choose some acceptable discount rate. The idea of a "rate of return" is a relatively familiar concept to decision-makers and, as a result, that the IRR is a more readily understandable summary statistic of a cost-benefit appraisal than the NPV [30].

Results and Discussion

The baseline net GHG removals by sinks were assumed to be zero. As regards the required planting area, it was estimated a project totalizing 7,000 hectares. The project area is assumed to be at any of the 7 municipalities of the Baturité Mountain, at the minimum level of 900 MASL.

The final estimate of biomass for the primary rain forest was 50.07 $t.ha^{-1}$ (Table 2), giving an average stock of carbon of 23 tons. ha^{-1}. Consequently, the average value of the CO_2 removed from the atmosphere was 92,13 $t.ha^{-1}$ for the first year.

The stock of accumulated CO_2 is not so high compared with other estimates in the Atlantic Rainforest biome. Saldarriaga et al. [33] in a study on the successional development in the Amazon region, found that bole biomass in four stands of mature forest was 286 $tCO_2\ ha^{-1}$. According to Lacerda et al, [34], Tropical Rain Forest Ibate -. São Paulo, reaches an average carbon sequestration and 96.15 tCO_2, in spite should remember that the study area is an exception in the middle of the semiarid landscape Northeast of Brazil

The total tCERs, considered the entire area is 27.093.603 tCO_2, and the average annual emission reduction corresponds to 903,120 tCO_2. In order to compare the estimation, was checked other projects developed in Brazil. The results show that the Florestal Santa Maria Project achieved ER of 997,444 tCO_2 [35] and The Cikel Brazilian Amazon REDD APD Project avoiding Planned Deforestation in 370000 tCO_2 [36]. It puts the Baturite mountain project in a middle position compared to the above two projects.

Cash flow

The simplified cash flow of activities related to CDM scenario, NZ ETS and VCS are presented in Table 3. The three scenarios presented differ in revenue. CMD market represents revenues of US $781,198.5 on the first verification (5 years of the project), derived from the price US $ 1,1 and quantity carbon accumulated in the rain forest (**710180** tCO_2). The NZ ETS market for being the highest in terms of price, represents a revenue of US $ 5,610,426 on the first verification. The VCS market with closing price of US $ 7,6*tCO_2, has revenues of US $ 6,050,091, position edit selfat an intermediate position compared to the other two scenarios evaluated. According to calculations, the project needs in its implementation year US$ 2,391,000, and the mean cost of maintenance for the following years is U$ 25,000 (Table 3).

Evaluating projects

The criteria for evaluation of projects (NPV, IRR) used in the economic analysis of the three scenarios, under a discount rate of 10% and a period of 30 years, showed that the NZ ETS and VCS scenarios are economically viable. The CDM project proved economically unviable as both the implementation (1 year) and in maintenance, had negative cash flow and hence the NPV and IRR criteria were unviable.

The IRR of the projects in the GHG program VCS and NZ ETS were higher than the discount rate, which makes them viable projects. Comparing these results with those of other forestry projects, the NZ ETS and OTC project to rain forest is greater than other forestry projects (Table 3) Net return.

The VCS scenario is feasible and is positioned at the intermediate level of profitability compared with the other two scenarios; moreover the VCS market is safe and has historically been the largest market in relation to the volume of transactions of emissions. The VCS presented a high growth in the year 2012. The growth of voluntary market is attributed to the explosion in the share of REDD market due to formal recognition in the international arena and approval of methodologies with few requirements for the projects [37].

Year	Biomass	Carbon stock	$tCO_2\ h^{-1}$	tCO_2. By the Project (7000 ha)
1	50,07	25,03513	92,12928	644904,9
2	51,32	25,6575	94,41962	660937,3
3	52,57	26,2873	96,73726	677160,8
4	53,85	26,9245	99,08218	693575,2
5	**55,14**	**27,56912**	**101,4544**	**710180,5**
6	56,44	28,22114	103,8538	726976,4
7	57,76	28,88055	106,2804	743962,9
8	59,09	29,54736	108,7343	761139,9
9	60,44	30,22155	111,2153	778507,1
10	**61,81**	**30,90313**	**113,7235**	**796064,5**
11	63,18	31,59208	116,2589	813812
12	64,58	32,28841	118,8214	831749,5
13	65,98	32,99211	121,411	849876,7
14	67,41	33,70317	124,0277	868193,7
15	**68,84**	**34,42159**	**126,6715**	**886700,2**
16	70,29	35,14737	129,3423	905396,2
17	71,76	35,88049	132,0402	924281,5
18	73,24	36,62097	134,7652	943356,1
19	74,74	37,36878	137,5171	962619,9
20	**76,25**	**38,12394**	**140,2961**	**982072,6**
21	77,77	38,88642	143,102	1001714
22	79,31	39,65624	145,935	1021545
23	80,87	40,43338	148,7948	1041564
24	82,44	41,21784	151,6817	1061772
25	**84,02**	**42,00962**	**154,5954**	**1082168**
26	85,62	42,80871	157,5361	1102752
27	87,23	43,61511	160,5036	1123525
28	88,86	44,42882	163,4981	1144486
29	90,50	45,24983	166,5194	1165636
30	92,16	46,07813	169,5675	1186973
Total estimated ERs				27.093.603 tCO_2
Total number of crediting years				**30**
Average annual ERs				903,120 tCO_2

Source: Autor.

Table 2: Biomass, Capture Carbon and Capture Dioxide Carbon average value.

Cost Year	0	1	2	3	4	5
Transaction Cost						
PDD	70000					
Registration						35509
Certification						21305
Validation	20000					
Land Acquisition	2,200,000					
Implementation Cost						
Activities to reduce deforestation and Operating cost.	100000					
Maintenance cost						
Activities to reduce reforestation and Operating Cost		20,000	20,000	20,000	20,000	**20,000**
Monitoring						10,000
Total	2,390,000	20,000	20,000	20,000	20,000	30,000
Revenues						
CDM						781,198
NZ ETZ						5,610,426
VCS						6,050,091
Profit						
CDM	-2,390,000	-20,000	-20,000	-20,000	-20,000	704384
NZ ETZ	-2,390,000	-20,000	-20,000	-20,000	-20,000	5533611
OTC	-2,390,000	-20,000	-20,000	-20,000	-20,000	5973276
Financial Results						
Discount Rate	0,1		0,1		0,1	
Total period crediting	30		30		30	
Net present Value	-$1,184,195		$6,588,178		$10,532,595	
Internal Rate of Return	4%		27%		28%	

Source: Adapted from Conservation International, CI Floresta (2011) and Baalman and White (2012)

Table 3: Cash Flow in US$ CDM, NZ ETS and VCS GHG program Cash flow in the three scenarios evaluated.

The market for CDM presented infeasibility attributed to the low market price (US$ 1.1). The sharp drop in market prices in Europe, in recent years, influenced the demand for carbon credits, causing a reduction in volumes sold and stagnation in the flow of transactions. While some politicians and nations are afraid to reverse carbon credits, several companies are voluntarily internalizing the cost of reducing emissions of their economic activities. As in the regulated market is affected by the low prices, the voluntary market continues to grow.

Conclusions

Latin America traditional offset supply countries a run for their volume, seeing 19 MtCO_2e transacted from the region's projects. Through its 8 MtCO2e transaction with KfW, Brazil's Acre state – along with sizable transactions from a few REDD+ projects in other locales – pushed Brazil over the top as the market's most popular project location in 2013 [27].

The APA Sierra Baturité represents a great potential as a carbon sink, sequestering in average 903,120 tCO_2 ha^{-1}. As possibilities to contribute to the mitigation of climate change through forestry projects are profitable. According to market prices and costs for 2013, it is concluded that the forestry project for carbon sequestration is unfeasible in Clean Development Mechanism and feasible in voluntary markets (NZ ETS and VCS).

The study presented only considers the GHG reduction of above ground biomass. However GHG standards are demanding project developers that include below ground biomass, deadwood and soil organic carbon.

Transaction costs vary according to the organization that does the validation and verification of carbon credits and tend to decrease according to the size and scope of the project. This fact prevents the individual participation of small farmers in projects for the generation of carbon credits similar to the scenarios presented in this study features.

The study presented can serve as a starting point to evaluate other projects that reduce emissions of greenhouse gases. However, Trigeorges [38] notes that under uncertainty, the future value of a variable is characterized by a distribution of variability. The recommendation for future work is to model a sensitivity analysis that considers the possibilities of risks and uncertainties.

References

1. Tubiello FN, Salvatore M, Rossi S, Ferrara A, Fitton N et al. (2013) The Faostat database of greenhouse gas emissions from agriculture. Environmental Research Letters 8: 1

2. Smith P, Martino D, Cai Z, Gwary D, Janzen HH et al. (2008) Greenhouse gas mitigation in agriculture. Philosophical Transactions of the Royal Society, B. 363: 789-813.

3. Baalman, White P (2014) Training Course. 302 GHG Accounting for Forest and Other Land Use Projects.

4. Brazil (2004) Provides for tax incentives to be granted to individuals and companies investing in Clean Development Mechanism - CDM that generate Certified Emission Reductions - CERs, authorizing the establishment of investment funds in CDM projects and other measures.

5. CDM A/R (2011) AR-ACM0003 Large-scale Consolidated Methodology: Afforestation and reforestation of lands except wetlands.

6. Lima Júnior VB, Rezende JLP, Oliveira AD (1997) Determination of discount

rate to be used in the economic analysis of forestry projects. Cerne Magazine 3: 45-66.

7. Oliveira TS (2005) History of human impacts and geoambientales aspects. In: Diversity and Conservation of Biota in Serra de Baturité, Ceara.
8. MMA (2002) Workshop Atlantic Forest of Northeast - Ministry of Environment, Water Resources and Legal Amazon.
9. UNFCCC (2005) CGE GHG Inventory Handbook (NAI) Land-use change and forestry sector.
10. FAO (2000) Forest Resources of Europe, CIS, North America, Australia, Japan and New Zealand (industrialized temperate/boreal countries), UN-ECE/FAO Contribution to the Global Forest Resources Assessment 2000, Main Report. Geneva Timber and Forest Study Papers 17: 445.
11. Brown S (2002) Measuring carbon in forests, current status and future challenges Environ. Pollut. 116: 363-372.
12. Salati E (1994) Issue x CO_2 sequestration - a new business opportunity for Brazil. In : Seminar x emission sequestration of CO_2 - a new business opportunity for Brazil, Rio de Janeiro.anal CVRD 15-37.
13. IPCC (2006) Guidelines for National Greenhouse Gas Inventories. Prepared by the National Greenhouse Gas Inventories Programme ed Eggleston HS, Buendia L, Miwa K, Ngara T and Tanabe K (Japan: Institute For Global Environmental Strategies)
14. Souza CL, Ponzoni FJ (1998) Vegetation indices for evaluating and bands TM / Landsat for timber volume estimation implanted forest of Pinus spp. In: Brazilian Symposium of Remote Sensing, 9, Santos. Proceedings. Santos : INPE.
15. Foody GM, Boyd DS, Cutler MEJ (2003) Predictive relations of tropical forest biomass from Landset TM data and their transferability between regions. Remote Sens Environ. 85: 463-74.
16. Brown S, Gillespie AJR, Lugo AE (1989) Biomass estimation methods for tropical forests with applications to forest inventory data. Forest Science 35: 881-902.
17. Watzlawick LF (2003) Estimation of biomass and carbon in Mixed Rain Forest and forest plantations from the satellite IKONOS II image data. 120. Thesis (PhD in Forest Sciences) - Sector of Agricultural Sciences, Federal University of Paraná, Curitiba.
18. Fernandes A, Salis S, Crispim S (2008) Carbon stocks of the tree layer of Cerrado in wetland Nhecolândia. Technical release 68. December: Corumba, MS.
19. Cifor (2011) Standards and methods available for estimating project-level REDD+ carbon benefits: Reference guide for project developers.
20. Kempff N (2011) Conservation planning In NKPDD.
21. Grace LR, Rodigheri HR, Conto AJ de (2000) Forest production costs concept and application. Columbus: 32.
22. Conservation International Peru (2011) Alto Mayo Conservation Iniciative. Project Description, Verified Carbon Standard.
23. Dudek DJ, Wiener JB (1996) Joint Implementation, Transaction costs, and climate change', Organization for Economic Cooperation and Development, OECD/GD 96: 173, Paris.
24. Undp (2014) The clean development mechanism: a user's guide. chapter 3: Developing the Project Design Document.
25. Social Carbon (2014) REDD Financial Feasibility Assessment Tool.
26. Florestas CI (2011) Sistemas Agroflorestais.
27. Peters S (2014) Turning over a New Leaf State of the Forest Carbon Markets A Report by Forest Trends' Ecosystem Marketplace & Bloomberg New Energy Finance.
28. Govt (2013) In: Forestry in the Emissions Trading Scheme.
29. Singh G (2009) Understanding Carbon Credits. Aditya Books, Ltda. 466pp
30. Pearce D, Atkinson G, Mourato S (2006) Cost-Benefit Analysis and the Environment. Recent Developments.
31. Santos E (2001) Financial Management of Small and Medium Enterprise. São Paulo: Atlas.
32. MMA (2002) Workshop Atlantic Forest of Northeast - Ministry of Environment, Water Resources and Legal Amazon.
33. Saldarriaga Jg, West DC, Tharp ML, Uhl C (1988) Long-term chronosequence of forest succession in the upper Rio Negro of Colombia and Venezuela. J. Ecol 76: 938-958.
34. Lacerda JSDe, Couto, Hilton TDZ, Hirota MM, Pasishnyk N, et al. (2009) Estimation of biomass and carbon in restored areas by planting native species. METRVM, USP- Esalq 1-23.
35. Plant Environmental Intelligence (2012) Florestal Santa Maria Project Last accessed: November 2014.
36. CKBV Florestal Ltda (2012) Cikel Brazilian Amazon REDD APD Project - Avoiding Planned Deforestation Last Accessed: September 2014.
37. Instituto Carbono (2013) Carbon market reveals uncertainties.
38. Trigeorgis L (1996) Real options: managerial flexibility and strategy in resource allocation. Cambridge, MA: MIT Press.

2

Farmers' Perception and Adoption of Agroforestry Practices in Osun State, Nigeria

Adedayo A. G[1] and Sobola Oluronke[2]

[1]*Department of Forestry and Wood Technology, Federal University of Technology Akure Ondo state, Nigeria*
[2]*Department of Forestry and Wildlife Management, Federal University Wukari, Taraba State, Nigeria*

Abstract

The study assessed Farmers' perception and adoption of agroforestry practices in Osun state, Nigeria. Multistage sampling technique was used for the study. Three local governments were randomly selected from each of the three senatorial districts in the state. Three farming communities were then randomly selected in each LGA why 10 respondents were purposively selected in each sampled community. Primary data were collected through semi- structured questionnaire to 270 respondents. In addition information was obtained from a key informant fin each sampled community with the use of interview schedule. The results of the study showed that there were divergent perceptions about agroforestry practice in the study area. 10% of the respondents in Osun west senatorial district were of the opinion that agroforestry practice is a scientific process that is difficult to practice, 62% perceived that it can improve farm productivity while 12% opined that the practice is not properly understood. Chi-square test ($p<0.05$) shows that there is significant association between respondents educational qualification and the adoption of agroforestry practices in all the senatorial districts of Osun state. Chi-square test ($p< 0.05$) also shows that there is a significant association between respondents; land ownership and the adoption of Agroforestry practices in all the senatorial districts of the study area. The adopted agro-forestry practices in the study area include retention of trees on farmland, planting of trees on boundaries, shifting cultivation and home gardening. . Challenges faced by farmers in the practice of agroforestry practice among others are; lack of technical knowhow, small land holdings, fire outbreak and poor extension service. More farmers can be encouraged to practice agroforestry through improved agroforestry extension service to the farmers, encouraging farmers to form cooperative groups, organizing training programmes for farmers in agroforestry and free distribution of tree seedlings to farmers.

Keywords: Farmers' perception; Adoption; Agroforestry practices; Home gardening; Nigeria

Introduction

A wealth of literature confirms that agroforestry practices may contribute to the welfare and livelihoods of farmers. Through the integration of trees on farm and in the landscape, agroforestry(AF), defined as a dynamic, ecologically based natural resource management system, diversifies and sustains production for increased social, economic and environmental benefits [1]. AF has thus been recognized as one of the strategies to introduce indigenous and exotic trees into cropping systems and impact on livelihoods of small-holder farmers in Sub-Saharan Africa [2].

Several studies have perceived AF practices in different ways; Lundgren and Raintree [3] viewed it as the set of land-use practices which involves the combination of trees, agricultural crops and/or animals on the same land management unit. Although cultivating trees in combination with crops and livestock is considered an ancient practice, factors such as the deteriorating economic situation in many parts of the developing world ;increased tropical deforestation; degradation and scarcity of land because of population pressures; and growing interest in farming systems, intercropping and the environment [4] have contributed to a rising interest in agroforestry since the 1970s. Most research on agroforestry has been conducted from the biophysical perspective, but socio-economic aspects in relation to perception of farmers are gaining attention [5].

In relation to the perceived benefit of agroforestry practices, its potential to uplift the socioeconomic conditions of the farmers has been identified. According to Maren and Carolyn [6], agroforestry enhanced the socioeconomic livelihood of rural farmers by enhancing income earning potentials and overall food and nutritional security as well as provision of fuel wood, fodder for animal consumption and employment. Earlier, Kandji et al. [7] pointed out that agroforestry systems improves the microclimate which in turn improves the adaptive capacity of land owners to climate change. Thus, Kandji et al. [7] suggested that the presence of trees in agricultural croplands can provide agricultural producers with an additional source of income that helps to strengthen their socioeconomic resilience. Integrating trees into systems where they can be planted close to each other and pruned or browsed intensively can help increase economic benefits. The farm-grown fodder increases milk production and can substitute for relatively expensive purchased dairy meal, thus increasing farmers' income [8].

Main agroforestry practices include improved fallows, taungya farming, home gardens, alley cropping, growing multipurpose trees and shrubs on farmland as well as boundary planting, [4,9]. Studies in several parts of Africa, including Nigeria have demonstrated the economic and agronomic returns of these practices [8]. Structures to retard the process of deforestation and environmental degradation have been adopted reluctantly and, even when adopted, the management and maintenance have been less than desirable. Thus, despite the

***Corresponding author:** Adedayo A.G, Department of Forestry and Wood Technology, Federal University of Technology Akure Ondo state, Nigeria; E-mail: gideonadedayo@yahoo.com

considerable progress in agroforestry research and dissemination, awareness and attitude of land users practicing AF systems is of utmost value before any recommendations should be made [1]. Faced with this situation, many researchers have called for the assessment of agroforestry systems (e.g. Franzel et al. [8]) from the perception of the farmers.

Given the potential of AF practices to improve agricultural land use systems and provide lasting benefits and alleviating adverse environmental effects at local and global levels, this study seeks to assess how farmers perceive Agroforestry practices in Osun state; ascertain the level of adoption of agroforestry practices by farmers in the study area; identify the various Agroforestry practices adopted by farmer in the study area identify Agroforestry tree species commonly planted by farmers in the study area; identify problems faced by local farmers, that influence the adoption of Agroforestry practices in the study area

Methodology

The study area

The study area is Osun state, Nigeria. The state was carved out of the old Oyo state on the 27th August 1991. It is located between longitudes 040301 E and 4051 E and latitude 70301 N and 70501 N, South-western Nigeria. It covers an area of approximately 14,875 square kilometers. The area is mainly agrarian, and it is also divided into three federal senatorial districts, each of which is composed of two administrative zones. The 1991 census puts the population of the state at 2.2million [10]. The State is made up of 30 local government areas with over 200 towns, villages and other settlements. The state has a considerable number of highly urbanized settlements some of which are Osogbo, Ile-Ife, Ipetumodu, Ilesa, Ikirun, Iwo, Ede, Ila-Orangun and Ikire. Others include, Ejigbo, Ilobu, Gbongan, Okuku, Inisa, Ijebu-Ijesa, Ipetu-Ijesha etc. The people of the state are mainly traders, artisans and farmers. Their other occupations include hand-woven textiles, tie and dye, leather work, calabash carving and mat-weaving. Osun State is bounded in the West by Oyo State, Ondo and Ekiti States in the East, Kwara State in the North and Ogun in the South. The State runs an agrarian economy with a vast majority of the populace taking to farming [10].

Method of data collection

Three local government areas were randomly selected from each of the 3 senatorial districts in the study area (Figure 1). Three farming communities were then selected in each LGA using random sampling technique to make a total of 27. In each of the twenty seven sampled farming communities, 10 respondents were purposively selected and semi-structured questionnaire was administered on them. A total number of 270 farmers were sampled to make up the total respondents of the study. The selected local government areas and the sampled communities are presented in Table 1.

Method of data analysis

The data collected was analyzed using descriptive statistics that include the use of frequency distribution and percentages Tables and bar charts. Chi- square test was used to test for the nature of association between respondent's age, family size, farm size, land ownership, farming experience, level of education and adoption of agroforestry practices in the study area.

Hypotheses tested

The following hypotheses were tested in the study.

- Ho: Age of farmers has no significant association with the adoption of Agroforestry practices in the study area
- Ho: Adoption of agroforestry practices is independent of the farm size of farmers in the study area
- Ho: Educational qualification of farmers has no significant association with the Perception/adoption of agroforestry practices in the study area
- Ho: Farmers land ownership status has no significant association with the adoption of agroforestry practices in the study area.
- Ho: Farmers farming experience has no significant impact on the adoption of agroforestry practices in the study area

Results and Discussion

Farmers perception of agroforestry practices in the study area

Table 2 shows that in Osun west senatorial district, 10% of the respondents were of the opinion that agroforestry practice is a scientific practice that is difficult to practice, 62% perceived that it can improve farm productivity, 12% opined that the practice is not properly understood while 16% were of the perception that agroforestry practice is a common practice among the local farmers. In Osun central senatorial district the highest percentage, 56% of respondents viewed agroforestry as a practice that can improve productivity, 19% opined that it is a common practice among local farmers, 13% were of the opinion that agroforestry is a scientific practice that is difficult to practice, while 11% perceived agroforestry practice as a practice that is not properly understood.

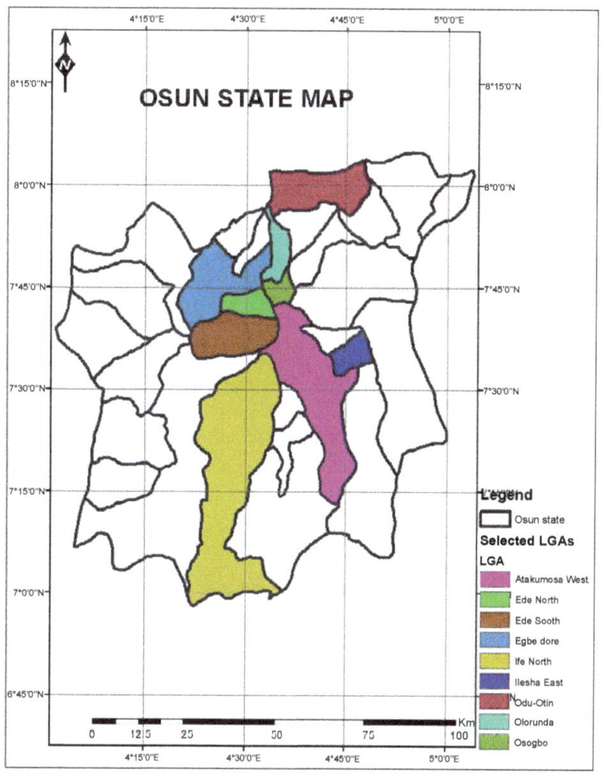

Figure 1: The map of Osun state showing the sampled local government area

Senetorial District	Local Government	Communities
Osun East	Ife North	Ipoye
		Olobo
		Oyere-aborishade
	Atakumosa West	Iloba
		Ila-Ijesa
		Itagunmodi
	Ilesha East	Ijayiregbe
		Isotun
		Ibala
OsunWest	Egbedore	Iragberi
		Iwoye
		Elemela
	Ede North	Elemo
		Eleeduogangan
		Abeere
	Ede South	Olodan
		Araromi
		Elewure
OsunCentral	Olorunda	Dagbolu
		Alabameta
		Oba-ile
	Odootin	Oyan
		Ekusa
		Ataka
	Oshogbo	Boredun
		Onigboyi
		Owode

Source: Field Survey, 2013

Table 1: The name of the communities selected in each of the senatorial district

From the foregoing, it therefore follows that there is a high degree of variation on the perception of the relevance of agroforestry technology to local farming situation in the study area. This agrees with Oladele and Fawole [11] who noted that farmers' perception of new technologies varies greatly in South- western Nigeria. Majority of the respondents in the study area are aware of the positive impact of agroforestry practices. The respondents were aware of the economic and environmental benefits of agroforestry practices and had favourable attitude towards those practices. It may be due to the fact that a significant proportion of the respondents were literate in the study area. Most of the respondents know that agro-forestry practices increases productivity although difficult to practice.

During focus group discussion agroforestry was viewed by the respondents to have a potential of solving their fuel wood need problems, improve the soil fertility and provide favorable climate for crops. Farmers had indigenous knowledge on the usage of trees in farming systems as well as some environmental uses, such as soil erosion control, provision of shade as well as climate moderation.

The responses are indication that planting of trees on their farmland will increase the overall output of the farmland. This benefit is expected to increase the overall household gains. Regmi [12] asserted that maintenance of mixed trees on the farmland made households resilient to cope with uncertainty and risks. At the time of insect or disease outbreak on one species they can meet their need from other species. Thus this practice has reduced the chances of complete crop failure. Similarly, Nepaune et al. [13] noted that agroforestry practices have the potential to maintain/improve surrounding condition of the forest and save time on collecting fodder and firewood from the forest. The saved time opened up avenues for other farming activities. A considerable number of the respondents pointed out that it is a common practice among local farmers as the practice of retaining trees on farm land has been a usual practice from time immemorial. It is noteworthy that these respondents have experienced improved productivity and saw increased role of farm trees to meet their need of livelihood. However they noted that it is quite difficult to practice, while some of the respondents also noted that the practice of agroforestry is not properly understood, this is an indication of lack of knowledge, which may be due to the problem of no access to agroforestry extension workers which has limited their information about agroforestry practices and as such limited their knowledge and perception about agroforestry innovations. This is in line with the work of Keilet al. [14] who considered information and knowledge about a given technology, as key to adoption of such technology.

Farmers' adoption of agroforestry practice in the study area

Table 3 shows farmers adoption of agroforestry practice in the study area. In Osun west senatorial district, 76% of respondents were involved in Agroforestry practice, while 24% affirm that they have never practice agroforestry. In Osun east, 80% of the respondents are involved in agroforestry practice while 20%have not adopted the practice. In Osun central, 72% of the respondents were involved compare to 28% which indicated that they had never practice agroforestry at any period of their farming experience. This means a higher percentage of farmers are involved in the practice of agroforestry in the study area. This implies that there is probably more awareness about the potentials of agroforestry practices in increasing productivity per unit area among the farmers in the study area. Akinbile et al. [15] noted that agroforestry has a way of instituting sustainable agricultural development in Nigeria. This is because agroforestry has the ability to combat the various environmental problems with the purpose of assisting farmers to maintain the fertility of their soils, ensure diversification of crop, wood and timber species per unit area and to stabilize, improve and conserve farmers' environment. The level of adoption of agroforestry by the framers therefore becomes very important since it determines the success of agroforestry practices. On the average an appreciable number of farmers (76%) indicated their adoption of Agroforestry while 24% indicated they did not adopt the practice. The main reason for the high level of adoption was because of multiple benefits the farmers gain from the crop-tree combination and also because agroforestry has been an age- long practice among the local farmers not only in the study area but all over the country.

Chi-square test ($p>0.05$) shows there is no significant association between respondents' age and the adoption of agroforestry practices in Osun West senatorial district while in Osun Central and Osun east senatorial districts ($p<0.05$) there is a significant association between respondents age and the adoption of agroforestry practices (Chi-square test of hypothesis 1). The reason for this might not be unconnected with the fact that there is no clear cut distinction in the ages of those that are involved in agroforestry practice in Osun West senatorial district. The different categories of ages of respondents are involved in agroforestry practice. However in Osun Central and Osun East senatorial districts more of the farmers that are involved in agroforestry practices are above 40 years of age.

Chi- square test ($p>0.05$) also shows that there is no significant association between respondents farm size and the adoption of agroforestry practices in the 3 senatorial districts of the study area (Chi-square test of hypothesis 2). This means farm size is not an important determinant in the adoption of agroforestry practices

Farmers perception	Osun West		Osun central		Osun East		Total	
	Freq	%	Freq	%	Freq	%	Freq	%
As a scientific practice that is difficult to practice	9	10	12	13	7	7	28	10
As a practice that can improve farm productivity	56	62	50	56	56	62	162	60
As a practice not properly understood	11	12	11	12	7	8	29	11
As a common practice among local farmers	14	16	17	19	20	22	51	19
Total	90	100	90	100	90	100	90	100

Source: Field Survey, 2013

Table 2: Farmers perception of agroforestry practice in osun state

Adoption	Yes	%	No	%	Total
Osun west	68	76	22	24	90
Osun east	72	80	18	20	90
Osun central	65	72	25	28	90

Source: Field Survey, 2013

Table 3: Farmers adoption of agroforestry practice

among local famers in the study area. Other factors like education and land ownership play more important role in determining the adoption of agroforestry practices among the farmers in the study area.

As such chi-square test (p<0.05) shows that there is significant association between respondents educational qualification and the adoption of agroforestry practices in all the senatorial districts of Osun state (Chi-square test of hypothesis 3). This shows that education is a determining factor in the adoption of agroforestry practice. When farmers are educated they have better access to information and innovations which help farmers to quickly adopt new technology.

In the same vein, chi-square test (p<0.05) shows that there is a significant association between respondents; land ownership and the adoption of Agroforestry practices in all the senatorial districts of the study area (chi-square test of hypothesis 4). This also shows that land ownership plays important role in adopting agroforestry since it involves trees planting. Adedayo (2004) showed that land ownership plays a significant role in the adoption of alley cropping among local farmers in Akure Local Government area of Ondo State, Nigeria. He noted further that tenant farmers are not usually allowed to plant trees as such they cannot adopt agroforestry practice because it involves tree planting.

Types of agroforestry practices adopted by farmers in the study area

The adopted agro-forestry practices in the study area include retention of trees on farmland, planting of trees on boundaries, scattered trees on farmland, shifting cultivation, taungya farming and home gardening. However, the level of adoption of each of the stated practices differs across the senatorial districts in the study area (Table 4). The results of the study revealed that, in Osun west senatorial district, planting of trees on boundary was the most adopted agro forestry practices (23%). 16% of the respondents in the same senatorial district adopted improved fallow, scattered trees on farm land and home garden.

In Osun central senatorial district, improved fallow, tree planting on boundary and home garden, carries the highest percentage. Basically 22% of the respondents were involved in improved fallow. This practice is followed by adoption of tree planting on boundary (20%) and home garden (18%). 13% of the respondents adopted the practice of scattered trees on farmland while 11% adopted tree retention practice. In Osun east senatorial district of the study area, the most adopted agro forestry practices were planting trees on farm boundaries (23%), home gardening (21%) and improve fallow (13%).

Agroforestry tree species commonly planted or retained by farmers in Study area

The result of the study shows that various agroforestry tree species are planted by farmers in the study area (Table 5). In Osun west senatorial district, 14% retained Millicia excelsa and Antiaris africana respectively. 13% of the respondents commonly plant Gliricidia sepium while 12% retain Khaya ivorensis during land preparation. 10% of the respondents retain trees such as Azadiracta indica, Treculia africana, Moringa oliefera and Antiaris africana on their farm lands.

Further, results revealed that commonly planted or retained agroforestry trees in Osun central senatorial district slightly differs from that of Osun west. Albizia zygia is the most common with 24% of the respondents retaining the tree on their farm land. Azadiracta indica and Gmelina arborea follows in a proportion of 19% and 12% respectively followed by Millicia excelsa (10%). Other agro forestry trees in Osun central include Khaya ivorensis and, Gliricidia sepium both of which are retained by 8% of the respondents.

Common agroforestry trees in Osun east are Moringa oliefera which is planted by 21% of the respondents. Others in the senatorial district are Gliricidia sepium (16%) and Albizia zygia (14%).Further, Cola spp, Citrus sinensis and Theobroma Cacao are also commonly planted tree crops in the study area. Specifically, in Osun east senatorial district of the study area 44% of the respondents plant Cola nitida, Citrus sinensis is planted by 51% of the respondents in Osun central while 11% and 12% plant the crop in Osun west and east senatorial district respectively.

Problems faced by local famers, which influence the adoption of agroforestry practices in the study area

Table 6 shows that the problems faced by local farmers which influence the adoption of agro forestry practices in the study area include land tenure problems, small land holdings, fire outbreak, lack of technical knowhow, non-availability of tree seedlings and trees casting shadow on crops among others. Among all these problems, lack of technical knowhow ranked highest. 19%, 24% and 14% of the respondents in Osun west, Osun central and Osun east senatorial districts of the study area respectively were of the view that lack of technical knowhow is the most difficult problem influencing the adoption of agroforestry practices.

This shows that poor technical knowhow is a strong barrier to

Practices	Osun West		Osun central		Osun East		Total	
	Freq	%	Freq	%	Freq	%	Freq	%
Retaining trees on farmland	11	12	10	11	9	10	30	11
Shifting cultivation	6	7	3	3	3	3	18	4
Improved fallow	14	16	20	22	12	13	46	17
Scattered trees on farmland	14	16	12	13	11	12	37	14
Planting trees on boundary	21	23	18	20	21	23	60	22
Taungya farming	10	11	11	12	9	10	30	11
Home gardening	14	16	16	18	19	21	49	19

Source: Field survey, 2013

Table 4: Types of agro-forestry practices adopted by farmers in the study area

Tree species	Osun West		Osun central		Osun East		Total	
	Freq.	%	Freq.	%	Freq.	%	Freq.	%
Tectona grandis	9	10	1	1	4	2	14	4
Gmelina arborea	0	0	11	12	3	1	14	4
Gliricidia sepium	12	13	7	7	14	16	33	12
Millicia excelsa	13	14	9	10	13	14	35	13
Azadiractha indica	9	10	17	19	8	4	34	11
Albizia zygia	5	6	20	24	13	14	38	15
Antiaris Africana	13	14	7	8	0	0	20	7
Khaya ivorensis	11	12	7	8	13	14	31	10
Treculia africana	9	10	6	2	3	3	18	5
Moringa oleifera	9	10	5	6	19	12	33	12
Cola nitida	9	10	1	1	40	44	50	19
Theobroma cacao	16	18	6	7	12	13	34	13
Elaeis guineensis	2	2	13	14	5	6	20	7
Mangifera indica	19	21	8	9	3	3	30	11
Chrysophyllum albidum	11	12	1	1	5	6	17	6
Afzelia africana	10	11	7	8	9	10	26	10
Parkia biglobosa	9	10	3	3	1	1	13	5
Citrus sinensis	10	11	46	51	11	12	67	25
Psidium guajava	4	4	5	6	4	4	13	5

Source: Field survey, 2013

Table 5: Types of agroforestry tree species commonly planted or retained by farmers in the study area

Problems faced by farmers	Osun West		Osun central		Osun East		Total	
	Freq	%	Freq	%	Freq	%	Freq	%
Fire outbreak	4	4	9	10	7	8	20	7
Poor extension service	22	24	19	21	1	1	42	15
Land tenure system	6	7	9	10	20	22	35	13
Small land holdings	5	6	5	6	9	10	19	7
Non availability of seed/seedlings	7	8	5	6	6	7	18	7
Lack of technical know-how	17	19	22	24	13	14	52	19
Long gestation period	4	4	6	7	15	17	25	9
Lack of incentives	12	13	7	8	16	18	35	13
trees casting shadow on crops	13	14	8	9	3	3	24	9

Source: Field survey, 2013

Table 6: Problems faced by farmers that influence the adoption of agroforestry practices

the adoption of agroforestry practices in the study area especially by the young farmers, who had little farming experience. Keil et al. [14] considered information and knowledge about a given technology as key to the adoption of new agricultural practices, especially when experiments about new technologies and innovations are carried out in the presence of farmers. Keil et al. [14] stated that farmers that are involved in on-farm experimentation of agroforestry technologies with the researchers are more likely to adopt the practice than those who are not

Poor extension service is another problem hindering adoption of agro forestry practices in the study area. 24% of respondents in Osun west, 21% in Osun central and 1% in Osun east respectively stated that they do not have access to extension services. Land tenure system and lack of incentives were also viewed to be part of the problems influencing adoption of agroforestry practices across the senatorial districts in the study area. 22% and 10% of the respondents in Osun East and Osun Central respectively stated that land tenure is a constraint to the adoption of agroforestry practices.

Land tenure is another major constraint to the adoption of Agroforestry practices in the study area because high proportion of respondents (46%) acquired their land through lease while 40% is through inheritance. Land under lease can only be used to grow agricultural crops such as cassava, maize, yam, cocoyam, okra, potatoes e.t.c. They are not allowed to plant trees that will grow over a long period of time. Adedayo [15,16] noted that tenant farmers and squatters find it difficult to plant trees on their farm land due to tenurial restrictions. Land owners always view tree planting by tenants as a way of claiming land ownership from them. As such they don't allow them to plant trees.

Other problems include long gestation period of trees, casting of shadow by trees and fire outbreak. In Osun west, 4%, 13% and 4% respectively stated that the above mentioned problems discouraged the adoption of agro forestry practices. In Osun east, 17% and 18% viewed long gestation period of trees and lack of incentives as problems affecting adoption of agro forestry practices.

Recommended solutions to the problems faced by farmers in the adoption of agroforestry practice in the study area

In view of the identified problems facing the adoption of agroforestry practices among farmers in the study area, the following recommendations can help to solve the problem to a great extent.

- Improved agroforestry extension services: The state forestry Department should ensure that improved agroforestry extension services are provided to the farmers in the study area. This will help to solve the problems of non-awareness and lack of knowledge about agroforestry practices.
- Training programme: The State Forestry Department should organize training programmes for farmers in the study area with practical demonstration of some agroforestry practices like pruning and mulching. This will help to expose them to some practical aspect of agroforestry. Such as, tree crop interactions, nursery establishment, seed pre-treatment and tree pruning activities,
- Distribution of tree seedlings to farmers: The state Forestry Department should endeavor to distribute tree seedlings to farmers freely. Where ever possible, short rotation tree seedlings should be given to farmers so as to encourage them to plant trees on their farm land.
- Farmers' cooperative groups should be encouraged: Farmers should be encouraged to form formal cooperative groups in the study area. This is to enable them benefit from the economies of large scale production and help them have access to research extension, credit facilities and eventually to draw down government/NGO interference, to better their lot.

The state government should help to establish adult literacy centres in the study area. These centres will generally help to educate illiterate farmers. Education will help to reduce ignorance among many of the farmers because many of them will be exposed to information and innovations that will help them to adopt agroforestry practices.

Conclusions

This study has shown that agroforestry practice is a common practice among farmers in the study area. Many of them (60% on the average) perceive agroforestry as a practice that can improve farm productivity. In spite of this, some of them (10%) perceived it as a scientific method that is difficult to practice, while 11% perceived it as a method not properly understood. As such not all the farmers in the study area have adopted agroforestry practice despite the fact that it is an age long practice. The study showed that there is a significant association between the age and educational qualification of respondents as well as respondents land ownership and the adoption of agroforestry practices in the study area. There is however no significant association between respondents' farm size and the adoption of agroforestry practices in the study area. The adopted agro-forestry practices in the study area include retention of trees on farmland, planting of trees on farm boundaries, scattered trees on farmland, shifting cultivation, taungya farming and home gardening. Many of the farmers are however faced with some problems which influence their adoption of agroforestry practices. These problems include; land tenure problems, small land holdings, fire outbreak, lack of technical knowhow, non-availability of tree seedlings and trees casting shadow on crops among others. Lack of technical knowhow ranked highest among all these problems followed by poor extension service and lack of incentives.

References

1. Mukadasi B, Nabalegwa W (2008) Extension for Agroforestry Technology Adoption: Mixed Intercopping of Crotalania (crotalaniagrahamjana) and maize (zea mays L.) in Kabale District Uganda. Environmental Research Journal 2: 131-137
2. Kwesiga F, Akinnifesi FK, Mafongoya PL, McDermott MH, Agumya A (2003) Agroforestry research and development in southern Africa during the 1990s: Review and challenges ahead. Agroforestry Systems 59: 173-186
3. Lundgren BO, Raintree JB (1982) Sustained Agroforestry. In Nestel B (ed.) Agricultural Research for Development: Potential and Challenges in Asia. ISNAR, The Hague, Netherlands
4. Nair PKR (1993) An introduction to Agroforestry. Kluwer Academic publishers, Netherlands.
5. Mercer DE, Miller RP (1998) Socioeconomic research in agroforestry: progress, prospects, priorities. Agroforestry Systems 38: 177–193
6. Maren Oelbermann and Carolyn E. Smith (2011). Climate Change Adaptation using Agroforestry Practices: A Case Study from Costa Rica. Stefano Casalegno (Ed.) Global Warming Impacts - Case Studies on the Economy, Human Health, and on Urban and Natural Environments, ISBN: 978-953-307-785-7, InTech, DOI: 10.5772/24172
7. Kandji ST, Verchot LV, Mackensen J, Boye A, Van Noordwijk M et al. (2006) Opportunities for linking climate change adaptation and mitigation through agroforestry systems. In: Garrity DP, Okono A, Grayson M, Parrott S (Eds) World Agroforestry into the Future (Edn) World Agroforestry Centre (ICRAF), ISBN 9290591846, Nairobi, Kenya.
8. Franzel S, Phiri D, Kwesiga F (2001) Assessing the adoption potential of improved fallows in eastern Zambia. In S Franzel, Scherr SJ (Eds.), Trees on the Farm: Assessing the adoption potential of Agroforestry Practices in Africa 37-64. Wallingford, UK: CAB International.
9. Sinclair FL (1999) A general classification of agroforestry practice. Agroforestry Systems 46: 161–180.
10. SEEDS (2007) State Economic Empowerment and Development Strategy (SEEDS). Osun State government.
11. Oladele OI, Fawole P (2007) Farmers perception of the Relevance of Agricultural technologies in South-Western Nigeria In : Journal of Human resources and Ecology 21: 191-194
12. Regmi BN (2003) Contribution of agroforestry for rural livelihoods: A case of Dhading District, Nepal. Paper presented at the International Conference on Rural Livelihoods, Forests and Biodiversity, Bonn, Germany 19-23.
13. Neupane RP, Sharma KR, Thapa GB (2001) Adoption of agroforestry in the hills of Nepal: A logistic regression analysis. Agricultural systems. Elsevier science limited 72: 177-196
14. Keil A, Zeller M, Franzel S (2005) Improved fallows in smallholder maize production in Zambia: do initial testers adopt the technology. Agroforestry Systems 64: 225-236
15. Akinbile LA, Salimonu KK, Yekinni OT (2007) Farmers Participation in Agroforestry Practices in Ondo State, Nigeria. Research Journal of Applied Sciences 2: 229-232
16. Adedayo AG (2004) Assessment of awareness and acceptability of Alley cropping among rural farmers in Akure South LGA, Ondo State, Nigeria. Journal of Tropical Forest Resources, 2: 99-108.

Soil Microbiological Properties and Enzyme Activities in Ginkgo-Tea Agroforestry Compared to Monoculture

Yaling Tian, Fuliang Cao*, Guibin Wang, Wangxiang Zhang and Wanwen Yu

College of Forest Resources and Environment, Nanjing Forestry University, 210037, Nanjing, Jiangsu Privence, China

Abstract

Studies on enzyme activities and microbiological properties in the soil were very important as they may give indications of the potential of the soil to support biochemical processes which were essential for maintaining soil fertility. The objective of this study was to investigate the difference of soil quality parameters, including organic carbon and nitrogen in the soil, microbial biomass, basal respiration, and the activities of soil enzymes (Catalase, polyphenoloxidase, dehydrogenase, urease, protease and invertase), between monoculture system and agroforestry system. Three management treatments were studied in this work: pure tea system (G_0), intercropped with grafted ginkgo seedlings (G_1 and G_2), which were in a twenty years old tea orchard. Three kinds of depths of soil (0-10 cm, 10-20 cm and 20-30 cm) were used for all treatments. All the parameters except polyphenoloxidase in soil surface in the agroforestry system showed significantly higher values as compared to those in G_0. The contents of soil organic C, total N, microbial biomass and the activities of enzymes were higher in the surface soil as compared to the soil from middle and lower layers. The activities of soil enzymes, such as catalase, dehydrogenase, urease, protease and invertase, and soil organic carbon, total nitrogen were significantly positively correlated. The results of this study suggested that growing teas in combination with ginkgo could be considered as a good forest management practice, which would enhance organic matter accumulation in the soil and improve the activities of soil enzymes, furthermore, could maintain soil productivity and sustainability.

Keywords: Agroforestry system; Ginkgo-tea; Enzyme activities; Soil biological properties

Introduction

Combining trees and crops in spatial or temporal arrangements had been used to improve the safety of food and nutrition and mitigate the pressure of the environment by offering sustainable and alternative products instead of monoculture production. In practices, agroforestry could improve soil quality, carbon sequestration, and water quality in cropping systems [1-3]. In fact, agroforestry could offer lots of social and environmental benefits to human beings, not only in landscapes but also in economies [4].

Tea (*Camellia sinensis*) was an important economic crop and planted widely on acid red soils in the tropical and subtropical regions in China [5]. Camellia sinensis needed full sun to part shade. They prefered a well drained, neutral to slightly acidic soil rich in organic matter (2 parts peat moss or compost to 2 parts loam to 1 part sand or perlite). Gingko (*Ginkgo biloba L.*) was a traditional economic tree species in China, and usually cultivated in agroforestry systems [6]. The practice of ginkgo agroforestry had been adopted in order to obtain more economic benefits during the initial stages of establishment.

Soil quality refereed to the capacities of soil to sustain biological productivity, maintain environmental quality, and promote plant and animal health [7-8]. Trees would uptake nutrients from deep soil layers and maintain soil fertility through the litter fall and other parts of plants [9]. The microbial parameters were usually used as indicators to evaluate the soil quality and the effects to the environment [10].

Soil microorganisms were involved in many soil biological processes [11], which played a crucial role in the circles of most major plant nutrients in soils [5]. Soil microbial biomass was not only the labile nutrient pool but also the medium for the transformation and cycling of organic matters and plant nutrients in soils [5].

In the soil system, soil enzymes had important biochemical functions in the whole process of organic matter decomposition. Enzyme could rapidly response to the changes in soil management practices [12], provide quantitative information on functional diversity of microbial activities [13], such as soil chemical processes, mineralization rates, and organic matter accumulation [11]. The measurement of soil enzyme activities in the ecosystem would help quantify and evaluate specific biological processes in the soil [8]. The combined measurements of enzyme activities and microbial biomass had been widely used over the last 10 years in the study of the microbiological responses to agricultural management [14-17].

Although there had been extensive studies on soil microorganism and soil enzymes [11,18-21], little had been reported on their roles in ginkgo-tea agroforestry system. Meanwhile, how to understand and maintain biodiversity had become an increasingly important issue for researchers, which was also the goal for resource management [8]. The objective of this research was to compare the effects of agroforestry and monoculture systems on soil organic carbon, total nitrogen, microbial biomass, basal respiration and enzyme activities.

Materials and Methods

Study site and soil sampling

The study was conducted in Sanfeng farm (31°40' N, 120° 42' E), located at the foot of Yu mountain in Changshu, Jiangsu Province,

***Corresponding author:** Dr. Fuliang Cao, College of Forest Resources and Environment, Nanjing Forestry University, 210037, Nanjing, Jiangsu Privence, China
E-mail: fuliangcaonjfu@163.com

China. The area was characterized by a subtropical wet monsoon climate with mean annual temperatures of 15°C and mean annual rainfall of about 1,500 mm, most of the rain falling between April and August.

This tea plantation was established in 1990 and three systems with different management treatments were employed in the study (Figure 1): (1) pure tea system (G_0), (2) intercropped with grafted ginkgo seedlings at a spacing of 10 m × 10 m (G_1), (3) intercropped with grafted ginkgo seedlings at a spacing of 6 m × 8 m (G_2). No inorganic nutrient amendments were applied, but herbicide and glyphosate were occasionally used. All the tea branches were pruned after tea picking seasons annually, with the tea rows 1.5 m width and 1m height.

The experimental design was completely randomized with a split plot for soil depths (0-10, 10-20 cm and 20-30 cm). There were two replicates for treatments and three sampling locations per treatment plot. Soil samples were collected with a soil auger and were placed in labeled plastic bags, which were sealed and transported to the laboratory in a cooler. All samples were maintained at field moist condition and were stored at 4°C until be analyzed. Three separate sub-samples were taken from each sample. One sub-sample was air-dried for soil carbon and nitrogen analysis. The second sub-sample was used to determine soil microbiological properties. The third sub-sample was used for enzyme assays after passing through a 2 mm sieve.

Soil analysis

Soil pH was determined in a 1: 2.5 soil/water ratio by a combination glass electrode [5]. Soil organic C was measured by oxidation with potassium dichromate. Total nitrogen was extracted with perchloric-concentrated sulphuric acid and determined by Kjeldahl digestion [6].

Soil microbial biomass C and N was determined by the chloroform fumigation extraction method with extraction coefficient of 0.45 and 0.54 for biomass C and N, respectively [22,23]. The K_2SO_4-extracted C of both fumigated and unfumigated samples was analyzed using a total organic C analyzer (Elementar liqui TOC, Germany). Basal respiration was determined by measuring CO_2 evolution. Twenty gram (oven-dry basis) of field-moist soil was incubated in 250 ml airtight glass vessels at 25° for 1 day. Vials containing 10 ml of 0.1 M NaOH were placed inside the flasks and the CO_2 evolved was determined by titration of carbonates with 0.1 M HCl [6,24]. Microbial quotient was defined as the ratio of microbial biomass C to soil organic C. The metabolic quotient was defined as the ratio of basal respiration to microbial biomass, i.e., the amount of CO_2-C produced per unit of microbial biomass carbon.

The activities of urease, sucrase, catalase, polyphenol oxidase, dehydrogenase and protease were determined by sodium phenolate colorimetry, sodium thiosulfate titration, potassium permanganate titration, iodimetry, TTC colorimetry and ninhydrin colorimetry, respectively [25].

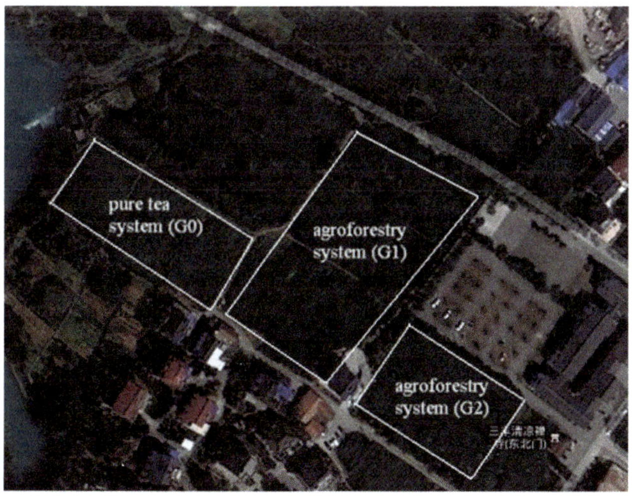

Figure 1: Satellite imagery obtained from Google Earth of the three sites in San-feng farm

Statistical analysis

Differences in soil properties for different soil depths and different systems were analyzed using one-way ANOVA. The least significant difference tests (Duncan's LSD) were used for pair-wise comparisons of treatment means. Differences were declared significant at the five percent level of significance ($p \leq 0.05$).

Results

Soil chemical properties

Soil samples collected from the three systems varied in chemical properties. As shown in Table 1, pH in upper layer (0-10 cm) of G_0 was significantly lower than others and increased with the soil depths.

There were significant differences among the three systems for organic C. Organic C was highest in G_2 of each layer, lowest in G_0 of upper layer and lowest in G_1 of both middle (10-20 cm) and lower layers (20-30 cm). Total N in upper layers showed significant differences between pure and agroforestry systems, which was 1.1 times higher in G_2 than that in G_0. There were no significant differences in middle layer but in lower layer, the orders of the differences in lower layer were: $G_2>G_1>G_0$. The accumulation for either organic C or total N was decreased with soil depths.

Microbial biomass C, N and microbial quotient

G_0 had the lowest soil microbial biomass C (C_{mic}) in comparison with G_1 and G_2 in both upper and middle layers, and there were no significant differences between G_1 and G_2 in the two layers. Meanwhile, in lower layer, soil microbial biomass C was significantly higher in G_2 than in G_0 (Table 2).

Similar significant differences were observed in microbial biomass N (N_{mic}) in upper layer. Microbial biomass N was significantly higher in G_2 than in other two systems both in middle and lower layers, and there were no significant differences between G_0 and G_1 (Table 2).

Systems	pH			Organic C (g•kg^{-1})			Total N (g•kg^{-1})		
	0-10cm	10-20cm	20-30cm	0-10cm	10-20cm	20-30cm	0-10cm	10-20cm	20-30cm
G0	5.21b	5.97	6.02	14.98c	11.26b	6.94a	1.68b	1.1	0.31c
G1	5.86a	6.12	6.37	16.5b	10.1c	5.02b	2.64a	1.4	0.82b
G2	5.86a	6.18	6.36	17.88a	12.08a	6.98a	2.84a	1.62	1.09a

Means within a column of the same position followed by a different lower-case letter were significantly different at p≤0.05.

Table 1: Soil pH, organic C and total N of the three systems

Systems	C_{mic} (μgCg^{-1}soil)			N_{mic} (μgNg^{-1}soil)			C_{mic}/C_{org}(%)		
	0-10cm	10-20cm	20-30cm	0-10cm	10-20cm	20-30cm	0-10cm	10-20cm	20-30cm
G0	264.2b	143.3b	66.7b	40.6b	27.4b	20.6b	1.75	1.28b	0.98b
G1	287.5a	161.8a	75.1ab	48a	30.3b	22.2b	1.74	1.60a	1.50a
G2	295.8a	166.8a	82.6a	52.4a	36.4a	26.5a	1.65	1.38b	1.21ab

Means within a column of the same position followed by a different lower-case letter were significantly different at p≤0.05.

Table 2: The contents of microbial biomass C, N and microbial quotient of the three systems

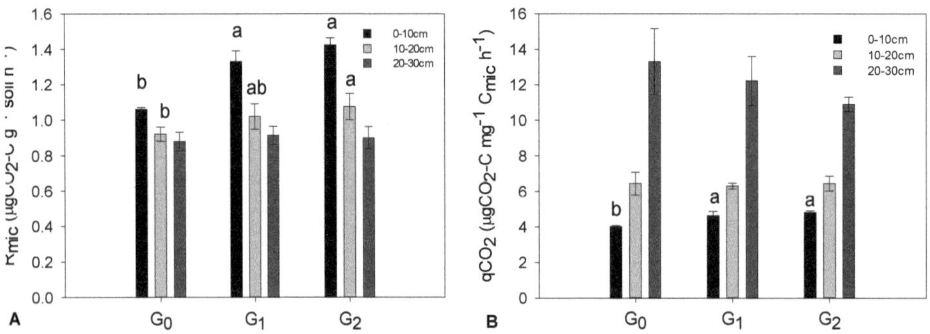

Figure 2: Basal respiration (A) and metabolic quotient (B) for the soil samples of three depths in the three systems. Different letters indicated significant differences at p ≤ 0.05. Bars represented standard errors.

Microbial quotient, the ratio of microbial biomass C to soil organic C (C_{mic}/C_{org}) was observed no significant differences among the three systems in upper layer. G_1 showed the highest amount of microbial quotient in middle layer. While in lower layer microbial quotient in G_1 was significantly higher than that in G_0, and there were no differences between G_0 and G_2 (Table 2).

Similar to organic C and total N, microbial biomass C and N and microbial quotient decreased with soil depths (Table 2).

Basal respiration and metabolic quotient

Basal respirations (R_{mic}) in upper layers of G_2 and G_1 were 1.31 and 1.25 times higher than those of G_0, respectively. And in middle layer, basal respiration of G_2 was 1.17 times higher than that of G_0. No significant differences were observed in lower layer among the systems (Figure 2A).

Similarly, metabolic quotient (qCO_2) in upper layers of G_2 and G_1 were 1.2 and 1.15 times higher than those of G_0, respectively. There were no significant differences in both middle and lower layers. Contrary to basal respiration, metabolic quotient showed a decreasing trend with soil depths (Figure 2B).

Enzyme activities

Catalase activities in G_1 and G_2 were significantly higher than G_0 in upper layer. Soil of G_1 in middle layer had 1.1 and 1.13 times the catalase activity as compared to that of G_0 and G_2, respectively. There were no significant differences in lower layer (Figure 3A). Polyphenoloxidase activities varied significantly among the three systems in upper layers, the order was: $G_1>G_0>G_2$. In addition, polyphenoloxidase activities showed the same trends in middle and lower layers, which were that G_1 had the highest activities and there were no significant differences between G_0 and G_2 (Figure 3B). Dehydrogenase activity was observed significantly higher in G_2 than in G_0 in upper layer, and with no significant differences between G_0 and G_1. There were no significant differences in middle layers among the three systems for dehydrogenase activity. G_2 showed the highest activity of dehydrogenase in lower layer compared to other systems (Figure 3C).

The trends of urease activity were observed similarly in upper and middle layers, with the orders: $G_2>G_1>G_0$. The differences in lower layers among three systems were not significant (Figure 3D). G_2 system had the highest protease activity in upper and middle layers, which were 1.56 and 1.40 times higher than G_0 and G_1, respectively. There were no significant differences in middle layer. Protease in G_1 showed the lowest activity than others in lower layer (Figure 3E). Invertase activities were significantly higher in G_1 and G_2 systems. G_1 had the lowest activity of invertase compared to G_0 and G_2 in middle layers. Significant differences were observed in lower layers, with the order: $G_2>G_0>G_1$ (Figure 3F).

Discussion

Chemical properties

The role of agroforestry in enhancing and maintaining long-term soil productivity and sustainability had been well documented [26]. Trees could enhance soil physical, chemical and biological properties by adding significant amount of above- and below-ground organic matters, and releasing and recycling nutrients in agroforestry systems [26,27]. Lee and Jose demonstrated that alley cropping systems involving pecan and cotton *(Gossypium hirsuitum)* in the southern United States had higher soil organic carbon and microbial biomass compared to monoculture cotton [28]. This study found that soil organic C and total N in ginkgo-tea agroforestry systems were significantly higher than that in pure tea system. The build-up of soil organic matter and nutrient turnover was affected by the input of litter fall. Soil organic matter held basic cations and was the source of energy for decomposers, which contributed to the increasment of the nutrients supply, such as N and K, in soil [29]. The higher organic carbon content of the upper layer in agroforestry systems could due to higher inputs of organic residues from ginkgo litter fall. The soil in tea orchard was very different than other soils [5]. Long-term root exudates and leaf litter could result in the decline of pH and the accumulation of Al [30,31]. In this study, as

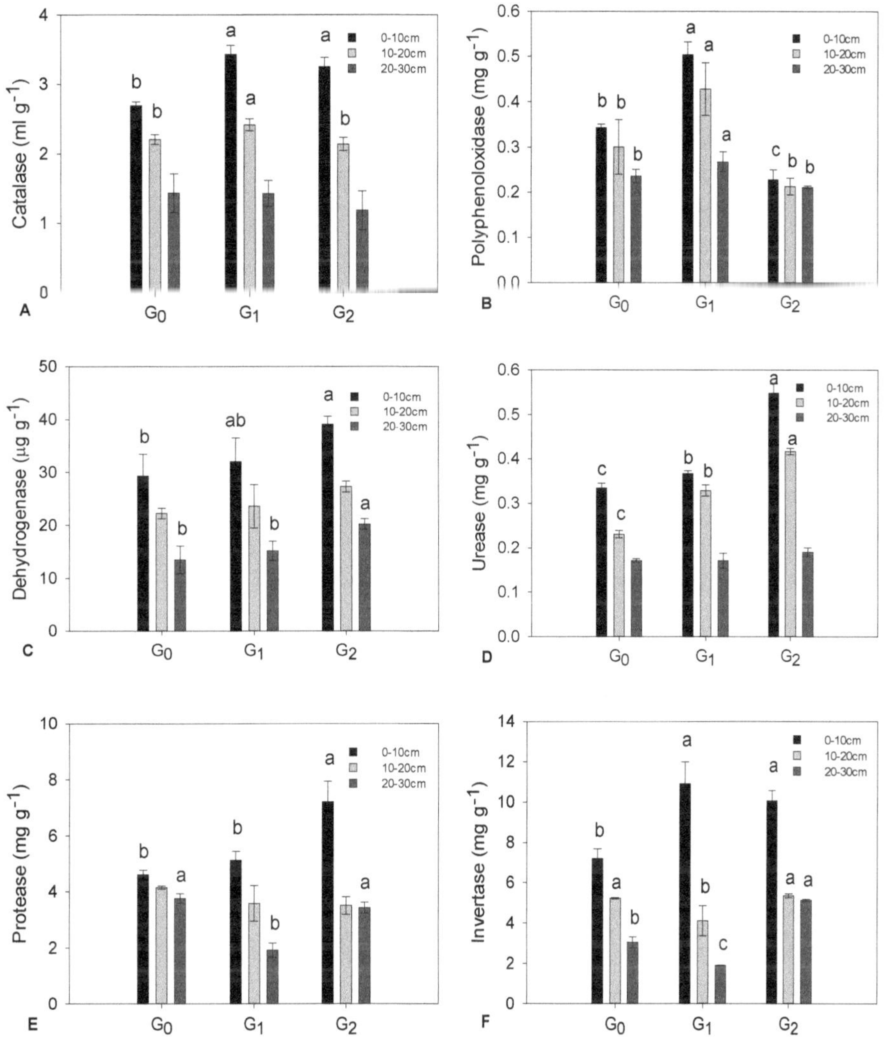

Figure 3: Enzyme activities of catalase (A), polyphenoloxidase (B), dehydrogenase (C), urease (D), protease (E) and invertase (F) for the soil samples of three depths in the three systems. Different letters indicated significant differences at $p \leq 0.05$. Bars represented standard errors.

intercropped with ginkgo, soil pH was significantly higher compared to monoculture, which suggested that ginkgo-tea agroforestry systems could improve soil pH, avoid soil exorbitance acidification and sustain soil productivity.

Microbiological properties

Soil microorganisms were important media in nutrient cycling and energy flow, and they were extremely sensitive to environmental changes [32]. The accumulation of soil organic C enhanced both the microbial biomass C and the proportion of microbial biomass C to soil organic C [5]. Fernandes found that the C biomass values ranged from 99 to 809 mg kg-1 in the top 10 cm and from 71 to 577 mg kg-1 in the 10-20 cm layer, whereas the N biomass values ranged from 11 to 101 mg kg-1 soil in the 0-10 cm layer and from 8 to 84 mg kg^{-1} in the 10-20 cm layer. These data indicated that C and N in the microbial biomass were concentrated in the first 10 cm of soil [33]. Our results provided additional evidence, which was that soil microbial biomass N decreased gradually with soil depth. The differences in soil microorganisms in upper layer among systems could be the results from different factors combined together, such as soil structure composition, root biomass and turnover, lignin content in crop residues, root and litter fall and microclimatic environment of community [34].

Microbial quotient had been used as an indicator of conversion efficiency of organic C into microbial C and losses of soil C during decomposition; it was also proposed to be used as a soil quality indicator to allow comparisons across soils with different organic matter contents [35]. In this study, there were no significant differences of microbial quotient among these systems. Meanwhile, it was found that microbial quotient of G_2 was lower than G_0 in upper layer, although the G_2 system had higher organic C, significantly. The reason for the phenomena should be that the decomposition of ginkgo leaf litter resulted in a large amount of organic C accumulated in soil of G_2 system and, consequently, lower microbial quotient in this soil. Microbial quotient represented between 0.82 and 1.77% in the areas under study. These percentages were in agreement with the results obtained by Wardle [36], and Anderson and Domsch [37].

Soil basal respiration was a widely used parameter for measuring the microbiological activity [5,34]. In this study, R_{mic} was significantly greater in G_2 system than that in pure tea system. A higher rate of R_{mic} may be due to the existence of a large pool of labile C substrates [38]. Metabolic quotient defined as the respiration produced per unit of microbial biomass, expressed as mg CO_2-C h^{-1} g^{-1} biomass-C, indicated the energy optimization as ecosystems develop, also used as a sensitive indicator of activity of microorganism, low values being presented in stable and mature systems [34]. Our results showed that there were significant differences between agroforestry systems and pure tea system. High soil basal respiration and low consumption of organic C in G_2 system could bring high metabolism efficiency and ample available organic C, and maintain soil fertility and sustainability, eventually.

Enzyme activities

Soil enzyme activity was critically important for soil quality and could provide indications of changes in metabolic capacity and nutrient cycling due to management practices [39]. Activities of soil enzyme under different vegetation types were significant different [40]. Numerous studies reported significantly higher activities of these enzymes in intercropping systems compared to monocultures [6,34,41]. In this study, the ginkgo-tea agroforestry systems revealed significantly higher enzyme activities compared to pure tea system.

Continuous monoculture would be detrimental to soil enzyme activities, for example, the catalase activities in the soil from a cucumber continuous monoculture system decreased significantly [39]. This study found that catalase activity was significantly higher in agroforestry systems than in pure tea system, which confirmed the previous studies. Polyphenoloxidase was an important oxidase in aromatic compounds cycling [42]. It was found that polyphenoloxidase activity of G_2 system was significantly lower than others, which disagreed with the studies showing that agroforestry system had higher activity [6,39]. But Zhou found that polyphenoloxidase activity was negatively related with soil humification [41], it could be hypothesized that soil in G_2 had the highest degree of humification. Dehydrogenase activity was considered as the reflection for the total range of oxidative activity of soil microflora and, consequently, may be assumed to be a good indicator of microbiological activity [43]. In our study, G_2 system showed relatively high values in dehydrogenase activity as compared to pure tea system.

Urease was the enzyme that catalyzes hydrolysis of urea to CO_2 and NH^3, which was a vital process in the regulation of N supply to plants after urea fertilization [35]. The results of this study were therefore consistent with other studies reporting that urease activity was significantly affected by different soil management systems [44,45]. Understanding of urease activity dynamics could reveal more effective ways of managing N fertilizers [35]. Reductions in urease activity of G_0 system might negatively affect tea growth and yield. Protease and urease were involved in the N cycle [5,34]. Similarly, protease in G_2 was significantly higher than in other systems. Soil invertase catalyzed the hydrolysis of sucrose to glucose and fructose, and was linked to the soil microbial biomass [42].

Relationships between soil properties and soil enzyme activities

In this study, enzyme activities strongly followed the distribution of soil carbon and nitrogen among the systems. The organic C content was highly correlated with catalase, dehydrogenase, urease, protease and invertase enzyme activities (r=0.95, 0.95, 0.87, 0.87, and 0.93, respectively). The greater correlations between enzyme activity and organic carbon were consistent with previously published results [19,46-48]. It could be hypothesized that perennial vegetation provided environmental conditions suitable for greater accumulation of organic C and total N. Increased enzyme activities were attributed to the increasing of organic matter and litter quality and quantity as well as the improvement of soil physical parameters. Increased enzyme activity was proportionally linked to microbial function leading to improved nutrient cycling and availability, which favored root growth, promotes beneficial plant–microbial interactions, and eventually increased the total soil carbon pool [11].

Depth effects

Soil organic C, total N, microbial biomass and enzyme activities were greater in the surface soil as compared to sub-surface soil, which agreed with published results [8,49]. These differences were attributed to the higher organic matter accumulation, favorable moisture and temperature in the surface soil as compared to sub-surface soil [8]. In addition, in this study, metabolic quotient in upper layer was significantly lower than middle and lower layers, which might be due to the relatively high level of organic C on the soil surface.

Conclusions

The objective of this study was to evaluate the changes of soil quality parameters including soil organic C, total N, microbiological properties and enzyme activities in pure tea system in comparison with ginkgo-tea agroforestry systems. In the study, three soil depths (0-10 cm, 10-20 cm and 20-30 cm) and two densities of ginkgo trees were investigated. The results obtained in this study showed that adoption of a ginkgo-tea combination could lead to the increased long-term sustainability of soil fertility, although the benefits to some soil properties may not be apparent immediately. All parameters decreased gradually with soil depths, except metabolic quotient. The contents of soil organic C, total N, microbial biomass and the enzyme activities in ginkgo-tea agroforestry systems were significantly higher than those in a pure tea system. Soil enzyme activities (i.e., catalase, dehydrogenase, urease, protease and invertase) were highly correlated with soil organic C and total N. Higher soil enzyme activities and the content of microbial biomass were enhanced by agroforestry that may lead to the increasing of other soil quality parameters such as organic matter content, soil sustainability and productivity, so that the soil and ecosystem functions would be improved.

Acknowledgment

The authors thank the staff of Sanfeng farm in Changshu, Jiangsu Province, China, for providing experiment sites, helping and supporting during the study. Also thanks to all the postgraduate students in the laboratory of silviculture in Nanjing Forestry University.

References

1. Lal R (2004) Soil carbon sequestration impacts on global climate change and food security. Science 304: 1623-1627.
2. Nair PKR, Kumar BM, Nair VD (2009) Agroforestry as a strategy for carbon sequestration. J Plant Nutr Soil Sc 172: 10-23.
3. Nii-Annang S, Grünewald H, Freese D, Hüttl R, Dilly O (2009) Microbial activity, organic C accumulation and 13C abundance in soils under alley cropping systems after 9 years of recultivation of quaternary deposits. Biol Fert Soils 45: 531-538.
4. Nair PKR (2007) The coming of age of agroforestry. J Sci Food Agr 87: 1613-1619.
5. Xue D, Yao H, Huang C (2006) Microbial Biomass, N Mineralization and Nitrification, Enzyme Activities, and Microbial Community Diversity in Tea Orchard Soils. Plant Soil 288: 319-331.

6. Wang GB, Cao FL (2011) Integrated evaluation of soil fertility in Ginkgo (Ginkgo biloba L.) agroforestry systems in Jiangsu, China. Agroforest Syst 83: 89-100.

7. Karlen DL, Mausbach MJ, Doran JW, Cline RG, Harris RF, et al. (2009) Soil Quality: A Concept, Definition, and Framework for Evaluation (A Guest Editorial). Soil Sci Soc Am J 61: 4-10.

8. Paudel BR, Udawatta RP, Anderson SH (2011) Agroforestry and grass buffer effects on soil quality parameters for grazed pasture and row-crop systems. Appl Soil Ecol 48: 125-132.

9. Singh B, Sharma K (2007) Tree growth and nutrient status of soil in a poplar (Populus deltoides Bartr.)-based agroforestry system in Punjab, India. Agro forest Syst 70: 125-134.

10. Schloter M, Dilly O, Munch JC (2003) Indicators for evaluating soil quality. Agr Ecosyst Environ 98: 255-262.

11. Udawatta RP, Kremer RJ, Garrett HE, Anderson SH (2009) Soil enzyme activities and physical properties in a watershed managed under agroforestry and row-crop systems. Agr Ecosyst Environ 131: 98-104.

12. Bandick AK, Dick RP (1999) Field management effects on soil enzyme activities. Soil Biol Biochem 31: 1471-1479.

13. Sinsabaugh RL, Carreiro MM, Repert DA (2002) Allocation of extracellular enzymatic activity in relation to litter composition, N deposition, and mass loss. Biogeochemistry 60: 1-24.

14. Wick B, Tiessen H, Menezes R (2000) Land quality changes following the conversion of the natural vegetation into silvo-pastoral systems in semi-arid NE Brazil. Plant Soil 222: 59-70.

15. Christopher DC (2006) Impact of cattle grazing and inorganic fertiliser additions to managed grasslands on the microbial community composition of soils. Appl Soil Ecol 31: 73-82.

16. Bastida F, Moreno JL, Hernández T, García C (2007) The long-term effects of the management of a forest soil on its carbon content, microbial biomass and activity under a semi-arid climate. Appl Soil Ecol 37: 53-62.

17. Meriles JM, Vargas Gil S, Conforto C, Figoni G, Lovera E, et al. (2009) Soil microbial communities under different soybean cropping systems: Characterization of microbial population dynamics, soil microbial activity, microbial biomass, and fatty acid profiles. Soil Till Res 103: 271-281.

18. Lizarazo L, Jordá J, Juárez M, Sánchez-Andreu J (2005) Effect of humic amendments on inorganic N, dehydrogenase and alkaline phosphatase activities of a Mediterranean soil. Biol Fert Soils 42: 172-177.

19. Mungai NW, Motavalli PP, Kremer RJ, Nelson KA (2005) Spatial variation of soil enzyme activities and microbial functional diversity in temperate alley cropping systems. Biol Fert Soils 42: 129-136.

20. Balota EL, Colozzi A, Andrade DS, Dick RP (2003) Microbial biomass in soils under different tillage and crop rotation systems. Biol Fert Soils 38: 15-20.

21. Tian L, Dell E, Shi W (2010) Chemical composition of dissolved organic matter in agroecosystems: Correlations with soil enzyme activity and carbon and nitrogen mineralization. Appl Soil Ecol 46: 426-435.

22. Brookes PC, Landman A, Pruden G., Jenkinson DS (1985) Chloroform fumigation and the release of soil nitrogen: A rapid direct extraction method to measure microbial biomass nitrogen in soil. Soil Biol Biochem 17: 837-842.

23. Vance ED, Brookes PC, Jenkinson DS (1987) An extraction method for measuring soil microbial biomass C. Soil Biol Biochem 19: 703-707.

24. Hernández T, García C, Reinhardt I (1997) Short-term effect of wildfire on the chemical, biochemical and microbiological properties of Mediterranean pine forest soils. Biol Fert Soils 25: 109-116.

25. Guan YS (1986) Soil enzyme and reach methods(in chinese). Beijing: Agricultural Press.

26. Jose S (2009) Agroforestry for ecosystem services and environmental benefits: an overview. Agroforest Syst 76: 1-10.

27. Nair P (1997) Directions in tropical agroforestry research: past, present, and future. Agroforest Syst 38: 223-246.

28. Lee KH, Jose S (2003) Soil respiration and microbial biomass in a pecan - cotton alley cropping system in Southern USA. Agroforest Syst 58: 45-54.

29. Budiadi, Ishii HT, Sabarnurdin MS, Suryanto P, Kanazawa Y (2006) Biomass cycling and soil properties in an agroforestry-based plantation system of kayu putih (Melaleuca leucadendron LINN) in East Java, Indonesia. Agroforest Syst 67: 135-145.

30. Nioh I, Isobe T, Osada M (1993) Microbial Biomass and Some Biochemical Characteristics of a Strongly Acid Tea Field Soil. Soil Sci Plant Nutr 39: 617-626.

31. Pandey A, Palni LMS (1996) The rhizosphere effect of tea on soil microbes in a Himalayan monsoonal location. Biol Fert Soils 21: 131-137.

32. Berg MP, Kniese JP, Verhoef HA (1998) Dynamics and stratification of bacteria and fungi in the organic layers of a scots pine forest soil. Biol Fert Soils 26: 313-322.

33. Fernandes SAP, Bettiol W, Cerri CC (2005) Effect of sewage sludge on microbial biomass, basal respiration, metabolic quotient and soil enzymatic activity. Appl Soil Ecol 30: 65-77.

34. Wang H, Huang Y, Huang H, Wang KM, Zhou SY (2005) Soil properties under young Chinese fir-based agroforestry system in mid-subtropical China. Agroforest Syst 64: 131-141.

35. Balota EL, Chaves JC (2010) Enzymatic activity and mineralization of carbon and nitrogen in soil cultivated with coffee and green manures. Rev Bras Cienc Solo 34: 1573-1583.

36. Wardle DA (1992) A comparative assessment of factors which influence microbial biomass carbon and nitrogen levels in soil. Biol Rev 67: 321-358.

37. Anderson TH, Domsch KH (1990) Application of eco-physiological quotients (qCO2 and qD) on microbial biomasses from soils of different cropping histories. Soil Biol Biochem 22: 251-255.

38. Islam KR, Weil RR (2000) Land use effects on soil quality in a tropical forest ecosystem of Bangladesh. Agr Ecosyst Environ 79: 9-16.

39. Zhou XG, Yu GB, Wu FZ (2011) Effects of intercropping cucumber with onion or garlic on soil enzyme activities, microbial communities and cucumber yield. Eur J Soil Biol 47: 279-287.

40. Michel K, Matzner E (2003) Response of enzyme activities to nitrogen addition in forest floors of different C-to-N ratios. Biol Fert Soils 38: 102-109.

41. Zhou LK (1987) Soil enzyme (in chinese). Beijing: Science Press.

42. Gu Y, Wang P, Kong CH (2009) Urease, invertase, dehydrogenase and polyphenoloxidase activities in paddy soil influenced by allelopathic rice variety. Eur J Soil Biol 45: 436-441.

43. Perucci P, Bonciarelli U, Santilocchi R, Bianchi AA (1997) Effect of rotation, nitrogen fertilization and management of crop residues on some chemical, microbiological and biochemical properties of soil. Biol Fert Soils 24: 311-316.

44. Klose S, Tabatabai MA (2000) Urease activity of microbial biomass in soils as affected by cropping systems. Biol Fert Soils 31: 191-199.

45. Longo RM, Melo WJ (2005) Urease activity in oxisols as influenced by vegetation cover and sampling time. Rev Bras Cienc Solo 29: 645-650.

46. Myers RT, Zak DR, White DC, Peacock A (2001) Landscape-level patterns of microbial community composition and substrate use in upland forest ecosystems. Soil Sci Soc Am J 65: 359-367.

47. Kremer RJ, Li J (2003) Developing weed-suppressive soils through improved soil quality management. Soil Till Res 72: 193-202.

48. Udawatta RP, Kremer RJ, Adamson BW, Anderson SH (2008) Variations in soil aggregate stability and enzyme activities in a temperate agroforestry practice. Appl Soil Ecol 39: 153-160.

49. Shamir I, Steinberger Y (2007) Vertical Distribution and Activity of Soil Microbial Population in a Sandy Desert Ecosystem. Microb Ecol 53: 340-347.

Estimation of Above-Ground Forest Biomass Using Landsat ETM+, Aster GDEM and Lidar

Lissette Cortés*, Jaime Hernández, Diego Valencia and Patricio Corvalán

Faculty of Forestry and Nature Conservation, University of Chile, Chile

Abstract

The measurement of above-ground biomass is important to understand carbon flow between trees and the atmosphere; remote sensing plays an important role in making this possible for extensive and hard-to-reach areas. This study compared above-ground forest biomass estimation models using data from different sources, including Landsat ETM+, Aster GDEM, ALS (LiDAR) and forest inventories. Two sets of predictors were established: the first included variables extracted from Landsat ETM+ and Aster GDEM, while the second included variables from Landsat in combination with LiDAR products (Digital Terrain Model, Digital Surface Model and Canopy Height Model). The Random Forest algorithm was used to build all models; this method explicitly returns the importance of each predictor and therefore allows the selection of the best set of variables. Estimations were made separately by forest cover for *Pinus radiata*, *Eucalyptus globulus* and second-growth *Nothofagus glauca*. Better results were obtained using the combination Landsat-LiDAR than those using Landsat-Aster GDEM data. Also, the results were better when applying the model to pine cover (pseudo R2 77.22%).

Keywords: Above-ground biomass; Landsat ETM+; LiDAR; Random forest.

Introduction

The study of forests is a growing area of research due to their importance in different natural processes; the carbon cycle is one of the main processes. This is due to the fact that terrestrial ecosystems are the greatest contributors to the global carbon balance [1]; they contain close to 60% of the carbon stored in vegetation and 50% of that stored in the soil [2]. Carbon in forests may be divided into three components: biomass in living plants; plant remains and soil [3].

Most studies related to carbon content in forests are performed on the above-ground biomass, defined as the weight or equivalent estimation of organic material growing above the ground [4]; the different methods of estimating this content may be classified by the type of information they utilize.

The method most used to estimate above-ground forest biomass is that of forest inventories. These use allometric equations which relate easily-measured field data such as tree height and diameter at breast height (DBH) with above-ground biomass [5]. Although this is the most precise method it has greater cost, especially when it is desired to estimate biomass in extensive or hard-to-reach zones. Given the operational difficulties, methods have been developed which combine inventories with information acquired by remote sensing [6,7] and environmental variables (i.e. slope, exposition, altitude and ground curvature) [8].

Passive remote sensors use solar energy to obtain spectral information about objects. Among the most used are the medium-resolution multispectral satellites Landsat ETM+ and Aster (30 and 15 meters, respectively) [9]. The information provided by these sensors allows relating reflectance, vegetation indices and/or transformations of different spectral bands to above-ground biomass. One of the advantages of using these images is that some of them may be acquired free or at low cost. The same is true for the Digital Elevation Models (DEM), SRTM ("Shuttle Radar Topography Mission") and Aster GDEM (Aster "Global Digital Elevation Model"), from which it is possible to derive topographic information to be correlated with above-ground biomass [10,11].

LiDAR ("Light Detection and Ranging") is an active remote sensor that obtains a cloud of points with X,Y, and Z coordinates which provide information about the structure of objects on the ground [12]. Measurements are obtained by estimating the distance between the sensor and its objective by the time between the emission and return of a laser pulse [13]. Data of this type have been used by various authors [14-16] to obtain metrics or high-resolution models related to forest heights. One of the models most used is that of Digital Surface Models (DSM), which represent the surface of the Earth including all objects on it, and the Canopy Height Model (CHM), which estimates the height of the trees. One of the main difficulties and disadvantages of this type of data, compared to multi-spectral images, is the high cost of acquiring and processing them.

There are two major types of statistical models, those which focus on the data and those which focus on algorithms [17]. The focus based on the data begins by assuming a stochastic model such as linear or logistic regression, whose fit is used both to predict responses (Y) to future data (X) and to extract information about the nature of the relation between variables X and Y. The approach based on algorithms focuses on prediction and uses estimations of its exactitude to validate the models, without considering the type of relation between the variables. The latter approach is generally called machine-learning models.

In the construction of statistical models for the estimation of above-ground biomass both approaches have been used. Initially techniques

***Corresponding author:** Lissette Cortés, Faculty of Forestry and Nature Conservation, University of Chile, Chile, E-mail: lcortes.serey@gmail.com

were employed such as multiple regression, k-NN classification and neural networks [18] on data from active and/or passive sensors. However, in recent years the use of machine-learning type algorithms based on classification and regression trees has increased. Among these, the Random Forest regression has been used to estimate forest parameters [10, 19-22], with better results than other methods.

About 22% of the surface area of continental Chile (16.6 million ha) is covered with native forests and forest plantations, concentrated mainly in the central part [23]. This makes the country an interesting place for biomass studies, to estimate the capacity for carbon storage, to implement management measures or to estimate the potential production of biofuel.

The objective of this study is to compare models of above-ground forest biomass estimation, combining satellite information obtained from the Landsat ETM+ sensor, Aster elevation data, information derived from LiDAR point clouds and field measurements.

Material and Methods

Test site

The study area is an experimental forest property of the Universidad de Chile named Pantanillos, located in central Chile (Figure 1), between coordinates UTM (WGS84) 744187 - 746042 meters east and 6074000 - 6070800 meters north.

Pantanillos has a surface area of approximately 400 ha, the majority of which is covered with plantations of *Pinus radiata* D. Don and in lesser quantity of *Eucalyptus globulus* Labill. It also has sectors of second-growth native forest, in which hualo (*Nothofagus glauca* Phil. Krasser) is the predominant species.

Field reference data

Field data were compiled in February, 2011, by establishing a systematic net of conglomerates every 200 m. Each conglomerate was formed by five sampling subunits, one central and four at 30 m distance in each of the cardinal directions. Each subunit consisted of concentric circular plots of 2, 4 and 8 m radius, which included trees with DBH ≥ 5, 10 and 20 cm, respectively (Figure 1). The following state variables were recorded in each plot: DBH (cm), species, total height (m), and other complementary descriptors. A total of 83 conglomerates were measured which encompass 330 plots (some plots were missing due to accessibility issues).

Estimation of the above-ground biomass with field data

The following allometric functions were used to estimate the above-ground biomass of individual trees:

E. globulus [24]:

$$BA = -92 + 2.3 * LN(DBH) \tag{1}$$

P. radiata [25]:

$$BA_F = e^{\left(1.028 * LN\left[(DBH+1.9)^2 * H\right] - 4.892\right)} \tag{2}$$

Other species [26]:

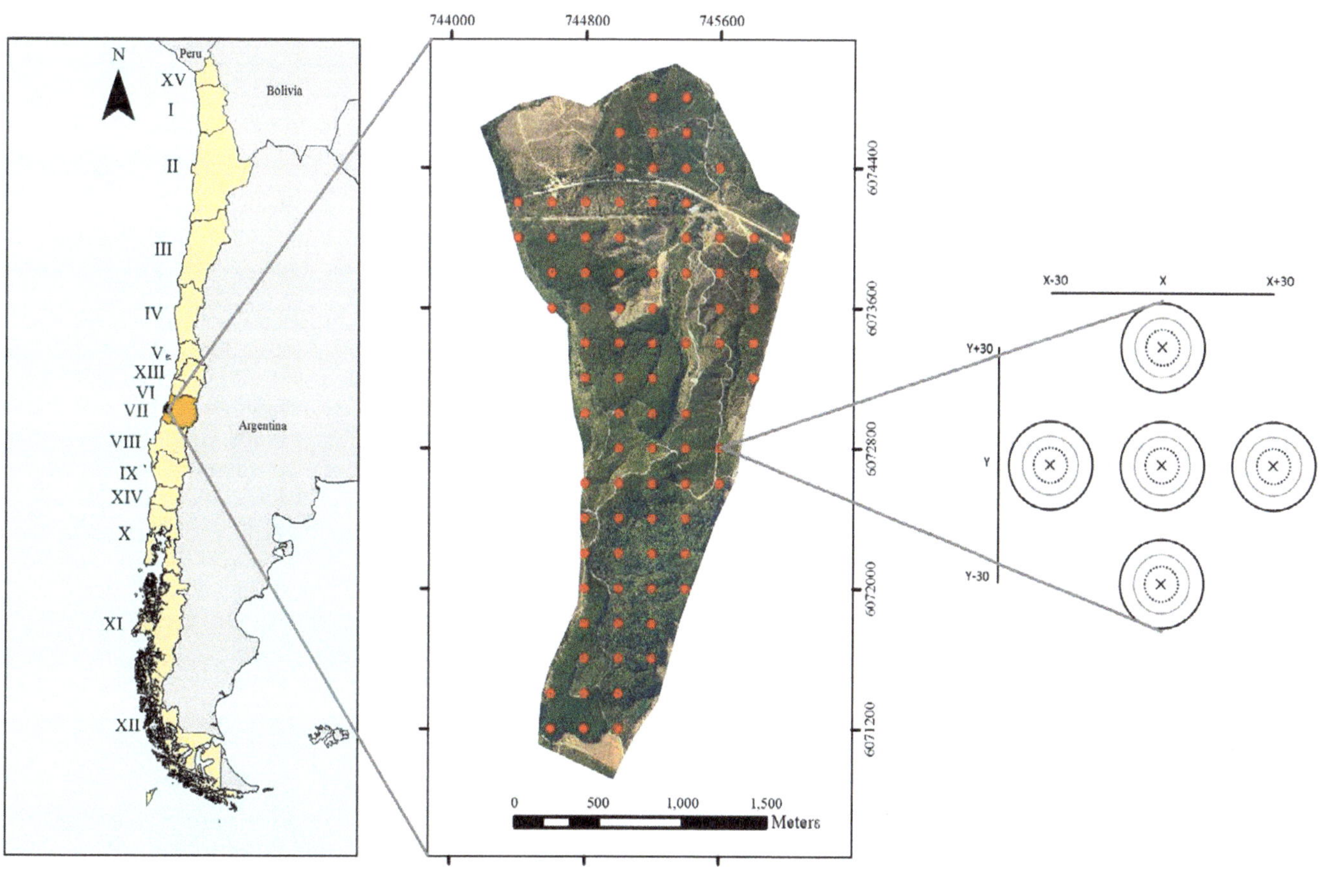

Figure 1: Location of the Pantanillos forest in the Región del Maule (left).Cluster distribution of forest inventory (center). Internal conformation of each cluster (right).

$$BA = e^{(-2.9149+1.9711*LN(DBH)+0.7442*LN(H))} \quad (3)$$

BA is the total above-ground biomass (kg), BA_F is stem biomass (kg), DBH (cm) and H is height (m).

For radiata pine BA_F is the above-ground biomass of the stem with bark; thus it was necessary to add the above-ground biomass component of branches and needles (non-stem biomass). This was done using the component equations of Corvalán and Hernández [27] for the percentage of stem above-ground biomass (BF%) and that of the non-stem biomass (BNF%).

The final equation for the stem above-ground biomass (BF%) was:

$$BF_{\%} = 51.71 * e^{(-98/X^{0.78})} + 33 \quad (4)$$

$$X = E * \sqrt{1000/N} * AB \quad (5)$$

Where X is the mean distance between trees, E is the age of the stand, AB is the basal area and N is the number of trees. Finally, BNF% was calculated as the complement of the stem biomass area, assuming a constant bark biomass of 10%.

$$BNF_{\%} = 100 - (BF_{\%} - 10\%) \quad (6)$$

Total above-ground biomass per hectare in each plot was calculated by summing the individual stem biomass of each tree multiplied by its respective expansion factor, plus the non-stem biomass expressed as an aggregate function of total stem biomass divided by total biomass.

Processing of satellite images

We obtained two Landsat ETM+ scenes from 8 December 2010 and 9 January 2011 (Path/row 233/85). Both images were rectified by polynomial functions using the UTM projection (WGS84 zone 18 south).

A radiometric correction was performed by conversion of Digital Number (DN) to radiance using the specific gain and offset of the sensor. We then applied an atmospheric correction using the method of Dark Object Subtraction proposed by Chávez [28], and a topographic correction using the method of Civco [29]. Reflectance was obtained by applying a simplified model with the correction parameters of Chuvieco [30]. Since the images were taken after 2003, that is, they contain the SLC-off error, both images were fused using the method of Scaramuzza *et al.* [31] (Figure 2). The reflectance bands were also used to obtain the Tasseled Cap transformation [32] and the vegetation indices shown in Table 1.

Processing of the aster elevation digital model

We obtained the free Aster GDEM Elevation Digital Model from the web site of the same name, with a spatial resolution of approximately 30 meters. From this we obtained topographical variables of aspect (in azimuth degrees), slope (in percentage) and land curvature (concave, convex or flat).

Processing of LiDAR data

LiDAR data were obtained in March 2011 with a mean density of the point cloud of 5 points per m², and a true three-band ortho image (visible RGB) with a spatial resolution of 0.5 m.

As a first step to obtain the Digital Terrain Model (DTM) from the LiDAR, we used the multi-scale curvature algorithm to identify the points at ground level [39]. The points not classified as ground were interpolated to obtain a Digital Surface Model (DSM) [40]. We also obtained a Canopy Height Model (CHM) as the difference between DTM and DSM, which shows the tree heights. The three models (DTM, DSM and CHM) were obtained at a spatial resolution of 1 m. As with the Aster GDEM, from the LiDAR DTM we obtained the topographic variables elevation, slope, aspect and land curvature.

Parceling

Parceling consisted in the identification of homogeneous units according to the dominant vegetation present. This was done by photo interpretation of the true ortho image complemented with field data. We identified 5 types of vegetation cover: *P. radiata* plantations, *E. globulus* plantations, second-growth native forests, mixtures of forest covers and a class called other, which included bare areas, buildings and areas outside the forest property.

Estimation of above-ground biomass

We used the method based on classification and regression trees called Random Forest (RF), available in the statistical software R-Project [41], under the name random forest package. This algorithm, proposed by Breiman [42], constructs a tree of classification or

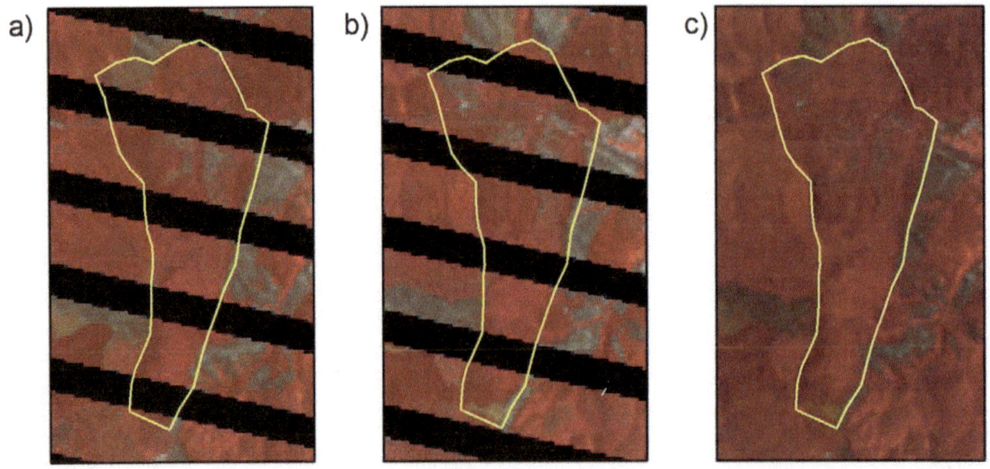

Figure 2: Correction of Landsat ETM+ SLC-off. a) Landsat ETM+SLC-off December/2010. b) Landsat ETM+SLC-off January/2011, c) Landsat ETM+ corrected.

Index	Equation	Reference
Difference Vegetation Index (DVI)	$DVI = \rho_{NIR} - \rho_R$	Roujean and Breon [33]
Simple Ratio (SR)	$SR = \frac{\rho_{NIR}}{\rho_R}$	Jordan [34]
Red Green Ratio (RGR)	$RGR = \frac{\rho_R}{\rho_G}$	Gamon and Surfus [35]
Red Green Index (RGI)	$RGI = \frac{\rho_G - \rho_R}{\rho_G + \rho_R}$	Coops *et al.* [36]
Normalized Difference Vegetation Index (NDVI)	$NDVI = \frac{\rho_{NIR} - \rho_R}{\rho_{NIR} + \rho_R}$	Rouse *et al.* [37]
Corrected Normalized Difference Vegetation Index (NDVIc)	$NDVI_c = NDVIv\left[1 - \frac{\rho_{MIR} - \rho_{MIR\,min}}{\rho_{MIR\,max} - \rho_{MIR\,min}}\right]$	Nemani *et al.* [38]

ρ_{NIR} = Near Infrared, ρ_R = Red, ρ_G = Green, ρ_{MIR} = Middle Infrared.

Table 1: Vegetation Indices utilized in the study.

regression using a process called bagging ("Bootstrap + Aggregation") [43,44]. In RF each node is separated using the best division chosen among a subset of randomly chosen predictor variables. This algorithm produces an internal error estimation, thus the results do not require cross-validation or independent validation [45].The error of estimation is calculated by leaving out of the bootstrap sample about one third of the reference cases not used in the construction of the tree. These data are so-called "out-of-bag" (OOB), and are predicted using the tree created by the bootstrap sample. On the average each datum is an OOB about 1/3 of the time; these predictions are averaged and the mean square error (MSE) is calculated [46].

RF provides a pseudo R^2 similar to the R^2 of a standard linear regression; it is called pseudo because it uses the values predicted by RF to calculate the MSE. Pseudo R^2 may take negative values, which are interpreted as very poor fits [47]. RF also calculates the relevance of the predictor variables, assigning a score that depends on the changes in the error when the values of this variable are permuted (called %IncMSE); if the effect is larger the variable is assigned greater importance [48].

We estimated above-ground biomass for *P. radiata*, *E. globulus* and second-growth native forest separately. Since RF includes an element of randomness, each estimation was repeated 100 times. To evaluate the sources of information, the root mean square error (RMSE) (equation 7) and percentage RMSE (equation) were calculated from the MSE returned by RF.

$$RMSE = \sqrt{\frac{1}{n}\sum_{i=1}^{n}\left(yi - \hat{\bar{yi}}_{OOB}\right)^2} \quad (7)$$

$$RMSE\% = \frac{RMSE}{\frac{1}{n}\sum_{i=1}^{n} y_i} * 100 \quad (8)$$

Where the y_i are the observed data.

Variables were extracted based on a point cloud established every 30 m, in which were also included the coordinates of each point and the point above-ground biomass calculated with field data when available. For the variables obtained with LiDAR data the extraction was performed based on an area of influence of 8 m radius based on the net of points, obtaining the mean value for the area. In the extraction of DCM and DSM the standard deviation was also estimated.

To compare the capacity of Landsat, Aster GDEM and LiDAR to estimate above-ground biomass, two sets of predictors which occupy combined sources of information were used (Table 2).

Results

With the Landsat + GDEM set using all the predictors the best fit was obtained in pine plantations, 56.02%. However, for eucalyptus and second-growth native forest the pseudo R^2 were low, with values of -5.73% and 2.5%, respectively. The selection of variables improved the fit and decreased the error (Table 3).The estimations in *P. radiata* and second-growth hualo selected variables related to topography and the spectral response, while in *E. globulus* only three predictors derived from Landsat were selected (Figure 3a).

In estimations using Landsat and LiDAR the fit values increased for *P. radiata* (pseudo R^2=75.72%) and second-growth hualo (pseudo R^2=13.90%) compared to those obtained using Landsat + GDEM, while *E. globulus* continued with negative pseudo R^2 (pseudo R^2=-0.92%).

Landsat +GDEM Set	Landsat+LiDAR Set
Landsat: Reflectivity (TM1 - TM5 y TM7) Vegetation Index Tasseled Cap (TC1-TC6)	Landsat: Reflectivity (TM1 - TM5 y TM7) Vegetation Index Tasseled Cap (TC1-TC6)
GDEM: Elevation (E) Slope (S) Aspect (Asp) Curvature (C)	LiDAR: Elevation (Li_E) Slope (Li_S) Aspect (Li_Asp) Curvature (Li_C)
	DSM (Li_DSM) Standard deviation DSM (Li_DSM_sd) CHM (Li_CHM) Standard deviation CHM (Li_CHM_sd)

Table 2: Sets of predictors used in the estimation of aboveground biomass.

		PR	EG	NG
Landsat + GDEM	Pseudo R^2	60.26 (± 0.55)	37.84 (± 2.19)	7.76 (± 1.24)
	RMSE	44.58 (± 0.31)	25.81 (± 0.83)	55.2 (± 0.37)
	RMSE%	51.59 (± 0.32)	60.83 (± 1.95)	75.24 (± 0.51)
Landsat + LiDAR	Pseudo R^2	77.22 (± 0.37)	37.92 (± 1.78)	30.95 (± 1.05)
	RMSE	33.75 (± 0.27)	25.79 (± 0.37)	47.76 (± 0.37)
	RMSE%	35.40 (± 0.28)	60.79 (± 0.87)	65.09 (± 0.50)

PR= Pinus radiata, EG= Eucalyptus globulus and NG= Nothofagus glauca.

Table 3: Fits and mean errors for estimations with selection of variables for the two proposed sets of predictors.

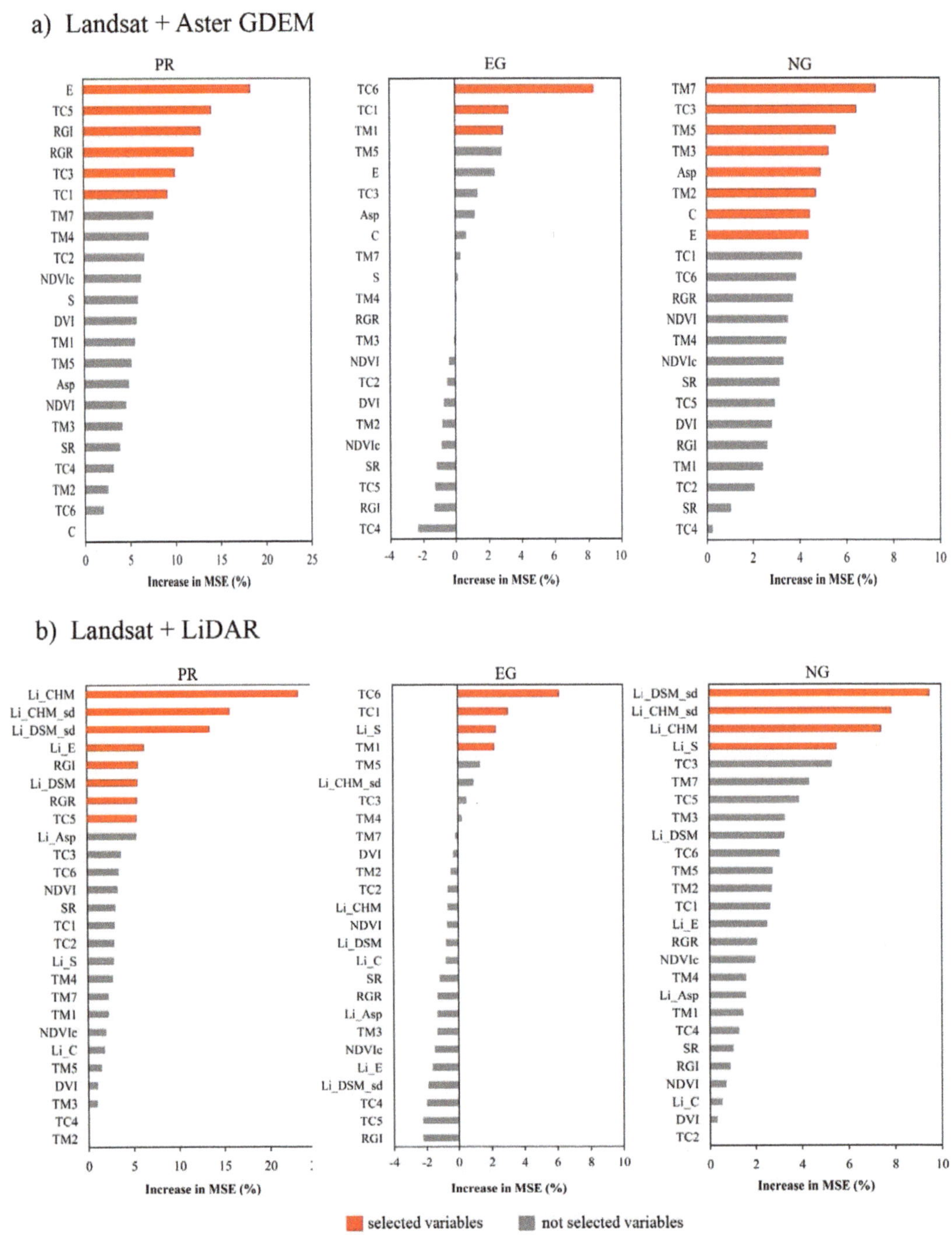

Figure 3: The graphs show the order of importance of the variables obtained in RF for the Landsat + GDEM and Landsat + LiDAR sets according to the dominant species. Red bars indicate the variables selected for the second estimation. PR= Pinus radiata, EG= Eucalyptus globulus and NG= Nothofagus glauca (native forest).

With selection of variables the fit in *P. radiata* was improved using variables of Landsat and LiDAR. For *E. globulus* a fit of only 37.92% was achieved (similar to that of Landsat + GDEM with selection of predictors) in which the variables of canopy structure were not relevant, in contrast to the estimations in second-growth hualo, in which only this class of variables was selected (Figure 3b).

Discussion

The set of predictors which used variables from Landsat and GDEM produced poorer fits than the second set, which incorporated information related to canopy height extracted from the LiDAR point cloud. These results concur with those of Lefsky *et al.* [49], who indicated that the metrics of canopy structure, mainly height and its standard deviation are highly correlated with the above-ground biomass of forests. Latifi *et al.* [19] also found this relation using a genetic algorithm for variable selection, although they also used other predictors extracted directly from the point cloud such as height percentiles.

In the estimations of pine and eucalyptus plantations and a second-growth native forest, both the fits and the type of predictors selected were considerably different for the two sets of predictors. The differences in fit among the three vegetation types may be due both to their horizontal and vertical structures. In pine plantations the horizontal structure is more homogeneous than in areas with native forest, which have not only a more complex and irregular canopy geometry but are also less dense. Clark *et al.* [50] found that plantations had higher correlations with metrics from LiDAR than forests with irregular structure. Our results agree with those of Popescu *et al.* [51], who obtained fits of 0.33 for deciduous forests (RMSE 44 ton/ha) and 0.82 for pine (RMSE 29 ton/ha), similar to our results.

Although eucalyptus plantations are homogeneous horizontally, the vertical structure of individual trees produced a large difference in the results. In pine plantations, the canopy structure allows a greater proportion of points to be intercepted in the canopy, which allows the creation of more precise models of the canopy surface, and thus also more precise DCM. On the contrary, in eucalyptus plantations trees are less leafy and canopies are sparser, thus many more points do not fall in the canopy of the trees but rather in the ground or the undergrowth, thus the DSM and DCM are less precise. LiDAR point clouds with higher densities may be suitable to compensate this problem but computing cost may also rise.

Conclusions

We conclude that the use of data which include information related to the canopy structure, such as the Digital Surface Models and Digital Canopy Models, produced better results (greater correlations) and lower errors. This was true in the estimations of pine and second-growth native forest but not for the estimations in *E. globulus*.

The differences in the fits of the estimations may be due to the precision obtained in the canopy models, given the vertical structure present in the different kinds of cover evaluated. In the estimations for *E. globulus* the utilization of LiDAR data did not improve the fit, which may indicate that the point density (5 points/m^2) is not dense enough to represent correctly the tree canopies, sub-estimating the height of the stand.

This study estimated biomass using digital models related to canopy height, which estimates the mean height of the vegetation in each pixel. However, it is also possible to use other variables extracted directly from the LiDAR cloud points which may reveal the structure of other strata (such as understory); using these variables could improve the fits estimations.

Acknowledgments

This study was financed by the BIOCOMSA Consortium (CORFO-INNOVA) as part of their project "Evaluation and monitoring of existing Aerial Biomass and optimization of supply networks to biofuel plants. We also thank the Laboratorio de Geomática y Ecología del Paisaje (GEP) of the Universidad de Chile.

References

1. Gibbs H, Brown S, Niles JO, Foley JA (2007) Monitoring and estimating tropical forest carbon stocks: making REDD a reality. Environmental Research Letters 2: 045023.
2. Dong J, Kaufmann RK, Myneni RB, Compton JT, Kauppi PE, et al. (2003) Remote sensing estimates of boreal and temperate forest woody biomass carbon pools, sources and sinks. Remote Sensing of Environment 84: 393-410.
3. Bravo F, Bravo-Oviedo A, Diaz-Balteiro L (2008) Carbon sequestration in Spanish Mediterranean forests under two management alternatives: a modeling approach. European Journal of Forest Research 127: 225-234.
4. Schlegel B, Gayoso J, Guerra J (2000) Medición de la capacidad de captura de carbono en bosques de Chile y promoción en el mercado mundial. Manual de procedimientos: muestreos de biomasa forestal. Chile 24.
5. Fehrmann L, Kleinn C (2006) General considerations about the use of allometric equations for biomass estimation on the example of Norway spruce in central Europe. Forest Ecology and Management 236: 412-421.
6. Labrecque S, Fournier RA, Luther JE, Piercey D (2006) A comparison of four methods to map biomass from Landsat-TM and inventory data in western Newfoundland. Forest Ecology and Management 226: 129-144.
7. Ji L, Wylie BK, Nossov DR, Peterson B, Waldrop MP, et al. (2012) Estimating aboveground biomass in interior Alaska with Landsat data and field measurements. International Journal of Applied Earth Observation and Geoinformation 18: 451-461.
8. Lu M (2006) The potential and challenge of remote sensing-based biomass estimation. International Journal of Remote Sensing 27: 1297-1328.
9. Wulder MA, White JC, Fournier RA, Luther JE, Magnussen S (2008) Spatially explicit large area biomass estimation: three approaches using forest inventory and remotely sensed imagery in a GIS. Sensors 8: 529-560.
10. Pierce AD, Farris C, Taylor AH (2012) Use of random forests for modeling and mapping forest canopy fuels for fire behavior analysis in Lassen Volcanic National Park, California, USA. Forest Ecology and Management 279: 77-89.
11. Main-Knorn M, Moisen GG, Healey SP, Keeton WS, Freeman EA, et al. (2011) Evaluating the Remote Sensing and Inventory-Based Estimation of Biomass in the Western Carpathians. Remote Sensing 3: 1427-1446.
12. Harding D, Lefsky M, Parker G, Blair J (2001) Laser altimeter canopy height profiles: methods and validation for closed-canopy, broadleaf forests. Remote Sensing of Environment 76: 283-297.
13. Lefsky MA, Cohen WB, Parker G, Harding DJ (2002) Lidar remote sensing ecosystem studies. Bioscencie 52: 19-30.
14. Hyde P, Dubayah R, Peterson B, Blair JB, Hofton M, et al. (2005) Mapping forest structure for wildlife habitat analysis using waveform lidar: Validation of montane ecosystems. Remote Sensing of Environment 96: 427-437.
15. Boudreau J, Nelson RF, Margolis HA, Beaudoin A, Guindon L, et al. (2008) Regional Aboveground Forest Biomass Using Airborne and Spaceborne LiDAR in Québec. Remote Sensing of Environment 112: 3876–3890.
16. Kellndorfer J, Walker W, Lapiont E, Kirsch K, Bishop J, et al. (2010) Statistical fusion of lidar, InSAR, and optical remote sensing data for forest stand height characterization: A regional-scale method based on LVIS, SRTM, Landsat ETM+, and ancillary data sets. Journal of Geophysical Research 115: 10.
17. Breiman L (2001) Statistical Modeling: The Two Cultures. Statistical Sciencie 16: 199-301.
18. Koch B (2010) Status and future of laser scanning, synthetic aperture radar and hyperspectral remote sensing data for forest biomass assessment. ISPRS Journal of Photogrammetry and Remote Sensing 65: 581-590.

19. Latifi H, Nothdurft A, Koch B (2010) Non-parametric prediction and mapping of standing timber volume and biomass in a temperate forest: optimizing variable selection on optical/LiDAR-derived predictors. Forestry 83: 395-407.

20. Gleason CJ, Im J (2012) Forest biomass estimation from airborne LiDAR data using machine learning approaches. Remote Sensing of Environment 125: 80-91.

21. Immitzer M, Atzberger C, Koukal T (2012) Tree Species Classification with Random Forest Using Very High Spatial Resolution 8-Band WorldView-2 Satellite Data. Remote Sensing 4: 2661-2693.

22. Leutner BF, Reineking B, Müller J, Bachmann M, Beierkuhnlein C, et al. (2012) Modelling Forest α-Diversity and Floristic Composition — On the Added Value of LiDAR plus Hyperspectral Remote Sensing. Remote Sensing 4: 2818-2845.

23. CONAF (2011) Catastro de los Recursos Vegetacionales Nativos de Chile. Monitoreo de Cambios y Actualizaciones.

24. Bennett LT, Weston CJ, Attiwill PM (1997) Biomass, nutrient content and growth responses to fertilisers of six-year-old Eucalyptus globules plantations at three contrasting sites in Gippsland, Victoria. Australian Journal of Botany 45:103-121.

25. Madgwick HAI (1994) Pinus radiata: Biomass, Form and Growth. Ed. HAI Madgwick, Rotura, New Zealand 428.

26. Gayoso J, Guerra J, Alarcón D (2002) Contenido de carbono y funciones de biomasa en especies nativas y exóticas. Documento Nº1. Proyecto FONDEF Medición de la capacidad de captura de carbono en bosques de Chile y promoción en el mercado mundial. Universidad Austral de Chile Instituto Forestal 53.

27. Corvalán P, Hernández J (2011) Tablas de estimación de biomasa aérea bruta en pie para plantaciones de Pino Insigne en Chile. Facultad de Ciencias Forestales y de la Conservación de la Naturaleza. Departamento Gestión Forestal y su Medio Ambiente; Santiago, Chile 100.

28. Chávez PS (1988) An improved dark-object subtraction technique for atmospheric scattering correction of multispectral data. Remote Sensing of Environment 24: 459-479.

29. Civco DL (1989) Topographic normalization of Landsat Thematic Mapper Digital Imagery. Photogrammetric Engineering and Remote Sensing 55: 1303-1309.

30. Chuvieco E (2010) Teledetección Ambiental. La observación de la tierra desde el espacio. 3ra Edición. Ariel Ciencia, España, Barcelona 590.

31. Scaramuzza P, Micijevic E, Chander G (2004) SLC gap-filled products, phase one methodology. Landsat Technical Notes.

32. Crist E, Cicone R (1984) Applications of the tasseled cap concept to simulated Thematic Mapper Data. Photogrammetric Engineering of Remote Sensing 50: 343-352.

33. Roujean JL, Bréon FM (1995) Estimating PAR absorbed by vegetation from Bidirectional Reflectance Measurements. Remote Sensing of Environment 51: 375-384.

34. Jordan CF (1969) Derivation of leaf area index from quality of light in the forest floor. Ecology 50: 663-666.

35. Gamon JA, Surfus JS (1999) Assessing leaf pigment content and activity with a reflectometer. New Phytologist 143: 105-117.

36. Coops N, Johnson M, Wulder M, White J (2006) Assessment of QuickBird high spatial resolution imagery to detect red attack damage due to mountain pine beetle infestion. Remote Sensing of Environment 30: 69-76.

37. Rouse J, Haas R, Schell J, Deering D (1973) Monitoring vegetation system in the great plains with ERTS. Third ERST Symposium 1: 309-317.

38. Nemani R, Pierce L, Running SW, Band L (1993) Forest ecosystem processes at the watershed scale: Sensitivity to remotely sensed leaf area index estimates. International Journal of Remote Sensing 14: 2519-2534.

39. Evans JS, Hudak AT (2007) A multiscale curvature algorithm for classifying discrete return LiDAR in forested environments. Geoscience and Remote Sensing 45: 1029-1038.

40. Isenburg M, Liu Y, Shewchuk J, Snoeyink J, Thirion T (2006) Generating Raster DEM from Mass Points via TIN Streaming. Geographic Information Science 186-198.

41. R Development Core Team (2011) R: A language and environment for statistical computing. R Foundation for Statistical Computing, Vienna, Austria.

42. Breiman L (2001) Random forests. Machine Learning 45: 5-32.

43. Bauer E, Kohavi R (1999) An empirical comparison of voting classification algorithms: Bagging, boosting, and variants. Machine Learning 36: 105-139.

44. Dietterich TG (2000) An experimental comparison of three methods for constructing ensembles of decision trees: Bagging, boosting, and randomization. Machine Learning 40: 139-157.

45. Prasad A, Iverson L, Liaw A (2006) Newer Classification and Regression Tree Techniques: Baggins and Random Forests for Ecological Prediction. Ecosystems 9: 181-199.

46. Grömping U (2009) Variable Importance Assessment in Regression: Linear Regression versus Random Forest. The American Statistician 63: 308-319.

47. Liaw A, Wiener M (2002) Classification and Regression by random Forest. R News 2: 18-22.

48. Reif D, Motsinger A, McKinney B, Crowe J, Moore J (2006) Feature selection using a random forests classifier for the integrated analysis of multiple data types. IEEE Symposium on Computational Intelligence and Bioinformatics and Computational Biology 1-8.

49. Lefsky MA, Cohen WB, Harding D, Parker G, Acker SA, et al. (2002) Lidar remote sensing of above-ground biomass in three biomes. Global Ecology & Biogeography 2: 393-399.

50. Clark ML, Clark DB, Roberts D (2004) Small-footprint lidar estimation of sub-canopy elevation and tree height in a tropical rain forest landscape. Remote Sensing of Environment 91: 68-89.

51. Popescu SC, Wynne RH, Scrivani JA (2004) Fusion of Small-Footprint Lidar and Multispectral Data to Estimate Plot- Level Volume and Biomass in Deciduous and Pine Forests in Virginia, USA. Forest Science 50: 551-565.

5

Evaluating *Pinus nigra* Arn. Ssp Salzmannii (Dunal Franco) Initial Seedling Growth in Different Mediterranean Mountain Areas

Manuel Esteban Lucas-Borja*

Higher Technical School of Agricultural and Forestry Engineering. Castilla La Mancha University, Albacete, Spain

Abstract

A detailed understanding of the laws and processes that determine the ecosystem dynamics is essential in order to develop well-adjusted forestry management plans in light of climate change. This includes a good understanding of natural regeneration processes since natural regeneration can become an important component of stand resilience. Moreover, climatic changes may reduce the success of natural regeneration and hence require adjustments to silvicultural practices. This work analyzes the effect of two different soil treatments and different stand densities on the initial *Pinus nigra Arn. ssp salzmanni* seedlings growth at six testing sites located in the Serranía de Cuenca (Spain). Preliminary results showed that soil preparation does not affect initial seedlings growth since differences between soil preparation techniques were always not significant (P>0.05). This work also showed that Spanish black pine initial seedlings growth tended to be higher at the most common and favourable habitats than in the marginal populations. Further studies are necessary to better understand the specific influences across sites and under changing climatic conditions.

Keywords: Soil preparation; Brushing; Scalping; Basal area; Spanish black pine.

Introduction

The Mediterranean forests have a rich biodiversity, are attractive for tourism and recreation, and have a great potential for high value forest products. In this context, sustainable forest management must encompass the maintenance of the ecosystem's integrity to support multiple users. In spite of the hot and dry season in the Mediterranean region, and water deficits during the growing seasons, natural forest regeneration problems can be overcome by selecting silvicultural characteristics of trees, which are best adapted to the specific ecological conditions of the Mediterranean eco-regions.

Spanish black pine (*Pinus nigra Arn*) is a taxonomically complex species found in the Mediterranean region (Greece/Turkey and the Iberian Peninsula), and there are various subspecies with similar adaptive capacity and low genetic differentiation [1]. Spanish black pine forests have been included in the Europe Union endangered habitats listing of natural habitats requiring specific conservation measures (as per Resolution 4/1996 in the Convention on the Conservation of European Wildlife and Natural Habitats), due to their environmental value (in particular landscaping and soil protection) and the difficulty of achieving successful natural regeneration. It is necessary to analyze the factors involved in its natural regeneration in order to advise forestry engineers and foresters.

Since the end of the 19th century, Spanish black pine stands in Spain have been managed under different systems that do not always promote forest stand persistence [2]. A shelterwood system with a 100 to 120 year rotation length and regeneration periods of 20 to 30 years regeneration period is now one of the most commonly used management methods [3,4]. Several authors agree that it is a shade-tolerant species in the early stages, although others suggest that it may invade full-lighted sites. Preto [5] recommend a crown coverage level from 20 to 40% in order to achieve an adequate natural regeneration. Debazac [6] observed its capacity to establish both in abandoned crops and under 90 years old pure stands. Amorini and Gambi [7] concluded that Spanish black pine needs a high proportion of sunlight. Other authors focus the question from a side-shade point of view; thus Pardé [8] concluded that the natural regeneration is difficult in openings and requirements in side shade are needed at least. Recently, some research works has been focused on the Spanish black pine seed rain, seed germination and seedlings survival under different stand densities and soil preparation techniques [9,10]. Data from these studies showed that Spanish black pine regeneration can be avoided in different forest stand in Cuenca Mountain due to the low number of seedlings survival and that climatic conditions had a strong influence in regeneration process. However, an adequate interval of basal area in combination with soil treatments (i.e. scalping) can improve seed germination and seedlings survival in mast years (low seed production years).

Initial seedling recruitment has always been difficult to achieve due to factors including summer drought, soil compaction, masting conditions, seed predation, inadequate overstorey density, and attacks by European pine shoot moth on seedlings and damage from grazing animals, and the combination of each [3]. However, available information on management history shows that the challenges of natural regeneration of Spanish black pine stands in the Cuenca Mountains are not new. During the last century forest managers tried to enhance regeneration by using soil treatment, artificial regeneration, and the introduction of new varieties for the species. Yet, results have been inconsistent [9]. Protecting areas of natural regeneration from browsing by large mammals has also been necessary, and is a major problem for regeneration of forest stands.

The stages of seeds and seedlings (i.e. seed germination, seedlings survival and early seedling growth) have been recognized as the most

***Corresponding author:** Manuel Esteban Lucas-Borja, Higher Technical School of Agricultural and Forestry Engineering. Castilla La Mancha University, Albacete, Spain, E-mail: ManuelEsteban.Lucas@uclm.es

important stages in natural regeneration governing the regeneration, structure and succession of vegetation [2]. Early natural regeneration can be viewed as the passing of a population through a series of environmental sieves that selectively filter out some individuals while allowing others to pass through to the next stage. Most Spanish black pine studies have been focused in seed germination or seedling survival stages [2,9-14] and little attention has been paid to the initial seedling growth. The objective of this study is to determinate in a general sense and under different field conditions (six experimental forests distributed across an altitudinal gradient) the influence of soil preparation on initial Spanish black pine seedling survival. The comparison among contrasting geographical ranges may contribute to an understanding of the role of environmental conditions in this natural regeneration stage and assist in forecasting Spanish black pine regeneration responses to global climate change. To accomplish this objective, initial seedlings growth were measured and modeller in different experimental sites and under two soil treatments. We hypothesized that soil preparation has a positive influence on initial seedling growth and dry weight although forest location can influence this trend.

Material and Methods

From 2000 to 2002 six experimental sites throughout the Serranía of Cuenca (Spain) were set-up in even-age forest stand dominated by Spanish black pine, representing the climatic gradient of this area. The selected sites were Forest of Public Utility n° 131 "Cerro Candalar" (site 1), Forest of Public Utility nº 133 "Ensanche de las Majadas" (sites 2 and 3), where forest of Spanish black pine are mixed with Scots pine (*Pinus sylvestris*), Forest of Public Utility nº 109 "Ensanche de Buenache" (site 4), where Spanish black pine is mixed with Spanish juniper (*Juniperus thurifera L.*) and Forest of Public Utility nº 106 "Los Palancares y Agregados" (sites 5 and 6), close to pine stands where Spanish black pine is mixed with Maritime pine (*Pinus pinaster*). According to the soil classification included in Soil Survey Staff (1999), *Calcic haploxerept and Lithic haploxeroll* is present in the forest of Public Utility n° 106, Los Palancares y Agregados (sites 5 and 6 respectively) whereas *Typic xerorthent and Lithic xerorthents* can be found in the others experimental sites. Further information related to the experimental sites can be viewed in Tables 1-3.

On each experimental site, arranged together in two 150 m parallel strips (north-south orientation) 30 rectangular plots were established (10 m length x 3 m width) in early February of 2000. Each plot was influenced by a similar Spanish black pine canopy cover and overstorey density, which can be observed in Table 1. With respect to soil treatments, three types of seedbed were once made in early January of 2000 before starting dissemination and randomly assigned to each plot: brushing (removal of ground vegetation), scalping (organic matter removed from up to 1-2 cm of mineral soil) and control (without treatment). Brushing and scalping were made with the straight blade of a Caterpillar D7. Removed vegetation and ground vegetation of control plot was characterize by *Eryngium campFestre L.*, *Thymus bracteatus L.* and *Geranium selvaticum L.*, for experimental sites number 1, 2, 3 and 4 and by *Eryngium campestre L., Geranium selvaticum L., Centaurea paniculata and Achillea odorata* for experimental sites number 5 and 6. At the end of the experiment in December-2002, all surviving seedlings had their height, crow volume (height x width x breadth) and basal diameter measured. Then and in order to obtain the seedlings dry weight, all seedlings were collected and carried out to the laboratory, where seedlings were dried using an oven (85 degrees during 48 hours) and weighted.

Hypothesized differences in initial seedling growth among experimental sites (random effect) and soil treatments (fixed effect) were tested using analysis of variance (ANOVA). Each representative forest stand can be considered as spatially independent. The post-hoc test applied was Fisher´s least significant difference (LSD) method. A significance level of α=0.05 was adopted throughout, unless otherwise stated.

Two models have also been adjusted by multiple linear regression (one with measurements taken in the field and another in the laboratory) with a view to estimating the biomass (total dry weight) of the seedlings measured. The structure of the models studied is as follows:

$$Pst = k \cdot D_c^{a1} \cdot d_b^{a2} \cdot h^{a3} \quad \text{and} \quad Pst = k \cdot L_r^{a1} \cdot d_b^{a2} \cdot h^{a3}$$

where Pst is the total dry weight (g), Dc is the diameter of the crown, L_r is the length of the whole root (mm), db is the diameter of the stem base (mm), h the total height of the aerial section (mm), and k, a_1 a_2 and a_3 are the coefficients estimated by the LSM.. Statistical analyses were made with Statgraphics 5.0® and JMP 7.0® software.

Results

Results showed that whereas experimental site was an important source of variation, soil preparation was not an important factor for any tested parameter related to initial seedling growth. As can be seen in Table 4, seedlings basal diameter and height were significantly influenced by experimental site but not by soil treatment. With respect to crow volume and dry weight, neither experimental site nor soil preparation were important factors (P>0.05).

Spanish black pine seedlings basal diameter tended to be greatest for the species located at Cerro Candalar (mean standard error: 0.96 ± 0.02 mm) and site 1 at Los Palancares y Agregados (mean standard error: 0.96 ± 0.01 mm). No differences were found for the rest of experimental sites (Ensanche de las Majadas site 1 and 2 (mean standard error): 0.87 ± 0.03 mm and 0.86 ± 0.03 mm, respectively; Ensanche de Buenache (mean standard error): 0.87 ± 0.02 mm and site 2 of Los Palancares y Agregados (mean standard error): 0.88 ± 0.03 mm). Soil preparation techniques did not have a significant impact in seedlings basal diameter although seedlings collected in soil preparation plots showed higher basal diameter (mean standard error: 0.93 ± 0.17 mm, 0.94 ± 0.15 mm and 0.84 ± 0.19 mm for scalping, brushing and control, respectively). In relation to seedling height, it was again highest in Cerro Candalar (mean standard error: 46.06 ± 1.03 mm). No differences were found between the rest of experimental areas (mean standard error: 39.01 ± 3.07 mm, 42.08 ± 1.90 mm, 43.24 ± 1.23 mm, 41.23 ± 1.67 mm, 38.72 ± 4.17 mm for Ensanche de Las Majadas site 1 and 2, Ensanche de Buenache and Los Palancares y Agregados site 1 and 2, respectively). Soil preparation generated higher seedlings height although differences were statistically not significant (mean standard error: 42.46 ± 2.73 mm, 42.47 ± 1.95 mm and 40.84 ± 2.09 mm for scalping, brushing and control, respectively). Crow volume factor was not influenced neither experimental site (mean standard error: 15.40 ± 3.17 mm, 14.40 ± 1.96 mm, 16.58 ± 2.24 mm, 17.96 ± 3.01 mm, 15.90 ± 2.67 mm, 16.41 ± 2.14 mm for Cerro Candalar, Ensanche de Las Majadas site 1 and 2, Ensanche de Buenache and Los Palancares y Agregados site 1 and 2, respectively) nor soil preparation (mean standard error: 15.75 ± 1.83 mm, 16.52 ± 2.78 mm and 16.13 ± 2.19 mm for scalping, brushing and control, respectively) (Figures 1 and 2).

In the case of seedlings dry weight (Figures 3 and 4) forest location and soil preparation did not affect significantly this parameter (Table 4). However, seedling dry weight tended to be higher in Cerro Candalar (mean standard error: 0.28 ± 0.09 g) and in site 1 at Los Palancares y Agregados (mean standard error: 0.27 ± 0.08 g) if we compare with the

Site	Experimental site	n/ha	B.A.	Dm	Hm	Ho	Principal Stand Composition
1	"Cerro Candalar"	399	12.59	25.54	12.41	13.56	100% Pn
2	"Ensanche de las Majadas"	265	11.79	24.19	13.12	14.89	75% Pn; 25% PS
3	"Ensanche de las Majadas"	387	13.52	26.44	12.72	13.21	70% Pn; 30% PS
4	"Ensanche de Buenache"	373	13.56	29.71	15.85	16.45	70% Pn; 30% JT
5	"Los Palancares y Agregados"	307	11.87	23.16	11.33	13.01	100% Pn
6	"Los Palancares y Agregados"	555	14.02	27.94	13.76	15.35	100% Pn

Table 1: General characteristics of the forest stand at each experimental sites.
n/ha: Number of trees per ha; B.A: Basal area (m^2/ha); *Dm*: Average stand diametrer (cm); *Hm*: Average stand height (m); *Ho*: Dominant stand height (m).
Pn: *Pinus nigra* Arn. ssp *salzmannii*,; Jt: *Juniperus thurifera* L.; Pp: *Pinus pinaster* ; Ps: *Pinus sylvestris*.

Site	Geographic coordinates	A	P	E	P	P_s	S	t_{mm}	t_m	T_{MM}
1	40º15'40" N, 1º56'40" W	1350	2.0	-	1137	139	3	-4.5	9.6	28.3
2	40º14'30" N, 1º58'10" W	1440	19.5	North	1137	139	3	-4.5	9.6	28.3
3	40º16'10" N, 1º58'40" W	1420	10.0	North	1137	139	3	-4.5	9.6	28.3
4	40º06'08" N, 1º55'10" W	1335	3.8	-	1031	111	3	-1.7	9.8	27.4
5	40º01'50" N, 1º59'10" W	1230	4.0	-	600	100	3	-0.5	11.9	30.5
6	40º01'20" N, 1º58'40" W	1200	4.0	-	600	100	3	-0.5	11.9	30.5

Table 2: Geographical location and physiographical and climatic characteristics of each experimental site (Meteorological station of Environment Ministery, Cuenca 1950-2007).
A: altitude above sea level (m). *p*: slope (%). E: exposure. P: mean annual precipitation (mm). P_s: mean summer precipitation (mm). S: drought period (months). t_{mm}: mean lowest temperatures of the coldest month (°C). t_m: mean annual temperature (°C). Tmm: mean highest temperatures of the hottest month (°C).

Site	Soil Type*	O	A					B or C				
		TH	TH	R.F.	C.	S.	SA.	TH	R.F.	C.	S.	SA.
1	Typic Xerorthent	3	21	15	22	10	68	39	17	24	24	52
2	Lithic Xerorthents	7	20	43	16	10	74	40	80	4	38	58
3	Lithic Xerorthents	5	25	35	20	8	72	**	90	16	8	76
4	Typic Xerorthent	6	44	48	16	12	72	>20	10	22	12	66
5	Calcic Haploxerept	3	8	13	18	10	72	33	3	30	4	66
6	Lithic Haploxeroll	5	18	42	4	6	90	28	50	6	4	90

Table 3: Summary of edaphic characteristics for each experimental site.
*Soil Type: Soil Survey Staff (1999). Tn: soil thickness (cm). RF: rock fragments (%). C: clay (%). S: silt (%). Sa: sand (%).
**Deeper than soil pit because rock fragments

Factor	n	Basal diameter	Height	Crown volume	Dry weight
Experimental site	182	*	*	not significant	not significant
F-ratio		2.49	3.03	1.13	2.23
P-value		0.02	0.01	0.75	0.06
Soil preparation	182	not significant	not significant	not significant	not significant
F-ratio		0.92	0.28	0.27	0.42
P-value		0.49	0.75	0.34	0.65

Table 4: Results of analyses of variance for Spanish Black pine seedlings at the end of the experiment (December 2002). For selected factors (experimental site and soil preparation) asterisk indicates differences statistically significant (P-value<0.05).

other experimental sites (Ensanche de las Majadas site 1 and 2 (mean standard error): 0.19 ± 0.05 mm and 0.17 ± 0.05 mm, respectively; Ensanche de Buenache (mean standard error): 0.20 ± 0.06 mm and site 2 of Los Palancares y Agregados (mean standard error): 0.22 ± 0.03 mm). Seedlings dry weight was higher but not statistically significant in scalped and brushed plots (mean standard error: 0.23 ± 0.04 g and 0.24 ± 0.03 g, respectively) than in control plots (mean standard error: 0.20 ± 0.02 g)

The model which best represents the total dry weight data was the following:

$$Pst = k \cdot L_r^{a1} \cdot d_b^{a2} \cdot h^{a3}$$

where Pst is the total dry weight (g), Lr is the total root length (mm), db the diameter at the stem base (mm), h is the total height of the aerial section (mm), and k, a_1 a_2 and a_3 are the coefficients estimated by the LSM. The previous model has been adjusted after logarithmic data transformation as we go on to show:

$$\ln(Pst) = a_0 + a_1 \ln(L_r) + a_2 \ln(d_b) + a_3 \ln(h)$$

This model accounts for 56.56% of the dry weight variability according to the multiple correlation coefficient corrected in terms of explicative variables, of which Lr accounts for 23.85%, h for 20.05% and db• for 12.66%. The inflation of variance factor takes the values of 1.16, 1.19 and 1.25, respectively, thus no significant multicolinearility exists. According to the inclusion order in the model however, an explained

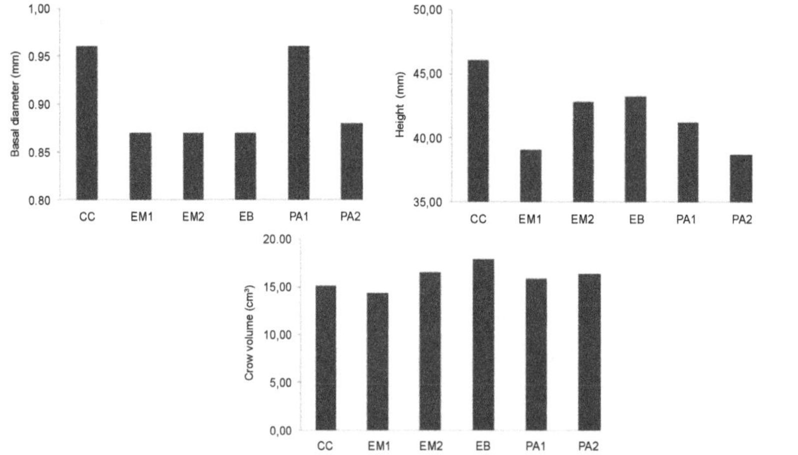

Figure 1: Mean basal diameter, total height and crow volume for seedlings collected at each experimental site. CC: Cerro Candalar, EM1: Ensanche de las Majadas (zone 1), EM2: Ensanche de las Majadas (zone 2), Ensanche de Buenache, PA1: Los Palancares y Agregados (zone 1), PA2: Los Palancares y Agregados (zone 2).

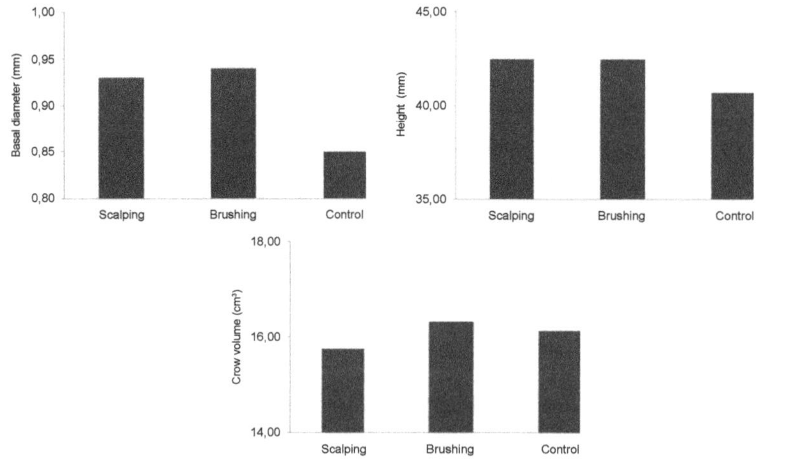

Figure 2: Mean basal diameter, total height and crow volume for seedlings collected at each soil preparation

variability of 27.31% was detected when comparing the variability explained by each variable collectively. Thus, the model that estimates the dry weights of those seedlings sampled in the laboratory would be as follows:

$$Pst = EXP(-0\ ,2353) \cdot L_r^{0,70823} \cdot d_b^{0,617602} \cdot h^{1,17668} \cdot EXP(0,135565/2)$$

With regard to the seedlings measured in the field, which represent the real population, data collected from the aerial section variables were used exclusively. The model that best represents the total dry weight data and which exclusively takes these variables into account was:

$$Pst = k \cdot D_c^{a1} \cdot d_b^{a2} \cdot h^{a3}$$

This model was adjusted by means of logarithmic transformation:

$$\ln(Pst) = a_0 + a_1 \ln(D_c) + a_2 \ln(d_b) + a_3 \ln(h)$$

This model accounts for 44.39% of the total variability, of which the variables Dc, db and h account for 6.52, 16.83 and 21.04%, respectively. Although the inflation of variance factor is low (1.178, 1.204 and 1.256, respectively), 22.36% of collectively explained variability has been detected. The model to estimate the dry weights of the seedlings sampled in the field would be the following:

$$Pst = EXP(-7,46813) \cdot D_c^{0,331751} \cdot d_b^{0,827746} \cdot h^{1,24616} \cdot EXP(0,173503/2)$$

Discussion and conclusion

Management strategies must be focused on ecosystem persistence and a sufficient degree of ecosystem stability. A detailed understanding of the laws and processes, which determine ecosystem dynamics, is an essential basis for developing well-adjusted management plans. This understanding must include the natural regeneration process because a comprehension of natural regeneration is fundamental for planning the stand regeneration in order to get the stand persistence. The plant dynamics must be the base upon which any forest management decision must be made. Water deficits adversely affect several physiological processes, resulting in loss of vegetative growth as well as in injury and mortality seedlings. Moreover, our previous research [9-11] indicates that site preparation, which exposes the mineral soil, enhances conditions for Spanish black seed germination and seedlings survival. Light to moderate partial cut would also favour establishment because the seedbed conditions created would be less extreme than those encountered in more open, exposed seedbeds. Authors concluded on

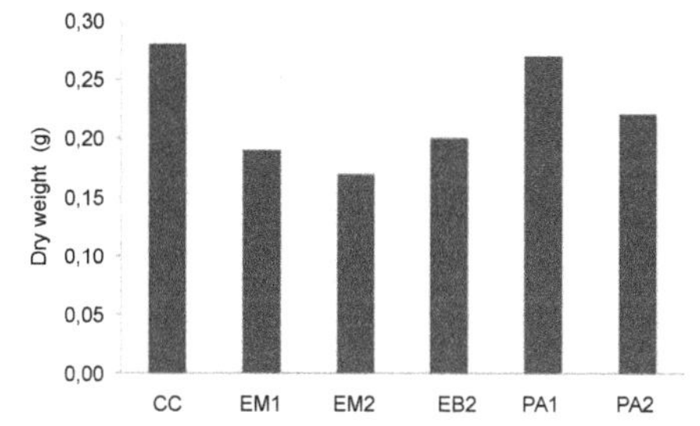

Figure 3: Dry weight for seedlings collected at each experimental site. CC: Cerro Candalar, EM1: Ensanche de las Majadas (zone 1), EM2: Ensanche de las Majadas (zone 2), Ensanche de Buenache, PA1: Los Palancares y Agregados (zone 1), PA2: Los Palancares y Agregados (zone 2).

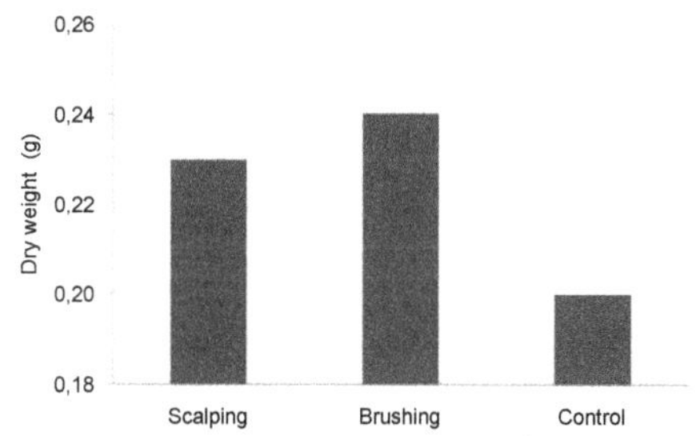

Figure 4: Dry weight for seedlings collected at each soil preparation.

these previous research works that the effect of the overstorey density in combination with soil preparation can be summarised as being positive one on the Spanish black pine seed germination and seedlings survival in Mediterranean forest areas [9-11].

However and contrary to Spanish black pine seed germination and seedling survival percentages previously recorded in the same forest areas [11], soil preparation does not affect initial seedlings growth since differences between soil preparation techniques were always not significant *(P>0.05)*. However, higher seedlings basal diameter, seedlings height and seedling dry weigh were always found in disturbed soils indicating that may be with longer period of study (more than one year) differences could be higher (statistically significant) and soil preparation could became an important factor in relation to initial seedling growth. Contrary to this, once Spanish black pine seedlings root can extend into the deeper soil to avoid competition from ground vegetation and to attain sufficient water and resources, herbal layer of control plots can have a beneficial component, generating lower soil temperatures and higher soil moisture [12]. Thus, soil preparation may negatively affect initial seedling growth. On this context, longer research period would be desirable and needed.

This work also showed that Spanish black pine initial seedlings growth tended to be higher at the most common and favourable habitats than in the marginal populations. As can be seen in Table 1, Spanish black pine populations dominated Cerro Candalar and Los Palancares y Agregados forest stand composition, indicating that these forest areas would be more appropriate and suitable for Spanish black pine initial seedling recruitment. Spanish black pine forests are mixed with Scots pine in Ensanche de las Majadas (sites 2 and 3) and with Spanish juniper in Ensanche de Buenache (site 4) indicating that some shifts in natural conditions occurs and others species and pines growth better, compete or even displace Spanish black pine. Thus, initial seedling growth appears to be site specific and further research would be desirable in order to obtain better conclusion on this topic.

In relation to seedling biomass models, the low proportion of variation explained by site and soil treatment with regard to seedlings biomass production, and the fact that root length is the best biomass predictor, suggests that biomass production at this age depends mostly on microsite conditions favorable for root development. Moreover, Good root development may allow maximum production at reduced fertility levels, and larger root systems may contribute to increased nutrient use efficiency [13], thus favoring initial seedling recruitment.

Higher overstorey density can promote higher seedlings height and basal diameter and negative effects of water deficit and drought period can be corrected (seedling protection by surroundings trees). On places where summer drought is not a problem with respect to initial seedling recruitment, higher sunlight reaching soil surface can be promoted with lower overstorey densities and thus, higher seedlings growth can be generated. At the present work, overstorey density effect over initial seedling growth was not tested (all experimental forest showed similar overstorey density). It would be desirable to evaluate it on the near future.

Knowledge of the processes and factors influencing nature regeneration is essential for its successful application. On this case, the optimal forest management plan should be a compromise between undisturbed forest floor for maximal early seedlings growth and soil preparation for promoting the number of germinated seeds and seedling survival (this last condition demonstrated by the author in other work developed in Cuenca Mountain). The nature of the relationship between natural regeneration of Spanish black pine and all factors involved appears to be site specific and further research work is needed to determine whether the relationships described above apply to other Spanish black pine stand across Spain. It is impossible to fully comprehend the dynamics of seedlings banks on the basis of short-tern studies. Therefore, long-term demographics studies on tree seedlings are necessary for understanding the processes of natural Spanish black pine regeneration, particularly on the new context of climatic change.

References

1. Vidakovic M (1991) Pinus nigra In: Vidakoviæ M. (ed.): Co- nifers: Morphology and variation. Grafièki Zavod Hrvatske, Zagreb 491-520.
2. Tíscar Oliver PA (2003) Condicionantes y limitaciones de la regeneración natural en un pinar oromediterráneo de Pinus nigra subsp. Salzmannii. Investigación agraria. Sistemas y recursos forestales, 12: 55-64.
3. Serrada Hierro R (2002) Consideraciones sobre el tratamiento de masas de pino laricio (Pinus nigra Arn.) en la zona sur del sistema ibérico. Comparación con el pino silvestre. En: GRANDE ORTIZ, M. A., GARCIA ABRIL, A and RODIGUEZ SOLANO, R. (ed.). Gestión Forestal sostenible de los pinares de Pinus nigra Arn. Fundación Conde del Valle de Salazar. Madrid.
4. Serrano Cuenca JL (2002) El Pinus nigra Arn. en la Serranía de Cuenca: Problemas y perspectivas. En España. En: GRANDE ORTIZ, M. A., GARCIA ABRIL, A and RODIGUEZ SOLANO, R. (ed.). Gestión Forestal sostenible de los pinares de Pinus nigra Arn. Fundación Conde del Valle de Salazar. Madrid.
5. Preto G (1983) Il pino nero nella Val de Bidente (Forlì). Aspetti della rinnovazione naturale. Annali dell'Istituto Sperimentale per la Selvicoltura 14: 49-99.
6. Debazac EF (1963) L'aire spontaneé du pin de salzmann en France. Rev. Forest. Fr. 10: 768-794.
7. Amorini E, Gambi G (1976) Un essempio di rinnovazione spontanea del pino nero sull'Appennino Tosco-Marchigiano.Annali dell'Istituto Sperimentale per la Selvicoltura 7: 1-25.
8. Parde J (1962) La régénération du pin noir d'Autriche en Lozère. Rev. Forest. Fr. 14: 931-938.
9. Lucas-Borja ME (2008) La regeneración natural de los montes de Pinus nigra Arn. ssp salzmannii en la Serranía de Cuenca; Bases para la gestión forestal del monte "Los Palancares y Agregados" (CU). PhD Thesis dissertation, Castilla la Mancha University, Spain.

10. Lucas-Borja ME, Silva-Santos P, Fonseca T, López Serrano FR., Tiscar Oliver, PA, et al. (2010) Modelling Spanish black pine postdispersal seed predation in Central-eastern Spain. For. Systems 19: 393-403.

11. Del Cerro A, Lucas-Borja ME, Martínez García E, López-Serrano FR, Andrés-Abellán M, et al. (2009) Influence of stand density and soil treatment on the Spanish black pine (Pinus nigra Arn. ssp. Salzmannii) regeneration in Spain. Invest. Agrar; Sist. Rec. For. 18: 167-180.

12. Del Cerro Barja A, Navarro Lopez R, Lucas Borja ME, Andres Abellan M, Garcia Morote FA, et al. (2006) Factores que influyen en la difícil regeneración de los montes de Pinus nigra Arn., en la Serranía de Cuenca. Revista Montes (2º trimestre): 33-39.

13. Tardieu F (1994) Growth and functioning of roots and of root systems subjected to soil compaction. Towards a system with multiple signalling?. Soil and Tillage Research, 30: 217-243.

14. Tiscar Oliver PA (2007) Dinámica de regeneración de Pinus nigr subsp. Salzmannii al sur de su área de distribución: etapas, proceos y factores implicados. Inves. Agrar.: Sist, Recur. Fores. 16:124-135.

6

Ethnobotanical Knowledge of Medicinal Plants among Tribal Communities in Orissa, India

LS Kandari[1]*, Ashish K. Gharai[2], Tripti Negi[3] and PC Phondani[4]

[1]*School of Natural Resources and Environmental Sciences, College of Agriculture and Environmental Sciences, Haramaya University, P. Box # 70, Dire-Dawa, Ethiopia*
[2]*Centre for People's Forestry, Tarnaka, Hyderabad, India*
[3]*Wildlife Institute of India, P.O. Box No. # 18, Chandrabani, Dehradun 248 001, Uttarakhand, India*
[4]*G.B. Pant Institute of Himalayan Environment & Development, Almora, Uttarakhand, India*

Abstract

Non-timber forest products (NTFPs) are an important source of Pharmaceutical industries for developing a variety of drugs in one hand and in other hand income for the traditional communities living in and around forests. The pressure on forests for the collection of NTFPs has motivated conservation biologists and forest managers to find ways to harvest NTFPs in a sustainable manner. Identification of plant species is the first and foremost criterion in planning for sustainable utilization of resources. The present study was conducted in five-forest division covering 12 villages having an average 345 households of Keonjhar district in Orissa on various NTFPs products. A total of 24 plant species of NTFPs are collected from the forest. The collected NTFPs were used in different purposes i.e. medicines, vegetables, essential oils, dye yielding, food items and other miscellaneous items by the tribal people. The seasonal variations of the availability of NTFPs indicates that there were seven NTFPs found in winter season, eight found in summer season, four in monsoon season and five found in around the year. It was observed that 88.9% villagers were engaged in agriculture followed by (10.14%) labor category and only 0.86% in business or service class. In this study, it is suggested that the tribal community should be involved in the established market chain for marketing of NTFP's. This will be also useful for the conservation and maintenance of NTFPs, which is a major livelihood option for poor traditional communities if we used in a sustainable way.

Keywords: Conservation; Harvesting; Livelihood; Marketing; Non-Timber Forest Products; Sustainable

Introduction

Forest plays an important role in enhancing livelihood requirements for rural community and in maintaining ecological balance. Over 53 million tribal people in India belong to 550 communities of 227 ethnic groups [1-3] and about 60% of the rural communities directly rely on forest for their day-to-day requirement. The dependence of communities on forest can be judged from the fact that an average tribal family draws half of its annual income from forest, 18% from agriculture, 13% from cattle and 18% from other employment [2,4]. Indigenous people have good knowledge of the floristic wealth of their surroundings and having close association with biodiversity rich areas. Apart from timber and firewood that are conceived as major forest produces, non-timber forest products (NTFP) include all products obtainable from forest. NTFPs indeed play a very significant role in the rural economy in terms of providing employment, income potential and life support sustenance [5]. For hundreds and thousands of years, indigenous communities have been utilizing NTFPs for various purposes i.e. edible, medicinal, food and other purposes and have been considered as the secondary production. However, it has acquired some importance in the recent years in policy strategies of the government owing to its significance in life support substance to numerically large number of forest dwelling communities. During the last few years, NTFPs have become adversely affected due to anthropogenic pressure (*i.e.*, encroachment, unsustainable harvesting, forest fire and overgrazing). Moreover, unlike timber that brings profits to government treasure, economic benefits provided by NTFP are accrued largely by the local people and forest dwellers [6,7].

In India Orissa is the second largest state producing NTFPs and 37% of its population depend on NTFPs for their daily day-to-day requirements. The Keonjhar district of Orissa, is harbours a rich diversity of ethnobotanical species, which generate considerable benefits from social and economic perspectives. Until now, people are preparing medicines from their available species of plants, which were used to treat common diseases. Ethnomedicinal uses of plants in different parts of Orissa are well studied [8,9]. However, Keonjhar district is often ignored in spite of the luxuriant growth of ethnomedicinal plants in and around different blocks of the district. There is also not taxonomically or ethnobotanically studies done on these plant. Therefore, the present study has been designed to report the ethnobotanical uses of different NTFPs by the local people. The objective of this study is to motivate to the forest dwellers and villagers to come forward for the utilization, cultivation and marketing of NTFPs from this region.

Study area

This ethnobotanical study was conducted in the Keonjhar district of Orissa state of India during year 2006-2007. The district of Keonjhar, lying between 21° 1' N and 22°10' N latitude and 85° 11' E to 86° 22' E longitude. The district presents a panorama of millennia, both from the geographical and anthropological point of view and spread over an area of 8,240 Sq. k ms Figure 1. The soil of the district is mostly

***Corresponding author:** LS kandari, School of Natural Resources and Environmental Sciences, College of Agriculture and Environmental Sciences, Haramaya University, P. Box # 70, Dire-Dawa, Ethiopia, E-mail: luxkandari@gmail.com

red throughout the district and in the south, there is a small patch of black cotton soil. The important minerals available in huge quantity in the district are Iron-ore, Manganese and Chromites. The climate of the district is characterized by an oppressively hot summer with high humidity. Summer generally commences in the month of March. Temperature begins to rise rapidly attaining the maximum in the month of May. During the summer, maximum temperature is 38.2°C. The weather becomes more pleasant with the advent of the monsoon in June and remains as such up to the end of October. The temperature in the month of December is lowest i.e. 11.7°C and sometimes it even drops down to 7°C. The average annual rainfall is 1534.5 mm.

Socio-economic setup

The study villages are located in interior pockets surrounded by hills and forests. Many villages are not physically accessible easily to the outsiders. The hills and uplands determine the quality of land and land based productivity. The multiple sources of economy of the people consist of animal husbandry, agriculture, forestry and wage labour. The modern village institutions in the name of Van Samrakshan Samiti (V.S.S), Self Help Groups (SHG), Village Development committees (VDC), Village Education Committee (VEC) etc. are supported and influenced by the traditional village institutions, which are guided by the culture and cultural practices of the target communities. There are eight types of STs, living in Keonjhar district, including 4 primitive tribal groups (PTGs) and 12 Schedule caset (SC) communities. The principal tribes were Bathudi, Bhuyan, Bhumij, Gond, HO, Juang, Kharwar, Kisan, Kolha, Kora, Munda, Oraon, Santal, Saora, Sabar and Sounti. These sixteen tribes constituted 96.12 % of the total tribal population of the district.

Methodology

A household survey was conducted using questionnaire and oral interviews with locals to study the uses of NTFPs, annual collection of NTFPs by household and market value of these NTFPs in the selected villages. The indigenous use pattern of each species was also documented during group discussion with the locals. Indigenous people actively engaged in indigenous knowledge were interviewed for recording their unique knowledge about traditional and indigenous

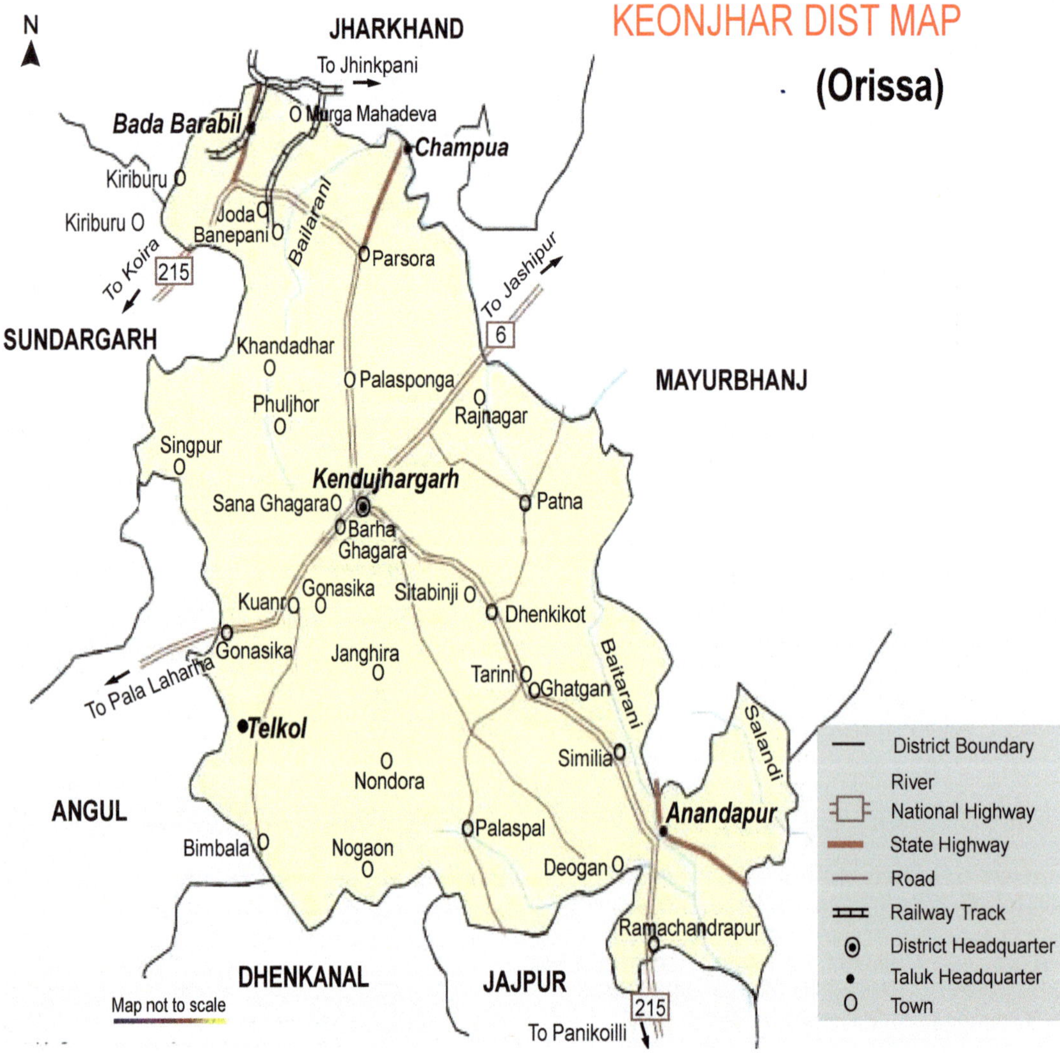

Figure 1: Location map of the study area.

uses of different plant species and their mode of usage. Information on plant local names, local uses and plant parts used and collection methods of each species was also captured. In addition to this, focused group discussions (FGD) were conducted with different age groups of people i.e., (6-12 years, 13-18 years, 19-45 years and above). Ethnobotanical data viz., local name, mode of preparation, medicinal uses were collected through questionnaire, interviews and discussions among the tribal and health practitioners in their local language. Our questionnaire allowed descriptive responses on the plant prescribed, such as part of the plant used, medicinal uses. Voucher specimens were collected identified, by referring standard local floras [10].

Results

A total of 24 plant species of NTFPs are collected from the study site and documented detail about ethnobotanical knowledge. Out of all these eight were collected daily, ten collected weekly, six on fortnight basis. The seasonal variations of the availability of NNTP's indicates that there were seven NTFP's found in winter season eight were found in summer season, four in monsoon and five around the year (Table 1). The commonly collected NTFP's were used in herbal medicines, vegetables & oil seeds, transtuff and dyestuff, food items, and other miscellaneous items. Among five groups of NTFP, drugs constitute the largest proportion of the total in all followed by food items (Figure 2). The production potential of NNTP's varies from region to region owing to the fact of differential natural conditions, socio-climatic conditions, rainfall, biodiversities and more importantly the ecological and environmental conditions. However, the mode of collection also varies from species to species within the season of availability. The collected NTFPs are from wild used by the local communities for curing a verity of ailments since long time (Table 2). The popularly collected NTFPs by majority of people reported in the study includes *Shorea robusta* (95%), followed by mushrooms (92%), *Mangifera Indica* (92%),

SL No	NTFP	Botanical names	Family	Season of availability	Mode of collection
1	Anla	*Emblica officinalis*	Euphorbiaceae	Winter	F
2	Ashwagandha	*Withania somnifera*	Solanaceae	Summer	W
3	Bela	*Aegele marmelos*	Rutaceae	Monsoon	W
4	Chironji	*Buchanania lanzan*	Anacardiaceae	Summer	D
5	Nimba	*Azadirachta indica*	Meliaceae	Summer	F
6	Bahada	*Terminalia bellerica*	Combretaceae	Winter	F
7	Basanga	*Adhotoda vasica*	Acanthaceae	Summer	D
8	Chakunda	*Cassia Tora*	Caesalpiniceae	Winter	D
9	Harida	*Terminalia chebula*	Combretaceae	Winter	F
10	Karanja	*Pongamia pinnata*	Fabaceae	Winter	F
11	Kendu	*Diospyros melanoxylon*	Ebenaceae	Summer	F
12	Kusum	*Schleichera oleosa*	Sapindaceae	Monsoon	D
13	Mahua flower	*Madhuca indica*	Sapotaceae	Summer	D
14	Mango	*Mangifera Indica*	Annacardiaceae	Summer	W
15	Padmachakra	*Gissampelos parcira*	Menispermaceae	Whole year	W
16	Palasi	*Butea monospema*	Eabaceae	Monsoon	D
17	Patalgarud	*Rauwolfia serpentine*	Apocynaceae	Whole year	W
18	Pita alu	*Dioscorea wallichii*	Dioscoreaceae	Whole year	W
19	Sal leaves	*Shorea robusta*	Dipterocarpaceae	Whole year	W
20	Siali fruit	*Boswellia serrata*	Burseraceae	Winter	W
21	Tamarind	*Tamarindus indica*	Papilionanceae	Summer	W
22	Tulasi	*Ocinum sanctum*	Lamiaceae	Winter	D
23	Tejpatta	*Cinnamomum tamala*	Lauraceae	Whole year	W
24	Mushrooms	*Mucuna pruriens*	Leguminosae	Monsoon	D

Note: F= Frequently, D= Daily, W= Weekly

Table 1: Season availability, mode of collection of NTFP's by household.

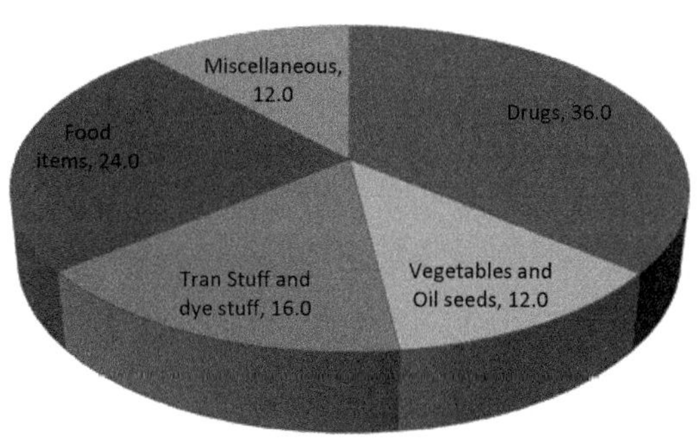

Figure 2: NTFPs used for different purposes by the local people.

SL No	Botanical names	Part used	Traditional Uses by Family
1	*Emblica officinalis*	**Fruits**	Fruit are used as ain rheumatism
2	*Terminalia bellerica*	Fruits	Fruit are used as a herbal medicine
3	*Terminalia chebula*	Fruits	Fruit are used as in stomach pain
4	*Pongamia pinnata.*	Fruit	Fruit are used as a herbal medicine
5	*Schleichera oleosa*	Seed	Oil seed, medicine and soap
6	*Mangifiers Indica*	Fruit	Fruit are used in vomiting and stomach pain.
7	*Rauwolfia serpentine*	Root	Roots are used as a herbal medicine
8	*Shorea robusta*	leaves	The leaves are used for making leaf plates used for edible.
9	*Tamarindus indica*	Fruit	Tamarind is mainly for the fruit and used in variety of dishes.
10	*Cinnamomum tamala*	Leaves	Used as spices.
12	*Buchanania lanzan*	Seeds	Seeds are used after grinding for cooking.

Table 2: Traditional use of NTFPs by the house holds.

Diospyros melanoxylon (80%), *Schleichera oleosa* (67%), *Boswellia serrata* (62%) and 54.6% *Madhuca indica* (Figure 3). *Shorea robusta* leaf and seed are used by tribal people for their livelihood. Leaf used for the plats ad the seed used for oil. *Diospyros melanoxylon* (Kendu leaves), also called Blackwood (Indian ebony), used in making *bidis* (Indian cigarettes). *Diospyros melanoxylon* are the most important NTFP, both from the point of view of the poor as well as the State. *Schleichera oleosa* (Kusum) *Boswellia serrata* seeds are also used for oil. However, most of the NTFPs used by the indigenous people for food and other food supplements, only these people used few species for medicinal purposes. It was observed that 88.9% villagers were engaged in agriculture followed by (10.14%) labour class and only 0.86% in business or service class (Table 3). Distribution of different age groups people engaged in the collection of NTFPs also vary, it was found that the age class between 18-45 years were highly engaged in collection of NTFPs compare to other age groups (Table 4). For collection of NTFPs, maximum distances of 6 to 10 k.ms were covered by 48.6% households for the collection of NTFPs (Table 5). The quantity of collection of NTFPs depends upon its marketability as well as its availability in the forest areas near to the village. Apart from *Diospyros melanoxylon*, which is collected by the department, rest of the NTFPs collected by the locals surreptitiously. However, NTFP collection is one of the main sources of livelihood for people who live below poverty line (BPL). In the district, tribal and non-tribal communities prefer to collect those NTFP species, which give them maximum returns.

Discussion

The study classifies NTFP's according to various purposes, frequency and mode of collection across the seasons, age group. In present study, sal leaves fetch a lucrative source of income to primary collector households. It can be inferred that the volume of procurement in the critical months indicates greater proportion to meet the pressing needs of the households when the distance and time factors do not stand on the way. Analysis of the data reveals that *Shorea robusta* leaves, as well as mushrooms are the main NTFPs collected in Keonjhar district. It is observed that household prefers collection of *Madhuca indica* and *Schleichera oleosa* knots as compared to other NTFPs. The forest dwellers economy is heavily dependent on precious

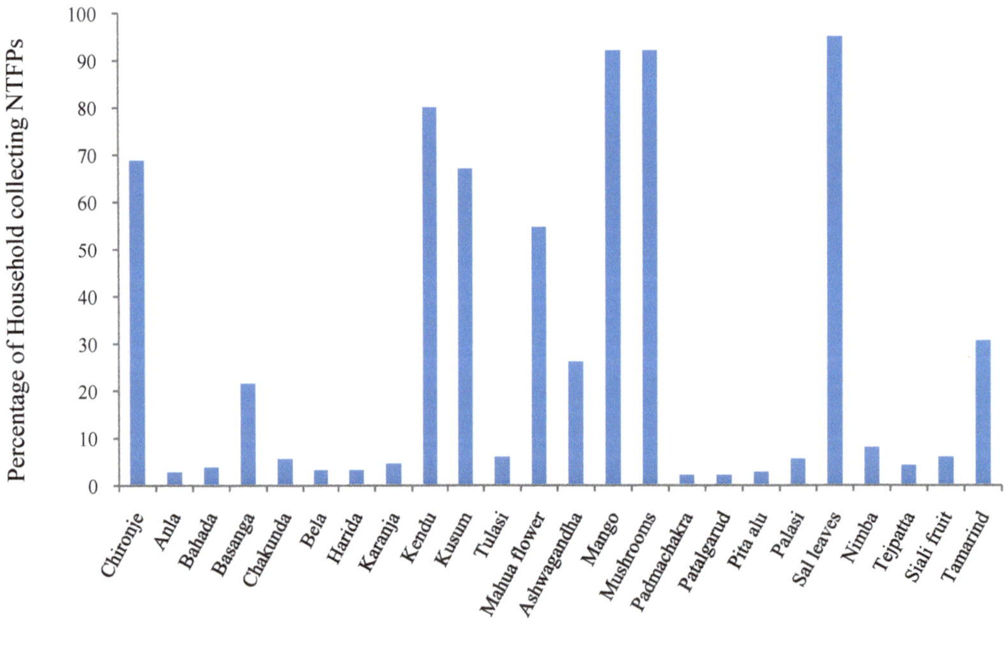

Figure 3: Percentage of households collecting NTFP from wild.

Forest Divisions	No of village covered	Total no of HHs	Population			Total	Primary Occupation		
			Male	Female	Children		Agriculture	Labour/NTFP	Business Service
Keonjhar	12	345	613	581	695	1889	307 (88.98)	35 (10.14)	3 (0.86)

Table 3: Distribution of population covered in the study in the Forest Division.

Forest division	Total HH	Age distribution (in Years)												Total		
		6 to 12			13 to 18			18 to 45			45 above			Persons engaged in NTFP collection		
		Male	Female	Total	Male	Female	Total	Male	Female	Total	Male	Female	Total	Male	Female	Total
Keonjhar	345	112	139	251	62	97	159	187	248	435	62	128	190	423	612	1035

Table 4: Distribution of different age group engaged in the collection of NTFPs.

Distance covered (in k.m.)	Response in percentage (N=345)
>2	5.79
3 to 5	18.55
6 to10	48.69
11 to 15	16.52
16 & above	10.43

Table 5: Coverage of average distance by households for the collection of NTFPs (In k m).

forest resources in general and NTFP's in particular. Another aspect of NTFP's procurement is the lack of full optimization of collection despite huge forest potential due to difficult terrines and inaccessible regions, constraints of market networks, poor and inadequate collection by government sponsored agencies, co-operatives and exploitative interest of businessmen, traders and middlemen, stringent forest laws etc. Despite of this, the forest dwellers are still depending on NTFP's to a substantial extent. Since agriculture yields are lower and uncertain in many areas, the dependence on NTFP's in terms of collection, processing and marketing is still high both for food security and for household needs [11]. Besides these, NTFPs, *Diospyros melanoxylon* leaves also contribute significantly to a household income despite its trade being governed by forest department. Bhattacharya and Hayat [12] conducted a study in sheopur kalan forest division of Madhya Pradesh and reported that several commercially important species were become extinct including, *Boswellia serrata, Sterculia urens* and *Phyllanthus emblica.* With changing political economy of forest resources around the world, benefits of NTFPs are increasingly discussed in valuing tropical forests [13]. The study shows that market forces namely demand for and supply of commodities determine the price of NTFP's.

The nature of the products also influences the market behavior and unorganized market bargaining between unequal ultimately determines the price behavior. Many traders at the road head and among the main trails are the critical link between these NTFPs and the market. They receive market information on species and price along with cash advances from large wholesalers; they are the first major point of transaction for local collectors. The data also observes that some of the agents are product- biased since they deal with particular products. In this process of commercialization of NTFPs, as observed in the ground, it is expected to dominate the market network.

Conclusion

In the state economy, it is observed that the economic contributions of NTFP can't be ignored. From the current study, it is revealed that collection of NTFP species varies according to their market demand and availability, therefore, in order to provide better price of NTFPs to the villagers/collectors, steps should be taken to promote sustainable collection of NTFPs and it becomes imperative to develop marketing facilities. In addition, a numbers of opportunistic buyers from nearby town periodically visit the villages in search of direct supplies thus; the establishment market facilities will also improve the market access of NTFP collectors. However, by developing capacity of local to adopt the value addition options for NTFPs so that their household income improves substantially. However, an attitudinal gradient in the mind of the forest dwellers has been observed which have shifted from a pure constructive dependence to destructive dependence. Therefore, efforts are urgently require for conservation and sustainable marketing of theses NTFPs involving the local tribal communities of the region.

Acknowledgement

The authors are grateful to the Director, Indian Institute of Forest Management, Bhopal for providing opportunity and constant support. The authors are also thankful to the officers of the Orissa forest department and the members of the selected JFMC in Keonjhar and all local people. The University of Haramaya, Ethiopia is highly also acknowledged.

References

1. Maikhuri RK, Gangwar AK (1993) Ethnobotanical notes of the Khasi ad Graro tribes of Meghalaya, north east India. Economic Botany 47: 345-357.
2. Maikhuri RK, Nautiyal S, Rao KS, Saxena KG (1998) Role of medicinal plants in the traditional health care system: A case study from Nanda Devi Biosphere Reserve. Himalaya Curr Sci 75: 152-157.
3. Phondani PC, Maikhuri RK, Rawat LS, Farooquee NA, Kala CP, et al. (2010) Ethnobotanical Uses of Plants among Bhotiya Tribal Communities of Niti Valley in Central Himalaya, India. Ethnobotany Reserch & Application 8: 233-244.
4. (1996) Ministry of Environment and Forest (MoEF) MoEF Annual Report.
5. Negi VS, Maikhuri RK, Rawat LS (2011) Non-Timber Forest Products (NTFPs): A Viable Option for Biodiversity Conservation and Livelihood Enhancement in Central Himalaya. Biodiversity Conservation 20: 545–559.
6. Lele S (1994) Sustainable use of biomass resources. A note on definitions, criteria and practical application. Energy for Sustainable Development 1: 42-46.
7. Panayotou T, Ashton P (1992) Not by timber Alone: Economics and ecology for sustaining Tropical Forests. Washington DC, Island press.
8. Pandey AK, Rout SD (2006) Ethnobotanical uses of plants by tribals of Similipal Biosphere Reserve (Orissa). Ethno 18: 102-106.
9. Rout SD, Panda T, Mishra N (2009) Ethno medicinal plants used to cure different diseases by tribals of Mayurbhanj district of North Orissa. Ethno Med 3: 27-32.
10. Saxena HO, Brahmam M (1996) The Flora of Orissa.
11. Kandari LS, Omprakash MD (2009) Enhancing livelihood through non-timber forest products (NTFPs) Collection: A case study of Keonjhar district, Orissa, India. Ann For 17: 251-257.
12. Bhattacharya P, Hayat SF (2004) Sustainable NTFP management for rural development: a case from Madhya Pradesh, India. International Forestry Review 6: 161-168.
13. Tewari DD (2000) Managing non-timber forest products as an economic resource. The J Inter Econ 11: 269–287.
14. Dobhal P, Sawan S, Sharma N (2007) Studies on medicinal plants of two villages of Chakrata forest division, Uttarakhand, India. Ann of For 15: 351-357.
15. Maikhuri RK, Rao KS, Chauhan K, Kandari LS, Prasad P, et al. (2003) Development of marketing of medicinal plants and other forest products-can it is a pathway for effective management and conservation. The Indian For 129: 169-178.
16. Rout SD, Panda SK (2010) Ethnomedicinal plant resources of Mayurbhanj district, Orissa. Indian J Trad Know 9: 68-72.

The Effect of Enclosures in Rehabilitating Degraded Vegetation: A Case of Enderta District, Northern Ethiopia

Tagel Gebrehiwot[1*] and Anne van der Veen[2]

[1]*University of Twente, P. O. Box 217, 7514 AE Enschede, The Netherlands*
[2]*Professor of Governance and Spatial Integrated Assessment, University of Twente, P. O. Box 217, 7514 AE Enschede, The Netherlands*

Abstract

Ethiopia is one of the African countries facing problems of environmental degradation. In particular, the problem is severe in the Tigray region of Northern Ethiopia highlands where, environmental degradation is threatening many parts of the region. Efforts to contain this problem and facilitate natural rehabilitation have been made at several levels. The establishment of area enclosures, and soil and water conservation works are two of the main strategies promoted to contain land degradation and restore the natural vegetation. This study investigates the effect of area enclosures on regenerating the degraded vegetation. A comparative analysis was performed by investigating two sites: enclosed and unprotected sites. A temporal vegetation change analysis and a vegetation index differencing technique were applied to investigate the change in vegetation that has occurred in the two study sites. Multi-temporal SPOT NDVI images for the period 2001 to 2009 were used for this purpose. The findings of the remote sensing analysis indicated that a consistent positive improvement in vegetation regeneration was observed in the enclosure area. This is indicative of the role of area closure, exclusion of livestock grazing and human interference from selected areas, as an important policy instrument to allow the recovery degraded vegetation provided that they are properly protected and managed.

Keywords: NDVI; Enclosure; Degradation; Rehabilitation

Introduction

Land degradation processes imply a reduction of the potential productivity of the land such as soil degradation and reduction of the quantity and diversity of natural vegetation [1]. Land degradation is one of the most common and serious environmental problems that affect two billion hectares (22.5%) of the world's agricultural land, pasture, forest and woodland [2]. Dry lands are environmentally fragile, and as a result susceptible to degradation. Consequently, the livelihood of millions of farmers living in the dry lands around the world is threatened by degradation of arable farm land.

In Africa, land degradation is a serious problem with a considerable impact on the economies of many countries. Ethiopia is one of the several African countries faced with problems of environmental degradation [3-5]. Land degradation has been identified as the most serious environmental problem in the Ethiopian highlands (lying above 1,500 m above sea level) that comprise nearly 44 per cent of the country's total area and about 88 per cent of the human population [6,7]. Currently, these areas are typified by degradation of vegetation cover and widespread soil degradation.

A long history of land clearing and sedentary agriculture has changed the vegetation cover and degraded the environment of the northern Ethiopian highlands. The conversion of forests, woodlands and shrub lands into croplands and cultivation of steep slopes by traditional ox-plough technology have resulted in loss of natural vegetation cover and severe soil erosion [8-10]. According to the World Bank [11] report, the rate of deforestation in the highlands is estimated to be 150,000 ha per year. Forests have shrunk from the original cover of 65 per cent of the country, and 90 per cent of the highlands, to currently 2.2 per cent and 5.6 per cent respectively. As a result, land degradation has reached an alarming state in the highlands [12,13].

Rapid population growth, over-exploitation of natural resources, limited arable land, and shortage of employment opportunities in the industrial and service sectors have increased pressure on forests and grazing land resources in Ethiopia [14-16]. The inevitable human-induced land degradation through the persistent need for food and firewood over the centuries led to the cutting down of forests on the mountain slopes.

To tackle the problems of deforestation and land degradation, the government has extensively carried out a number of rehabilitation programs supported by the World Food Program's Food-For-Work scheme to prevent further soil erosion and other forms of environmental degradation. Soil and water conservation activities, and the establishment of area enclosures have been two of the main strategies promoted to control land degradation and restore the natural vegetation.

Accordingly, establishing area enclosures in various parts of the region has been a priority in the past decades. However, despite the emerging and promising ecological importance of area enclosures in the study region, very little scientific studies have been carried out to evaluate the effect of enclosures in the rehabilitation of degraded lands. But neither has potential use of remote sensing in monitoring vegetation regeneration after enclosures been evaluated. Important exception includes IAO (2008). IAO (2008) examined the change in vegetation cover that has occurred in area enclosure in Enderta district using two period Landsat images. However, their study failed to show the long-term temporal changes in vegetation cover that has occurred in the area. This apparent limitation provides us with the starting point

***Corresponding author:** Tagel Gebrehiwot, University of Twente, P. O. Box 217, 7514 AE Enschede, The Netherlands, E-mail: t.g.gidey@utwente.nl; ealzabeth@yahoo.com

of this paper [17].

The aim of this paper is to investigate the role of area enclosure in restoring degraded vegetation in a randomly chosen district that is adversely affected by heavy grazing and clearing of vegetation. A comparative analysis was performed by investigating two sites, enclosed and unprotected areas, in order to evaluate the effect of enclosures. Existing literature indicates satellite observations have proven to be a valuable data source providing timely and spatially continuous data with improved details.

Accordingly, the use of large-scale multi-temporal remotely sensed image databases has become a practical tool for monitoring changes in vegetation and assessing natural resource [18]. Thus, a range of remotely sensed vegetation indices have been used to monitor vegetation condition. In this paper, the temporal changes in vegetation that have occurred in the two study sites were examined by applying remotely sensed vegetation indices, Normalized Difference Vegetation Index in particularly, and vegetation index differencing technique. The goal of this research is thus to determine whether area enclosures are effective in regenerating vegetation cover over the course of a 9 year period from 2001 to 2009.

Description of the program

An enclosure refers to a specific land unit that is protected from the activities of a particular class of animals using appropriate barriers such as fencing to control the influence of animals [19]. Area enclosures in the Ethiopian context can be defined as the degraded land that has been excluded from human and livestock interference for rehabilitation. In principal, human and animal interference is restricted for a certain period of time to encourage natural regeneration.

Area enclosures in the Tigray region have been developed through a participatory process with local communities and due consideration was given to ways of gaining their confidence to increase their participation in natural resource management. Community members are involved in site selection and decision-making for the enclosure areas. Local communities reach an agreement before to strictly protect area enclosures from any form of grazing, harvesting grass, and tree cutting. This is supported by local rules and regulations developed by the village council. Communities are also given clear responsibilities to make decisions on issues pertaining to protection, access, management and utilization of the resources. Moreover, the village council is involved in organizing and encouraging participation in woodlots development and also nominate and finance the guards who protect the area enclosures.

Materials and Methods

The Study area

The present study focused on Tigray region, which is one of the regions in northern highlands of Ethiopia most affected by environmental degradation. This is because it has been settled for long and human population density is among the highest in the country [20,21]. Impoverishment of the ever increasing highland population and their search for subsistence income has led to massive deforestation [22,23]. In the past decades, Tigray has lost nearly all of its forests. The environmental degradation caused by deforestation, vegetation clearance and subsequent soil erosion has led to periodic crop failure even under normal rainfall conditions. Moreover, over exploitation of land resources and misuse of land by the age-old traditional system, which pays no attention to conservation methods have led to the loss of natural vegetation cover. As a result, low agricultural productivity and land degradation are pervasive [24-26].

To empirically detect the change in vegetation induced by government policy intervention of area enclosures, a study was conducted in a randomly chosen district, Enderta district. Geographically, it lies between latitudes 130-140 N and longitudes 390-40030' E in the southern part of Tigray region (Figure 1). The climate is predominantly semi-arid with irregular rainfall and frequent drought periods. The mean annual rainfall of the district (locally called woreda) is estimated to be less than 532 mm. Exploitative land use practices such as relentless land cultivation have resulted in most of the land being severely degraded. This area was previously densely forested and dominated by Juniper procera and Olea Africana [27]. As in many parts of the Ethiopian highlands, vegetation clearance to provide additional areas for cultivation, fuel wood and house construction has been a common practice in the woreda.

The study was conducted in two randomly drawn study sites: enclosed and unprotected sites (Figure 1). The area enclosure site covered a total area of 49,255 hectares enclosed for the enhancement of natural vegetation since 2001, while the unprotected land covered 16,932 hectares. The study sites were selected from the same woreda in order to prevent the influence of agro-ecology on the analysis of vegetation cover. Moreover, the two study sites have homogeneous land cover, mainly covered by natural vegetation particularly scrubs which are relatively better for vegetation change analysis than the mixed land uses.

Government official data on area enclosures indicates that by 2001 a total of 262,704 ha of degraded land in Tigray region have been enclosed for rehabilitation purposes. Areas enclosed in several districts of Tigray reached 952,000 ha by 2009, covering about 17.8 per cent of the total land (Figure 2).

SPOT-NDVI datasets

A time series of geo-referenced SPOT Vegetation ten day maximum value composite (MVC) NDVI images for the period 2001-2009 were acquired from VGT4Africa of the DevCoCast project website (http://www.vgt4africa.org). The SPOT Vegetation products have 1km spatial resolution and have a 10-day temporal resolution. The satellite data recorded by SPOT Vegetation instrument are already pre-processed in order to ensure the quality of the NDVI product [28]. The SPOT NDVI

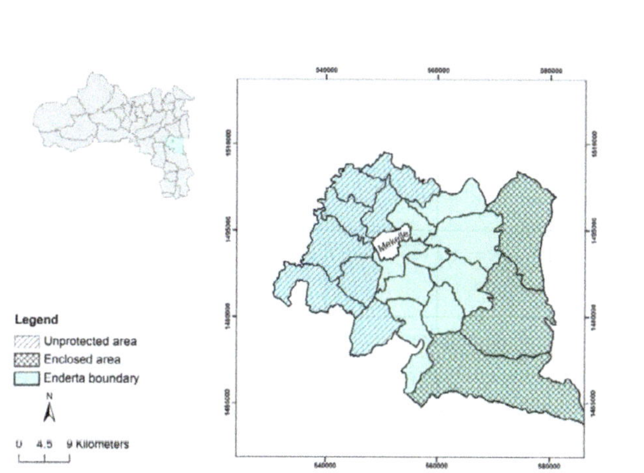

Figure 1: Geographic locations of the study sites in Enderta *woreda*, Southern Tigray, Ethiopia.

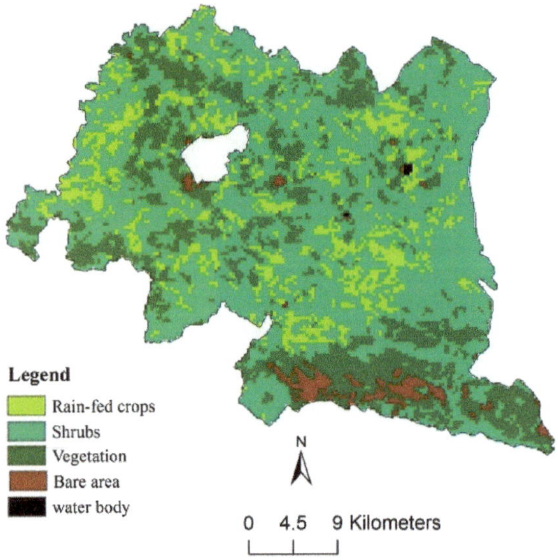

Figure 2: Land cover map of Enderta *woreda*. The land cover map is extracted from the European Space Agency (ESA) Global Land Cover map 2009 with spatial resolution of 300 m.

products are useful for monitoring of long-term vegetation changes and seasonality studies.

According to the seasonal calendar of the study area, all images were taken in the third decade of February. The main rainy season extends from mid-June to mid-September. This period is not an ideal time principally because vegetation differences between the two study sites cannot be captured during the wet period due to moisture differences in the landscape. The NDVI differences between the units during the rainy season will be uniform more or less. In addition, temporal variations in the NDVI reflect the vegetation's response to weather variability and hence the dynamic change of vegetation cover that could occur due to seasonal variations cannot be prevented. In contrast, during the dry season (December to February) there is no effective rainfall, therefore only evergreen trees, bushes and shrubs can be captured with the remotely sensed images. Consequently, temporal variations in the NDVI reflect the presence of vegetation cover. Against this backdrop, February was selected as the ideal month for this study to examine the vegetation change thus preventing the dynamic change of vegetation cover that could occur due to seasonal variations.

Method

Pre-processing of data

Before the SPOT-Vegetation NDVI products were ready for use, some pre-processing was done. Thus, ILWIS GEONETCast toolbox was used to sub-map the original images into the study sites. Then a 10-day composite NDVI map for the third decade of February was produced for both study sites for the period 2001 to 2009 to investigate the temporal change in vegetation.

NDVI-based vegetation change detection

Numerous studies have demonstrated that the Normalized Difference Vegetation Index (NDVI) has been widely applied for monitoring of the dynamic variation of biophysical parameters, land degradation, deforestation, change detection, monitoring famine early warning systems, and drought detection [1,29-36]. The vegetation condition and biomass are often correlated with NDVI [37-39]. Accordingly, for the purpose of this study NDVI, which is the most common vegetation index applied in vegetation change studies, was applied to investigate the role of area enclosure in the restoration of vegetation on the degraded lands.

Observed NDVI trend is an important input and appears reasonable from monitoring of long-term vegetation change perspective. Accordingly, multi-temporal NDVI maps have been successfully used in many vegetation change studies to determine inter-annual variability of vegetation state [40-43]. It is determined as:

$$NDVI = \frac{NIR - RED}{NIR + RED} \quad (1)$$

where NIR is the reflectance in the near-infrared wavelengths and RED is the reflectance in the red wavelengths. In order to identify the temporal changes in vegetation that have occurred in the two study sites, multi-temporal NDVI maps were constructed as they aid to quantitatively analyse the temporal effects of the phenomenon [44]. The consistency of the change in vegetation was further examined by computing NDVI anomalies.

Change detection analysis was subsequently applied. Change detection is the process of identifying differences in the state of an object or phenomenon by observing it at different times [45]. Several methods are available for detecting change in satellite images. According to Hubert-Moy et al. [46], the choice of method is dependent on the area characteristics. In this study, a vegetation index differencing technique was applied for the change detection analysis, which has the advantage of emphasising differences in the spectral response of different features by reducing impacts of topographic effects [44]. Accordingly, vegetation differencing technique was applied to investigate the direction of change and examine whether vegetation conditions in 2009 was worse or better as compared to the average of the previous years. In order to perform the differencing, we first produced the average NDVI map for the years 2001-2008. The average NDVI map was then subtracted from the 2009 SPOT NDVI map to obtain the difference image.

Results and Discussion

The multi-temporal analysis indicated that the temporal NDVI maps for area enclosure demonstrate a noticeable change during the period of 2001-2009, while no change was observed for the unprotected area (Figures 3 and 4). A snapshot of the vegetation condition on the area enclosure clearly shows a consistent increase in the NDVI values between the period 2001 and 2009 as indicated in Figure 3. Figure 3 further shows that the vegetation condition in the enclosed area was very poor at the beginning of the study period and remarkable change was observed over time indicating the effect of protection time on vegetation cover. The enclosed site became greener over the 9 year time period of the study as deduced by the multi-temporal maps. The positive increase in NDVI values indicates high green leaf biomass, canopy closure, or leaf area [47,48]. This change is attributed to its closure from livestock interference and indiscriminate tree felling, which actually fostered regeneration of vegetation cover in the area.

Moreover, the dominant vegetation type of the enclosure area is described as dense mixed bush land dominated by Euclea shimperi locally known as "Kilio". The cover types are in the average range between 1m and 3m in height and with canopy cover of between 50 and 80 per cent. In the month of February, it also consists of dry grasses with some green herbaceous species such as Rumex abyssinicus locally known as "Hohot", Aloe barbadensis locally known as "ika" and Ocimun lamiifolium. On this basis, the increasing positive change

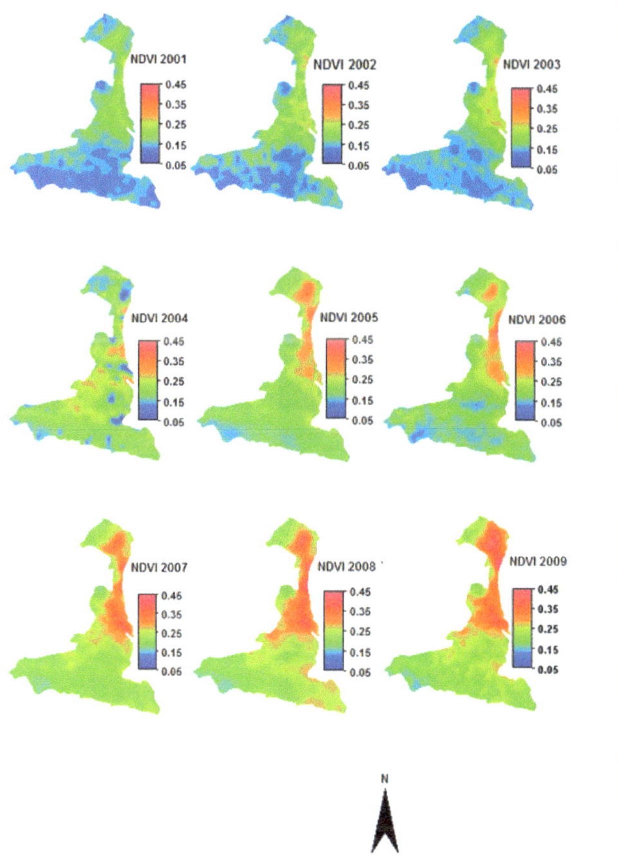

Figure 3: Temporal NDVI maps on area enclosure for the period 2001-2009

in the multi-temporal NDVI values observed in the area enclosure is strong evidence that it is mostly related to improved vegetation as a result of the protection of the area from human and livestock interference, which are amongst the main causes of deforestation. The establishment of area enclosures guarantees to improve the recovery process of vegetation in the degraded areas.

On the contrary, Figure 4 shows a decrease in the NDVI values between the period 2001 and 2009 in the unprotected area. The temporal NDVI maps for the unprotected area shows a better vegetation condition at the beginning of the study period compared to the enclosure area. However, the vegetation starts to decline dramatically since the year 2003. This is mainly attributed to the increasing population, which is growing at a very rapid rate of about 3 per cent annually and their persistent increased demand for cropland, subsistent income and fuel wood that led to vegetation clearance at an alarming rate in the area. Moreover, majority of the people in the Enderta woreda are engaged in subsistence mixed farming. Accordingly, dependence on land for livelihood, absence of alternative employment opportunities, low productivity of cultivated land and the associated poverty have driven vegetation clearance. The unprotected area is currently under a great pressure of wood collection as many people are collecting and transporting wood for sale to nearby urban areas. The findings of the multi-temporal analysis on the two study sites showed the significance of area enclosures in the restoration of vegetation on degraded lands.

The temporal change in vegetation conditions that have occurred in both sites were further analysed by determining NDVI anomalies, which is computed by subtracting the long-term average NDVI (2001-2009) from the corresponding NDVI values in each year (Figure 5). The result of the temporal analysis reveals a consistent positive increase in NDVI anomaly in the area enclosure, while the NDVI anomaly for the unprotected area indicates a declining trend and exhibits variability over time accentuated with positive and negative anomalies. In the period 2001 to 2009, a linear trend line fit (R2=0.73) every year there is 8.03 NDVI increase in the area enclosure, while the NDVI anomaly in the unprotected area tends to decline by 3.62 (R2=0.57) annually indicating a declining trend in vegetation cover. The findings prove that there is strong evidence that enclosure areas appear to be successful in regenerating natural vegetation on degraded drylands. Although the comparison is done for the dry season, we also verified the results by examining the effect of other exogenous factors mainly rainfall.

One climatic station namely Mekelle airport, which is appropriate and nearby the selected study sites was considered to investigate the effect of rainfall. Rainfall for the main rain season at Mekelle airport station tended to decline annually by 8.9 mm y-1 for the period 2001 to 2009 (Figure 5). During the nine years of rainfall recording, mean seasonal rainfall amounts were below the mean in four years. The rainfall pattern over the study area exhibits variability's over time with a high coefficient of variation (25%) accentuated with positive and negative anomalies. Moreover, the study woreda has experienced both dry and wet years over the last 9 years. From the analysis, we found that seasonal rainfall in the study woreda does not show an increasing trend over the period 2001-2009, and thus the effect of rainfall on the observed vegetation was not significant. Consequently, it can be stated that the observed increasing trend in vegetation health in the enclosed area is attributed to the protection of the area from human and livestock interference, which is an outcome of the conservation policies. Similarly, in Tanzania positive trends in NDVI values in woodland and

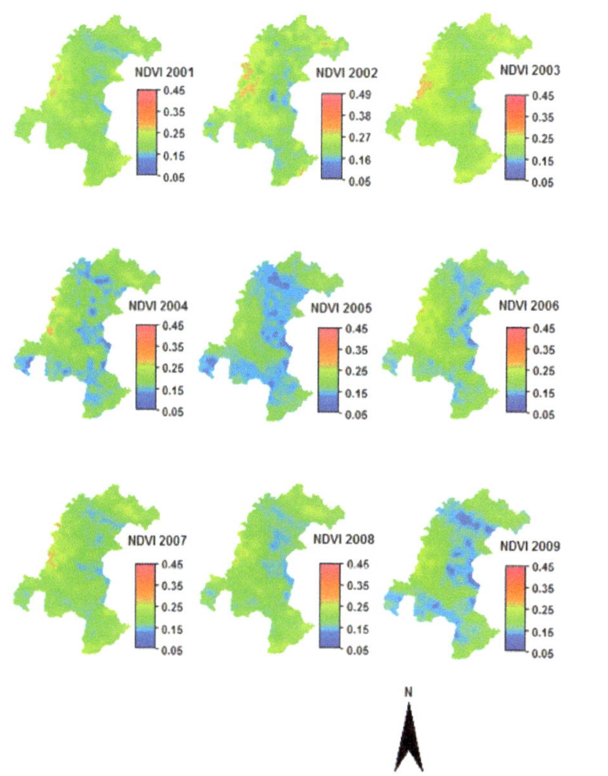

Figure 4: Temporal NDVI maps on unprotected area for the period 2001-2009

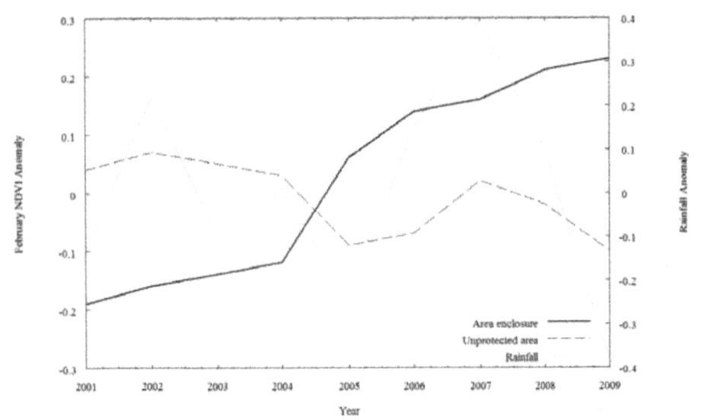

Figure 5: Average NDVI anomalies for area enclosure and unprotected areas vis-à-vis seasonal rainfall (June – September) in Mekelle Airport station for the period 2001-2009.

forest pixels have been associated to changes in conservation policies [18].

Furthermore, vegetation index differencing technique was applied to detect the direction of change in vegetation over the period 2001 to 2009. According to Dobson et al. [49], the basic premise for image differencing is that subtraction results in an image dataset where values less than or greater than zero indicate areas of change. Accordingly, the vegetation index difference for both study sites revealed that a positive change in the vegetation cover is observed in 2009 compared to the 8 years average for the enclosed area, while no change was observed in the unprotected area (Figure 6). The vegetation change detection further indicated that 11 per cent (5,030 hectare) of the protected area showed a substantial increase over the last 9 years when compared with the average NDVI of the previous 8 years. 57 per cent (26,924 hectare) of the protected area showed a moderate change and 32 per cent of the area does not show a change over the period. Feoli (1994), as cited in Aerts et al. [50], reported a direct correlation between vegetation cover and the length of protection time. However, the results of the vegetation index difference for the unprotected are indicated that no change in 2009 compared to the 8 years average (Figure 6). This again indicated that areas that are protected from human and animal intrusion have a good chance of regeneration than unprotected areas.

The vegetation change analysis therefore showed evidence of vegetation restoration of previously degraded lands due to establishment of enclosure. It was also evident that despite the study sites experience uniform rainfall amounts, the enclosure area had improved NDVI reading than the unprotected area. Therefore, the key difference was due to the 'with' and 'without' government intervention to protect the areas.

The findings of the remote sensing analysis provide evidence that a consistent positive improvement in vegetation regeneration is observed in the area enclosure site. This indicates the role of area enclosures, exclusion of livestock grazing and human interference from selected areas, as an important instrument to allow regeneration of degraded vegetation with concomitant benefits for reversing land degradation. Establishment of enclosures is therefore a good policy instrument provided that they are properly protected and managed. It is a cheap method for the restoration of degraded lands as natural processes lead to the revival of the vegetation without any human intervention and financial investment [51,52].

Similar findings have been reported in studies by IAO [17], Palik et al. [53] and Hobbs and Harris [54]. Aerts et al. (2004) also regarded area enclosure as a powerful tool for environmental rehabilitation. This general finding further echoes the findings of Kindeya [55]; Mengistu et al. [52], Muluberhan et al. [56] and Nyssen J. et al. [57] where similar encouraging results on area enclosures in Tigray region have been reported. Descheemaeker et al. [58] also reported area enclosures as an important measures to combat land degradation and increase biomass production. The Nyssen et al. [57] study on the impacts of environmental rehabilitation in Tigray also revealed that sheet and rill erosion rates have decreased, spring discharge, vegetation cover and crop production have improved.

As it is noted earlier in section 1.1, the involvement of local communities in the rehabilitation program has also contributed a lot to the success of the area enclosure program. Devolution of natural resource governance to the lowest level; to those who are close to the problems of natural resource is also one of the important instruments introduced in the environmental rehabilitation measures. Involving community members in site selection and decision-making for the enclosure areas has created a sense of ownership and the community's commitment for effective protection and sustainable management of the resources. Enforceable community laws are also crucial for the success of conservation efforts in the region. The involvement of the community members in designing conservation strategies is showing a great success in achieving the goals set for the projects.

By and large, the strategy of area enclosure is vital to halt and reverse land degradation. The establishment of area enclosures is indeed one of the best solutions to stop further land degradation and to promote reclamation. Thus, based on the findings of our analysis we conclude that the strategy of area enclosure is an effective policy measure to control land degradation and restore natural vegetation on the degraded lands.

Conclusion

A long history of land clearing and sedentary agriculture has changed the vegetation cover and degraded the environment of the northern Ethiopian highlands. The conversion of forests, woodlands and shrub lands into croplands and cultivation of steep slopes by traditional ox-plough technology have resulted in loss of natural vegetation cover. The study region is one of the regions in northern Ethiopia most affected by environmental degradation. In the past decades, Tigray has lost nearly all of its forests. The environmental degradation caused by deforestation, vegetation clearance and subsequent soil erosion has led

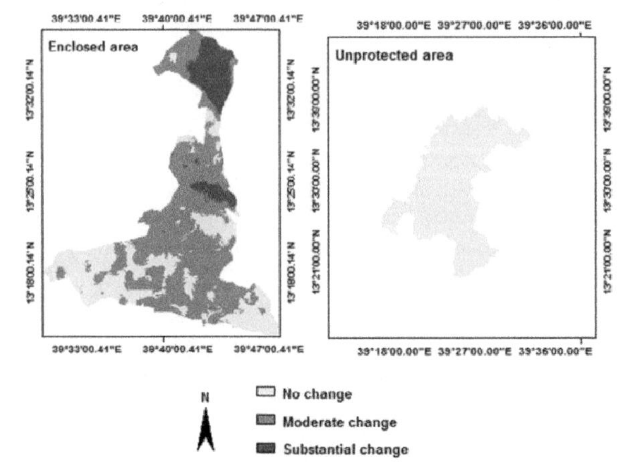

Figure 6: Classified NDVI difference map for area enclosure and unprotected area, Enderta *woreda*

to periodic crop failure even under normal rainfall conditions.

In order to reverse the problems of deforestation and land degradation, the establishment of area enclosures have been promoted to as strategy control land degradation and restore the natural vegetation. Subsequently, establishing area enclosures has been a priority in various parts of the region for the past decade. Hence, investigating the effect of area enclosures in regenerating the natural vegetation on the degraded areas was the primary objective of this study. The findings of the remote sensing analysis indicated that a consistent positive improvement in vegetation regeneration was observed in the enclosure area. This indicates the role of area closure, exclusion of livestock grazing and human interference from selected areas, as an important policy instrument to allow the recovery degraded vegetation provided that they are properly protected and managed.

The strategy of area enclosure coupled with community based resource management practice is a key factor for the observed restoration of natural vegetation cover. Concomitant to this finding, our paper demonstrates the contribution of remote sensing to applications in policy evaluation. Finally, we recommend that further study in other parts of the region is required to gain an informed understanding of the roles played by area enclosures in the restoration of the degraded vegetation in this environmentally very dynamic region of the northern part of Ethiopia.

However, like all studies, ours is not without limitation. First, our analysis is restricted to the examination of multi-temporal maps and the finding is not supported by ground truth data for both study sites. Though is multi-temporal NDVI images have been successful to determine inter-annual variability of vegetation state, the generalization of the results might be limited due to the lack of field data. This limitation should be kept in mind when evaluating the conclusions of our study.

Acknowledgement

The authors gratefully acknowledge the Flemish Institute for Technological Research NV (GEONETCast - VGT4AFRICA) for their kind help in providing the SPOT images used in this study. The comments of the anonymous reviewers are also gratefully acknowledged

References

1. Sommer S, Hill J, Megier J (1998) The potential of remote sensing for monitoring rural land use changes and their effects on soil conditions. Agr Ecosyst Environ 67: 197-209.
2. Al Dousari AM, Misak R, Shahid S, (2000) Soil compaction and sealing in AL-Salmi area, Western Kuwait. Land Degrad. Dev. 11: 401-418.
3. World Bank (1987) Ethiopia; recent economic development and future prospects for recovery and growth.
4. Alemnehm D (1991) Environment, famine and politics in Ethiopia: A view from the village, Lynne Rienner, Boulder/London.
5. FAO (2007) State of the world's forests 2007. Food and Agriculture Organization of the United Nations, Rome. vol 2010.
6. Kruger H , Berhanu F, Michael YG, Kefeni K (1996) Creating an inventory of indigenous SWC measures in Ethiopia, in: Sustaining the Soil: Indigenous Soil and Water Conservation in Africa. IIED, London. 170-180.
7. Haileslassie A, Priess J, Veldkamp E, Teketay D, Lensschen J.P et al. (2005) Assessment of soil nutrient depletion and its spatial variability on smallholders' mixed farming systems in Ethiopia using partial versus full nutrient balances. Agr. Ecosyst. Environ 108: 1-16.
8. McDougall I, Morton WH, Williams NAU (1975) Age and rate of trap series basalts at Blue Nile Gorge, Ethiopia. Nature 254:207-208.
9. Virgo KJ, Munro RN (1978) Soil and erosion features of the Central Plateau region of Tigrai, Ethiopia. Geoderma 20: 131-157.
10. Aune JB, Bussa MT, Asfaw FG, Ayele A.A (2001) the ox ploughing system in Ethiopia: can it be sustained? Outlook Agri 30: 275-280.
11. World Bank (2001) African Development Indicators. Washington
12. Herweg K, Stillhardt B (1999) The variability of soil erosion in the highlands of Ethiopia and Eritrea. Average and extreme erosion patterns. Soil Conservation Research Report 33. Berne, Switzerland, University of Berne.
13. Nyssen J, Poesen J, Moeyersons J, Deckers MH, Lang A et al. (2004) Human impact on the environment in the Ethiopian and Eritrean highlands- state of the art Earth-Sci Rev 64: 273-320.
14. EIU (1985) Quarterly economic review of Uganda, Ethiopia, Somalia, and Djibouti: Annual supplement. Economic Intelligence Unit (EIU), London.
15. Hurni H (1988) Ecological issues in the creation of famine in Ethiopia. Paper presented at the National Conference on Disaster Prevention and Preparedness Strategy for Ethiopia, 5-8 December, Addis Ababa.
16. Hurni H, Kebede T, Zeleke G, (2005) Implications of Changes in Population, Land Use and Land Management for surface runoff in the upper basin area of Ethiopia. Mt. Res. Dev 25: 147-154.
17. IAO (2008) Land Evaluation in Enderta district, Tigray, Ethiopia. Ministero degli Affari Esteri, Istituto Agronomico per l'Oltremare (IAO), Università degli Studi di Firenze, Facoltà di Agraria, IAO 28th Edition, Florence.
18. Pelkey NW, Stoner CJ, Caro TM (2000) Vegetation in Tanzania: Assessing long term trends and effects of protection using satellite imagery. Biol Conserv 94: 297-309.
19. Young S (1958) Exclosures in big game management in Utah. J. Range Manage 11: 186-190
20. Fattovich R (1997) The peopling of the Tigrean Plateau in ancient and medieval times (ca. 4000 B.C. – A.D. 1500): evidence and synthesis, in: Bard, K.A. (Ed.), The environmental history and human ecology of northern Ethiopia in the late Holocene: Preliminary results of a multidisciplinary project. Naples Insituto Universitario Orientale, Italy, pp. 81-105.
21. Pender J, Gebremedhin B (2008) Determinants of agricultural and land management practices and impacts on crop production and household income in the highlands of Tigray, Ethiopia. J Afr Econ 17: 395-450.
22. Chadhokar P, Solomon A (1988) Importance of revegetation in soil conservation in Ethiopia, in: Rimwanich (Eds.), Constraints and solutions to application of conservation practices. Bangkok, Thailand 1203-1213.
23. Nyssen J (1997) Vegetation and soil erosion in Dega Tembien (Tigray, Ethiopia). Bulletin du Jardin botanique national de Belgique. 66: 39-62.
24. Bekele S, Holden S (1998) Resource degradation and adoption of land conservation techniques in the Ethiopian Highlands: A case study in Andit Tid, North Showa. Agric. Econ 18: 233-247.
25. Sonneveld BGJS, Keyzer MA (2002) Land under pressure: soil conservation concerns and opportunities for Ethiopia. Land Degrad Dev 14: 5-23.
26. Holden S, Shiferaw B (2004) Land degradation, drought and food security in a less-favoured area in the Ethiopian highlands: a bio-economic model with market imperfections. Agric. Econ 30: 31-49.
27. Darbyshire I, Lamb H, Umer M (2003) Forest Clearence and Regrowth in Northern Ethiopia During the Last 3000 Years. The Holocene 13:537-546.
28. Jacobs T, Piccard I, Bydekerke L, Gontier E (2006) S10 NDVI, in: Bartholomé, E. (Ed.), VGT4Africa user manual. First edition. European Communities, Luxembourg, pp. 97-105.
29. Tucker C (1979) Red and photographic infrared linear combination for monitoring vegetation. Remote Sens Environ 8: 127-150.
30. Chavez Jr P.S, MacKinnon D.J (1994) Automatic detection of vegetation changes in the Southwestern United States using remotely sensed images. Photogramm. Eng. Rem 60:571- 583.
31. Myneni R, Asrar G (1994) Atmospheric effects and spectral vegetation indice. Remote Sens Environ 47: 390-402.
32. Bannari A, Morin D, Bonn F, Huete A.R (1995)A review of vegetation indices. Remote Sens. Rev13.95-120.
33. Lawrence RL, Ripple WJ (1998) Comparison among vegetation indices and band wise regression in a highly disturbed, heterogeneous landscape: Mount St. Helens, Washington. Remote Sens Environ 64: 91-102.

34. Huete A, Didan K, Miura T, Rodriguez E.P, Gao X (2002) Overview of the radiometric and biophysical performance of the MODIS vegetation indices. Remote Sens. Environ 83: 195-213.

35. Anyamba, A, Tucker CJ (2005) Analysis of Sahelian vegetation dynamics using NOAA-AVHRR NDVI data from 1981-2003. J Arid Environ 63: 596-614.

36. Prasad AK, Sarkar S, Singh R, Kafatos M (2007) Inter-annual variability of vegetation cover and rainfall over India. Adv Space Res 39: 79-87.

37. Hunt ER (1994) Relationship between woody biomass and PAR conversion efficiency for estimating net primary production from NDVI. Int. J. Remote Sens 15: 1725-1730.

38. Gamon JA, Field CB, Goulden ML, Griffin KL, Hartley AE, et al. (1995) Relationship between NDVI, canopy structure, and photosynthesis in three Californian vegetation types. Ecol. Appl 5: 28-41.

39. Nilsson H (1995) Remote sensing and image analysis in plant pathology. Annu Rev Phytopathol 33: 489-528.

40. Marsh SE, Walsh JL, Lee CT, Beck LR, Hutchinson CF (1992) Comparison of multi-temporal NOAA-AVHRR and SPOT-XS satellite data for mapping land cover dynamics in the West African Sahel. Int J Remote Sens 13: 2997-3016.

41. Di L Rundquist D.C, Han L (1994) Modelling relationships between NDVI and precipitation during vegetative growth cycles. Int. J. Remote Sens15: 2121-2136.

42. Reed BC, Brown JF, Vanderzee D, Lovel TR, Merchant JW (1994) Measuring phenological variability from satellite imagery. J Veg Sci 5: 703-714.

43. Lyon J, Yuan D, Lunetta RS, Elvidge CD (1998) A change detection experiment using vegetation indices. Photogramm. Eng Rem S 62: 143-150.

44. Lu D Mausel P, Brondízio E, Moran E (2004) Change detection techniques. Int J Remote Sens 25: 2365-2407.

45. Singh A (1989) Digital change detection techniques using remotely sensed data. Int J Remote Sens 10: 989-1003.

46. Hubert-Moy L, Cotonnec A, Le Du, L, Chardin A, Perez P (2001) A comparison of parametric classification procedures of remotely sensed data applied on different landscape units. Remote Sens. Environ 75: 174-187.

47. Sader SA, Winne JC (1992) RGB-NDVI color composites for visualizing forest change dynamics. Int J Remote Sens 13: 3055- 3067.

48. Myneni RB, Keeling C, Tucker C, Asrar G, Nemani R et al. (1997) Increased plant growth in the northern high latitudes from 1981 to 1991. Nature. 386: 698-702.

49. Dobson J.E, Bright E.A, Ferguson R.L, Field D.W, Wood L.L et al.(1995) NOAA Coastwatch Change Analysis Program (C-CAP) Guidance for Regional Implementation, NOAA Technical Report NMFS123, NOAA Coastal Ocean Program, U.S. Department of Commerce, Washington, D.C.

50. Aerts R, November E, Behailu M, Deckers J, Hermy M, et al. (2002) Forest rehabilitation: one approach to water conservation in Central Tigray. Eth. J. Water Sci. Technol. 6: 34-37.

51. Bendz M (1986) Hill Side Closures in Welo. Ethiopian Red Cross Society Mission Report, Vaxjo, Sweden.

52. Mengistu T, Teketay D, Hulten H, Yemshawd Y (2005) The role of enclosures in the recovery in the recovery of woody vegetation in degraded dryland hillsides of central and northern Ethiopia. J. Arid Environ 60: 259-281.

53. Palik BJ, Goelbel PC, Kirkman LK, West L (2000) Using landscape hierarchies to guide restoration of distributed ecosystems. Ecol. Appl 10: 189-202.

54. Hobbs RJ, Harris JA (2001) Restoration ecology: repairing the earth's ecosystems in the new millennium. Restor. Ecol 9: 239-246.

55. Kindeya G (1997) Area Enclosures as an Approach in the Management of dryland Biodiversity: A case study of Tigray Region. Mekelle University, Ethiopia.

56. Muluberhan HA, Houhoulis PF (2006) The role of area enclosures and fallow age in the restoration of plant diversity in northern Ethiopia. Afr J Ecol 44: 507-514.

57. Nyssen J, Munro N, Haile M, Poesen J, Descheemaeker K (2007) Understanding the environmental changes in Tigray: a photographic record over 30 years. VLIR-Mekele University Zala-Daget Project.

58. Descheemaeker K, Nyssen J, Rossi J, Poesen J, Haile M et al. (2006) Sediment deposition and pedogenesis in exclosures in the Tigray highlands, Ethiopia. Geoderma. 132:291-314.

8

Spatial Relationships between Canopy Structure and Understory Vegetation of an Old-Growth Douglas-Fir Forest

Bo Song[1]*, Jiquan Chen[2] and Thomas M Williams[1]

[1]*Belle W. Baruch Institute of Coastal Ecology and Forest Science, Clemson University, Georgetown, USA*
[2]*Department of Earth, Ecological & Environmental Science, University of Toledo, Toledo, USA*

Abstract

Canopy structure and spatial distributions of understory vegetation are known to be closely correlated in forest community. The primary objective of this study is to examine the spatial relationships between the canopy heterogeneity and understory vegetation at multiple scales in an old-growth Douglas fir forest. A 12 ha plot located in the T.T. Munger Experimental Forest of Washington was stem-mapped, and the understory vegetation was surveyed along two 400 m transects by species and life form groups. Canopy structure of the plot was modeled utilizing a stem map and a crown geometry model. The three-dimensional canopy characteristics derived from the models were then used to assess the effects of canopy structure on understory vegetation at multiple resolutions using correlation and wavelet analysis techniques. Correlation results showed significant associations of understory species with the three-dimensional canopy structure. The majority of dominant herb species were highly associated with canopy opening, and they were more severely affected by lower canopy layers (heights of 10-35 m) than by dominant-codominant layers (heights of 40-50 m). The saplings of western hemlock had a significant negative association with lower canopy layers, while vine maples and the saplings of Pacific silver fir were positively correlated with dominant-codominant canopy layers. Wavelet analysis indicated that the relationship between canopy structure and understory vegetation was highly scale-dependent, *i.e.*, understory variables responded differently to the heterogeneity of canopy patterns at different scales.

Keywords: Forest canopy, Spatial heterogeneity, Wavelet analysis, Scale, Temperate forest.

Introduction

Canopy structure is a key variable influencing many aspects of forest ecosystems [1-7]. Numerous studies have indicated that patterns of understory vegetation and physical environment (e.g., temperate, soil nutrients and moisture availability) were directly related to canopy openness [2,6-10]. For example, forest canopies provide microhabitats for plants, including mosses, herbs, and shrubs [11-13], and influence the composition of understory species, seedling regeneration, and microclimate conditions [2,6,7,10,14-17]. As the upper interface between forests and the atmosphere, forest canopies provide critical buffering from external disturbance for forest interior and create a microclimatic regime (e.g., sun flecks) below the canopy that drives understory processes. A high degree of structural diversity of the canopy provides high microsite variability, and consequently, diverse understory vegetation [2-4,13].

Studies have also demonstrated the importance of the three-dimensional structure of forest canopies in determining microclimatic conditions in the understory [1,4,5,16,17]. Vertical structures, such as gap aspect ratio, tree height, and branch architecture, affect light regime and moisture level in the understory [9,10,18,19]. As a result, the differences in crown geometry and tree architecture can greatly influence the development of understory vegetation and successional dynamics [20,21]. Horizontal structural features lead to spatial variation in the understory microclimate and are important for developing and maintaining a structurally heterogeneous stand and diverse understory [2,8,13,21]. Radiation that reaches the understory is also affected by the spatial location of trees, stand age, density, and gap size [22-24]. Thus, understory conditions likely vary in a substantial manner due to different heights of canopy and horizontal locations because of the heterogeneity of the canopy structure. These height and location dependent variations in canopy structure are important for understanding the relationships between canopy and understory vegetation.

The primary objective of this study is to examine quantitatively the relationships between three-dimensional canopy structure and distribution of understory plants by investigating the variations in canopy structure at different heights and transect locations. We are particularly interested in examining whether the relationships between canopies at different heights and understory vegetation vary and if they are scale-dependent along transects.

Methods

Study area

The old-growth Douglas-fir (*Pseudotsuga menziesii*) forest at the Wind River Canopy Crane Research Facility (WRCCRF) is located within the T. T. Munger Research Natural Area of the Wind River Experimental Forest, 45°48'N, 121°58'W, on the Gifford Pinchot National Forest in southern Washington. The site is in the Wind River valley at 355 m elevation, deep within the southern Washington Cascade Range. A seasonally intermittent stream runs through the study site. The weather in the area is characterized as a temperate wet

***Corresponding author:** Bo Song, Belle W. Baruch Institute of Coastal Ecology and Forest Science, Clemson University, Georgetown, USA
E-mail: bosong@clemson.edu

winter, dry summer climate with 2,528 mm of annual precipitation, with less than 10% occurring between June and September. Average annual snowfall is 2,330 mm. Mean annual temperature is 8.7°C. Soils are medial, mesic, Entic Vitrands that are deep (2-3 m), well drained, loams and silt loams, generally stone free, and derived from volcanic tephra [25].

This study area was located in a 500-year-old temperate forest stand dominated by Douglas fir and western hemlock (*Tsuga heterophylla*), with the tallest trees averaging 55 to 65 m (maximum 67 m). Other canopy species include western redcedar (*Thuja plicata*), western white pine (*Pinus monticola*), Pacific silver fir (*Abies amabilis*), grand fir (*Abies grandis*), and noble fir (*Abies procera*). Understory trees include Pacific yew (*Taxus brevifolia*) and Pacific dogwood (*Cornus nuttallii*). Dominant understory shrub species are vine maple (*Acer circinatum*), salal (*Gaultheria shallon*), and dwarf Oregon grape (*Berberis nervosa*).

Data collection and analysis

A 12ha (400 × 300 m) plot was established at the study site. Trees with diameters at breast height (DBH, 1.3 m above the ground) greater than 5 cm were measured and identified to species. Their cardinal coordinates (x, y) and the elevation (z) were measured using a computerized total surveying station (WILD TC600 Total Station, Leica Geosystems AG, Heerbrugg, St. Gallen, Switzerland).

Two 400 m line transects with east-west orientation were established in the plot (Figure 1). Understory plants and their life forms (*i.e.*, shrubs, herbs, seedlings, and saplings) were sampled systematically

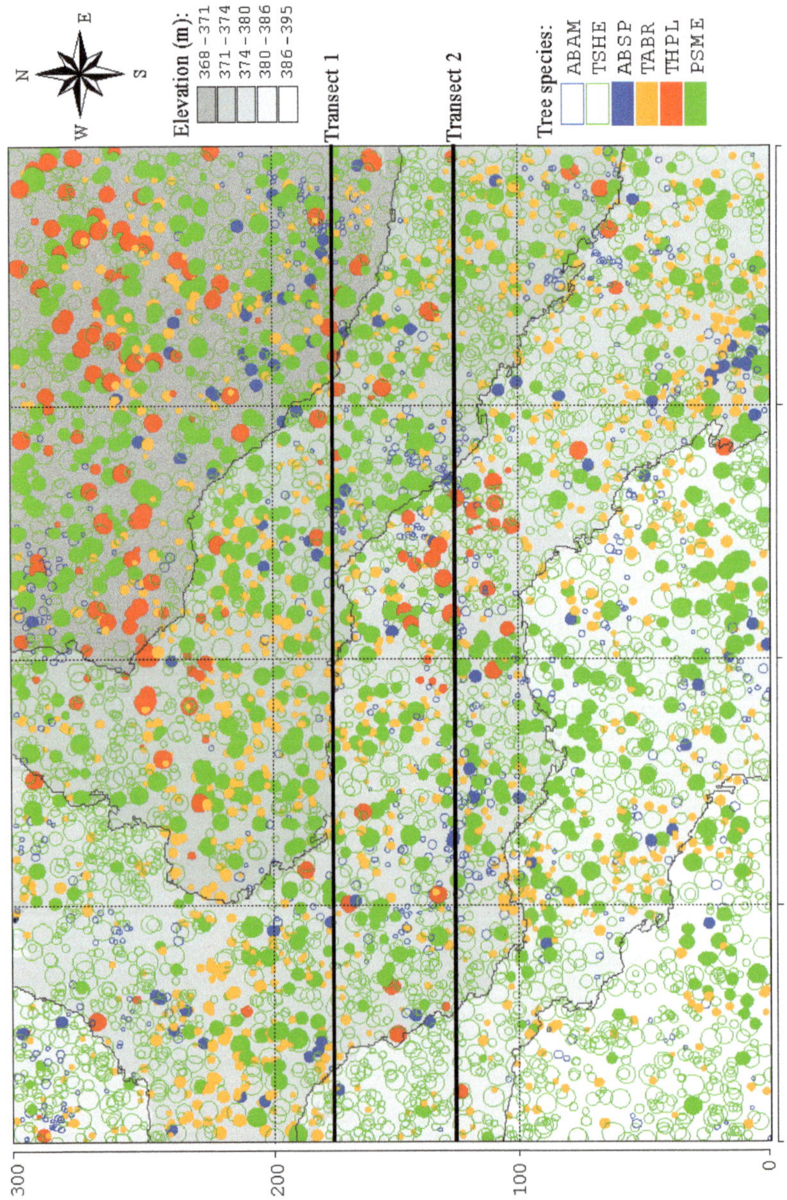

Figure 1: Location and layout of two 400 m transects within a stem-mapped 12 ha (400 x 300 m) old-growth Douglas-fir forest in the Wind River Canopy Crane Research Facility in southern Washington. The species codes are the same as in Table 1. The sizes of circles represent the differences in tree size.

using 2 2 m quadrates every 5 m along these two 400 m transects during the summer of 1996. Percent cover was recorded for each species and life form groups. For the shrubs and saplings, the coverage of different height classes (0-1 and 2-7 m) was also recorded.

Canopy structure was modeled utilizing a stem map and a crown geometry model, and was tested with field-collected data [26]. The outputs of the model include the total projection of canopies for all trees in the plot, and canopy crosscuts at different heights (5-50 m with an interval of 5 m) for all species. Canopy openings along the two transects were also calculated using circular quadrates (radius = 5 m) located at the centers of the 2 × 2 m understory-sampling quadrates.

Frequency and average percent cover of all taxa present along the transects were calculated and tabulated. To assess species dominance, two additional diversity measures were calculated for each species: Shannon-Weiner Diversity Index (H-diversity) and evenness (Table 1). Abundance of each species was also plotted against distance along the transects to help assess distributional patterns and possible spatial relationships with canopy structure. Pearson correlation coefficients were used to examine the association between canopy structure (both horizontal and vertical) and understory vegetation.

To evaluate and compare the patterns of canopy structure and understory vegetation at multiple scales along the transects, we used wavelet analysis for each transect. Wavelet analyses, including wavelet transform and wavelet variance have been applied in several studies of ecological patterns at a range of temporal and spatial scales [24,27-29]. The advantages of using wavelet analysis include: 1) the ability to detect multi-scale patterns, 2) retain location information along the transect, and 3) no need to assume stationary in the data (unlike semi-variogram analysis, [27,29,30]. The wavelet transform is defined as

$$W(a, x_j) = \frac{1}{a} \sum_{i=1}^{n} f(x_i)\, g(\frac{x_i - x_j}{a}) \qquad (1)$$

Where $f(x_i)$ is the data pattern function, $g((x_i\text{-}x_j)/a)$ is a window function (*i.e.*, wavelet) for a given range of scale *a*, centered at location x_j along a transect, and *n* is the number of sample quadrates. We used a Mexican-hat wavelet to quantify peaks and troughs in the data. The resulting image was used to identify distribution patterns of canopy structure and understory vegetation at different scales along the transect and to determine whether they correspond with each other. In addition, we calculated the wavelet variance as:

$$V(a) = \frac{1}{n} \sum_{i=1}^{n} W^2(a, x_j) \qquad (2)$$

Wavelet variance indicates whether one scale contributes more or less than others to the overall pattern along a transect [27]. Pearson correlation coefficients were used to quantify the relationship between wavelet transforms of different variables at multiple scales. A program developed by Li and Loehle [31] was used to calculate *W(a, x)* and *V(a)*.

Results

Canopy structure and the composition of understory vegetation

The old-growth Douglas-fir forest was dominated by western hemlock (TSHE) and Douglas-fir (PSME) (Figure 1). Western hemlock was found in all canopy layers, with an average DBH of 30.8 cm, ranging from 4.8 to 132.1 cm, and height ranging from 4 to 57 m. Douglas-firs were generally taller than 45 m, with an average DBH of 106.2 cm. Most grand fir (),noble fir (), and red cedar (THPL) were in the dominant-co dominant layers (Figure 1), with averages DBH of 45.9 (firs) and 61.4 (cedars) cm. Pacific silver firs (ABAM) and Pacific yews (TABR) were found within the understory layer (Figure 1), with averages DBH of 11.1 and 14.3 cm, respectively. The canopy structure of this plot seemed very heterogeneous and complex, and appears as a mosaic of various patches (Figure 1). The average canopy opening was 19.6%.

Twenty-seven herbs and fourteen shrub species were recorded in a total of 162 plots along the two transects (Table 1). The abundance-frequency relationship for herb species was of negative-exponential shape (Figure 2a). Most herbaceous species occurred infrequently (< 15% of plots) and had low coverage (< 5%). No herb species with high frequency and abundance was found (Figure 2a). The three most frequent herb species were twinflower (*Linnaea borealis*), queen's cup (*Clintonia uniflora*), and vanilla-leaf (*Achlys triphylla*), which appeared in more than 55% of the plots, although their cover ages were low (< 6%). There were some herb species, such as northern bedstraw (*Galium boreale*) and sword fern (*Polystchum munitum*), with very low frequency, but high abundance. Contrary to the herb species, the abundance-frequency relationship for shrub species was of positive-exponential shape (Figure 2b). There were several shrub species with high frequency and abundance, such as Salal and dwarf Oregon-grape, yet no shrub species with high frequency and low coverages.

The canopies were generally closed, but embedded with many openings (Figure 3). The average canopy opening along Transect 1 was 20.3%, with an 18.9% standard deviation, whereas the average opening for Transect 2 was 14.7%, with a 15.4% standard deviation. Transect 1 crossed several relatively large openings, and a trail crossed Transect 1 at around the 325 m location, while Transect 2 had continuous closed canopy for an extended distance (from 180 to 270 m) and relatively aggregated openings.

Spatial associations between understory vegetation and canopy structure

There were obvious spatial associations between understory vegetation and the heterogeneity of canopy pattern along both transects (Figures 3 and 4). The covers (%) of vanilla-leaf, three-leaved anemone (*Anemone deltoidea*), queen's cup, and twinflower were high when there were large canopy openings. Covers of these herbs were close to zero under closed canopies. The covers of the lower canopies (10-35m high) were negatively associated ($P < 0.005$) with covers of vanilla-leaf, queen's cup, inside-out flower (*Vancouveria hexandra*), and three-leaved anemone, while vanilla-leaf, queen's cup, inside-out flower, and three-leaved anemone were positively associated with each other ($P < 0.001$).

The spatial patterns of total abundance and total richness of all herbs were positively correlated with canopy openness ($P < 0.001$), and both were negatively correlated with the cover of lower canopies (10-35m high, $P < 0.005$). The covers and richness of herb species were larger for Transect 1 than those for Transect 2 (Figure 3a, 3b). Pathfinder (*Adenocaulon bicolor*), northern bedstraw, sweet-scented bedstraw (*Galium triflorum*), and nodding trisetum (*Tricetum cernuum*) only appeared on Transect 1, and only within the largest canopy opening on Transect 1. Hooker's fairy bells (*Disporum hookeri*), broad-leaved starflower (*Tiarella trifoliata*), and inside-out flower were also associated with the largest canopy opening on Transect 1. Inside-out flower had another peak at location 325 m of Transect 1, adjacent to a trail.

The total abundance of all shrubs was higher when there were large

			Frequency	Average	Spatial Heterogeneity	
Code	**Scientific Name**	**Common Name**	(%)	**Cover**(%)	**H-diversity**	**Evenness**
Herbs						
LIBO	*Linnaea borealis*	Twinflower	62.3	2.21	3.78	0.82
CLUN	*Clintonia uniflora*	Queen's cup	56.8	1.83	4.17	0.92
ACTR	*Achlys triphylla*	Vanilla-leaf	56.2	5.99	3.51	0.78
VISE	*Viola sempervirens*	Trailing yellow violet	47.5	0.88	4.13	0.95
TROV	*Trillium ovatum*	Western trillium	45.1	0.66	4.15	0.97
VAHE	*Vancouveria hexandra*	Inside-out flower	29.6	2.61	3.09	0.80
ANDE	*Anemone deltoidea*	Three-leaved anemone	24.1	0.76	3.49	0.95
PTAQ	*Pteridium aquilinum*	Bracken fern	22.8	6.19	2.98	0.83
XETE	*Xerophyllum tenax*	Bear-grass	19.1	8.89	3.00	0.87
CHUM	*Chimaphila umbellata*	Prince's-pine	13.6	0.73	2.95	0.96
GOOB	*Goodyera oblongifolia*	Rattlesnake-plantain	11.7	0.84	2.59	0.88
DIHO	*Disporum hookeri*	Hooker's fairybells	11.1	1.81	2.41	0.84
TITR	*Tiarella trifoliata*	Three-leaved foamflower	9.9	2.90	1.90	0.69
CHME	*Chimaphila menziesii*	Menzies 'pipsissewa	6.2	0.60	2.21	0.96
ADBI	*Adenocaulon bicolor*	Pathfinder	4.3	5.21	1.67	0.86
TRLA	*Trientalis latifolia*	Broad-leaved starflower	3.7	0.83	1.64	0.92
CASC	*Campanula scouleri*	Scouler's harebell	2.5	1.25	1.31	0.95
SMST	*Smilacina stellata*	Star-flowered false	2.5	0.75	1.24	0.90
POMU	*Polystchum munitum*	Sword fern	1.9	8.83	0.71	0.65
HIAL	*Hieracium albiflorum*	White flowered hawkweed	1.9	1.17	1.00	0.91
GABO	*Galium boreale*	Northern bedstraw	1.2	12.50	0.69	1.00
BLSP	*Blechnum spicant*	Deer fern	1.2	7.00	0.34	0.49
GATR	*Galium triflorum*	Sweet-scented bedstraw	1.2	3.25	0.67	0.96
COCA	*Cornus canadensis*	Dwarf dogwood	1.2	0.50	0.69	1.00
PTAN	*Pterospora andromedea*	Pine drops	1.2	0.50	0.69	1.00
VECU	*Veronica cusickii*	Cusick's speedwell	1.2	0.50	0.69	1.00
TRCE	*Tricetum cernuum*	Nodding trisetum	0.6	2.50	0.00	0.00
Shrubs						
GASH	*Gaultheria shallon*	Salal	87.0	18.54	4.41	0.89
BENE	*Berberis nervosa*	Dwarf Oregon-grape	72.2	20.65	4.36	0.92
VAPA	*Vaccinium parvifolium*	Red huckleberry	44.4	13.00	3.80	0.89
ACCI	*Acer circinatum*	Vine maple	32.1	17.38	3.30	0.84
ROGY	*Rosa gymnocarpa*	Baldhip rose	16.7	4.54	2.43	0.74
VAAL	*Vaccinium alaskaense*	Alaskan blueberry	8.6	9.89	2.16	0.82
RUUR	*Rubus ursinus*	Trailing blackberry	5.6	1.22	2.03	0.92
VAME	*Vaccinum membranaceum*	Black huckleberry	3.7	1.92	1.63	0.91
COCO	*Corylus cornuta*	Beaked hazelnut	2.5	14.25	0.35	0.25
SYAL	*Symphoricarpos albus*	Common snowberry	2.5	1.25	1.09	0.79
RHPU	*Rhamnuspurshiana*	Cascara	1.2	0.75	0.64	0.92
VAOV	*Vaccinum ovalifolium*	Oval-leaved blueberry	0.6	17.00	0.00	0.00
HODI	*Holodiscus discolor*	Ocean spray	0.6	8.50	0.00	0.00
CONU	*Cornus nuttallii*	Pacific dogwood	0.6	5.00	0.00	0.00
ACCI (2 - 7 m)	*Acer circinatum*	Vine maple	14.8	34.8	2.77	0.87
Seedlings						
TSHE	*Tsuga heterophylla*	Western hemlock	82.7	2.1	3.86	0.79
ABAM	*Abies amabilis*	Pacific silver fir	26.5	0.7	3.52	0.94
PSME	*Psudotsuga menzeisii*	Douglas-fir	17.3	0.5	3.33	1.00
TABR	*Taxus brevifolia*	Pacific yew	5.6	0.5	2.20	1.00
TSHE	*Tsuga heterophylla*	Western hemlock	40.1	27.2	3.64	0.87
ABAM	*Abies amabilis*	Pacific silver fir	35.8	16.0	3.39	0.83
TABR	*Taxus brevifolia*	Pacific yew	3.1	8.9	1.43	0.89
ABGR	*Abies grandis*	Grand fir	1.2	8.8	0.60	0.86

Table 1: Frequency, average cover, H-diversityfor understory vegetation found along two transects in an old-growth Douglas-fir forest at Wind River Canopy Crane Area. The codes are the abbreviations of species scientific names.

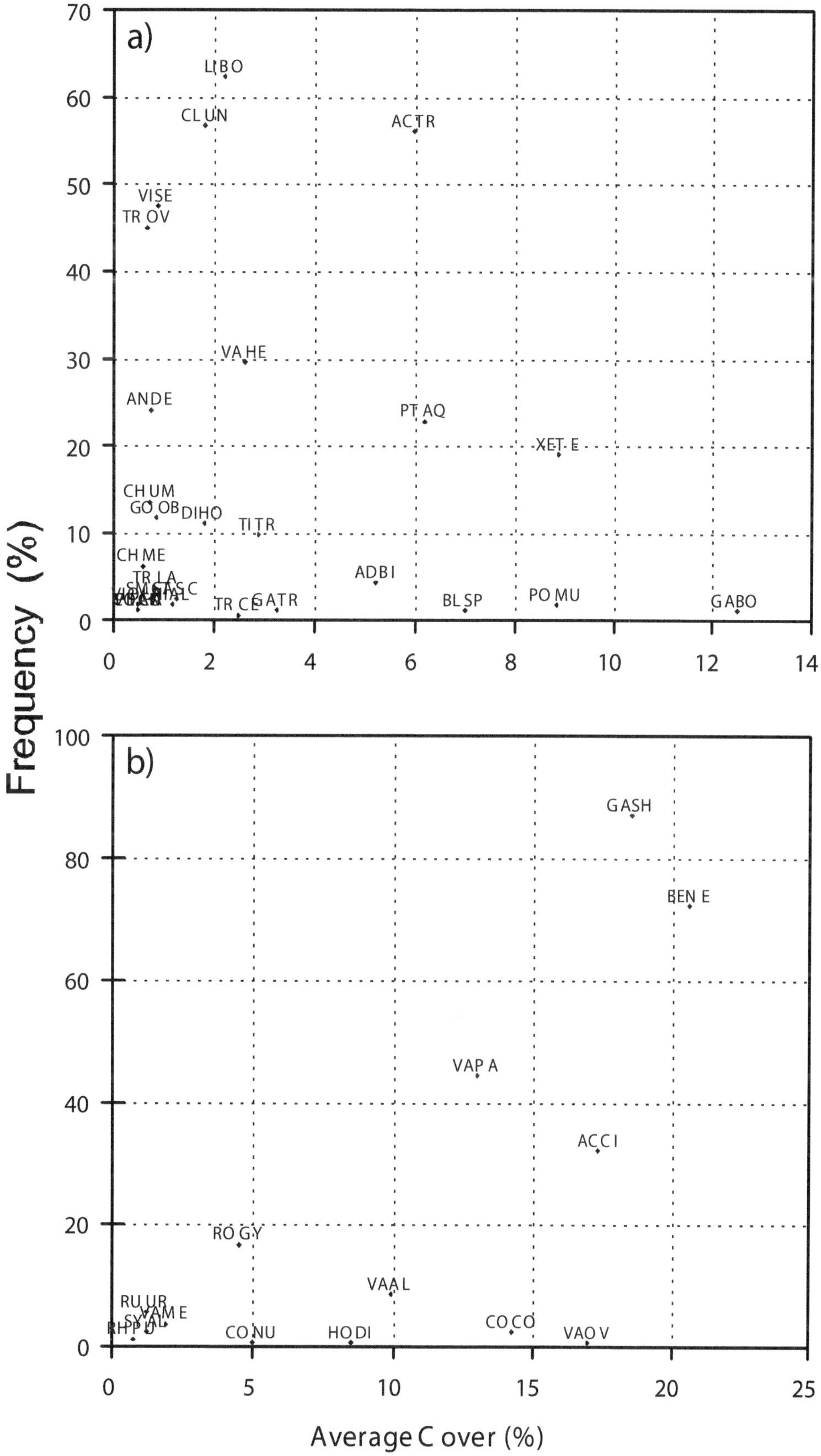

Figure 2: Relationships between species frequency and average cover for both herbaceous (a) and shrub (b) species. See Table 1 for the species codes.

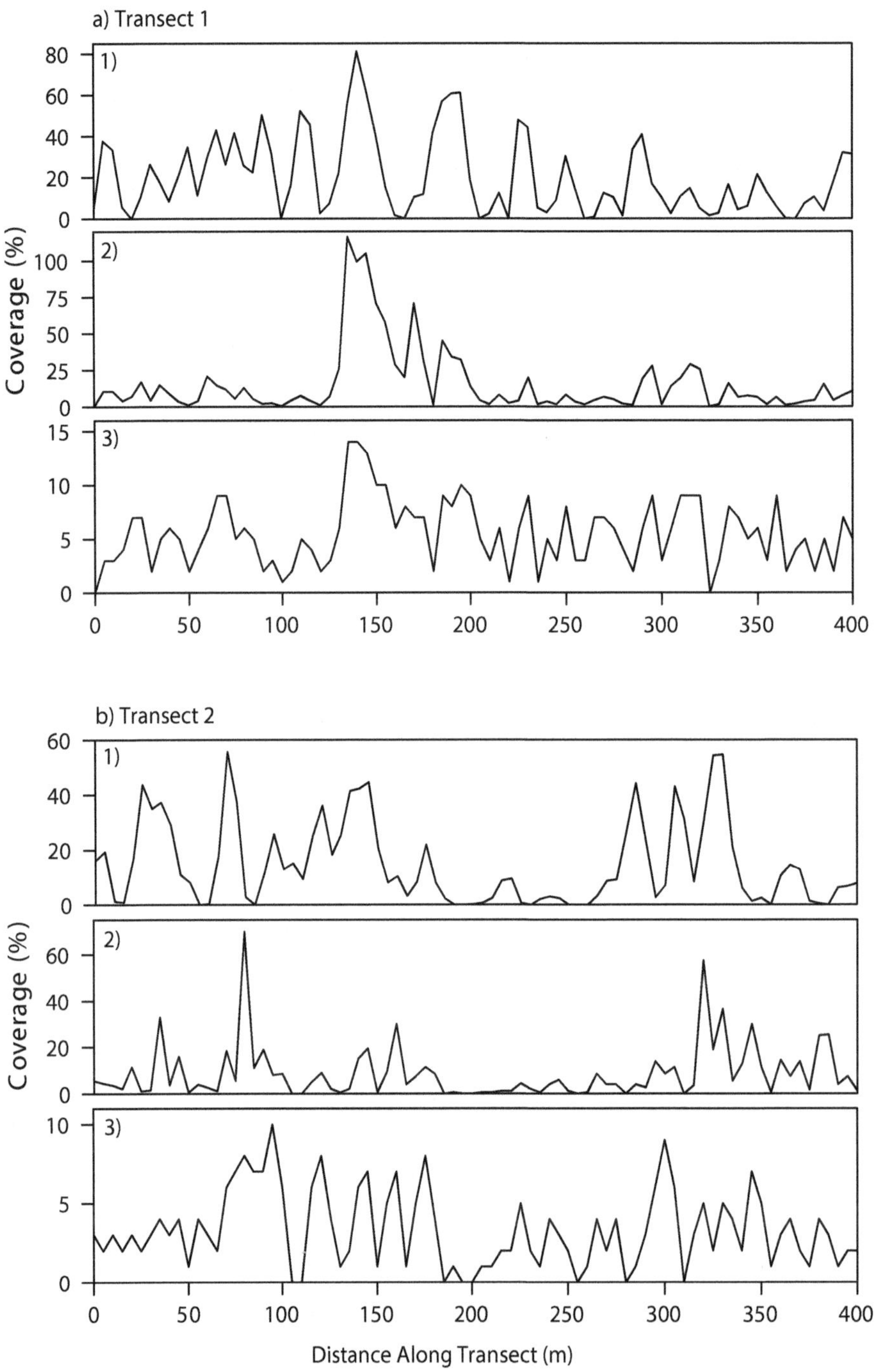

Figure 3: Distribution patterns along two 400 m transects for 1) model derived canopy opening (= [canopy opening area/(πr^2) * 100], r = 5 m, %); 2) total herb cover (%); and 3) herb richness. The canopy openings were estimated from a canopy model, while herb total cover and richness were observed from 2×2 m quatrats along the transects.

canopy openings around both transects (Figure 4a and 4b). Within a relatively large connected canopy opening across Transect 1 (Area 1-2 in Figure 4a), which was located at the edge of a trail, the covers of shrub species salal, Alaskan blueberry (*Vaccinium alaskaense*), and red huckleberry (*Vaccinium parvifolium*) reached their maximum. Salal had significant positive correlation ($P < 0.05$) with Alaskan blueberry, red huckleberry, and bear-grass (*Xerophyllum tenax*), while the latter three species tended to co-occur along the transects. Salal was negatively correlated ($P < 0.05$) with Douglas-fir canopies of all heights. Vine maples (2-7 m high) were positively correlated with the dominant-co dominant layers of the over story canopy (canopy heights at 40, 45, and 50 m, with $P < 0.005$, 0.005, and 0.001, respectively). The highest cover of vine maples occurred at the same locations where the dominant canopy cover reached a maximum.

The saplings of Pacific silver fir were also positively correlated with dominant-co dominant layers of over story canopies (canopy heights at 45 and 50 m, with $P < 0.01$ and 0.001, respectively), while western

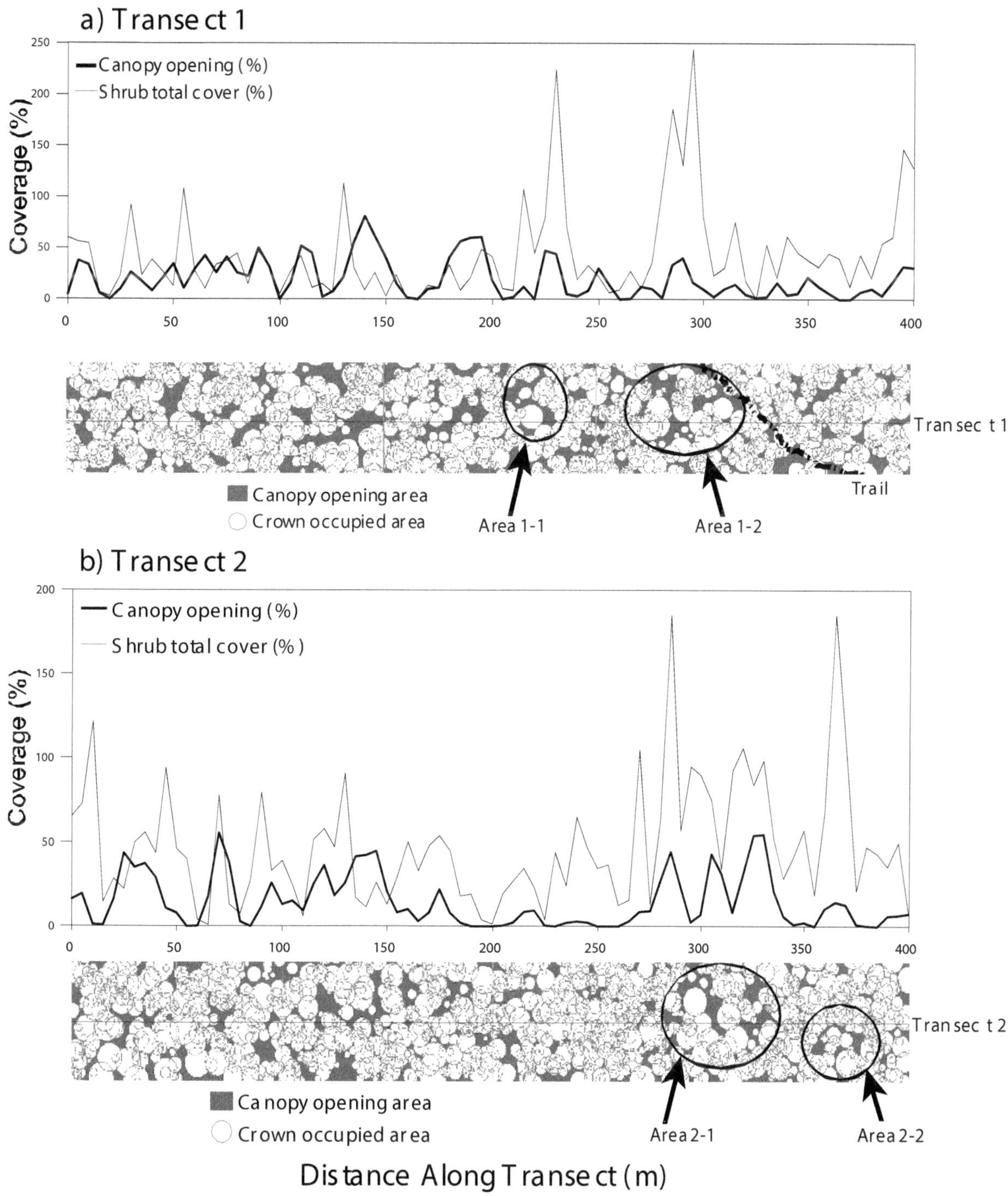

Figure 4: Spatial distributions of model derived canopy opening (%) and total shrub cover (%) along Transects 1 and 2 (a and b, respectively). The canopy projections were also shown, which were centered along these two transects, with buffer zones of 50 m (25 m on each side).

hemlock saplings were negatively associated with lower canopies (canopy height at 10, 15, and 20 m, with P < 0.01, 0.01, and 0.001, respectively).

Spatial scales determined by wavelet analysis

A high percentage of canopy openings were found at two locations (about 135 and 185 m) along Transect 1, as indicated by wavelet transforms (Figure 5a). Richness and total cover of herbs on Transect 1 reached their maximum at similar locations and scale ranges, clearly corresponding with high values of canopy openings (Figure 5a-5c). The correspondence of patterns among the variables on Transect 2 was not as obvious as on Transect 1 (Figure 5d-5f). There was a higher correspondence between wavelet patterns of total herb richness (Figure 5e) and model derived canopy openings (Figure 5d) when the distance along the transect was shifted to the left (Figure 5e), suggesting a possible lag effect.

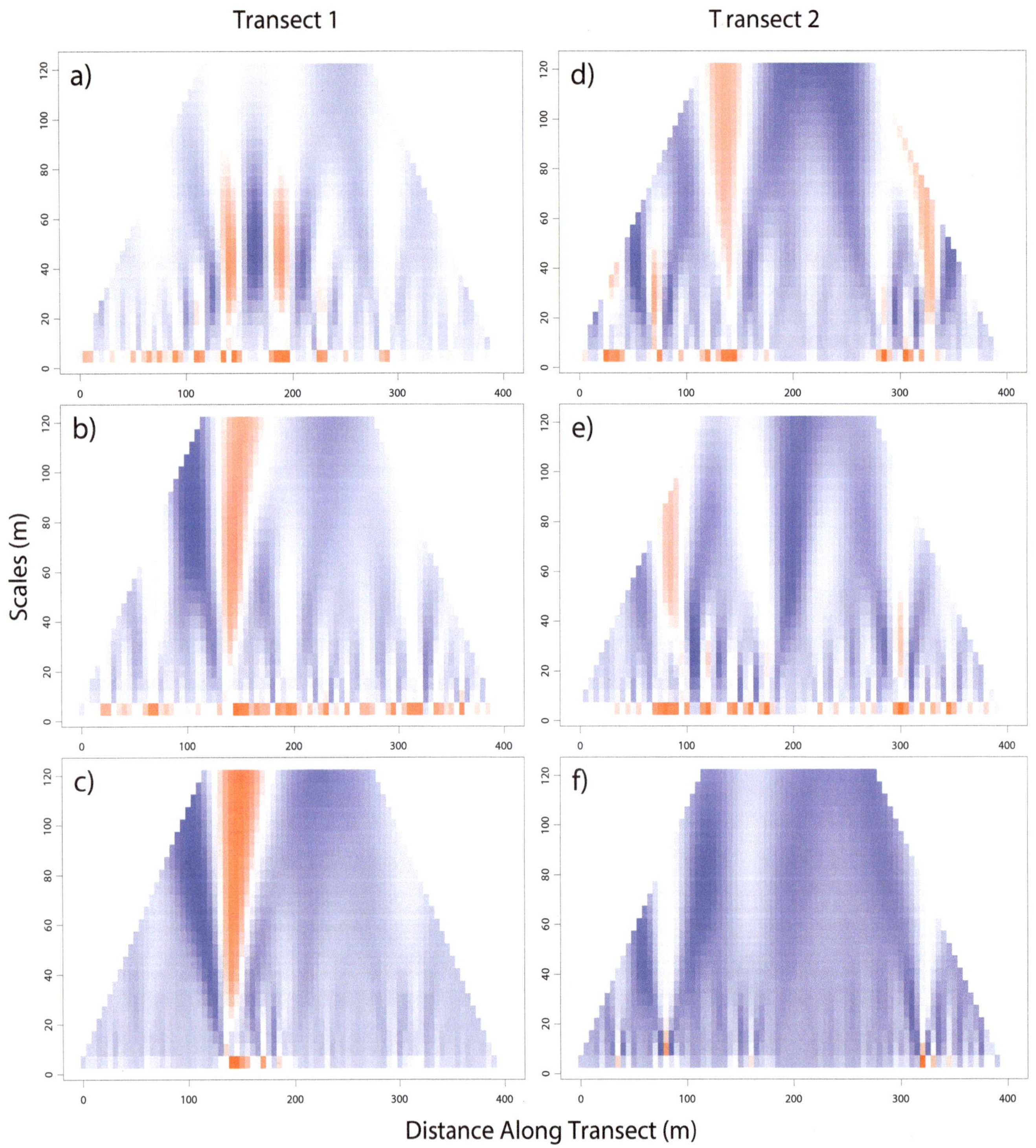

Figure 5: Wavelet transforms of model derived canopy openings (a, d), herbaceous species richness (b, e), and total cover of herbaceous species (c, f) along Transect 1 and Transect 2, respectively. Different colors and shadings in a wavelet transform image represent changes of a variable (e.g., percent cover) at different locations (m, X-axis) and different scales (m, Y-axis). Red colors represent higher values of the variable, whereas dark blue represents lower values.

Along Transect 1, the correlations between wavelet transforms of canopy openings and total cover and richness of herbs were highest at scales of 25-40 m (Figure 6a). Along Transect 2, the correlation between canopy openings and total herb cover reached its maximum near the 45 m scale (Figure 6b), while at a similar scale, there was no correlation between canopy openings and total herb richness (Figure 6b). However, when moving along the transect of total herb richness to both east-west directions (negative if to the left and positive if to the right) at 5 m intervals, we found that the correlation between the two variables was maximized when this lag was equal to -15 m, with significant increases starting at around the 45 m scale (Figure 6b).

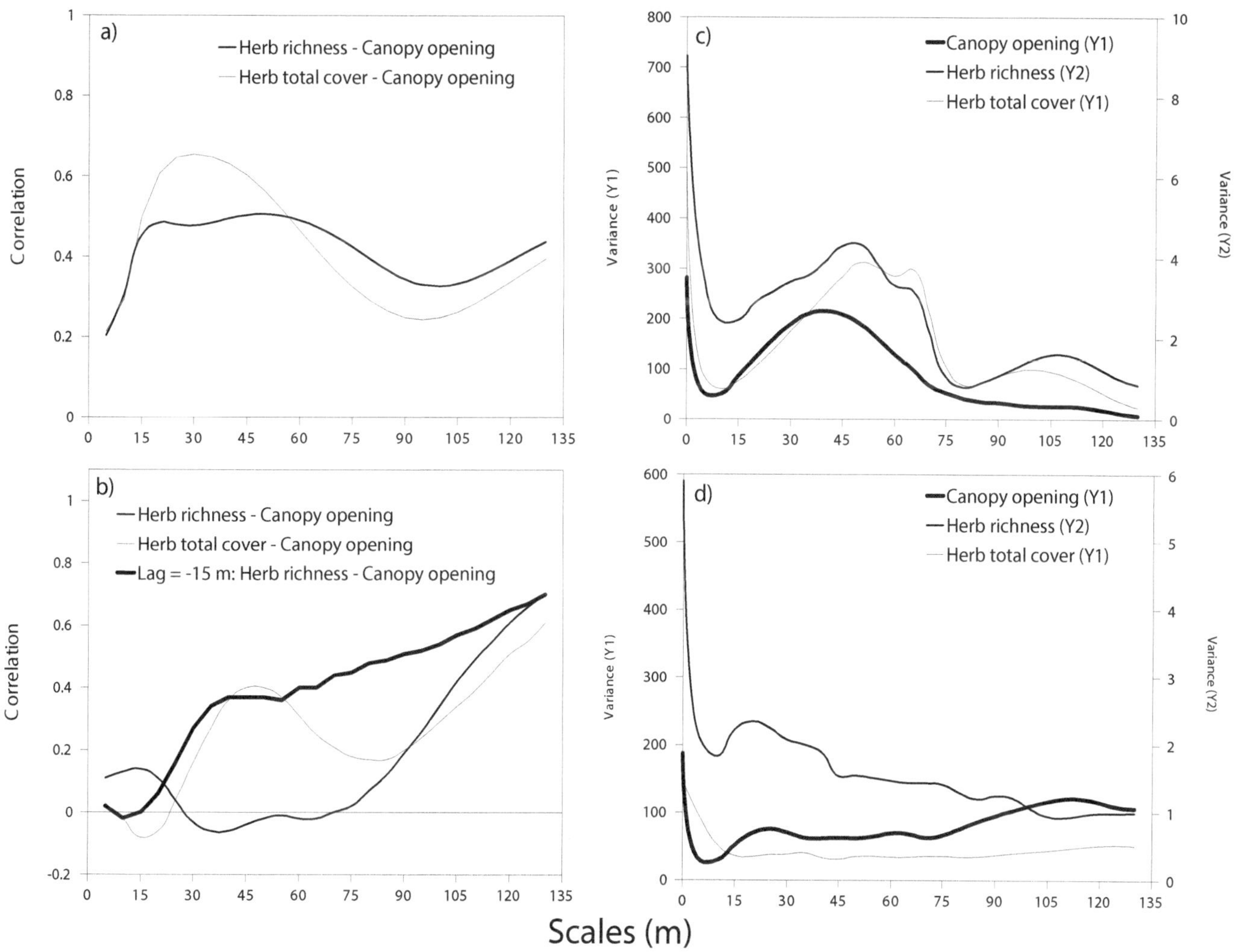

Figure 6: a) and b) are correlations of wavelet transforms among different variables along Transects 1 and 2, respectively, at different scales. c) and d) are the variances of wavelet transforms of different variables along Transects 1 and 2, respectively, at different scales. X-axis represents the scale (m) in all four graphs.

There was one peak in the wavelet variance for canopy openings on Transect 1 at about40 m scale, indicating that the dominant patch size of canopy openings tended to be about 40 m on Transect 1 (Figure 6c). There were two peaks on Transect 2 at scales around 17 and 115 m (Figure 6d). Further examining the canopy opening distribution along Transect 2 (Figure 3b) suggested that these two peaks may reflect the patchiness of canopy openings at these two scales. The wavelet transforms (Figure 5d) also reflected this trend that the first dominant patchiness was at around the 17 m scale and then this patchiness dissolved into larger scale dominance at around the 115 m scale.

Along Transect 1, the wavelet variances of total cover and richness of herbs were similar, reaching their peaks at scales around 45 m (Figure 6c). This showed that patches of high richness and total cover of herbs occurred mainly at the 45 m scale on Transect 1. There were similarities among wavelet variances of canopy openings, the total herb cover, and herb richness on Transect 1. While there was not an obvious peak in the wavelet variance of total herb cover along Transect 2, the wavelet variance of herb richness peaked at around the 17 m scale. The result indicated that the high herb richness on Transect 2 occurred primarily at this scale, and it corresponded to the dominant patchiness of canopy openings.

Discussion

One of the contributions of this study was a unique view of three-dimensional canopy structure (See authors' previous publications for details, [5,26]. example of horizontal structure shown in Figure 1, and examples of vertical structure shown in Figures 7 and 8, both horizontally and vertically, in quantitatively characterizing the influence of canopy structure on understory vegetation at various scales. There existed vertical heterogeneity at different canopy heights along both transects (Figures 7 and 8). The canopy cover can be very different at different heights, e.g., in some locations on the transects, there were high canopy cover at lower canopy layer but low canopy cover at dominant- co dominant layer (Figure 7a, figure 7b, Figure 8a); while there were opposite situations in other locations on the transects (Figure 7c). In some locations, there were high canopy covers at intermediate canopy layer yet low canopy covers at lower canopy layers and dominant-co dominant layer (Figure 8b), suggesting that although the effect of the dominant- co dominant layer is important, the influences of the lower and intermediate canopies should not be ignored. Lower canopies can also play a critical role in the growth and spatial distribution of understory vegetation [16,26].

The similarity of wavelet variances among canopy openings, herb

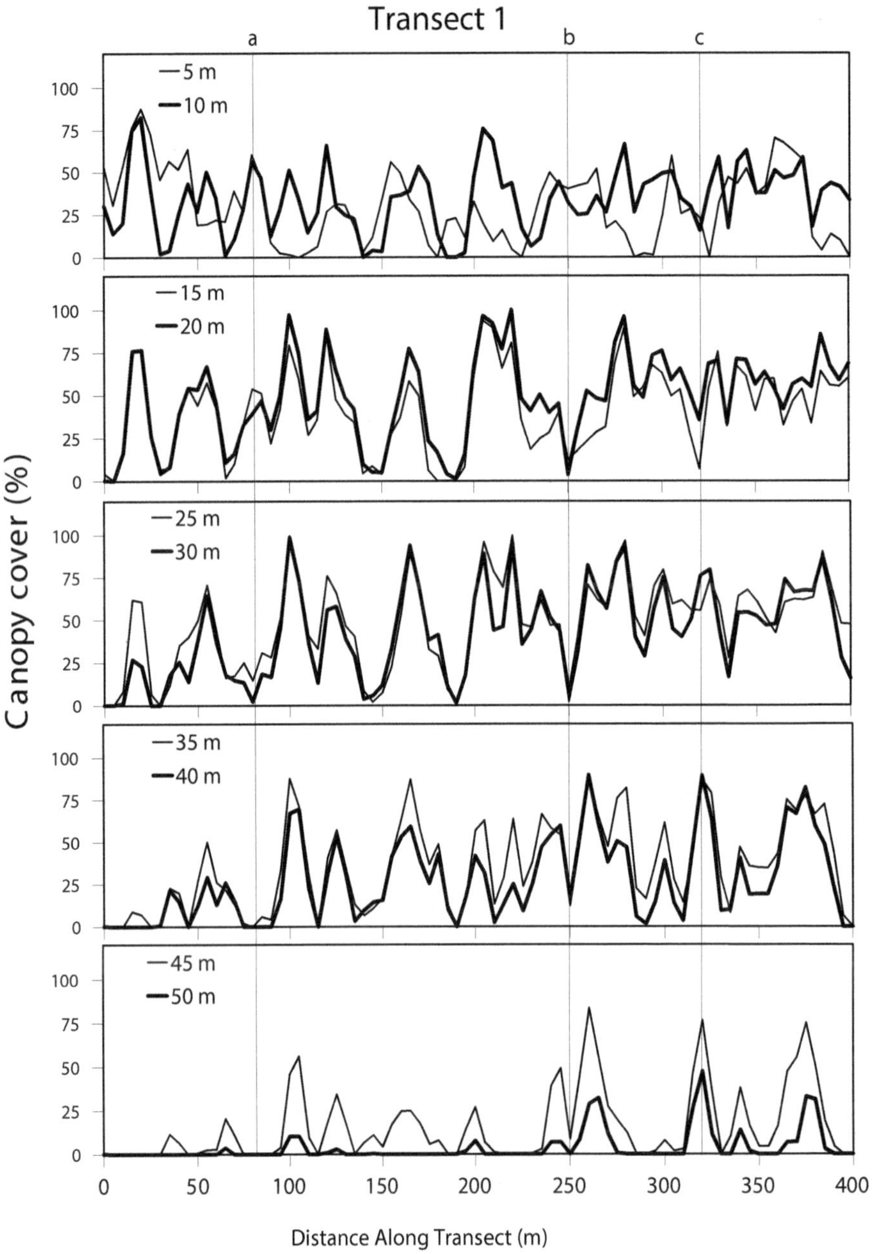

Figure 7: Along Transect 1, the canopy projection (%) at different tree heights (from height at 5 – 50 m, with an interval of every 5 m). a and b showed the locations where there were relatively high canopy covers (%) at lower canopy layers yet close to zero canopy cover at dominant-codominant layer, c showed the situation opposite to a and b.

total cover, and richness on Transect 1 reflected their correspondence at some scales (≈40-45m, Figure 6c), while on Transect 2 there was only correspondence between canopy openings and herb richness at scale around 17 m. It was not clear why the overall correlation between wavelet transforms of canopy openings and herb richness was the highest when the distance (horizontal-axis) of herb richness was shifted 15 m to the left along Transect 2, yet no lag effect was found along Transect 1. The reason perhaps is that the sun orientation is more influential when patch sizes (of both canopy openings and herbs) are smaller, thus less overlap and less direct (no lag) correlation. It was also not clear why the correlations of wavelet transforms of both total herb cover and richness with canopy openings were higher at the 45 m scale (Figure 6a and Figure 6b). It may be related to the geographical location of the site and how light penetrates this old-growth Douglas-fir forest in which the heights of the dominant-co dominant layer are around 40 - 45 m.

Reproduction and growth of understory species are greatly restricted by light availability [26,32] and microclimate [9,10,15], which are determined by canopy structure. As a result, the distribution of understory species depends on the structure of the canopy [24,33,34]. The results of this study showed that patterns in composition and diversity of understory species had close associations with horizontal and vertical canopy structures, although their responses to different layers of the canopy varied.

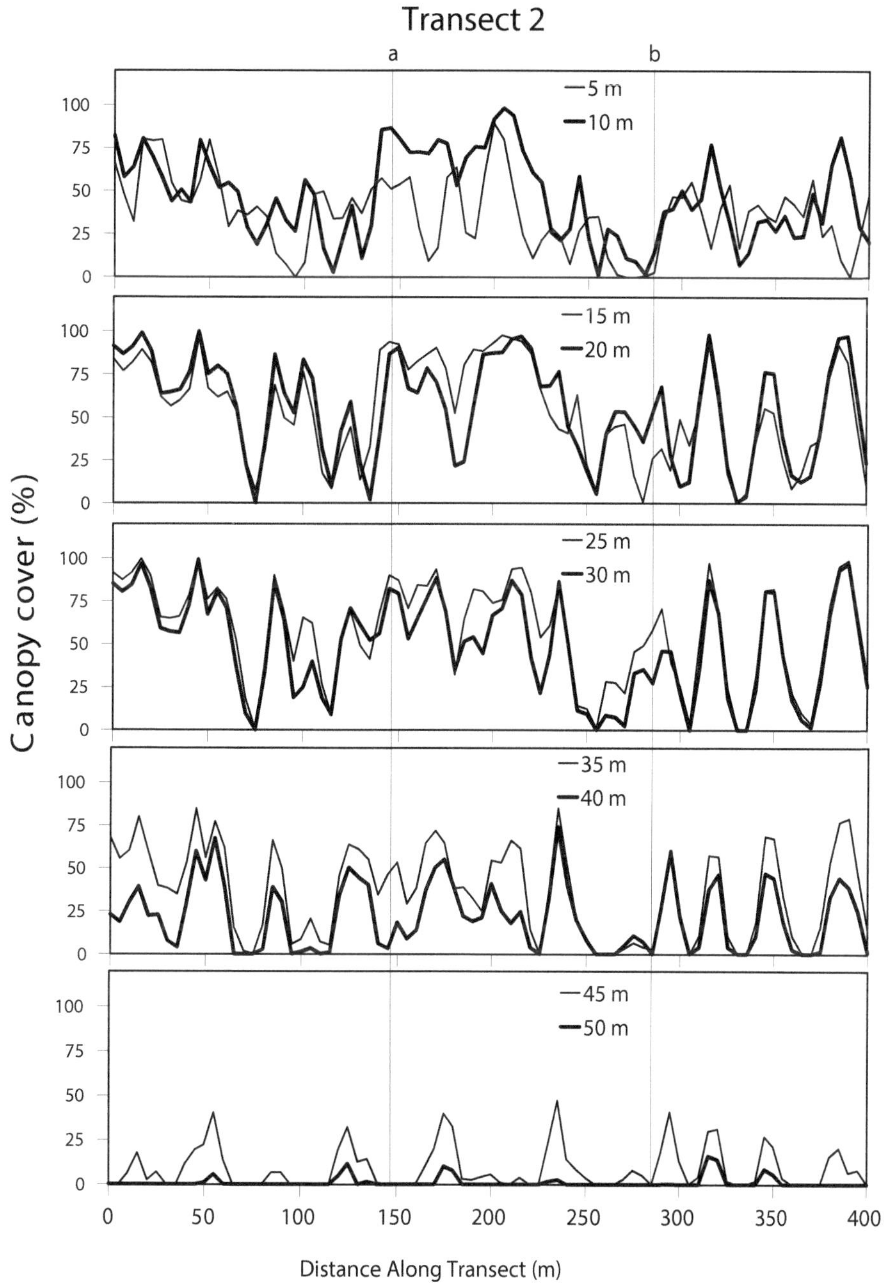

Figure 8: Along Transect 2, the canopy projection (%) at different tree heights (with the same heights and intervals as in Fig. 7). a showed the similar situation as in Fig. 7a and b, b showed where there was high canopy covers in the intermediate canopy layers (20 – 30 m height) yet close to zero canopy covers at both low (5 – 10 m height) and dominant-codominant canopy layer.

It was interesting that the herb and shrub species in this old-growth Douglas-fir forest had the opposite trends in terms of the relationship between abundance and frequency (negative-exponential shape for herbs and positive-exponential shape for shrubs). Although the positive relationship between abundance and frequency for the shrubs is quite expected, the negative relationship for the herbs is worth an explanation. The latter relationship likely results from the limited number of gap patches and light availability in these patches in the understory. In this old-growth forest, light limits the growth of understory and large gap openings are few. Whenever such gaps are available, they are quickly occupied by the light demanding species, such as those locating at the bottom right corner of Figure 2a.

It is well known that species diversity increases and composition changes near forest or road edges because these edges provide a unique environment for many understory species [35]. In this stand, because of the trail that crossed Transect 1, it seemed more disturbed than Transect 2 it. Adjacent to this trail there were relatively large connected canopy openings cutting through both the north and south sides of the transect. These openings created high heterogeneity not only in canopy structure but also in the understory vegetation. Some understory herb and shrub species responded to these openings, suggesting that these species are associated with disturbance. The total abundance and richness of all herbs along Transect 1 were larger than the ones along

Transect 2 because there were more canopy openings along Transect 1 than Transect 2.

In summary, in this study, canopy structure was quantified horizontally and vertically (See authors' previous publications for details, [5,26]. Also example of horizontal structure shown in Figure 1, and examples of vertical structure shown in Figures 7 and 8), allowing quantitative characterization of the influences of canopy on understory vegetation at various scales [1,16,24]. Most dominant herb species that were positively associated with canopy openings were more negatively affected by lower canopy layers than by the dominant- co dominant layers. The spatial distributions of herb species were mainly affected by canopy openings and disturbance. The abundance of total shrubs had a positive relationship to canopy openness along the two sampled transects, although the interpretation on the associations between canopy structure and individual shrub species were not immediately clear. Wavelet analysis is a useful tool in characterizing the pattern changes and in comparing variables of interest across scales and locations along transects. The relationship between canopy structure and understory vegetation was highly scale-dependent, *i.e.*, understory variables responded differently to the heterogeneity of canopy patterns at different scales.

Acknowledgements

This research was supported by grants from USDA Forest Service, Agreement No. PNW94-0541 and a Challenge Fellowship at Michigan Technological University.

References

1. Chen J, Bradshaw GA (1999) Forest structure in space: a case study of an old growth spruce-fir forest in Changbaishan Natural Reserve, PR China. Forest Ecology and Management 120: 219-233.
2. McGuire JP, Mitchell RJ, Moser EB, Pecot SD, Gjerstad DH, et al. (2001) Gaps in a gappy forest: plant resources, longleaf pine regeneration, and understory response to tree removal in longleaf pine savannas. Can. J. For. Res. 31: 765-778.
3. Ozanne CMP, Anhuf D, Boulter SL, Keller M, Kitching RL, et al. (2003) Biodiversity meets the atmosphere: A global view of forest canopies. Science 301: 183-186.
4. Parker GG, Harmon ME, Lefsky MA, Chen JQ, Van Pelt R, et al. (2004) Three-dimensional structure of an old-growth Pseudotsuga-tsuga canopy and its implications for radiation balance, microclimate, and gas exchange. Ecosystems 7: 440-453.
5. Song B, Chen JQ, Silbernagel J (2004) Three-dimensional canopy structure of old-growth Douglas-fir forest. Forest Science 50: 376-386.
6. Chávez V, Macdonald SE (2010) The influence of canopy patch mosaics on understory plant community composition in boreal mixedwood forest. Forest Ecology and Management 259: 1067-1075.
7. Hollett Jr. BP, Jackson MT (2013) Composition Gradients in Forest Canopy and Groundlayer Vegetation within a White River Floodplain Forest. Proceedings of the Indiana Academy of Science 93: 313-322.
8. Spies TA, Franklin JF (1989) Gap characteristics and vegetation response in coniferous forests of the Pacific Northwest. Ecology 70: 543-545.
9. Hart SA, Chen HYH (2006) Understory vegetation dynamics of North American boreal forests. Critical Reviews in Plant Sciences 25: 381-397.
10. Barbier S, Gosselin F, Balandier P (2008) Influence of tree species on understory vegetation diversity and mechanisms involved - A critical review for temperate and boreal forests. Forest Ecology and Management 254: 1-15.
11. Schroeer AE, Hendrick RL, Harrington TB (1999) Root, ground cover, and litterfall dynamics within canopy gaps in a slash pine (*Pinus elliottii Engelm.)* dominated forest. Ecoscience 6: 548-555.
12. Quinby PA (2000) First-year impacts of shelterwood logging on understory vegetation in an old-growth pine stand in central Ontario, Canada. Environmental Conservation 27: 229-241.
13. Brosofske KD, Chen J, Crow TR (2001) Understory vegetation and site factors: implications for a managed Wisconsin landscape. Forest Ecology and Management 146: 75-87.
14. Chen J, Franklin JF, Spies TA (1993) Contrasting microclimates patterns among the clearcut, edge, and interior area of old-growth Douglas-fir forest. Agricultural and Forest Meteorology 63: 219-237.
15. Chen J, Saunders SD, Crow T, Brosofske KD, Mroz G, et al. (1999) Microclimatic in forest ecosystems and landscapes. Bioscience 49: 288-297.
16. Van Pelt R, Franklin JF (1999) Response of understory trees to experimental gaps in old-growth Douglas-fir forests. Ecological Applications 9: 504–512.
17. Van Pelt R, Franklin JF (2000) Influence of canopy structure on the understory environment in tall, old-growth, conifer forests. Can. J. For. Res. 30: 1231-1245.
18. Nielsen BO, Ejlersen A (1977) The distribution pattern of herbivory in a beech canopy. Ecological Entomology 2: 293-299.
19. Canham CD, Finzi A, Pacala SW, Burbank DH (1994) Causes and consequences of resource heterogeneity in forests: interspecific variation in light transmission by canopy trees. Can. J. For. Res. 24: 337-349.
20. Kabakoff RP, Chazdon RL (1996) Effects of canopy species dominance on understory light availability in low-elevation secondary forest stands in Costa Rica. Journal of Tropical Ecology 12: 779-788.
21. Stewart GH (1988) The influence of canopy cover on understory development in forests of the western Cascade Range, Oregon, USA. Vegetatio 76: 79-88.
22. Reifsnyder WE, Furnival GM, Horowitz JL (1971) Spatial and temporal distribution of solar radiation beneath forest canopies. Agricultural Meteorology 9: 21-37.
23. Massman WJ (1982) Foliage distribution in old-growth coniferous tree canopies. Can. J. For. Res. 12: 10-17.
24. Kembel SW, Dale MRT (2006) Within-stand spatial structure and relation of boreal canopy and understorey vegetation. Journal of Vegetation Science 17: 783-790.
25. Shaw DC, Franklin JF, Bible K, Klopatek J, Freeman E, et al. (2004) Ecological setting of the wind river old-growth forest. Ecosystems 7: 427-439.
26. Song B, Chen J, Densanker PV, Reed DD, Bradshaw GA, et al. (1997) Modeling canopy structure and heterogeneity across scales: from crowns to canopy. Forest Ecology and Management. 96: 217-229.
27. Bradshaw GA, Spies TA (1992) Characterizing canopy gap structure in forest using wavelet analysis. Journal of Ecology 80: 205-215.
28. Brosofske KD, Chen J, Crow TR, Saunders SC (1999) Vegetation responses to landscape structure at multiple scales across a Northern Wisconsin, USA, pine barrens landscape. Plant Ecology 143: 203-218.
29. Keitt TH, Fischer J (2006) Detection of scale-specific community dynamics using wavelets. Ecology 87: 2895-2904.
30. Mi XC, Ren HB, Ouyang ZS, Wei W, Ma KP (2005) The use of the Mexican Hat and the Morlet wavelets for detection of ecological patterns. Plant Ecology 179: 1-19.
31. Li BL, Loehle C (1995) Wavelet analysis of multiscale permeabilities in the subsurface. Geophysical Research Letters 22: 3123-3126.
32. Martens SN, Breshears DD, Meyer CW (2000) Spatial distributions of understory light along the grassland/forest continuum: effects of cover, height, and spatial pattern of tree canopies. Ecological Modelling 126: 79-93.
33. Okland RH, Eilertsen O (1996) Dynamics of understory vegetation in an old-growth boreal coniferous forest, 1988-1993. Journal of Vegetation Science 7: 747-762.
34. Goldblum D (1997) The effects of treefall gaps on understory vegetation in New York State. Journal of Vegetation Science 8: 125 132.
35. Fraver S (1994) Vegetation responses along edge-to-interior gradients in the mixed hardwood forests of the Roanoke River Basin. North Carolina Conservation Biology 8: 822-832.

9

Information, Multi-faceted Forest Ownership and Timber Supply

Francesca Rinaldi* **and Ragnar Jonsson**

European Commission, Joint Research Centre (JRC), Institute for Environment Sustainability (IES), Forest Resources and Climate Unit, Via E. Fermi, 2749 I-21027 Ispra, Italy

Abstract

Existing decision support systems (DSS) do not account for forest owner heterogeneity, nor do they explicitly model the reaction of forest owners to policy. Hence, current DSS are suitable for response analysis, but much less useful for policy impact assessment or forecasting. The current study presents a theoretical model of harvesting behavior which provides the basis for a simulation model, Expected Value Asymmetries (EVA), useful for analyzing how timber supply and forest characteristics are affected when forest owners differ as to responsiveness to information, risk aversion, and patience as regards postponement of harvesting revenues. The simulation results clearly indicate that the model is well adapted for considering forest owner heterogeneity when assessing the impact of policy on the inter-temporal development of forest resources and timber market conditions. Finally, it is outlined how EVA could integrate forest owner specific harvesting behavior in an augmented Decision Support System (DSS), thus addressing the inability of DSS operational at pan-European level to model the interaction between policy and forest management decisions.

Keywords: Forest policy; Forest owner; Timber supply; Information; DSS

Introduction

A sizeable proportion of the forests in the European Union (EU) is in private hands [1], and private forest ownership is increasing in the forest-rich sub-regions of Central-East and Northern Europe [2]. Thus, considering the attitudes and objectives of private forest owners is important for ensuring that policy instruments are effective [3]. It is also important to recognize that forest owners are far from a homogeneous group as regards objectives and attitudes (see, e.g., [4-8]).

Hence, efficient policy impact analysis requires the ability to model the interaction between policy and forest management decisions, also acknowledging the heterogeneity of forest owners. Existing Decision Support Systems (DSS) operational at pan-European level are somewhat lacking in this respect. In the instances where forest owner heterogeneity is accounted for, simplistic heuristic approaches are used. An example is the EUwood study (Hamburg University 2010) where forest owner harvesting behavior was linked to forest holding size, so that small forest holdings were assumed to result in a smaller percentage of the theoretical wood supply being available. Another prominent example is the EFSOS II study [9] (UNECE/FAO), which provided so called policy scenarios—depicting the outcome for the European forest sector when the priority of forest management is given to different objectives—for the assessment of trade-offs facing policymakers. Just as was the case in the EUwood study, the pan-European forest resource assessment model EFISCEN [10] was used both to provide estimates of potential wood supply, in EFSOS II ingested as a constraint on the production of wood products in the economic forest sector model EFI-GTM [11], and to project the future state of the forests given assumptions of future wood demand and pre-specified management regimes.

The analysis in these two studies neither considered the reaction of forest owners to policy and the effect of forest owner heterogeneity on the inter-temporal development of forests, nor the logical feedback existing from the forest sector model to the forest resource assessment model. The latter constitutes a relevant problem, since multiple errors would propagate over time if the 'actual' quantity of harvested wood were set equal to this upper bound, without taking into account that the satisfaction of the demand might have required a lower harvesting level. Hence, this kind of modelling endeavors has its value for response analysis, but is not very useful for policy impact assessment or forecasting.

Further, information is a commonly used policy instrument, not the least in the context of forest policy [12]. The importance of information as a policy tool is illustrated by considerable efforts spent by Member states of the EU on information, training and capacity building of private forest owners to foster wood mobilisation (Standing Forestry Committee ad hoc Working Group II 2008). Information as a policy tool certainly works by modifying the expectations of forests owners/managers about the future, e.g., concerning demand and prices, so that informed harvesting decisions are taken using updated expectations [13]. Consequently, policy impact analysis presupposes the ability to model the reaction of forest owners to policy in the form of information releases.

Thus, the current study aims to provide a theoretically founded framework to incorporate forest owner heterogeneity, including the asymmetric response to information, in a DSS. Departing from earlier studies using the Fisherian two-period consumption-saving-harvesting model [14-20], in particular Rinaldi and Jonsson [13], we develop a theoretical model of harvesting behavior when forest owners differ as to the receptiveness to information. This theoretical model is then used as the basis for a more complex simulation tool—Expected Value Asymmetries (EVA)—allowing for a higher degree of heterogeneity in terms of forest owners' preferences. EVA is subsequently used to demonstrate how forest owner heterogeneity affects both the inter-temporal development of forests and timber market conditions. Further, it is discussed how EVA, by working as an intermediary, could provide feedback from an economic forest sector model to a forest

***Corresponding author:** Francesca Rinaldi, European Commission, Joint Research Centre (JRC), Institute for Environment Sustainability (IES), Forest Resources and Climate Unit, Via E. Fermi, 2749 I-21027 Ispra, Italy; E-mail: francesca.rinaldi@jrc.ec.europa.eu

resource assessment model, while at the same time ensuring that forest owner specific harvesting behavior is integrated in the DSS.

The paper proceeds as follows: the next section presents the theoretical model, with particular emphasis on the modelling of uncertainty and information arrival. Then the policy simulation tool EVA is presented. Hereafter, simulation results are presented and discussed. The subsequent section outlines how EVA could be part of a DSS, completing the feedback loop between an economic model of the forest sector and a forest resource assessment model while at the same time ensuring that forest owner harvesting behavior is accounted for. The paper concludes with summary and conclusions.

Theoretical Model

The theoretical model used here is simple and stylized. First of all, in order to derive analytical solutions, we consider a two-period economy, so that the optimal harvesting problem of each forest owner reduces to how much should be harvested immediately and how much in a(n) (indeterminate) future period. Further, for what concerns policy making, we simply assume that some unspecified policy is initiated at the beginning of period one, and forest owners perceive it as new information (i.e., a signal) regarding timber demand and prices. Policy design is obviously an issue of considerable importance. However, it is beyond the scope of this paper, whose focus specifically is the modelling of forest owners' harvesting behavior.

The representation of policy through signals is particularly interesting, allowing the policy to be perceived as "good (news)" or "bad (news)" from the forest owner perspective. Further, it accounts for the circumstance that policy-effects might be perceived as long lasting and/or time contingent, and also for the specific degree of confidence of individual forest owners/managers (in the policy maker). The theoretical model exclusively focuses on direct economic risk (as opposed to biological risk), a choice that is at least partially supported by empirical evidence showing that private forest owners consider direct economic risk, such as price changes, much more important than indirect economic risk resulting from biological damage [21].

We consider a two-period economy in which two distinct groups, *I* and *NI*, of risk-averse private landowners want to maximize the utility derived from individual final wealth, by choosing how much to harvest in each period. Within each group landowners are identical, further we also assume that they all have the same utility index *U*, with exponential form and constant risk-aversion *r*, that is $U=-exp(-rw^j)$, where w^j represents wealth at time *2* for an individual belonging to group $j=I, NI$. Besides having identical preferences, all the forest owners are also initially endowed by identical forests characterized by the same area, age, tree species and growth function.

Each landowner harvests in both periods, in particular, we denote by x^j_i the quantity harvested at time *i* by a generic landowner belonging to group *j*. Given the amount harvested at time *1*, x^j_1, the stock available at time *2* is uniquely defined by the growth function and the initial forest endowment Q. In addition, since we are assuming that utility derives from final wealth only, and that the rotations beyond the second period are ignored, it also follows that such a stock will be entirely harvested at time *2*. For simplicity we will consider a linear growth function. Obviously such an assumption violates the concavity requirement, however, our analysis does not crucially rely on this specification, which instead allows for a reduction in the notational and computational complexity of the model. Therefore, we assume:

$$x^j_2=k(Q-x^j_1)\ 0\leq x^j_1\leq Q \tag{1}$$

where *Q* is the initial forest stock, and *k* a positive constant.[1]

The demands for wood in both period, D_1 and D_2, are inelastic and unknown. We let p_i be the price at which wood harvested at time *i* is sold on the wood market. Notice that, since the forest must be harvested completely at the end of the second period, and since demand is inelastic, our framework focuses only on the realization of the time 1 equilibrium through the equilibrium price p_1. Finally we assume that timber revenues $p_1x^j_1$ are invested at time *1* on a risk free bond with gross return $(1+b)$, at time *2*.

Uncertainty

When the landowner makes his/her harvesting decision, the demand D_1 and the timber prices are not known. However, all landowners know that p_1 is derived as an equilibrium price, while they hold beliefs concerning the distributions of D_1 and the dynamic of prices. In particular, both D_1 and p_2 are driven by a stochastic common component ε_a that can be thought of as a long-run economic shock affecting both periods (an example could, e.g., be a realization of the bioeconomy, creating a long-term increase in the demand for wood), and two time contingent shocks, ε_D and ε_2, respectively. The variance of ε_a, σ_a^2, can be thought of as a measure of economic risk. The two shocks ε_D and ε_2 are mutually independent, and, in addition, they are also independent from ε_a. Therefore, they represent specific time contingent shocks exclusively affecting time *1* (ε_D), and time *2* (ε_2), respectively. Hence, σ_D^2 and σ_2^2 can be thought of as measures of time specific risk. To make examples, one can imagine a temporary variation in exchanges rates affecting timber demands, or a provisional reduction in the activity of a major sawmill in the area in question.

Hence, $D_1=D(1+\varepsilon_a)+\varepsilon_D$ and $p_2=p_1(1+m)+\varepsilon_a+\varepsilon_2$, where $m, D\in R^{++}$, $\varepsilon_a \sim N(0, \sigma_a^2)$, $\varepsilon_D \sim N(0, \sigma_D^2)$, $\varepsilon_2 \sim N(0, \sigma_2^2)$. The constant *D* can be thought of as base demand level, and the percentage variation from it (ε_a) depends on the overall long-run economic development. In addition, the forest sector (and henceforth the demand) can be hit also by a time-contingent shock ε_D, which will have no residual effect at time *2*. The price at time 2, depends on the equilibrium price at time *1*, the long-run economic shock, ε_a, and the time specific shock for time 2, ε_2. Notice that, since the forest will be harvested at the end of the second period, only the demand in the first period is relevant for the analysis.[2]

For concreteness, we assume that $m>b$, for example one can imagine that *m* is a fixed inflation rate and, in absence of any additional information, forest owners/managers might assume that future timber prices will adjust for inflation.

Information Arrival

The model assumes that while uncertainty definitely resolves at time *2*, with the realization of the timber price, p^2, some news concerning the forest sector already arrives at the beginning of time *1*, through a signal *s*. In particular,

$$s=\alpha\varepsilon_a+\varepsilon_D+\varepsilon_s \tag{2}$$

where ε_s is a noise distributed according to $\varepsilon_s \sim N(0, \sigma_s^2)$. Only the members of *I* are able to observe the signal, while the others ignore its existence. For simplicity, we normalize the total number of landowners to 1, and we assume that a fraction β of the entire group observes

[1] In particular, the possibility of obtaining corner or interior solutions to the maximization problem does not depend on this assumption.

[2] Obviously one could alternatively use more complex specifications $D_1=D(1+a\varepsilon_a)+\varepsilon_D$ and $p_2=p_1(1+m)+b\varepsilon_a+\varepsilon_2$, where a, b$\in\Re^{++}$, however this would not add much to the results, while it would largely reduce tractability.

the signal, while the remaining $1-\beta$ does not. The signal is essentially the perceived effect of some unspecified policy. In particular, this construction implicitly assumes that forest owners and managers regard any type of policy as affecting demand for wood and prices, and henceforth their expected values. For concreteness, here we will consider a policy viewed as beneficial that could possibly increase demand and prices, that is, $s>0$.

σ_s^2 is a measure of the precision of the signal, since higher σ_s^2 induces also higher variability of the signal and, therefore, lower precision (we refer the reader to [7] for the comparative statics analysis on σ_s^2). However, notice that σ_s^2 could also be interpreted as the (inverse) degree of confidence into the policy maker.

$\alpha[0,1]$ represents the degree to which the signal is specific to the entire time horizon. If $\alpha=1$, the signal is also partially informative about the long-run shock ε_a, and, consequently, about p_2. On the contrary, if $\alpha=0$, the signal exclusively brings information contingent to time *1*, and it cannot be used to infer the timber price realization at time *2*. In concrete, one can imagine that, when the signal is announced, each forest owner (in group *I*) evaluates its degree of specificity.

The informativeness of the signal with respect to the demand and the timber price, respectively is defined as usual as $\gamma^D=cov(s, D_1)/var(s)$ and $\gamma^P=cov(s, p_2)/var(s)$. Given our assumptions, s, $p_2|s$ and $D_1|s$ are also normally distributed. In particular, standard Bayesian updating leads to:

$$p_2|s{\sim}N(p_1(1+m)+s\gamma^P,\ (1-\alpha\gamma^P)\sigma_a^2+\sigma_2^2) \qquad (3)$$

$$D_1|s{\sim}N(D+s\gamma^D,\ (1-(\alpha/D)\gamma^D)D^2\sigma_a^2+(1-\gamma^D)\sigma_D^2) \qquad (4)$$

If not explicitly mentioned otherwise, in the following we will assume that $\alpha\neq 0$. Obviously, we also assume $Q-D-s\gamma^D>0$.

Landowners' heterogeneity, information and equilibrium pricing

At the beginning of period *1*, all the forest owners are endowed with a forest stock Q, and growth of the forest is described by (1) above. Each landowner invests the revenues $p_1x^j_1$ from timbers harvested at time *1*, x^j_1, in a risk free bond with gross return $1+r$. Hence, within each group, the maximization problem consists in choosing optimal current timber supply x^j_1, in order to maximize the utility from final wealth w^j.

Given harvesting levels x^j_1, x^j_2, final wealth at time *2* is:

$$w^j=p_2x^j_2+p_1x^j_1(1+b) \qquad (5)$$

where x^j_2 is given by (1).

Before the harvesting and investment decisions are taken, members of group *I* receive some news concerning the forest sector by means of a signal $s=\alpha\varepsilon_a+\varepsilon_D+\varepsilon_s$. This information is then used within the group to update the priors on D_1 and p_2, and henceforth to choose optimal contingent harvesting x^I_1. More generally, both groups choose optimal harvesting at time *1*, given the available information Ω^j. In particular Ω^j is represented by the signal for landowners belonging to group *I* while it is an empty set for the others, so that $\Omega^I=s$ and $\Omega^{NI}=\varnothing$. The maximization problem becomes:

$$\underset{x_1^j}{Max}\ E[p_2x^j_2+p_1x^j_1(1+b)\ |\Omega^j]\ j=I,NI \qquad (6)$$

under the constraint given by (1).

Since p_1 is taken as given and $p_2|\Omega^j$ is normally distributed, so is $w|\Omega^j$. Further, the utility index is a standard constant risk aversion exponential utility, so that the expected utility of w^j given information Ω^j is a strictly increasing transformation of the kernel $E[w^j|\Omega^j]-0.5\,rVar[w^j|\Omega^j]$.

First Order Conditions

The first order conditions associated to the maximization problem (6) yield:

$$x^I_1=Q-\frac{k\left[p_1(1+m)+\gamma^P s\right]-p_1(1+b)}{rk^2\left[\left(1-\alpha\gamma^P\right)\sigma_a^2+\sigma_2^2\right]} \qquad (7)$$

$$x^{NI}_1=Q-\frac{kp_1(1+m)-p_1(1+b)}{rk^2\left[\sigma_a^2+\sigma_2^2\right]} \qquad (8)$$

Clearly, an increase in the expected growth rate of timber prices *m*, or a reduction in the risk free rate *b* reduces the short-run timber supplies from both groups, x^I_1 and x^{NI}_1. Vice versa, an increase in the time *1* price p_1, or in the risk aversion coefficient *r*, or in the growth index *k*, increases both x^I_1 and x^{NI}_1, while an increase in the informativeness parameter α reduces x^I_1. The effect of an increase in risk (σ_2^2 and σ_a^2) on the short-run supply of group *NI* is clear, leading to an increase in the time *1* supply x^{NI}_1. The effect on x^I_1 is generally the same; however, there exist combinations of parameters for which an increase in risk induces a reduction in x^I_1. Finally, when the signal is received, the informed group displays a lower supply at time *1* than the uninformed one, ceteris paribus.

Expected Equilibrium

The equilibrium price is computed by equalizing total timber supply at time *1*, to the expected time *1* demand, given the available information, that is $E[D_1|\Omega^j]$. Specifically, assuming that the correct expected demand is indeed $D+\gamma^Ds$, or, equivalently that the signal is not biased on purpose, the expected time *1* equilibrium price p^*_1 solves the equation $\beta x^I_1+(1-\beta)x^{NI}_1=D+\gamma^Ds$, that is:

$$p_1^{I*}=\left(Q-D-\gamma^D s\right)\times\frac{rk^2\left[\left(1-\alpha\gamma^P\right)\sigma_a^2+\sigma_2^2\right]\left[\sigma_a^2+\sigma_2^2\right]}{\left[(1+m)k-(1+r)\right]\left[\sigma_a^2+\sigma_2^2-(1-\beta)\alpha\gamma^P\sigma_a^2\right]}+\dots-\frac{\beta ks\gamma^P\left(\sigma_a^2+\sigma_2^2\right)}{\left[(1+m)k-(1+r)\right]\left[\sigma_a^2+\sigma_2^2-(1-\beta)\alpha\gamma^P\sigma_a^2\right]} \qquad (9)$$

Notice that, if the entire population was made by *I* forest owners, time *1* expected demand would still be $E[D_1|s]=D+\gamma^Ds$., while it would be *D* if only the *NI* group was present. In these two particular cases, the expected equilibrium prices for the groups would be p^{I*}_1 and by p^{NI*}_1 respectively:

$$p^{I*}_1=(Q-D-\gamma^Ds)\ \frac{rk^2\left[\left(1-\alpha\gamma^P\right)\sigma_a^2+\sigma_2^2\right]}{(1+m)k-(1+r)}-\frac{\gamma^P sk}{(1+m)k-(1+r)} \qquad (10)$$

$$p^{NI*}_1=(Q-D)\ \frac{rk^2\left[\sigma_a^2+\sigma_2^2\right]}{(1+m)k-(1+r)} \qquad (11)$$

p^{NI*}_1 is also the equilibrium price that would realize if no signal was sent, in particular notice that $p^{I*}_1<p^{NI*}_1$.

Neither the price p^{I*}_1 nor p^{NI*}_1 realizes at the equilibrium, since the overall population of forest owners is in fact made up by the two groups.

In particular, $p^{I*}_1<p^*_1<p^{NI*}_1$, so that the heterogeneity of landowners' population implies that the *I* (*NI*) group is asked to satisfy less than the entire demand, but at a higher (lower) price than p^{I*}_1 (p^{NI*}_1). An

increase in β, that is, in the proportion of informed landowners reduces the equilibrium price as expected, in the extreme case of $\beta=1$, $p^{I*}_1 = p^*_1$. This finding has concrete implications since it directly implies that increased information penetration reduces the (equilibrium) price for timber (Figure 1).

Methodology

Expected value asymmetries: a harvesting behavior model

In this Section, we present a prototype tool based on the framework introduced above for general harvesting behavior forecasting and policy simulation. Our model, Expected Value Asymmetries, or shortly EVA, expands the framework analyzed above by considering five possible categories of forest owners, differing in terms of risk aversion and patience as regards postponing harvesting revenues, as detailed below. With the exclusive purpose of facilitating the analysis, the results presented here are based on the assumption that the initial growing stocks and growth rate are identical among the five categories. In what follows we will first detail the way we have expanded the framework introduced in the previous Sections, also providing a characterization of the five categories, and then present some result obtained from the analysis conducted in EVA.

As already mentioned, forest owners differ in their purposes, degree of information, and risk attitudes concerning their forests. To reflect this heterogeneity, we have assigned to each category of forest owners two coefficients: risk aversion, r, and degree of patience with respect to postponing harvesting income, q. In particular, we assign weights, $1-q$ and q, to monetary wealth from timber sale at time 1 and time 2, respectively, and we modify (5) assuming that individuals want to maximize the utility of final *weighted* wealth $w_q^j=(q)p_2x^j_2+ +(1-q)p_1x^j_1(1+r), 0\leq q\leq 1$.

The use of the pair $(1-q,q)$ allow us to express a degree of patience with respect to monetary outcomes realized in the future (as opposed to present ones), in particular, if $q=0$, the landowner only values monetary outcomes realized from current timber sale, while the opposite is true if $q=1$. When $q=0.5$ instead, monetary outcomes from current and future timber sale are equally valued. A higher degree of patience ($0.5\leq q<1$) signalizes that the forest owner assigns the forest additional value beside the one provided by timber sales, and therefore is keen on carrying a larger forest stock to the future, even if this might not be optimal from a pure monetary perspective. This might be the case for example if the forest is part of an inheritance to be left to the children, or if has high ecological value, or if it provides non-monetary amenity services.

In creating forest owner typologies, we draw on earlier studies creating and analyzing private forest owner typologies [4-8]. We build in particular on Ingemarson et al. [7] in considering the following five categories of forest owners: Urban Dweller (UD), Economic Man, Elderly Couple, Green Values, and Multi-objective. In choosing the number and characteristics of forest owner categories, particular attention was made so as to allow sufficient variation as well as a continuum as regards objectives, and associated risk aversion and patience with respect to monetary outcomes. Here follows a brief description of the types:

I. Urban Dweller, being detached from his/her forest holding, has no clear objective with respect to his/her forest holding, and is not interested in, or receptive to, information/policies regarding forests. In light of the theoretical framework presented above, he/she belongs to the *NI* group. Since in essence the forest is not valued, risk aversion (r^{UD}) is low, for the same reason, the forest is left un-harvested in the first period, so that the patience coefficient q^{UD} is set to 1.

II. Economic Man has as overriding objective the maximization of the monetary value of the forest, no matter if it comes from present or future timber sale, equivalently, the patience coefficient q^{EM} is at intermediate level. Risk aversion (r^{EM}) is also at intermediate level, and he/she is informed about possible policies affecting the timber market.

III. Elderly Couple's main purpose is to leave the holding in as favorable (economic) condition for their children as possible. Consequently, the value of wealth from current timber sale is low (equivalently, patience (q^{EC}) is high) and risk aversion (r^{EC}) concerning future timber market conditions is high. For the same reasons, the Elderly Couple also tries to collect information on policies affecting the timber market.

IV. Green Values' main objective is promoting non-timber values, e.g., biodiversity and recreation. The value of monetary wealth from current timber sale, and consequently the opposite of patience ($1-q^{GV}$), is relatively small, since postponing final felling sustains non-timber values for a longer period. Risk aversion (r^{GV}) is fairly low. Information concerning timber markets is ignored, so that Green Value belongs to the *NI* group.

V. Multi-objective values both monetary and amenity (non-timber) benefits of forest holdings. Risk aversion (r^{MO}) is at intermediate/high levels and, due to amenity valuation, patience q^{MO} with respect to future timber sales is fairly high. Information about the future evolution of timber markets is collected and valued.

Next, we rank the five categories according to risk aversion (r) and the degree of patience with respect the future (q). From the lowest to the highest risk aversion (r) we have: Urban Dweller, Green Values, Economic Man, Multi-objective, and Elderly Couple ($r^{UD}<r^{GV}<r^{EM}<r^{MO}<r^{EC}$). Similarly, from the highest coefficient q to the lowest: Urban Dweller, Elderly Couple, Multi-objective, Green Values, and Economic Man ($q^{UD}>q^{E}>q^{MO}>q^{GV}>q^{EM}$).

Since all groups maximize the utility form economic wealth only, we assume that the risk aversion coefficient (r) ranges over 2 and 4, as typically assumed in standard economic theory [22]. Therefore we assume: (Table 1).

Next, we derive for each group the first order condition corresponding to the maximization problem (6), wherein (5) has been replaced by the expression for final weighted wealth (w_q^j). We then solve for the equilibrium (i.e. for the optimal short run timber supply of each group and the equilibrium price) by setting to zero the system formed by the five first order conditions derived before and the market clearing condition that equalizes total period 1 supply to the expected demand.

It should be noted that the ranking and the numerical assignment to the various parameters are informed estimates only, based on a number of studies of forest owner attitudes [4-8]. For example, the high degree of the patience of the Elderly Couple is derived from Ingemarson et al. [7], where the category, or objectives cluster, "traditionalists", which

$r^{UD}=2$	$r^{GV}=2.5$	$r^{EM}=3$	$r^{MO}=3.5$	$r^{EC}=4$
$q^{UD}=1$	$q^{GV}=0.55$	$q^{EM}=0.5$	$q^{MO}=0.6$	$q^{EC}=0.9$

Table 1: Preferences parameters per category: r^{UD}, r^{GV} r^{EM}, r^{MO}, r^{EC} risk aversion coefficients for Urban Dweller (UD), Green Values (GV), Economic Man (EM), Multi-objective (MO) and Elderly Couple (EC); q^{UD}, q^{GV} q^{EM}, q^{MO}, q^{EC} patience coefficients for Urban Dweller (UD), Green Values (GV), Economic Man (EM), Multi-objective (MO) and Elderly Couple (EC).

inspired our Elderly Couple, is the one that to the largest extent expects children or other relatives to take over the forest estate. Accordingly, we assume that Elderly Couple, wishing to hand over a well-stocked forest holding, are less concerned with wealth from current timber sales.

In addition, a number of simplifications are implicitly assumed, such as inelastic demand and the absence of any budget or consumption requirements for the forest owners. However, as already pointed out, our purpose is not to launch a fully operative tool for policy-analysis, but rather to demonstrate some lines along which our theoretical framework could be elaborated for more applied research. In addition, there exist studies providing an empirical basis for the quantification of some of the parameters, e.g., the distribution of growing stock on different owner categories (see, e.g., Favada et al. [9]).

Simulation Results and Discussion

In this section we present some of the research questions that could be addressed by the module EVA. We start by considering questions related to how forest owner heterogeneity affects the state of the forest resources and the timber market, and then we reassess the situation, considering also the effects of policy (unspecified, but with market implications).

Forest owner heterogeneity and forest development

Let us assume that the growing stock at time 1 is equally partitioned among the five categories, so that 20% of the growing stock (the one owned by Urban Dweller) is unmanaged, in the sense that no harvesting activities take place. Further, we assume that 20% of the entire growing stock is interesting from a biodiversity perspective (BIO in Figure 2), and it is entirely owned by a single category, while the remaining 60% is mainly valuable for timber (WOOD in Figure 2). We are interested in analyzing the composition of the growing stock at time 2, depending on whether the BIO area is owned by Multi-objective or Economic Man.

The answer is provided in Figure 3a and 3b below: clearly, the unmanaged proportion (31.43%) is the same as this area is owned by the Urban Dweller in both examples, while if the BIO proportion is owned by Multi-objective, it will also represent a higher percentage of the total growing stock at time 2. Indeed, the higher degree of risk-aversion of the Multi-objective with respect to the Economic Man would on one side induce a higher short-run harvest level; however, for this particular parameters' assignment, this effect is more than compensated by the opposite one generated by the higher degree of patience of Multi-objective, which instead reduces short-time harvesting in favor of a higher growing stock available at time 2.

Though the quantitative findings strictly depend on the chosen values of the parameters, the results derived through the EVA model are interesting per se, in a qualitative sense. Indeed, in essence, the model simply computes the equilibrium, allowing for forest owner heterogeneity. The graphs above show that, independently from the numerical specifications adopted, such heterogeneity makes the distribution of forest land on owner types non-trivial. Hence, given that private forest owners are a heterogeneous group; our results highlight the necessity to consider forest ownership when producing policy analysis and forest resource outlooks.

Forest owner heterogeneity and the timber market

Next we compare how timber market characteristics (namely timber equilibrium price and market shares in the sense of shares of total timber supply) are affected by the heterogeneity in forest ownership. Specifically, we consider two alternative partitions of the forest growing stock: one that assigns 20% of the growing stock to Elderly Couple, 30% to Economic Man, and the remaining 50% equally between Multi-Objectives and Green Value, and one that assigns 30% of the growing stock to Elderly Couple, 20% to Economic Man, and the remaining 50% equally between Multi-Objectives and Green Value.

The equilibrium price that realizes in the second case is some 9% lower than the one realized under the first partition. This results from the intrinsic value of wood being lower for the Elderly couple, whose degree of patience with respect to the future is higher. Figure 4a and 4b below compare the market shares under the two possible partitions.

Forest owner's heterogeneity and policy

When policy is considered, the EVA model accounts for the fact that different categories might have different degrees of confidence in the policy maker ($1/\sigma_s^2$ in the theoretical framework). Specifically, we rank those groups that are informed about forest policy according to their degree of confidence as follows (from the lowest to the highest): Elderly Couple, Multi-objective and Economic Man (as already mentioned, Green Values are assumed to be not receptacle for information that concerns timber market developments). In particular, $\sigma^2_{EM}=0.02$, $\sigma^2_{MO}=0.04$, and $\sigma^2_{EC}=0.06$.

Going back to the first example discussed in this Section, we assume that the BIO percentage is indeed owned by Economic Man, and that the policy maker would like to reduce the proportion of unmanaged stock and increase the one of BIO at time *2*. The EVA model could then, e.g., be used for reassessing the composition of the growing stock, should the policy maker act in order to inform and train the Urban Dweller so that he/she will also eventually decide to harvest at time *1*. Henceforth, the Urban Dweller is induced to assign value to his monetary wealth from harvesting at time 1, according to the patience coefficient $q^{UD}=0.48$, and his degree of confidence into the policy maker is pretty high with respect to the other categories, so that $\sigma^2_{UD}=0.03$. The result of such a simulation is shown in Figure 5 below: the growing stock at time *2* is reduced by 8.55%, however such a reduction is mainly due to the decrease of unmanaged stock (-58.28%) since both the BIO and the WOOD proportions increase (by 79.5% and 20.4%, respectively).

Next, we move to the third example considered above, and we analyze the effects on timber markets of two alternative policies: (a) one aimed at maintaining the actual market conditions and (b) one aimed at increasing the short-run supply of wood. The first policy consistently modifies the market shares (Figure 6a below), leaving the total supply unaffected, while the equilibrium price drops by some 14 percent. The second policy generates an additional minimal adjustment of the market shares (Figure 6b below) and a further reduction (3.8%) of the equilibrium price, while the short-run supply increases by 4.3% (and consequently the growing stock in the second period reduces 3.8%). Notice that this result might appear counterintuitive, however the contradiction is only illusory since the policy maker essentially acts as risk-reducer from a forest owner perspective.

In particular, the arrival of information (independently from its content) reduces the perceived variance as it appears in formula (3) and (4), and the results of our simulation also confirm this finding. Indeed, the introduction of the policy aimed at maintaining the actual market conditions reduces the first period's individual supplies of all three informed groups (meaning of each landowner representing one of the three groups), as it appears in Figure 7, since the perceived variance of the second period's price is reduced. Indeed, the absolute percentage variation in the short run supply is higher the higher is the coefficient of risk aversion (-13.1% for the Elderly Couple, -6.7% for Multi-Objective

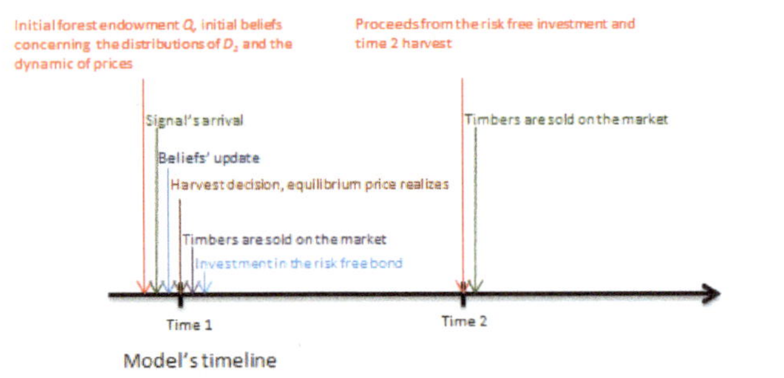

Figure 1: Expected value asymmetries: a harvesting behavior mode.

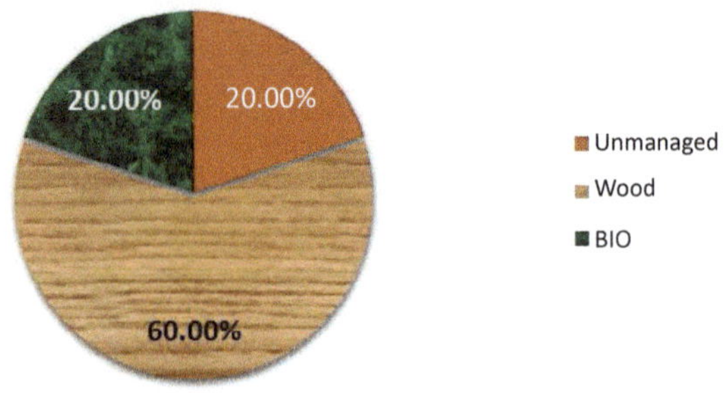

Figure 2: Initial growing stock composition.

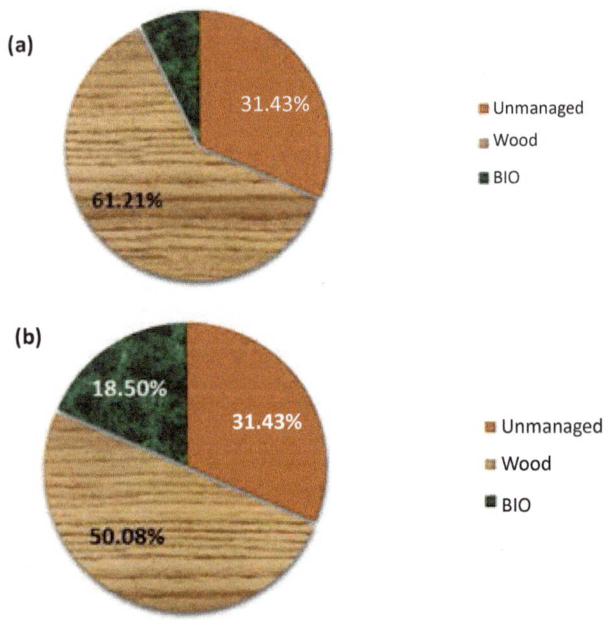

Figure 3: Time 2 growing stock composition (a) BIO area owned by Economic Man, (b) BIO area owned by Multi-Objective.

and -1.5% for the Economic Man). Next, when the second policy is put in place, the content of the information becomes relevant, and the informed landowners raise both the expected value of the demand and the one of the future price (for given equilibrium price).

Hence the overall result is a priori ambiguous, since one should sum the effect of the reduced variance (i.e., reduction of short-run supply) to the one of increased expected value, which in itself is uncertain since the increase in the expected demand (future price) increases (decreases) the short-run supply. In our simulation these contrasting forces are represented by a period one supply which is lower than the one without but higher than the one under the first policy considered. Again, the simulation results, in quantitative terms, depend on the chosen parameter values. Nevertheless, in a qualitative sense, they demonstrate the necessity to take into account the heterogeneity of forest owners when issuing policy initiatives.

A Suggested Augmented DSS

In this section it is outlined how EVA can integrate forest owner heterogeneity in an augmented DSS, linking a forest sector model with a forest resource assessment model.

EVA as part of a decision support system

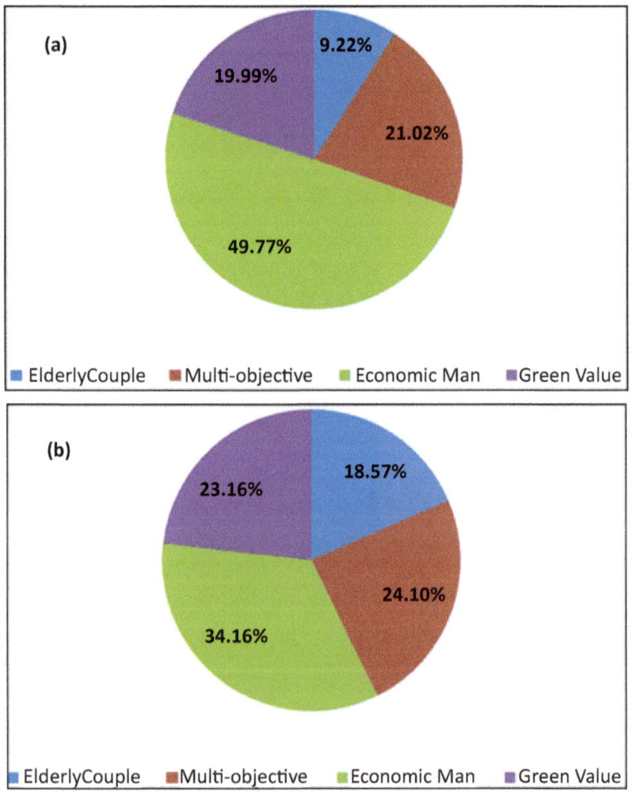

Figure 4: Market shares, (a) When the largest percentage (30%) of the growing stock is owned by Economic Man, (b) When the largest percentage (30%) of the growing stock is owned by Elderly Couple.

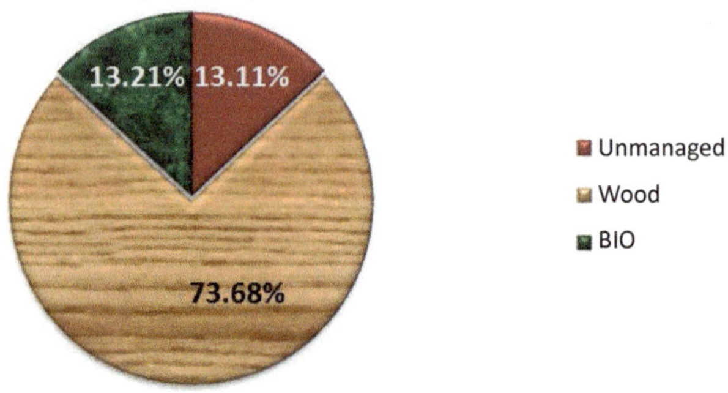

Figure 5: Time 2 Growing stock composition after policy.

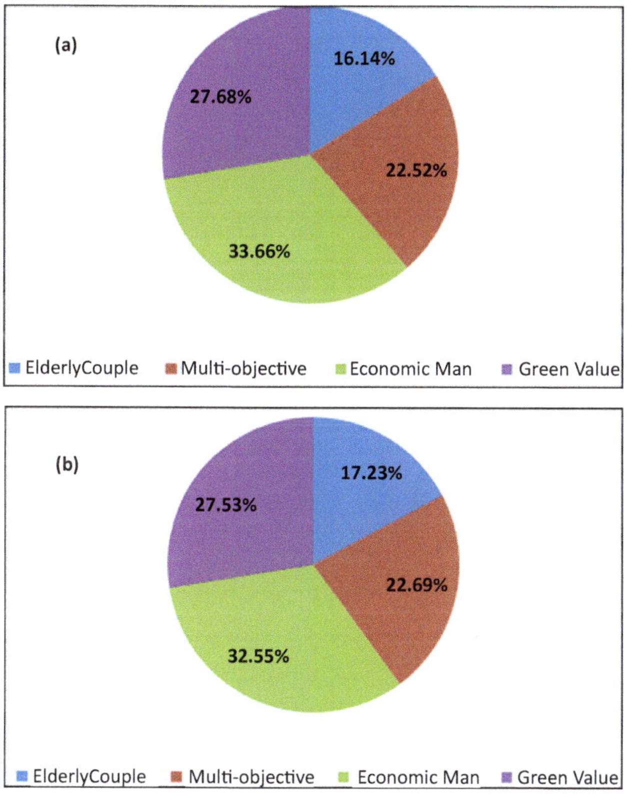

Figure 6: Market shares, (a) After policy to keep total supply constant (30% of growing stock owned by Elderly Couple), (b) After policy aimed at increasing short-run timber supply (30% of growing stock owned by Elderly Couple).

The EVA model can work as an intermediary between a forest resource assessment model, (FRAM), such as, e.g., the European Forestry Dynamics Model (see https://gna.org/projects/efdm/) or EFISCEN [10], and an economic model for the forest sector (EMFS), such as, e.g., EFI-GTM [11]. In particular, EVA can complete the loop between the two models, using as inputs the maximum harvestable level from the FRAM and the demand for wood primary products from the EMFS, and returning as output the amount effectively harvested by different forest-owner categories, to be ingested in the FRAM.

To explain the linkages between FRAM and EMFS more concretely, let us consider a FRAM belonging to the class of area-matrix state transition models as first developed in Sallnäs [23], such as the aforementioned European Forestry Dynamics Model or EFISCEN, meaning that a separate area matrix is assigned to each forest type according to region, owner type, species, site class (Figure 7). The FRAM derives the total maximum harvestable level—given legal constraints, biodiversity considerations, etc.—ingested by the EMFS as upper bound to the supply of primary products (Figure 8).

Simultaneously, this information is also sent, at cell-level, to EVA along with information regarding forest growth rate and owner type, also at cell-level. EMFS derives the forest sector equilibrium and sends the demand for primary products to EVA. Given this information, for each cell EVA computes the amount to be harvested in order to maximize the utility from wealth of the respective forest owner under the resource constraint provided by FRAM and the requirement that the entire demand derived in FRAM has to be satisfied. EVA's results are then fed back into FRAM completing the loop from EMFS to FRAM (Figure 8).

As knowledge concerning the distribution of forest land on different categories of forest owners is generally quite scant, 'expert judgment' is needed to allocate the forest resource on different owner types. Further, as already noted, the numerical assignment to the various parameters in EVA are to be regarded as informed estimates only. This lack of information makes frameworks as the one outlined above most suited for policy scenario analysis, since scenario analysis is better equipped to deal with indeterminacies caused by ignorance than, e.g., extrapolating methods [24,25].

Summary and Conclusions

Effective forest policy instruments require policy makers able to foresee what the harvesting reaction of private forest owners will be. This in turns depends on private forest owners' specific attitudes and objectives. However, private forest owners do not constitute a homogenous group, as already noted in several studies. Existing DSS do not account for forest owner heterogeneity, nor do they explicitly model the reaction of forest owners to market and/or policy signals

The current study seeks to address these shortcomings by (i) explicitly considering different types of forest owners, and by (ii) focusing on the role of information as a policy tool. A theoretical model of timber supply that studies how information releases affect harvesting behavior when forest owners differ in the awareness or receptiveness to this information is developed. Under these conditions, an asymmetry arises, which affects equilibrium pricing and, consequently, also harvesting decisions.

The theoretical model put forward permits information releases to influence forest owners' expectations of timber demand and equilibrium pricing. Consequently, informed landowners display a lower short-run timber supply than uninformed ones, whereas increased information penetration lowers timber price. Thus, the framework has concrete policy implications, as it demonstrates how information can be used as a tool for influencing timber markets.

This theoretical framework is then expanded, allowing for a higher degree of heterogeneity in terms of forest owners' preferences: we assume five forest owner types, differing as to (i) the responsiveness to market related information, (ii) risk aversion, and (iii) patience as regards postponing harvesting revenues. This harvesting behavior model, EVA, can, as a form of 'artificial laboratory', be used for policy analysis. Further, by working as an intermediary, EVA could provide feedback from an economic forest sector model to a forest resource assessment model, while at the same time ensuring that forest owner specific harvesting behavior is integrated in a DSS.

Simulation results derived through EVA indicate that the model is

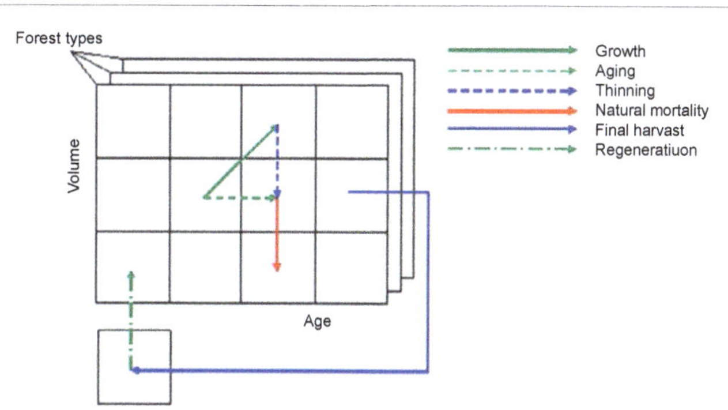

Figure 7: An area matrix state transition model.

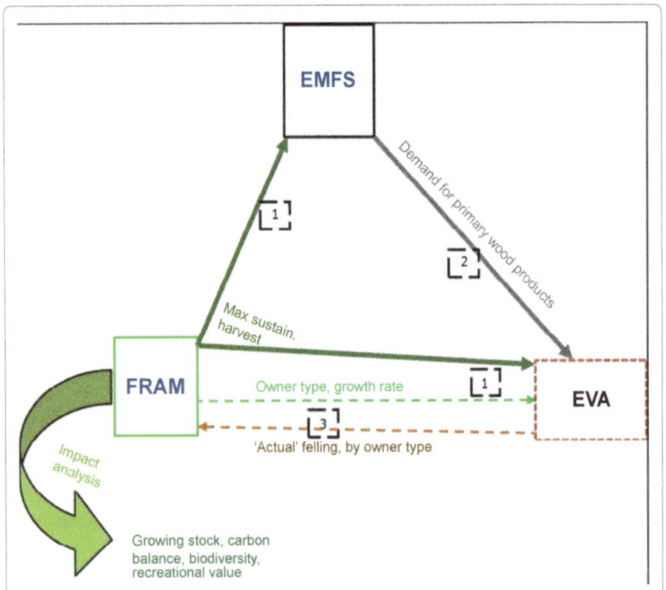

Figure 8: EVA as part of a DSS.

well adapted for modelling how forest owner heterogeneity and policy in the form of information release interact to affect the inter-temporal state of forest resources as well as the timber market. The numerical assignments to the various parameters of the EVA model are informed estimates only, derived from studies creating forest owner typologies. Further, in the simulation the distribution of forest land on the different categories is completely arbitrary.

Hence, though there are some studies providing empirical support for the quantification of some of the parameters, more empirically based knowledge, with a greater geographical coverage - concerning risk preferences, receptiveness to market-related information, and the distribution of forest land on different types of forest-owners - would enhance simulation models like EVA and contribute to the implementation of effective policy instruments. Imperfect information makes EVA and modelling frameworks as the one outlined above mostly suited for policy scenario analysis, since scenario analysis is better equipped to deal with indeterminacies caused by ignorance than, e.g., extrapolating methods [24,25]. This, however, is a caveat valid for most, if not all, modelling activities.

List of Symbols

- U: utility function;
- r: risk-aversion coefficient
- w^j: wealth at time 2 for an individual belonging to group j
- I: index identifying the group of informed landowners
- NI: index identifying the group of uninformed landowners
- x^j_i: quantity harvested at time i by a generic landowner belonging to group j
- x^j_1: quantity harvested at time 1 by a generic landowner belonging to group j
- x^j_2: quantity harvested at time 2 by a generic landowner belonging to group j
- x^{NI}_1: quantity harvested at time 1 by a generic landowner belonging to group NI
- x^I_1: quantity harvested at time 1 by a generic landowner belonging to group I
- k: forest growth rate
- Q: initial forest endowment
- D_1: time 1 demand for wood
- D_2: time 2 demand for wood
- p_i: price at which wood harvested at time i is sold on the wood market
- p_1: price at which wood harvested at time 1 is sold on the wood market
- p_2: price at which wood harvested at time 2 is sold on the wood market
- b: risk-free rate
- ε_a: long-run economic shock
- ε_D: time 1-contingent shock
- ε_2: time 2-contingent shock
- σ_a^2: variance of the long-run economic shock ε_a
- σ_D^2: variance of time 1-contingent shock ε_D
- σ_2^2: variance of time 2-contingent shock ε_2
- m: price variation coefficient
- s: signal
- ε_s : noise of the signal
- σ_s^2: variance of the noise ε_s
- β: fraction of the owners' population that observes the signal
- α: degree to which the signal is specific to the entire time horizon
- γ^P: informativeness of the signal with respect to the timber price
- γ^D: informativeness of the signal with respect to the timber demand
- Ω^j: time 1 available information a generic landowner belonging to group j
- p^{I*}_1: expected equilibrium price at time 1 for a landowner belonging to group I
- p^{NI*}_1: expected equilibrium price at time 1 for a landowner belonging to group NI
- p^*_1: expected market time 1-equilibrium price
- q: degree of patience with respect to postponing harvesting income
- w^j_q: *weighted* wealth at time 2 for a landowner belonging to group j with patience coefficient q
- r^{UD}: risk-aversion coefficient for the "*Urban Dweller*" category
- r^{GV}: risk-aversion coefficient for the "*Green Value*" category
- r^{EM}: risk-aversion coefficient for the "*Economic Man*" category
- r^{MO}: risk-aversion coefficient for the "*Multi Objective*" category

- r^{EC}: risk-aversion coefficient for the "*Elderly Couple*" category
- q^{UD}: patience coefficient for the "*Urban Dweller*" category
- q^{GV}: patience coefficient for the "*Green Value*" category
- q^{EM}: patience coefficient for the "*Economic Man*" category
- q^{MO}: patience coefficient for the "*Multi Objective*" category
- q^{EC}: patience coefficient for the "*Elderly Couple*" category
- σ^2_{UD}: inverse of the degree of confidence in the policy maker patience coefficient for the "*Urban Dweller*" category
- σ^2_{EM}: inverse of the degree of confidence in the policy maker patience coefficient for the "*Economic Man*" category
- σ^2_{MO}: inverse of the degree of confidence in the policy maker patience coefficient for the "*Multi Objective*" category

σ^2_{EC}: inverse of the degree of confidence in the policy maker patience coefficient for the "*Elderly Couple*" category

The authors declare no conflict of interest. The opinions expressed herein are those of the authors and do not necessarily reflect the views of the European Commission.

References

1. Pulla P, Schuck A, Verkerk PJ, Lasserre B, Marchetti M, Green T (2013) Mapping the distribution of forest ownership in Europe. European Forest Institute, EFI Technical Report 88.
2. Forests in Europe (2011) FOREST EUROPE Liaison Unit. ISBN 978-82-92980-05-7.
3. Bliss JC, Martin AJ (1990) How tree farmers view management incentives. Journal of Forestry 88: 23-29.
4. Karppinen H (1998) Values and objectives of non-industrial private forest owners in Finland. Silva Fennica 32: 43-59.
5. Kline JD, Alig RJ, Johnson RL (2000) Fostering the production of nontimber services among forest owners with heterogeneous objectives. Forest Science 46: 302-311.
6. Boon TE, Meilby H, Thorsen BJ (2004) An empirically based typology of private forest owners in Denmark - Improving the communication between authorities and owners. Scandinavian Journal of Forest Research 19: 45-55.
7. Ingemarson F, Lindhagen A, Eriksson L (2006) A typology of small-scale private forest owners in Sweden. Scandinavian Journal of Forest research: 249-259.
8. Favada IM, Karppinen H, Kuuluvainen J, Mikkola J, Stavness C (2009) Effects of timber prices, ownership objectives, and owner characteristics on timber supply. Forest Science 55: 512-523.
9. UNECE/FAO (2011) European Forest Sector Outlook Study II.
10. Schelhaas MJ, Eggers J, Lindner M, Nabuurs GJ, Pussinen A, et al. (2007) Model documentation for the European forest information scenario model (EFISCEN 3.1.3). Technical Report 26 (Also Published as Alterra Report 1559). European Forest Institute, Joensuu.
11. Kallio AMI, Moiseyev A, Solberg B (2004) The Global Forest Sector Model EFI-GTM — The Model Structure. European Forest Institute, Joensuu, Finland.
12. Serbruyns I, Luyssaert S (2006) Acceptance of sticks, carrots and sermons as policy instruments for directing private forest management. Forest Policy Econ. 9: 285- 296.
13. Rinaldi F, Jonsson R (2013) Risks, Information and Short-Run Timber Supply. Forests 4: 1158-1170.
14. Johansson PO, Löfgren KG (1985) The economics of forestry and natural resources. Basil Blackwell, Oxford. 312.
15. Koskela E (1989) Forest taxation and timber supply under price uncertainty: Perfect capital markets. Forest Science 35: 137-159.
16. Koskela E (1989) Forest taxation and timber supply under price uncertainty: Credit rationing in capital markets. Forest Science 35: 160-172.
17. Ollikainen M (1990) Forest taxation and the timing of private nonindustrial forest harvests under interest rate uncertainty. Canadian Journal of Forest Research 20: 1823-1829.
18. Ollikainen M (1993) A mean-variance approach to short-term timber selling and forest taxation under multiple sources of uncertainty. Canadian Journal of Forest Research 23: 573-581.
19. Uusivuori J (2002) Nonconstant Risk Attitudes and Timber Harvesting. Forest Science 48: 459-470.
20. Gong P, Löfgren KG (2003) Risk-aversion and the short-run supply of timber. Forest Science 49: 647-656.
21. Lönnstedt L, Svensson J (2000) Non-industrial private forest owners' risk preferences. Scandinavian Journal of Forest Research 15: 651-660.
22. Bodie Z, Kane AJ, Marcus A (2008) Investments, 7th edition. New York: McGraw Hill/Irwin.
23. Sallnäs O (1990) A matrix growth model of the Swedish forest. Studia Forestalia Suecica 183. Swedish University of Agricultural Sciences, Faculty of Forestry: Uppsala, Sweden : 23
24. Raskin P, Gallopin G, Gutman P, Hammond A, Swart R (1998) Bending the curve: Toward Global Sustainability. Stockholm Environmental Institute, PoleStar Series Report 8.
25. Raskin P, Kemp-Benedict E (2004) Global Environment Outlook Scenario Framework: Background Paper for UNEP's Third Global Environment Outlook Report (GEO-3). United Nations Environment Programme (UNEP), Nairobi, Kenya.

Forest Management Influence on the Carbon Flux of *Cupressus lusitanica* Plantation in the Munessa Forest, Ethiopia

Yonas Yohannes[1]*, Olga Shibistova[2,3], Zeleke Asaye[4] and Georg Guggenberger[2]

[1]*Wolkite University, Wolkite, P. O. Box 07, Wolkite, Ethiopia*
[2]*Institute of Soil Science, Leibniz Universität Hannover, Herrenhäuser Str. 2, 30419 Hannover, Germany*
[3]*VN Sukachev Institute of Forest, SB-RAS, Akademgorodok, 660036 Krasnoyarsk, Russian Federation*
[4]*Adami Tulu Agricultural Research Center, Oromia Agricultural Research Institute, P.O.Box:35, Zeway, Ethiopia*

Abstract

The effect of plantation forests on the global carbon balance is controversially discussed in recent times. As soil respiration is a decisive component in the carbon exchange between terrestrial ecosystems and atmosphere, effects of forest management measures (e.g. thinning) in the context of driving parameters of soil CO_2 efflux is a key issue in optimizing carbon friendly land management. In the present study, we report the effects of thinning, soil temperature and soil moisture, and biotic parameters on soil CO_2 efflux rate. Soil CO_2 efflux was measured by using an Infrared Gas Analyzer. We selected thinned and un-thinned stands within six years old *Cupressus lusitanica* plantation forest. Soil respiration rate ranged from 1.47 to 6.92 µmol $m^{-2}s^{-1}$ (thinned) and 1.31 to 5.20 µmol $m^{-2}s^{-1}$ (control stand). Generally higher soil respiration rates were measured during wet than in dry season. Seasonal variability of soil CO_2 efflux was significantly ($p<0.05$) correlated with soil moisture, but poorly correlated with soil temperature. Soil respiration increased with increasing soil moisture and reached maximum at 31% but after this threshold it start to decline. In general, soil CO_2 efflux rate in the first and second year after thinning was 24% and 14% higher in the thinned stand. Increased soil temperature at the thinned stand contributed minor to the larger soil CO_2 efflux, the more important reason appeared to be the trees' direct response. Higher fine root production together with larger microbial concentrations representing different groups infers a higher autotrophic respiration by roots and associated mycorrhizal fungi as well as by heterotrophic respiration. Despite the higher CO_2 losses with soil respiration, the organic C and total N concentrations in soil rather tended to increase, indicating higher organic matter input to soil at the thinned stand.

Keywords: *Cupressus lusitanica*, Soil CO_2 efflux, Soil moisture, Soil temperature, Thinning

Introduction

Recent atmospheric studies have suggested a substantial terrestrial sink for CO_2 could be attributable to the African continent [1,2]. However, there is still great uncertainty of the magnitude and variability of the African carbon stocks and fluxes [3]. The African ecosystems contribute 20% of global net primary production, along with the same proportion on global forest degradation and deforestation induced CO_2 emission [4], of which 0.24 Pg C yr^{-1} for sub-Saharan Africa [5].

One component that needs to be addressed in the terrestrial carbon budget is the efflux of CO_2 from the soil surface to atmosphere as being ultimately attributable to the ecosystem biological activity [6]. Contributing more than 50% to the ecosystem respiration across variety of biomes [7,8], soil respiration determines the ecosystem carbon balance and thus an ecosystem sink or source activity [9]. Simulations of the soil CO_2 efflux for biogeochemical models are one of the main challenges due to high temporal variability and spatial heterogeneity of soil respiration at all scales with respect to environmental conditions and long-term site disturbance history. Soil temperature and soil moisture has been recognized to be the two most influential abiotic parameters leading to seasonal and diurnal variations of soil respiration in different ecosystems [10-12].

One of the economically viable strategies for sequestering atmospheric carbon and mitigation the climate change scenario in tropics could be development of plantations on suitable land areas [13]. The Global Forest Resource Assessment reported that in 2010 planted forests worldwide accounted 264 million hectare, 7.2 million hectare of which were located in sub-Saharan Africa. Among other African countries, Ethiopia has one of the longest forest plantation histories [14] with the current total plantation forests area of 511 thousand hectare [15]. Forest plantations in Ethiopia are mainly represented by monocultures of fast growing exotic tree species mainly Eucalyptus sp. and *Cupressus lusitanica* [16]. Productivity of plantations needs to be high to meet increasing demand of wood products and to enhance stand quality and ability of effective carbon sequestration in various compartments of a forest ecosystem [17].

Forest management practices may determine carbon exchange between the atmosphere and the forest [18]. One measure to increase the formation in tree biomass, mainly stems, is thinning. This practice affects physiological and hence growth responses of the remaining trees and micro-climatic changes within the modified stand [19], influencing the environmental condition beneath the canopy, including the soil microclimate, and root dynamics [20]. The aboveground productivity and the carbon transfer to soil via litter fall and rhizo deposition are decisive for carbon availability for microbial decomposition processes in soil. Hence, it is likely that changes in aboveground productivity will have consequences for belowground process and vice versa [21]. All

***Corresponding author:** Yonas Yohannes,Wolkite University, Wolkite, P. O. Box 07, Wolkite, Ethiopia, E-mail: yonyoh4@gmail.com

these changes may be reflected in the soil respiration. Most research on thinning treatment on plantation forests, including Ethiopia [22] has been done in order to identify management guidelines to produce optimal stand growth [23]. Few studies have now emerged that focused on the carbon implications of thinning while targeting volume production [24]. As principally soil CO_2 flux studies in African intact and plantation forests are scarce [25-28], to our knowledge no study investigated the impact of thinning on soil respiration in east African plantation forests.

The Munessa forest is one of the largest plantation forest in Ethiopia established in the 1970`s. It covers about 6973 ha of which *Cupressus lusitanica* (Mill.) is the dominant planted tree species that account for 60% of the total plantation area [29]. The objectives of this study were (i) to compare the effects of thinning treatment on soil CO_2 efflux rate between intact and thinned stands and (ii) to compare indicators for soil microbial biomass between intact and thinned stand. We hypothesized that thinning treatment would increase soil CO_2 efflux rate by stimulating the productivity of the remaining trees [30] and changing the soil microclimate conditions and as a result of improved growing conditions of remaining trees and change in soil microclimate, soil microbial biomass will be larger under thinned stand.

Materials and Methods

Study site and silvicultural treatment

The study was conducted in 6-years-old *Cupressus lusitanica* stand established in June 2003 located at the Munessa forest (7°25` 44´´N and 38°51´ 05´´E), West Arsi zone, Oromia regional state, Ethiopia. The slope is gentle and the elevation at the study site is 2126 m above sea level. The soil texture is predominately a slightly acidic and nutrient rich clay-loam that was evolved from volcanic parent material [31], and was classified as Mollic Nitisols according to the WRB system [32]. The chemical and physical properties of the surface soil of the experimental plots are presented in Table 1.

The study site is characterized by a bimodal rainfall pattern with a minor rainy season occurring from March to May and a major rainy season from July to November [33-35]. Own meteorological records since 2001 show that 80% of the annual precipitation fell in the major rain season, and no clear indication of a minor rainy season is given. Also the precipitation pattern during the observation time showed pronounced seasonality with a wet period from June to October 2009 and from February to October 2010 showing a monthly precipitation of up to 200 mm (Figure 1). In contrast, November 2009 to January 2010 and the last six months were characterized by dry conditions. Minima of air temperature occurred at f the transition of the rainy season to the dry season with an average monthly temperature of *c.* 14°C and increased towards the transition of the dry season to the rainy season peaking in an average monthly temperature of *c.* 19°C.

For the present study, we worked on two fenced stands (40m×40m) located 20m apart from each other. The experimental stands were selected by a silvicultural research team with the objective to investigate the influence of thinning operation on the growth performance of potential crop trees of *C. lusitanica.* The stand was initially planted at spacing of 2.5m×2.5m. In January 2008, selective thinning has been done in one of the fenced areas by removing the competitors trees next to trees assigned as potential crop. No thinning operation was carried out on the other fenced area and was considered as a control. After thinning the number of trees left in the thinned stand was 98 as compared to 144 in the control stand.

Soil respiration measurement

Before the start of the field campaign, within each stand sixteen randomly spots were selected in order to estimate the number of sampling points required for soil CO_2 efflux measurement. At each selected spot, circular PVC soil collars (20 cm in diameter and 5 cm long) were installed about one year and six months of lag time after the management impact. An insertion depth of about 1-2 cm into was chosen for soil CO_2 efflux measurement to minimize severing of roots and the mycorrhiza. Finally, the outer wall of the collars was sealed with fine sand to avoid leaking of the chamber-soil system.

The number of individual flux measurements needed for various degrees of precision at various confidence levels was computed following probability equation described by Snedecor and Cochran [36]:

$$n = (t_\alpha s / D)^2 \quad (1)$$

where t_α is Student's t with degrees of freedom at the 0.05 probability level, s is the standard deviation with values obtained at this study, and D is the specified error limit. The result in Table 2 shows the required number of measurement points in each of the stands under different degrees of precision. Clearly, large numbers of flux measurements are ideal but we learnt from the 16 collar measurements before the start of the sampling campaign, it was feasible to restrict our measurements on 20 collars per stand. This decision led us our soil CO_2 efflux estimation to fall within an interval of 20% of the population mean at 95% confidence interval (Table 2).

At each stand, we installed the 20 collars under the canopy of four individual randomly selected trees at a distance of about 0.7 to 1.3 m from the bole. This design was chosen because previous studies had shown higher variability and higher soil respiration rates were observed in collars established at close vicinity to tree stems [37,38]. A sampling design of installing collars close to the vicinity to tree stems gives more reliable estimates of fluxes from a given stand [39]. Olsthoorn et al. [40] has shown that such sampling design was also more error free in the assessment of fine root density. Soil collars were left in place throughout the measurement period. Herbaceous understory vegetation was avoided during collar set-up. However, when, in the following two year's period, any vegetation grew inside the collars, it was clipped back.

Soil CO_2 efflux was measured on a weekly basis from May 2009 through May 2011 with the exception of November and December 2009 due to instrument failure. Instantaneous soil respiration was measured using a LI-8100-103 soil survey chamber (LI-COR, Lincoln, NE, USA) connected to an Infrared Gas Analyzer Li-8100 (LI-COR, Lincoln, NE, USA). Soil temperature (°C) at a depth of 0.1 m was recorded adjacent to each collar simultaneously with soil CO_2 efflux measurements using a thermocouple probe (Li-8100-201) connected to the Li-8100. The

Stand	Soil depth	Sand*	Silt*	Clay*	pH§	CEC¶	BS''
		—— g kg^{-1} ——				$mmol_{(+)}$ kg^{-1}	%
Thinned	0-10 cm	210 ± 15	395 ± 62	395 ± 67	6.0 ± 0.5	638 ± 92	100
	10-25 cm	189 ± 8	371 ± 65	440 ± 70	5.7 ± 0.6	449 ± 69	100
Control	0-10 cm	203 ± 20	376 ± 20	421 ± 20	6.3 ± 0.5	568 ± 24	100
	10-25 cm	190 ± 26	329 ± 7	481 ± 27	5.7 ± 0.7	447 ± 74	100

*measured by the pipette method (33); § analyzed potentiometrically in 1 M KCl [1.2.5 (m/v)]; ¶ cation exchange capacity determined with the $BaCl_2$ compulsive exchange method (34); '' base saturation. Values are mean and standard deviation of three replicated samples

Table 1: Basic characteristic of the soils under each stand.

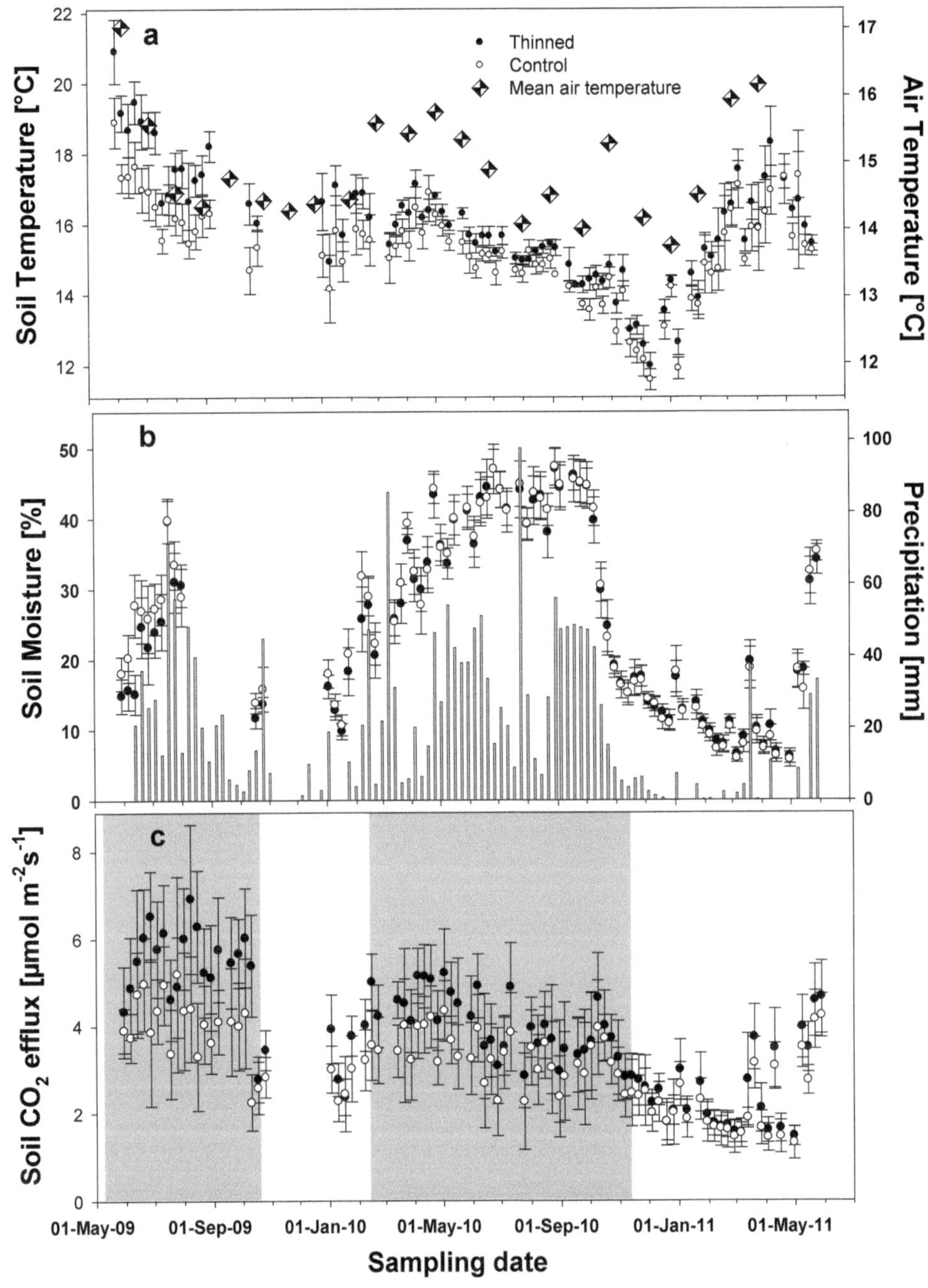

Figure 1: Seasonal patterns of air and soil temperatures (a), precipitation and soil moisture (b), and soil CO_2 efflux rate (c). Each data point for soil respiration, soil temperature and soil moisture is a mean of twenty measurements. Error bars indicate standard deviation. For the air temperature the monthly average temperatures measured by the Kuke weather station are presented. Data gaps are due to rain events or instrument failure. Periods with light grey background indicate wet periods as calculated by the approaches of Gibbs and Maher [52].

Interval about the full population mean (%)	99% confidence (α=0.01)		95% confidence (α=0.01)		90% confidence (α=0.01)		80% confidence (α=0.01)	
	Thinned	Control	Thinned	Control	Thinned	Control	Thinned	Control
± 10	31.95	34.10	24.27	25.91	20.37	21.74	15.85	16.92
± 20	26.85	28.65	20.40	21.77	17.12	18.27	13.32	14.22
± 30	22.88	24.41	17.38	18.55	14.59	15.57	11.35	12.11
± 40	19.73	21.05	14.99	15.99	12.58	13.42	9.79	10.44
± 50	17.18	18.34	13.05	13.93	10.96	11.69	8.52	9.10

Mean and standard deviation of soil CO_2 efflux on the measurement day was 5.34 ± 1.1 and 5.10 ± 0.8 µmol $m^{-2}s^{-1}$ for thinned and control stand respectively

Table 2: Number of sampling point required within a precision of ± 10 to ± 50% of the full population mean and at different confidence interval between 80-99% based on 16 measurements per stand in May 2009.

volumetric soil water content at 0.06 m depth was also measured with a handheld theta probe (ML2, Delta-T Device Ltd, Cambridge, UK) at three replicates around the collars immediately following each soil CO_2 efflux measurement. For each of the stands mean CO_2 efflux rates were calculated from the 20 chamber measurements obtained during individual sampling events.

Chemical and biological properties of the soil

After the soil respiration measurements were completed for a given sampling period, the soils to the depth from 0-10 cm and 10-25 cm below each PVC collar were collected to determine organic carbon (OC) and total nitrogen (TN), fine root density and microbial lipid concentration (PLFA analysis). Immediately after collection the soil samples were sieved with 2 mm mesh size and all visible fine roots were picked out. Root samples were then washed with distilled water and oven-dried at 65°C for 48 h for biomass determination.

After homogenization of each soil sample, aliquots (*c.* 2 g) were taken for organic C and total organic N analysis. After air-drying, soil samples were finely grinded with a steel ball mill (Mixer Mill, Retsch MM 200) and dried overnight at 105°C. Soil organic carbon and nitrogen contents were analyzed using a Vario EL III elemental analyzer (Elementar Analysensysteme GmbH, Germany).

Likewise, directly after homogenization of the soil, subsamples of about 10 g were collected separately for phospholipids fatty acid (PLFA) and neutral lipid fatty acid (NLFA) analysis. After removing all visible plant materials, the soil samples were then packed in small glass vials and kept frozen until analyzed. Lipid fatty acid extraction was carried out using frozen soil samples (1.5 g) extracted overnight with a chloroform-methanol citrate buffer mixture (1:2:08) by the modified method described by Bligh and Dyer [41]. The fatty acid 19:0 (nonadecanoic acid methyl ester) was added to the samples as an internal standard. Lipids were separated into neutral, glyco-, and phospholipids using solid phase extraction with silicic acid column (Bond Elut LRC-Si, Varian Agilent Technologies, Santa Clara, CA). Thereafter, the neutral and phospholipids were subjected to a mild-alkali methanolysis, and the resulting fatty acid methyl esters were separated by gas chromatography using an Agilent 7890A GC-MS (Varian Agilent Technologies, Santa Clara, CA). The fatty acids were quantified by comparison of the peak areas with those of the standard peak. Standard nomenclature was used to refer to the PLFAs according to the designation described in Zelles [42]. Phospholipids fatty acids i15:0, a15:0, i16:0, 18:1ω7c and cy19:0 were used to represent bacterial biomass [42], while the PLFAs 18:2ω6 and 18:1ω9 were considered to have fungal origin [43,44], and the NLFA 16:1ω5 was used as marker for arbuscular mycorrhizal fungi [45]. The sum of the mentioned PLFAs in addition to PLFAs being not assigned as biomarkers (16:1ω9, 16:1ω7c, 16:1ω5, 16:0, 10Me16:0a and 10Me16:0b) were used to represent microbial biomass [46,47] and expressed as nmol PLFA g^{-1} dry soil. At our study site, *C. lusitanica* is infected by arbuscular mycorrhiza.

Statistical analysis

Individual collars were used as the statistical units for analyzing the spatial variation. Statistical differences in the temporal variability of the soil temperature and soil moisture between thinned and control stands were analyzed using student *t*-test, at 95% confidence interval. A t-test or, if normality was not fulfilled after transformation, a Mann-Whitney Rank Sum test were used to analyze the difference between thinned and control stand in concentrations of fine roots (i.e., fine root biomass), OC and TN, total PLFAs and 16:1ω5 NLFA. For each collar from thinned and control stands the relationships between soil CO_2 efflux rate and soil chemical and biological parameters (fine root biomass, OC and TN, PLFAs) was examined with linear regression. The statistical analysis (α 0.05 level) was performed using R 2-11.0 [48] or SigmaPlot version 11 (Systat Software Inc., San Jose, CA, USA).

The effects of thinning, season, soil moisture and soil temperature (and their interaction) on soil CO_2 efflux variability were tested by linear mixed effect (LME) model. The model was set up with respiration rate at a collar as dependent variables with treatment (thinned and control) and season (wet and dry) as fixed effects. We nested the random effects of collar numbers within the experimental plot. Because of repeated measurements on the same sampling unit, correlation between the residuals within the same collar was expected. To account for this possible autocorrelation, a first-order continuous autoregressive structure (CAR(1)) error process was incorporated. Plotting the residuals of soil CO_2 efflux as a function of soil moisture showed a clear curvature and hence the quadratic effect of soil moisture is also included in the LME model. We used maximum likelihood to calculate parameter estimates and associated standard errors for each fixed effect, with the package nlme using program R Version 2.10.1 [48-50]. We examined residual plots to ensure that variance was homoscedastic and after fitting we have found out that the residual variation of soil CO_2 efflux increased with increased fitted values. Thus we used a log transformation of the soil CO_2 efflux data to improve homogeneity of variance [50]. We also used nonlinear regression analysis with the objective to describe the dependence of soil respiration on soil moisture or with soil temperature. More details about further statistical approaches can be found in Yohannes et al. [51].

Results

Seasonal variability in soil CO_2 efflux under thinned and control stands

For the period of May 2009 to May 2011, Figurel shows the temporal patterns of air temperature and soil temperature at 0.10 m depth (Figure 1a), precipitation and soil moisture at 0.06 m depth (Figure 1b), as well as the soil CO_2 efflux rate (Figure 1c) for the thinned and control stands. The seasonal pattern of the soil temperature followed that of the air temperature (Figure 1a), however, with smaller fluctuations. Likewise, soil moisture exhibited pronounced seasonal variations over the 2 years of observations, mirroring the precipitation pattern (Figure 1b). The highest soil water content of up to 49% were observed towards the end of the long rain period in August 2010, whereas the soil water contents<10% were registered at the end of the dry period in March - April 2011.

Average soil temperatures during the measurement period were 15.9 ± 1.7°C and 15.2 ± 1.5°C at the thinned and control stand, respectively. In general, soil temperature within the thinned stand were significantly higher as compared with the control stand (p=0.005), with higher between stands differences than within the stands (Figure 2). In contrast, there were no significant differences in volumetric soil water content between thinned and control stands (p>0.05). Considering soil moisture, the matrix of within-stand coefficient of variation versus the between-stand coefficient of variation lined up almost equally along the 1:1 line (Figure 2), indicating that the thinning treatment unlike soil temperature had no effect on change in soil moisture regime.

Across the measurement campaign, mean soil CO_2 efflux rate varied from 1.5 to 6.9 μmol m^{-2} s^{-1} and from 1.3 to 5.2 μmol m^{-2} s^{-1} at the thinned and control stand, respectively (Figure 1c) with seasonal minimum in January 2010 and April 2011 and maximum in June 2009 and June 2010. Soil respiration rates over the period of observation were significantly ($p<0.001$) higher in the thinned than the control stand (Figure 1c), and differences in the flux density appeared to be more pronounced during wet periods. The coefficient of variation in the soil CO_2 efflux rate was larger between stands than within stand (Figure 2), indicating the potential influences of stands specific biotic and/or abiotic factors on soil CO_2 efflux variability. The coefficient of variation in the seasonal soil CO_2 efflux rate ranged from 9% to 40% (thinned stand) and 12% to 48% (control stand). In general, the seasonality of soil CO_2 efflux pattern reflects the changes in soil moisture and different significantly between wet and dry periods (Table 3).

A non linear regression shows a significant ($p<0.05$) effect of soil moisture on soil CO_2 efflux rates that explained about 50% and 48% of the variability under thinned and control stand respectively (Figure 3). Soil CO_2 efflux rate as a function of soil moisture was fitted by Gaussian curve, which suggested the existence of thresholds in dependency of soil CO_2 efflux rate on soil moisture (Figure 3). Soil respiration rates coincidently increased with increasing volumetric soil water content

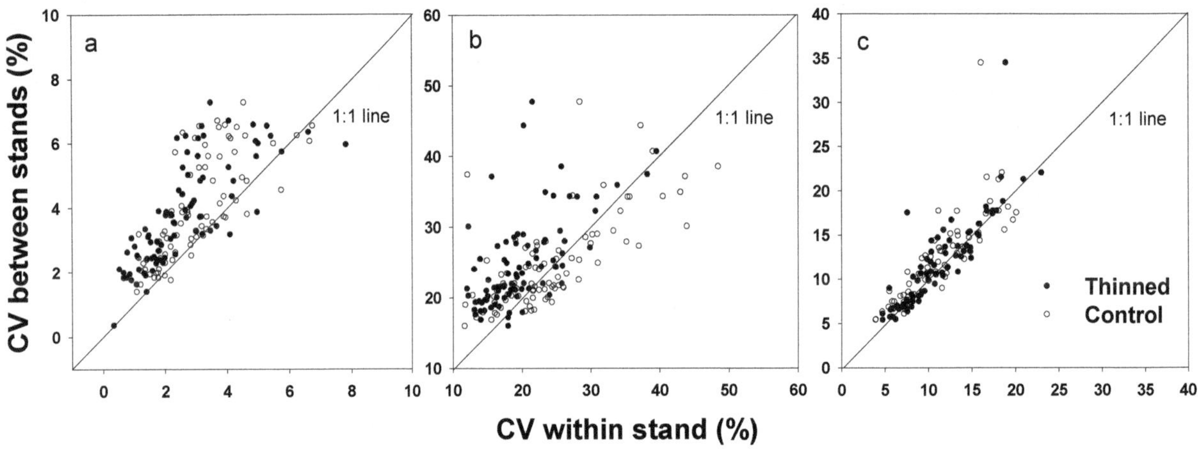

Figure 2: Coefficients of variation between stands versus CV within-stands in the thinned and control plantation stands: soil temperature (a), Soil CO_2 efflux (b), and soil moisture (C). Please note the difference on the scale of both-axis between the figures.

Factor	Coefficient	Standard error	df	F	p
Intercept	-1.0582	0.1008	1	2376.6198	<0.0001
Thinning	0.1259	0.0456	1	13.8562	0.0098
ST	0.0651	0.0057	1	1162.9606	<0.0001
SM	0.0896	0.0056	1	2651.6588	<0.0001
Season	0.0662	0.0135	1	920.3257	<0.0001
SM^2	-0.0015	0.0001	1	2125.6025	<0.0001
ST:SM	0.0004	0.0003	1	1.6307	0.2017

df, degree of freedom; *SM*, soil moisture; *ST*, soil temperature

Table 3: Summary statistics of linear mixed effect model describing the relationship of soil respiration to thinning treatment, covariates soil temperature and volumetric soil moisture content, and interaction term.

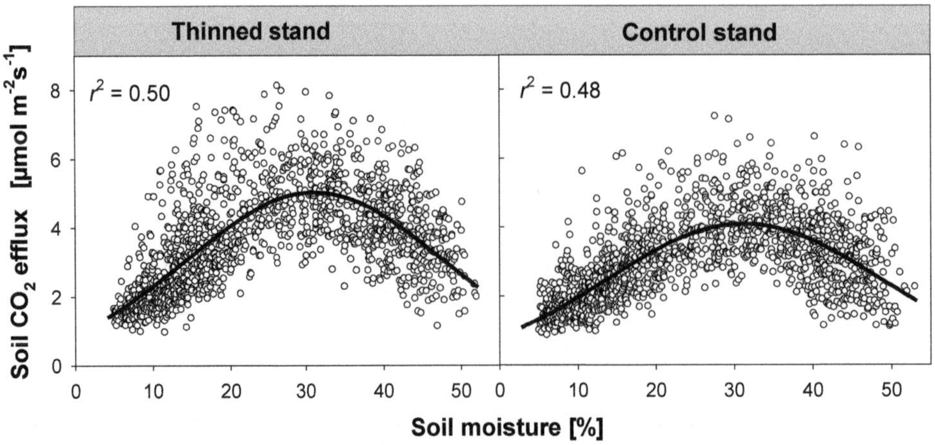

Figure 3: Relationship between soil CO_2 efflux rate (*SR*) and volumetric soil water content (*SM*) at 0.06 m soil depth. The parameters of the Gaussian function $SR=a*\exp(-0.5*((SM-x_0)/b)^2)$ are: a=5.01, b=16.80, x_0=30.97 (thinned stand); a=4.08, b=17.47, x_0=31.18 (control stand).

up to 31%, but the opposite was the truth after soil water content exceeded the abovementioned threshold.

Compared with soil moisture, soil temperature had a minor but significant impact on seasonal variability of soil CO_2 efflux. Soil temperature explained 17% and 10% of the temporal variability of the soil CO_2 efflux rate in thinned and control stands, respectively (Figure 4).

Soil CO_2 efflux as related to biological parameters and stand management

The variance components in the LME model analysis indicate that the nested effect of collars around individual tree attributed to 19% of the total variance of the random effect. This shows the likely effect of inherent characteristics (biotic and abiotic) of the soil underneath the collar to the overall soil CO_2 efflux variability. Organic carbon concentration in soil collected from each collars in 0–10 and 10–25 cm soil layers differed between the thinned and control stands, with mean values of 114 ± 27 and 40 ± 13 g C kg^{-1} soil for the former stand and 93 ± 20 and 21 ± 6 g C kg^{-1} soil for the latter (n=20, $p<0.05$ in both soil layer). Similarly the mean TN contents from 0-10 and 10-25 cm soil depth were 10 ± 2 and 4 ± 2 g N kg^{-1} soil for the thinned and 9 ± 2 and 2 ± 0.5 g N kg^{-1} soil for the control stand, being significantly different ($p<0.05$) at the 10-25 cm soil depth.

The concentrations of microbial biomarker (nmol total PLFA g^{-1} dry soil) used as a proxy for microbial biomass assessment were 22% larger ($p<0.05$) in the soil from 0-10 cm depth sampled from the thinned stand than in those under the control (Figure 5). The soils from the thinned stand also were characterized by larger concentrations of PLFAs being attributed to bacteria (nmol bacterial PLFA g^{-1} dry soil) and fungi (nmol fungal PLFA g^{-1} dry soil), being 19% ($p<0.05$) and 39% ($p=0.025$), respectively, more than in the soils from the control stand. At both soil depths, soils under the thinned stand contained also larger a concentration of the NLFA16:1ω5 arbuscular mycorrhizal fungi biomarker ($p<0.05$) (Figure 5).

The mean (n=20) fine root biomass was 145 ± 25 (0-10 cm soil depth) and 105 ± 10 (10-25 cm soil depth) at the control stand, being11% ($p=0.023$) and 12% ($p<0.001$) respectively smaller as for the thinned stand (not shown).

There was no significant relationship between soil CO_2 efflux

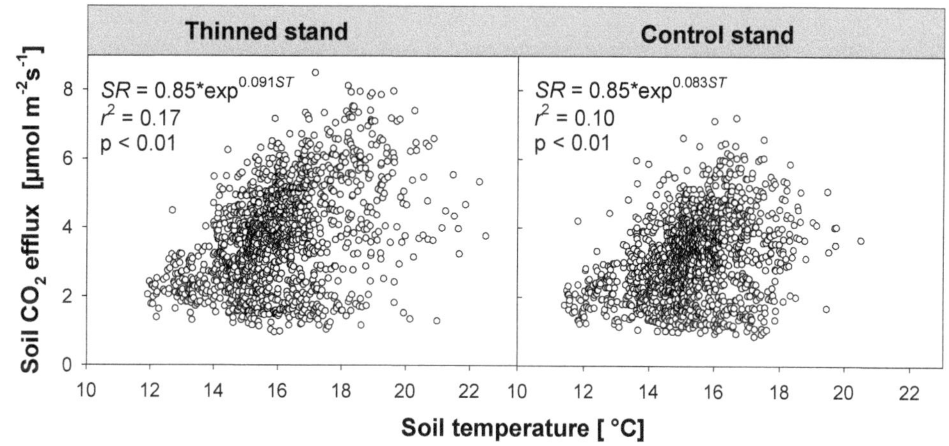

Figure 4: Relationship between soil CO_2 efflux rate (*SR*) and soil temperature (*ST*) at 0.1 m soil depth.

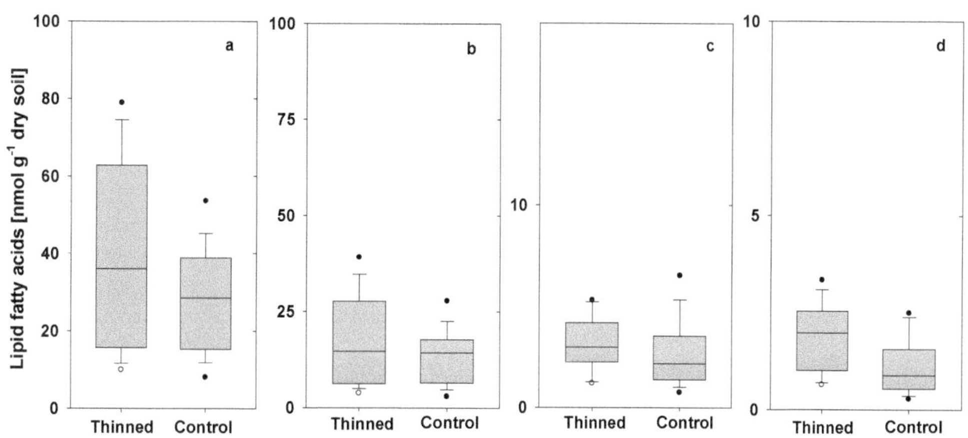

Figure 5: Total microbial (a), bacterial (b) and fungal markers (c) PLFA concentrations (nmol PLFA g^{-1} dry soil) and total AMF marker (d) concentration (nmol NLFA g^{-1} dry soil) collected from thinned and control stand from 0-10 cm (top graph) and 10-25 cm soil depth (bottom graph), n=20. Box plots indicate the distribution by percentiles. The lower and higher part of the box indicates the 25th and 75th percentiles, respectively. The value of the error bars are the 10th and 90th percentile, and the 0th percentile (median) is given by horizontal line within the box. Please note the difference on the scale of Y-axis.

with OC, fine root biomass, and PLFAs / NLFA analyzed in the soil samples taken directly after the measurement underneath each collars. However, the microbial fatty acids content in 0-10 cm soil depth were positively related with fine root biomass (Pearson's $r=0.74$ at $p<0.05$ and $r=0.69$ at $p<0.05$), OC ($r=0.52$ at $p<0.05$ and $r=0.18$ at $p<0.05$), and with TN ($r=0.59$ at $p<0.05$ and $r=0.10$ at $p<0.05$); values in parenthesis are for thinned and control stands, respectively.

Discussion

Seasonal variability of soil CO_2 efflux in relation to soil moisture and soil temperature

Not surprisingly, the soil respiration in both stands showed a similar seasonal pattern, with the higher rates observed in wet periods and the lower rates during dry periods. This result is consistent with our previous study conducted in adjacent natural forest [51]. Our finding is also in agreement with results reported for other tropical forests with seasonality in precipitations [25-28]. However, a positive effect of soil moisture on soil respiration could only be identified when it was below 31% volumetric soil moisture content. Soil water contents above this threshold led to a decrease in soil respiration (Figure 4). It should be also noticed that, the threshold for the evergreen *Cupressus lusitanica* stand reported here, is equal to the threshold of evergreen tree species (*Podocarpus falcatus* and *Prunus africana)* examined in the adjacent natural forest and exceeded that for deciduous *Croton macrostachys* [52]. The decline in soil CO_2 efflux rate at extremely wet conditions could possibly result from physiological stress of heterotrophic microorganisms due to oxygen deficiency that decreases decomposition rates of soil organic matter [53]. Alternatively, periods with the most precipitation are associated with conditions of low photosynthetic active radiation. This may decrease carbon assimilation rates and, consequently, the rates of associated autotrophic soil respiration [54].

Although the effect of soil temperature on the seasonal pattern of soil respiration has been widely confirmed by many researchers [10], our result indicated that soil respiration was only slightly positively related to soil temperature. This suggests that soil temperature was not the key determinant of soil respiration at the study site. The apparent weak contribution of soil temperature unlike to soil moisture is partly due to the relatively small temporal temperature fluctuations in this ecosystem, being not sufficient enough to drive seasonal variations in soil respiration [28,51]. Further, at times of high soil temperature soil CO_2 efflux was restrained by low soil moisture (Figure 4).

Soil CO_2 efflux in relation to CN and biotic factors

Regression analysis of the soil CO_2 efflux rates with both fine root biomass, soil OC or TN, and the different microbial parameter sampled underneath the respective collars showed no correlation. The lack of any relationship is surprising as soil CO_2 efflux from forest soil derived both from respiration of plant roots and soil microorganisms, positive relationship between soil respiration rate and fine root biomass and/or microbial parameters have been frequently reported [55,56]. In young stands of trembling aspen and paper birch, King et al. [57] found that increased fine root biomass was accompanied by an increase of soil CO_2 efflux by 39%. Fine roots influence soil physical and chemical environment via exudation of carbon-rich substances [58]. This argument highlights the decisive role of fine roots to change soil OC concentration which is one of the main determinants of soil CO_2 efflux variability [37]. In our case, we speculate that the lack of any relationship can be due to the overriding effect of abiotic parameters in the control of the soil CO_2 efflux rate at this particular date. At that time the averaged volumetric soil moisture underneath the individual collars was 33%, and with that above the threshold level of c. 31% as identified in Figure 3. The negative impact of the high soil moisture on soil respiration rates may have leveled off effects of different concentration of the organic substrate or the microbial biomass [59].

Thinning effects on soil CO_2 efflux

An averaged soil CO_2 efflux rate at the thinned stand was 13% larger than at the control stand. Thinning alters the abiotic stand parameters such as increasing the soil temperature or improving the light conditions due to loss of competitors. Soil temperatures increased with thinning intensity [60-62], and are probably caused by higher insolation to the soil surface under a more open canopy. As soil temperature affects the soil CO_2 efflux rates at the investigated stands to a certain extent (Figure 4), the increase in soil temperature might have contributed to the larger soil respiration in the thinned stand. In contrast, soil moisture, the major abiotic driving factor at the stands under study, did not get al. tered by the management impact.

A major reason of any thinning operation is the improvement of the light, nutrient, and water supply to increase stand productivity and obtaining better timber from potential crop trees [63]. After three years of thinning treatment, the basal area increment (at diameter breast height, 1.3 m from the ground) of the experimental trees in the thinned stand was on average 20% larger than trees under study in the control stand. As the soils are highly fertile and water supply is also not limited (in addition to the precipitation, there is lateral water input to the stand due to its downslope position), the major reason for this increased growth rates was rather the improved conditions for incoming photosynthetic active radiation. Larger biomass production requires more nutrient acquisition by the trees, thus leading to larger fine root biomass and larger colonization by arbuscular mycorrhiza, as it can be concluded from the larger concentrations of 16:1ω5 NLFA in the thinned stand soils. Hence, both components of the autotrophic continuum may have enhanced soil respiration at the thinned stand due to higher energy demand, thus increasing the autotrophic respiration [61].

However, fine roots, after their death, are also a source of heterotrophic respiration. Larger concentrations of biomarkers indicative for total microbial biomass (all PLFAs), bacteria (i15:0, a15:0, i16:0, 18:1ω7c and cy19:0 PLFAs) and fungi (18:2ω6 and 18:1ω9 PLFAs) suggesting fuelling of heterotrophic activity by higher substrate supply, i.e. by rhizodeposition, at the thinned stand. Such positive relations between fine roots and microbial biomass parameters have been frequently reported [64]. Also Hwang and Son [65] found larger fine root input to soil after thinning and argued that this was enough organic matter input to offset potential losses of soil OC by soil respiration. This might also explain our observation that despite the higher soil CO_2 efflux rates, OC and TN concentrations tended to be even larger at the thinned stand.

In general, our results concur with the findings of other studies reporting that thinning treatments led to elevated soil CO_2 efflux rates in plantation forests [66-69]. However, there are also reports showing the opposite effect of thinning on the soil CO_2 efflux rate [63,70]. Varying results on the effect of thinning treatment on soil CO_2 efflux variability could possibly arise from differences in thinning intensity, the timing of the post-thinning monitoring period, tree species and age of the study trees and sampling design [70-72]. In our study collars were placed randomly around trees as it has been shown that measurement of fluxes in a closer distance to the nearest tree give more reliable estimates of greenhouse gas emissions from a given stand than measuring further from the tree stem [39]. Oshashi et al. [67] designed their sampling points in a similar way and also found larger soil respiration rates of the thinned stand as compared to the control.

In the present study, selective thinning was applied in which only competitors' trees were removed that found next to trees assigned as potential crop trees. Concilio et al. [69] also carried out selective thinning and measured a 43% and 14% higher soil respiration rate in the thinned areas as in the control of a mixed-conifer forest and a hardwood forest, respectively. According to Son et al. [73] soil respiration was higher at a lightly thinned than at heavily thinned stand. A heavy management impact probably reduces the carbon translocation from plant to soil more, and with that root respiration and the supply of readily available OC source for heterotrophic respiration, as potentially the decomposition of the old organic matter is accelerated by more favorable abiotic conditions. Therefore, differences in thinning intensity might be also one of the reasons for the inconsistence reports among similar studies.

Further, the period when the soil respiration measurements have been done after the thinning might have an influence on the result. In a nine-year old Japanese cedar forest Ohashi et al. [67] reported that 3-4 years after thinning soil respiration was higher in a thinned stand than those of intact stand, but there was no difference 5 years after the thinning. In our result, the differences in average soil CO_2 efflux rate between thinned and control plot was larger in the first measurement year than the second. The largest soil CO_2 efflux during the early periods since thinning can be attributed to an increased heterotrophic respiration resulted from an accompanied change in the microclimate that favors decomposition of dead tree roots and slashed materials from the thinned trees. At our site, it is also likely that decomposition of dead roots from thinned trees influenced total soil respiration, nevertheless decomposition rate in the tropic is very fast [74] and less likely to prolonged more than a year. Therefore change in soil CO_2 efflux between thinned and un-thinned stand particularly after two years after thinning treatment could emanates from both autotrophic and heterotrophic respirations.

Conclusions

Thinning of the *C. lusitanica* stand was carried out in order to increase productivity of the trees and raise the economic value of the timber. In this study we investigated an important part of the C cycle, the soil CO_2 efflux, as influenced by thinning and in connection to the weather variables. The soil respiration rate was on average 13% larger in the thinned stand as it was estimated by the LME model. Soil moisture could have been identified as the key abiotic driver. Due to the pronounced seasonality in precipitation, soil moisture overrides the effect of soil temperature on soil CO_2 efflux rate that was found to be minor. While soil moisture did not differ between the thinned and control stands, temperature in soils (10 cm depth) of the former was significantly higher due to higher insolation. This higher soil temperature likely contributed to a certain extent to the higher soil CO_2 efflux at the thinned stand. However, it is most likely that the increased in soil respiration is the result of the trees' response on the silvicultural treatment. Higher fine root concentration of the thinned stand may suggest both, a higher autotrophic respiration and a higher heterotrophic respiration, as is also indicated by higher concentrations of biomarkers representing the microbial biomass. Despite the higher C losses as soil CO_2 efflux, the OC and TN concentrations in soil rather tended to increase. Hence, fostering tree growth by thinning obviously increased the organic C input to soil, counter balancing the C losses.

Acknowledgments

This work was supported by Deutsche Forschungsgesellschaft (DFG) within the project package (PAK 188). We thank Deksiso Bulcha, Getu Tadesse, Temesgen Yohannes, and Awol Assefa for their assistance and support in collecting data in the field. We also thank Christian Rumpf, Elke Eichmann-Prusch, Joanna Weiss, Leopold Sauheitl, Pieter Wiese, Roger-Michael Klatt, Silke Bokeloh and Ulrike Pieper for their laboratory assistance in PLFA and soil analysis. Likewise we are grateful to Frank Schaarschmidt for his advice in statistical analysis.

References

1. Stephens BB, Gurney KR, Tans PP, Sweeney C, Peters W, et al. (2007) Weak northern and strong tropical land carbon uptake from vertical profiles of atmospheric CO2. Science 316: 1732-1735.
2. Ciais P, Piao S-L, Cadule P, Friedlingstein P, Chédin A (2009) Variability and recent trends in the African terrestrial carbon balance. Biogeoscience 6: 1935-1948.
3. Williams CA, Hanan NP, Neff JC, Scholes RJ, Berry JA, et al. (2007) Africa and the global carbon cycle. Carbon Balance Manag 2: 3.
4. Laporte NT, Stabach JA, Grosch R, Lin TS, Goetz SJ (2007) Expansion of industrial logging in Central Africa. Science 316: 1451.
5. Ciais P, Bombelli A, Williams M, Piao SL, Chave J, et al. (2011) The carbon balance of Africa: synthesis of recent research studies. Philos Trans A Math Phys Eng Sci 369: 2038-2057.
6. Raich JW, Schlesinger WH (1992) The global carbon dioxide flux in soil respiration and its relationship to vegetation and climate. Tellus 44B: 81–99.
7. Janssens IA, Lankreijer H, Matteucci G, Kowalski AS, Buchmann N, et al. (2001) Productivity overshadows temperature in determining soil and ecosystem respiration across European forests. Global Change Biol 7: 269-278.
8. Shibistova O, Lloyd J, Zrazhevskaya G, Arneth A, Kolle O, et al. (2002) Annual ecosystem respiration budget for a Pinus sylvestris stand in central Siberia. Tellus 54B: 568-589.
9. Valentini R, Matteucci G, Dolman AJ, Schulze ED, Rebmann C, et al. (2000) Respiration as the main determinant of carbon balance in European forests. Nature 404: 861-865.
10. Davidson EA, Belk E, Boone RD (1998) Soil water content and temperature as independent or confounded factors controlling soil respiration in a temperate mixed hardwood forest. Global Change Biol 4: 217-227.
11. Reichstein M, Rey A, Freibauer A, Tenhunen J, Valentln R, et al. (2003) Modelling temporal and large-scale spatial variability of soil respiration from soil water availability, temperature and vegetation productivity indices. Global Biogeochem Cyle 17: 1-15.
12. Rustad LE, Huntington TG, Boone RD (2000) Controls on soil respiration: implications for climate change. Biogeochemistry 48: 1-6.
13. Dabas M, Bhatia S (1996) Carbon Sequestration through Afforestation: Role of Tropical Industrial Plantations. Ambio 25: 327-330.
14. Pohjonen V, Pukkala T (1990) Eucalyptus globulus in Ethiopian forestry. For Ecol Manage 36: 19-31.
15. FAO (2010) Global forest resources assessment, 2010 – Main report. FAO Forestry Paper 163. Rome, Italy.
16. Bekele-Tesemma A, Birnie, A, Tengnaes B (1993) Useful Trees and Shrubs for Ethiopia: Identification, Propagation and Management for Agricultural and Pastoral Communities. Regional Soil Conservation Unit/SIDA, RSCU, Nairobi.
17. Richter DD, Markewitz D, Trumbore SE, Wells CG (1999) Rapid accumulation and turnover of soil carbon in a re-establishing forest. Nature 400: 56-58.
18. 1 Maier CA, Albaugh TL, Allen HL, Dougherty PM (2004) Respiratory carbon use and carbon storage in mid-rotation loblolly pine (Pinus taeda L.) plantations: the effect of site resources on the stand carbon balance. Global Change Biol 0: 2335-2350.
19. Della-Bianca L, Dils RE (1960) Some effects of stand density in red pine plantation on soil moisture, soil temperature and radial growth. J Forestry 58: 373-377.
20. Tian DL, Yan WD, Fang X, Kang WX, Deng XW, Wang GJ (2009) Influence of thinning on soil CO2 efflux in Chinese fir plantations. Pedosphere 19: 273-280.
21. Wardle DA, Bardgett RD, Klironomos JN, Setälä H, van der Putten WH, et

al. (2004) Ecological linkages between aboveground and belowground biota. Science 304: 1629-1633.

22. Pukkala T, Pohjonen V (1993) Yields of Cupressus lusitanica in Ethiopia. Tiivistelmä: Cupressus lusitanican toutos Etiopiassa. Silva Fennica 27: 195-207.

23. Galloway G, Ugalde L, Vasquez W (2001) Importance of density reductions in tropical plantations: experiences in Central America. Forests Trees Livelihoods. 11: 217–232.

24. Hoover C, Stout S (2007). The carbon consequences of thinning techniques: Stand structure makes a difference. J Forestry 105: 266-270.

25. Epron D, Nouvellon Y, Roupsard O, Mouvondy W, Mabiala A, et al. (2004) Spatial and temporal variations of soil respiration in Eucalyptus plantation in Congo. For Ecol Manage 202: 149-160.

26. Epron D, Bosc A, Bonal D, Freycon V (2006) Spatial variation of soil respiration across a topographic gradient in a tropical rain forest in French Guiana. J Trop Ecol 22: 565-574.

27. Werner C, Kiese R, Butterbach-Bahl K (2007) Soil-atmosphere exchange of N2O, CH4, and CO2 and controlling environmental factors for tropical rain forest sites in western Kenya. J Geophysical Res. 112: doi:10.05029/02006jd007388.

28. Nsabimana D, Klemedtson L, Kaplin BA, Wallin G (2009) Soil CO2 flux in six monospecific forest plantations in Southern Rwanda. Soil Biol Biochem 41: 396-402.

29. Teshome T, Petty JA (2000) Site index equation for Cupressus lusitanica stands in Munessa forest, Ethiopia. For Ecol Manage 126: 339-347.

30. Balboa-Murias MA, Rodríguez-Soalleiro R, Merino A, Álvarez-González JG (2006) Temporal variations and distribution of carbon stocks in aboveground biomass of radiate pine and maritime pine pure stands under different silvicultural alternatives. For Ecol Manage 237: 29–38.

31. Fritzsche F, Zech W, Guggenberger G (2007) Soils of the Main Ethiopian Rift Valley escarpment: A transect study. Catena 70: 209-219.

32. FAO, ISRIC, ISSS (1998) World Reference Base for Soil Resources, vol. 84. Food and Agriculture Organization of the United Nations, Rome, Italy.

33. Gee GW, Bauder JW (1986) Particle-size analysis. In: A. Klute (Ed) Methods of Soil Analysis Part 1, (2nd ed) Agronomy series No. 9. American Society of Agronomy, Madison, Wisconsin: 383-411.

34. Gillman GP, Sumpter EA (1986) Modification to the compulsive exchange method for measuring exchange characteristics of soils. Aust J Soil Res 24: 61-66.

35. Griffits JF (1972) Ethiopian highlands. In: Griffits JF (Ed), Climate of Africa. World Survey of Climatology. 10: 369-387.

36. Snedecor GW, Cochran WG (1967) Statistical methods, Iowa State University Press, Ames, Iowa. USA. 1- 274.

37. Wiseman PE, Seiler JR (2004) Soil CO2 efflux across four age classes of plantation loblolly pine (Pinus taeda L.) on the Virginia Piedmont. For Ecol Manage 192: 297–311.

38. Saiz G, Green C, Butterbach-Bahl K, Kiese R, Avitabile V, et al. (2006) Seasonal and spatial variability of soil respiration in four Sitka spruce stands. Plant Soil 287: 161-176.

39. Butterbach-Bahl K, Rothe A, Papen H (2002) Effect of tree distance on N2O and CH4 fluxes from soils in temperate forest ecosystems. Plant Soil 240: 91-103.

40. Olsthoorn AFM, Klap JM, Voshaar JHO (1999) The relation between fine root density and proximity of stems in closed Douglas-fir plantations on homogeneous sandy soils: implications for sampling design. Plant Soil 211: 215–221.

41. BLIGH EG, DYER WJ (1959) A rapid method of total lipid extraction and purification. Can J Biochem Physiol 37: 911-917.

42. Zelles L (1999) Fatty acid patterns of phospholipids and lipopolysaccharides in characterization of microbial communities in soil: a review, Biol Fertil Soils 29: 111-129.

43. Frostegård Å, Bååth E (1996) The use of phospholipid fatty acid analysis to estimate bacterial and fungal biomass in soil. Biol Fertil Soils 22: 59-65.

44. Zelles L (1997) Phospholipid fatty acid profiles in selected members of soil microbial communities. Chemosphere 35: 275-294.

45. Olsson PA (1999) Signature fatty acids provide tools for determination of the distribution and interactions of mycorrhizal fungi in soil. FEMS Microbiol Ecol 29: 303-310.

46. Federle TW (1986) Microbial distribution in soil-new techniques. In: Megusar F, Gantar M, (Eds) Perspectives in Microbial Ecology, Slovene Society for Microbiology, Ljubljana, pp 493–498.

47. Frostegård Å, Bååth E, Tunlid A (1993) Shifts in the structure of soil microbial communities in limed forests as revealed by phospholipid fatty acid analysis. Soil Biol Biochem 25: 723-730.

48. Dean CB, Nielsen JD (2007) Generalized linear mixed models: a review and some extensions. Lifetime Data Anal 13: 497-512.

49. Ihaka R, Gentleman R (1996) R: a language for data analysis and graphics. J Comput Graphical Statistics 5: 299-314.

50. Pinheiro J, Bates D (2000) Mixed-Effects Models in S and S-PLUS. Berlin: Springer Verlag.

51. Yohannes Y, Shibistova O, Abate A, Fetene M, Guggenberger G (2011) Soil CO2 efflux in an Afromontane forests of Ethiopia as driven by seasonality and tree species. For Ecol Manage 261: 1090-1098.

52. Gibbs WJ, Maher JV (1967) Rainfall Deciles as Drought Indicators. Bureau of Metrology, Bulletin, No. 48. Common wealth of Australia, Melbourne: 1-37.

53. Linn DM, Doran JW (1984) Effect of water filled pore space on CO2 and NO production in tilled and non tilled soils. Soil Sci Soc of Am J. 48: 1267-1272.

54. Moyano PE, Kutsch WL, Rebmann C (2008) Soil respiration fluxes in relation to photosynthetic activity in broad-leaf and needle-leaf forest stands. Agric For Meteorol 148: 135–143.

55. Pregitzer KS, Zak DR, Maziasz J, DeForest J, Curtis PS, et al. (2000) Interactive effects of atmospheric CO2, and soil-N availability on fine roots of Populus tremuloides. Ecol Appl 10: 18-33.

56. Oscar JV-B (2007) Relationships among litterfall, fine-root growth, and soil respiration for five tropical tree species. Can J Forestry Res 37: 1954-1965. http://www.nrcresearchpress.com/doi/abs/10.1139/x07-057#.Ue-SDG0TnIU

57. King KS, Pregitzer DR, Zak J, Sober Isebrands JG, Dickson RE, et al. (2001) Fine-root biomass and fluxes of soil carbon in young stands of paper birch and trembling aspen as affected by elevated atmospheric CO2 and tropospheric O3. Oecologia 128: 237-250.

58. Kuzyakov Y, Chen W (2001) Photosynthesis controls of rhizosphere respiration and organic matter decomposition. Soil Biol Biochem 33: 1915–1925.

59. Fierer N, Schimel J (2002) Effects of drying–rewetting frequency on soil carbon and nitrogen transformations. Soil Biol Biochem 34: 777–787.

60. Thibodeau L, Raymond P, Camire C, Munson AD (2000) Impact of precommercial thinning in balsam fir stands on soil nitrogen dynamics, microbial biomass, decomposition, and foliar nutrition. Can J Forestry Res 30: 229-238.

61. Ma S, Chen J, North M, Erikson H, Bresee M, et al. (2004) Short-term effects of experimental burning and thinning on soil respiration in an old-growth, mixed conifer forest. Environ Manage 33: 148-159.

62. Yang A-R, Son Y, Noh NJ, Lee SK, Jo W, et al. (2011) Effect of thinning on carbon storage in soil, forest floor and coarse woody debris of Pinus densiflora stands with different stand ages in Gangwon-do, central Korea. For Scie Technol 7: 30-37.

63. Tang J, Qi Y, Xu M, Misson L, Goldstein AH (2005) Forest thinning and soil respiration in a ponderosa pine plantation in the Sierra Nevada. Tree Physiol 25: 57-66.

64. Wardle DA (1992) A comparative assessment of factors which influence microbial biomass carbon and nitrogen levels in soil. Biological Review 67: 321-358.

65. Hwang J, Son Y (2006) Short-term effects of thinning and liming on forest soils of pitch pine and Japanese larch plantations in central Korea. Ecol Res. 21: 671-680.

66. Messina MG, Schoenholtz SH, Lowe MW, Ziyin W, Gunter DK, et al. (1997) Initial responses of woody vegetation, water quality, and soils to harvesting intensity in a Texas bottomland hardwood ecosystem. For Ecol Manage 90: 201-215.

67. Ohashi M, Gyokusen K, Saito A (1999) Measurement of carbon dioxide

evolution from a Japanese cedar (Cryptomeria japonica D Don) forest floor using an open-flow chamber method. For Ecol Manage 123: 105-114.

68. Kowalski S, Sartore M, Burlett R, Berbigier P, Loustau D (2003) The annual carbon budget of a French pine forest (Pinus pinaster) following harvest. Global Change Biol 9: 1051-1065.

69. Concilio A, Ma S, Li Q, LeMoine J, Chen J, et al. (2005) Soil respiration response to prescribed burning and thinning in mixed conifer and hardwood forests. Can J For Res 35: 1581-1591.

70. Sullivan BW, Kolb TE, Hart SC, Kaye JP, Dore S, et al (2008) Thinning reduces soil carbon dioxide but not methane flux from southwestern USA ponderosa pine forests. For Ecol Manage 255: 4047-4055.

71. Jónsson A, Sigurdsson BD (2009) Effects of thinning and fertilization on soil respiration in a cottonwood plantation in Iceland. Biogeosciences Discussion 6: C2821-C2824.

72. Nilsen P, Strand LT (2008) Thinning intensity effects on carbon and nitrogen stores and fluxes in a Norway spruce (Picea abies (L.) Karst.) stand after 33 years. For Ecol Manage 256: 201-208.

73. Son Y, Jun Y, Lee Y, Kim R, Yang S (2004) Soil carbon dioxide evolution, litter decomposition, and nitrogen availability four years after thinning in a Japanese larch plantation. Communication in Soil Sci plant Analysis 35: 1111-1122.

74. Trumbore SE, Da Costa SE, Nepstad DC, De Camargo PB, Martinelli LA, et al. (2006) Dynamics of fine roots in Amazonian tropical ecosystems and the contribution of roots to soil respiration. Global Change Biol 12: 217–229.

Essential Oil Variability in Natural Hahadjerine Population of *Cupressus dupreziana* in Tassili n'Ajjer (Algeria)

M Ramdani[1]*, T Lograda[1], P Chalard[2], G Figueredo[3], JC Chalchat[2] and A Zeraib[1]

[1]*Laboratory of Natural Resource Valorization, Sciences Faculty, Ferhat Abbas University, 19000 Setif, Algeria*
[2]*Ecole nationale Supérieure de Chimie de Clermont Ferrand, Laboratoire de Chimie des Hétérocycles et des glucides, EA987, 63174 Aubière Cedex 01*
[3]*LEXVA Analytique, 460 rue du Montant, 63110 Beaumont, France*

Abstract

Essential oils extracted from dried leaves of *Cupressus dupreziana* A. Camus, an endemic species in the Tassili n'Ajjer (Sahara Central Algeria), were analysed by gas chomatography coupled to mass spectrometry (GC-MS). Terpinoid analyses were performed on 13 trees in the natural population of Hahadjerine in order to determine the intra-population variability. 39 trepenoids were identified; the averages of the principal components were trans-totarol (24.4%), Manoyl oxide (21.2%), α-pinene (15%) and Δ^3-carene (11.3%). The terpenoid markers used made it possible to determine the individual patterns of chemotypic variability. This variability confirms that genetic factors are not solely responsible for the decrease in the numbers of this species.

Keywords: *Cupressaceae*; *Cupressus dupreziana*; Genetic variability; Tassili n'Ajjer; Terpenoïds

Introduction

Tassili n'Ajjer is an ecosystems significant and fragile, situated in the south arid zones of Algeria (Sahara central). The Tassili n'Ajjer (National Park) protected area covers a total area of 72.000 km^2. The climate is hyper-arid, characterised by extreme meteorological variability and uncertainty. Mean annual rainfall ranges from 20mm to 100mm, with marked variations across years and seasons. Precipitation may be absent for several years at a given location. The absolute temperatures may range from -7°C to 50°C. The biodiversity inventory is far from complete and data on the distribution and status of most taxa require urgent updating.

The *Cupressus*, family Cupressaceae, is represented in Algeria by one widely endemic species found in the south of the country, namely in Tassili n'Ajjer. *C. dupreziana* A. Camus is a rare plant, which occurs in very dry regions, and is listed by the International Union for the Conservation of Nature and Natural Resources as an endangered species [1]. The climatic conditions, as well as human activities, seem to be the main factors reducing the diversity of this species.

A critical botanical review of this species has been made by Barry et al. [2]. Phytochemical investigations of *C. dupreziana* have been carried by [3-7]. The study on seven French cultivars of *Cupressus* gives an interesting essential oil composition including α-pinene, Δ^3-carene, sabinene, limonene and α-cadinol [8].

The dominant compound of *C. sempervirens* essential oil is the α-pinene [9]. The dominance of α-pinene, in *C. sempervirens*, is confirmed in addition to the sabinene and terpinene-4ol [10]. The α-pinene and Δ^3-carene are the major components of *C. dupreziana* [11-12]. The analysis of the samples of botanical garden of Algiers shows that *C. dupreziana* and *C. sempervirens* have the same major compounds, α-pinene and Δ^3-carene [11]. Several genetic studies on the species are currently being conducted [13-16].

The aims of study were linked to provide data that have always been rare; provide information on the chemical composition and search for variability within populations; fill gaps on natural populations of this species, because most previous studies have used botanical gardens samples.

Materials and Methods

Plant material

The aerial parts of *C. dupreziana* were collected from the Tassili n'Ajjer in Hahadjerine locality in april 2009. A voucher specimen is deposited in the Herbarium of the Sciences Faculty of Ferhat Abbas University (Algeria). Leaves and branches were dried at room temperature for 7 days, and used for analyses. The study is based on the analysis of a random sample of green branchlets of 13 trees from Hahadjerine population in Tassili n'Ajjer (Figure 1).

Essential oil analysis

The essential oils were extracted by hydrodistillation of dried plant material using a Clevenger-type apparatus for 3 h. The oils were stored in sealed glass vials at 4-5°C prior to analysis. Yield based on dry weight of the sample was calculated. The essential oil were analysed on a Hewlett-Packard gas chromatograph Model 5890, coupled to a Hewlett-Packard MS model 5871, equipped with a DB5 MS column (30m X 0.25mm; 0.25µm), programming from 50°C (5 min) to 300°C at 5°C/mn, 5 min hold. Helium as carrier gas (1,0ml/min); injection in split mode (1: 30); injector and detector temperature, 250 and 280°C respectively. The MS working in electron impact mode at 70 eV; electron multiplier, 2500 V; ion source temperature, 180°C; mass spectra data were acquired in the scan mode in *m/z* range 33-450.

The compounds assayed by GC in the different essential oils were identified by comparing their retention indices with those of reference

***Corresponding author:** M Ramdani, Laboratory of Natural Resource Valorization, Sciences Faculty, Ferhat Abbas University, 19000 Setif, Algeria
E-mail: Ramdani_ghalia@yahoo.fr

compounds in the literature and confirmed by GC-MS by comparison of their mass spectra with those of reference substances [17-19].

Statistical analysis

16 terpenoïds were obtained in sufficiently large quantities to be able to perform statistical analysis [20]. Data were first subjected to Principal Components Analysis (PCA) to examine the relationships among the terpenes compounds and identify the possible structure of the population. Cluster analysis (UPGMA) was carried out on the original variables and on the Manhattan distance matrix to seek for hierarchical associations among the populations. Statistical analyses were carried out using STATISTICA 9 software.

Results

This study included the natural Hahadjerine population, with 13 trees, in Tassili n'Ajjer. The average oil yield of the different trees was found to be 0.3%. The composition of the oils from the trees of Hahadjerine population differed only quantitatively. The general chemical profiles of the tested oils and the percentage content of the individual compounds are summarized in Table 1.

39 compounds were separated by GC-MS and characterized, with varied concentrations, particularly trans-totarol (19.1-33.5%), Manoyl oxide (14.1-26%), α-pinene (12.4-19.7%) and Δ^3-carene (8-17.7%). The average of the remaining compounds is low, such as germacrene-D (5.3%), Ferruginol (5.2%), cis-totarol (4.5%), isopimaradiene (2%) and 2-hydroxy-12-methoxy-19 -norpodocarpa-4,8,11,13-tetra-3-one (1.3%). Analyses revealed the presence of intra-population variability. The components identified (α-pinene, Δ^3-carene, germacrene-D, manoyl oxyde, cis and trans-totarol and ferruginol) show significant terpinoids variability (Figure 2). The compounds found in the Hahadjerine cypress generally resembled those previously reported to occur in some of cypress species [22-25]. In particular, α-pinene and Δ^3-carene were present in high relative levels. The only exception was the tree 11 which contains a low rate of Δ^3-carene, similar rate observed in several North American species of Cupressus [26].

The principal component analysis (PCA) performed on the correlations between the 16 variables presented three axes comprising

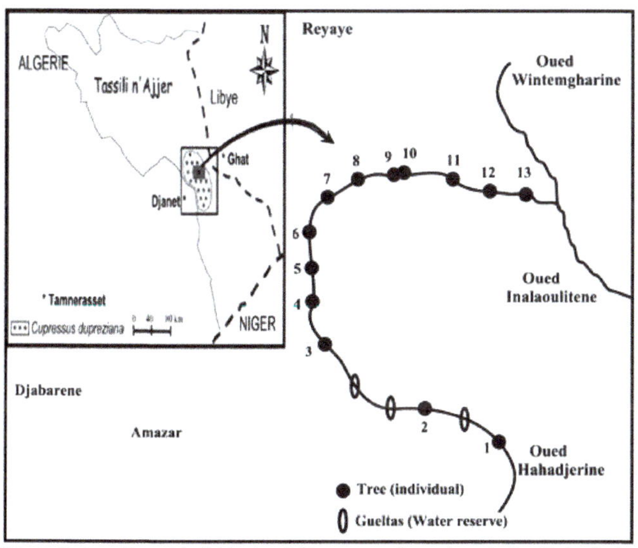

Figure 1: Location of individuals of *Cupressus dupreziana* in the Hahadjerine Wadi of Tassili n'Ajjer.

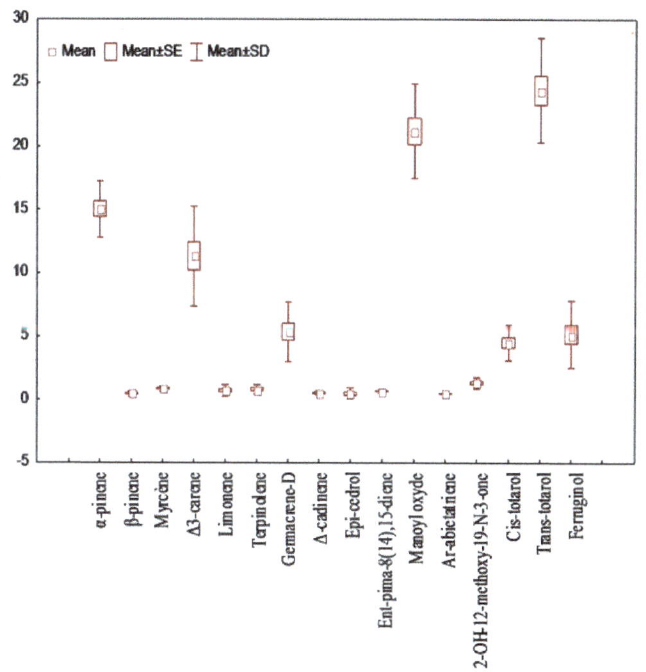

Figure 2: Chemical Variability of main compounds of *Cupressus dupreziana*.

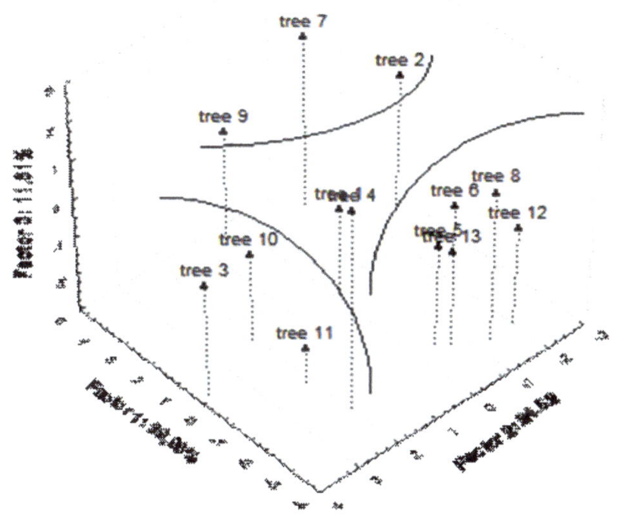

Figure 3: Ordination of the first three principal axes of Hahadjerine population.

76.19% of the total variation present in the original data. This analysis clustered populations in several groups, but the separation of populations is not clear. The ordination of population's means obtained for the three vectors is shown in (Figure 3).

All individuals of this population have showed high α-pinene and trans-totarol levels and low quantitative variations in all their components. Mono and sesquiterpenoids variability reflects the heterogeneity of the genetic structure of Hahadjerine population [21].

Genetic analyses were carried out using 16 terpinoids including

Compounds	Trees													Moy
	1	2	3	4	5	6	7	8	9	10	11	12	13	
α-pinene	13.0	8.1	9.7	9.5	11.1	10.6	11.7	9.8	11.7	9.8	14.2	11.2	15.8	11.2
Fenchene	0.2	0.2	0.2	0.1	0.2	0.2	0.4	0.2	0.3	0.2	0.1	0.2	0.2	0.2
Sabinene	0.2	0.0	0.2	0.1	0.1	0.1	0.2	0.1	0.2	0.2	0.1	0.0	0.1	0.1
β-pinene	0.4	0.3	0.4	0.4	0.4	0.4	0.4	0.3	0.4	0.4	0.5	0.0	0.0	0.3
Myrcène	0.7	0.5	0.6	0.6	0.6	0.6	0.9	0.6	0.9	0.6	0.7	0.5	0.5	0.6
Δ^3-carene	8.6	8.6	7.5	6.1	7.1	7.7	13.6	7.9	12.6	7.3	4.1	6.6	10.4	8.3
Para cymene	0.1	0.4	0.1	0.0	0.0	0.0	0.2	0.6	0.2	0.1	0.0	0.0	0.1	0.1
Limonene	0.7	0.1	0.5	0.4	0.4	0.4	1.2	0.1	1.2	0.7	0.4	0.3	0.6	0.5
β-phellandrene	0.1	0.0	0.1	0.1	0.0	0.0	0.2	0.0	0.2	0.1	0.1	0.1	0.2	0.1
Terpinolene	0.5	0.6	0.9	0.6	0.3	0.3	1.1	0.4	1.0	0.7	0.5	0.4	0.5	0.6
Terpinene-4-ol	0.0	0.2	0.2	0.0	0.0	0.0	0.4	0.0	0.4	0.2	0.1	0.1	0.1	0.1
α-terpineol	0.0	0.0	0.1	0.0	0.0	0.0	0.5	0.0	0.5	0.2	0.0	0.0	0.0	0.1
Terpinen-4-yle acetate	0.0	0.1	0.1	0.0	0.0	0.0	0.3	0.0	0.3	0.1	0.0	0.0	0.1	0.1
α-terpenyle acetate	0.0	0.0	0.0	0.0	0.0	0.0	0.2	0.0	0.2	0.0	0.5	0.0	0.0	0.1
β-caryophyllene	0.2	0.1	0.3	0.2	0.3	0.3	0.3	0.2	0.3	0.3	0.0	0.3	0.4	0.2
α-humulene	0.2	0.1	0.3	0.2	0.2	0.2	0.2	0.2	0.3	0.3	0.0	0.3	0.3	0.2
γ-muurolene	0.2	0.3	0.1	0.1	0.3	0.3	0.4	0.3	0.3	0.1	0.0	0.3	0.5	0.2
Germacrene-D	2.5	1.7	5.0	4.4	2.9	2.5	3.5	2.3	4.5	4.7	9.9	4.3	4.7	4.1
α-muurolene	0.0	0.0	0.0	0.0	0.0	0.0	0.1	0.0	0.1	0.1	0.2	0.2	0.2	0.1
γ-cadinene	0.2	0.2	0.1	0.1	0.3	0.2	0.4	0.2	0.3	0.1	0.1	0.2	0.3	0.2
Δ-cadinene	0.3	0.4	0.2	0.2	0.5	0.4	0.5	0.4	0.4	0.3	0.4	0.5	0.7	0.4
Caryophyllene oxyde	0.1	0.0	0.0	0.0	0.2	0.2	0.2	0.2	0.2	0.1	0.0	0.1	0.1	0.1
Epi-cedrol	0.2	0.4	0.2	0.1	0.0	0.3	1.1	0.3	1.0	0.4	0.1	0.3	0.3	0.4
Torrilenol	0.2	0.2	0.2	0.2	0.3	0.3	0.3	0.3	0.3	0.2	0.1	0.3	0.3	0.2
β-acorenol	0.0	0.1	0.0	0.0	0.0	0.1	0.4	0.1	0.4	0.1	0.0	0.1	0.2	0.1
Epi-α-cadinol	0.0	0.1	0.0	0.0	0.1	0.1	0.2	0.0	0.2	0.0	0.2	0.1	0.2	0.1
Epi-α-muurolol	0.0	0.0	0.1	0.0	0.0	0.0	0.2	0.0	0.0	0.0	0.4	0.1	0.1	0.1
α-cadinol	0.0	0.0	0.2	0.2	0.4	0.3	0.2	0.3	0.2	0.2	0.8	0.2	0.3	0.3
Eudesma-4(15),7-dien-3-β-ol	0.1	0.1	0.1	0.0	0.3	0.3	0.2	0.3	0.2	0.1	0.1	0.2	0.2	0.2
Eudesma-4(15),7-dien-1-β-ol	0.2	1.2	0.5	0.1	0.2	0.2	0.1	0.2	0.1	0.1	0.2	0.2	0.2	0.3
Ent-pima-8(14),15-diene	0.4	0.2	0.4	0.7	0.6	0.7	0.5	0.7	0.4	0.4	0.6	0.7	0.5	0.5
Manoyl oxyde	13.2	10.7	10.2	17.7	18.7	20.0	15.6	20.3	12.1	12.2	15.6	21.3	21.0	16.0
Isopimaradiene	1.3	1.0	0.9	1.6	1.8	2.0	1.5	2.2	0.9	0.9	1.5	2.3	1.8	1.5
13-epimanoyl oxyde	0.2	0.1	0.1	0.2	0.2	0.2	0.2	0.3	0.1	0.1	0.2	0.2	0.2	0.2
Ar-abietatriene	0.4	0.2	0.3	0.4	0.4	0.4	0.3	0.5	0.3	0.3	0.4	0.4	0.4	0.4
2-hydroxy-12-methoxy-19-norpodocarpa-4, 8,11,13-tetra-3-one	0.8	0.9	0.7	1.0	1.2	1.4	1.0	1.4	0.6	0.4	1.0	1.7	1.0	1.0
Unk 01	2.7	1.3	5.5	5.2	3.2	3.2	2.6	3.3	3.2	3.6	4.3	3.0	3.0	3.4
Totarol	14.5	10.6	20.5	25.6	22.8	21.7	15.0	19.1	14.3	16.2	21.0	20.8	17.7	18.4
Ferruginol	3.7	0.6	5.6	0.4	6.0	5.6	0.4	4.9	4.1	5.2	5.7	4.7	4.5	3.9
Unk 02	1.4	1.0	2.2	4.3	3.2	3.1	1.7	2.7	1.4	2.1	4.1	3.4	0.0	2.3
Unk 03	2.2	35.0	14.0	5.2	7.0	5.0	6.7	8.5	11.7	24.2	6.2	6.7	0.0	10.2
Total														87.4

Table 1: Chemical composition of essential oils of Hahadjerine population of *Cupressus dupreziana*.

some compounds that have been shown in other species of *Cupressus* to be under the control of single locus with two alleles. The dendrogram based on UPGMA clustering (Manhattan distance), shows the presence of many groups (Figure 4) that confirms result obtained from ACP analyses.

The first group formed by individuals (1, 13, 2, 7 and 9), rich in myrcene, Δ^3-carene, limonene, terpinolene and manoyl oxyde. The second group is divided into two groups, one formed by individuals (3, 10 and 11) rich in cis-totarol and germacrene-D, while the other includes individuals (4, 5, 6, 8, and 12) characterized by components manoyl oxyde and trans-totarol. Aggregation of Hahadjerine population trees into small groups is an indication of terpenoids variability in this population. The diversity of the terpinoids contents

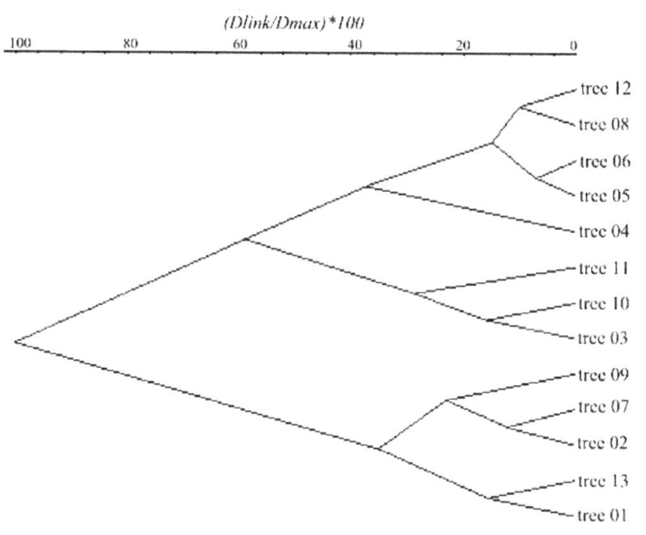

Figure 4: Dendrogram based on Manhattan similarity distance.

reflects the existence of considerable genetic variability [27,28]. The UPGMA analysis of terpene traits confirms this variability, but no clear conclusions can be transmitted between individuals' geographic distribution and genetic structure.

In brief, essential oils analysis carried out on 13 individuals of *C. dupreziana* showed both intra-population variability in their terpenoid content, with abundance of trans-totarol, Manoyl oxide, α-pinene, Δ^3-carene and germacrene-D.

Acknowledgements

The works was supported by Algerian MESRS and Laboratoire de Chimie des Hétérocycles et des glucides de Clermont Ferrant. The authors thank Lahbib tergui guide in Tassili n'Ajjer.

References

1. IUCN (2006) Conifer Specialist Group (1988) *Cupressus dupreziana* var. *dupreziana*. Red List of threatened Species. Retrieved on 10 June 2006. Listed as Critical (CR A2c. C1 v2.3).
2. Barry JP, Belin B, Celles JC, Dubost D, Faurel I, et al. (1970) Essai de monographie du *Cupressus dupreziana* A. Camus. Cyprès endémique du Tassili des Ajjers (Sahara central). Bul Soc Hist Nat Afr Nord 61: 95-176.
3. Piovetti L, Diara A (1977) Sesquiterpenes from *Cupressus dupreziana*. Phytochemistry 16: 103-106.
4. Piovetti L, Combaut G, Diara A (1980) Monoterpenes et sesquiterpenes oxygenes du *Cupressus dupreziana*. Phytochemistry 19: 2117-2120.
5. Piovetti L, Francisco C, Pauly G, Benchabane O, Bernard-Dagan C, et al. (1981) Volatile constituents of *Cupressus dupreziana* and sesquiterpenes of *C. sempervirens*. Phytochemistry 20: 1299-1302.
6. Pauly G, Yani A, Piovetti L, Bernard-Dagan C (1983) Volatile constituents of the leaves of *Cupressus dupreziana* and *Cupressus sempervirens*. Phytochemistry 22: 957-959.
7. Ramdani M, Rached O, Laouer H, El Koli M, Lograda T (2007) Chemical Composition and Antibacterial Activity of *Cupressus dupreziana* A. Camus Natural Product Communications 2: 945-949.
8. Christelle PL, Xavier F, Louisette LC, Loiseau AM, Fellous R, et al. (2003) Chemical composition of cypress essential oils: Volatile constituents of leaf oils from seven cultivated *Cupressus* species. J Essent Oil Res 15: 242-247.
9. Sacchetti G, Maietti S, Muzzoli M, Scaglianti M, Manfredini S, et al. (2005) Comparative evaluation of 11 essential oils of different origin as functional antioxidants, antiradicals and antimicrobials in foods. Food Chemistry 91: 621-632.
10. Tapondjoua AL, Adlerb C, Fontemc DA, Boudaa H, Reichmuth C (2005) Bioactivities of cymol and essential oils of *Cupressus sempervirens* and *Eucalyptus saligna* against *Sitophilus zeamais* Motschulsky and *Tribolium confusum* du Val. Journal of Stored Products Research 41: 91-102.
11. Chanegriha N, Baaliouamer A, Meklati BY, Chretien JR, Keravis J (1997) GC and GC/MS leaf oil analysis of four Algerian cypress species. J Essent Oil Res 9: 555-559.
12. Ramdani M, Lograda T (2008) Foliar Sesquiterpene Variations in Natural Populations of Cupressus dupreziana in Tassili n'Ajjer (Algeria). Asian Journal of Plant Sciences 8: 59-63.
13. Pichot C, Borrut A, El-Maataoui M (1998) Unexpected DNA content in the end osperm of *Cupressus dupreziana* A. Camus seeds and its implications in the reproductive process. Sexual Plant Reproduction 11: 148-152.
14. Pichot C, Liens B, Nava JL, Bachelier JB, El Maâtaoui M (2008) Cypress surrogate mother produces haploid progeny from alien pollen. Genetics 178: 379-383.
15. Pichot C, El-Maataoui M, Raddi S, Raddi P (2001) Surrogate mother for endangered *Cupressus*. Nature 412: 39.
16. Ramdani M, Lograda T, Chalard P, Chalchat JC, Figueredo G (2011) Chemical Variability of Essential Oils in Natural Populations of *Cupressus dupreziana*. Nat Prod Commun 6: 87-92.
17. Rösch P, Popp J, Kiefer W (1999) Raman and surface enhanced Raman spectroscopic investigation on *Lamiaceae* plants. Journal of Molecular Structure 480-481: 121-124.
18. Adams R.P (2001) Identification of Essential oil components by Gas chromatography/Mass spectroscopy; Allured/Carol Steam. IL. USA.
19. Swigar AA, Silverstein RM (1981) Monoterpenes, Infrared, Mass, NMR Spectra and Kovats Indices, Aldrich Chem. Co. Milwaukee, WI, USA.
20. Baradat Ph, Yazdani R (1988) Genetic expression for monoterpens in clones of *Pinus sylvestris* grow on different sites. Scand J For Res 3: 25-36.
21. Hemmateenejad B, Javadnia K, Elyasi M (2007) Quantitative structure-retention relationship for the Kovats retention indices of a large set of terpenes: A combined data splitting-feature selection strategy. Anal Chim Acta 592: 72-81.
22. Zavarin E, Lawrence L, Thomas MC (1971) Compositional variations of leaf monoterpenes in *Cupressus macrocarpa, C. pygmaea, C. goveniana, C. abramsiana* and *C. sargentii*. Phytochemistry 10: 379-393.
23. Kim YK, Cool LG, Zavarin E (1994) cis-Calamenene related sesquiterpenoids from *Cupressus Bakari* foliage. Phytochemistry 36: 961-965.
24. Cool LG, Zi-Ling Hu, Zavarin E (1998) Foliage terpenoids of Chinese *Cupressus* Species. Bioch. Syst. Ecol 26: 899-913.
25. Cool LG (2005) Sesquiterpenes from *Cupressus macrocarpa* foliage. Phytochemistry 66: 249-260.
26. Rafii Z, Cool LG, Zavarin E (1992) Variability of foliar Mono- and Sesquiterpenoids of *Cupressus Bakari*. Bioch. Syst. Ecol 20: 123-131.
27. Forrest GI (1980) Genotypic variation among native Scots pine populations in Scotland based on monoterpene analysis. Forestry 53: 101-128.
28. Raddi S, Sümer S (1999) Genetic diversity in natural *Cupressus sempervirens* L. Populations in turkey. Bioch Syst Ecol 27: 799-814.

12

Effect of Disturbance on Seed Germination and Seedling Growth of *Cassia fistula* (L.), *Albizia lebbeck* (L.) Benth. and *Dalbergia sissoo* Roxb. of Similipal Biosphere Reserve, Odisha, India

Mishra R.K*[1], Pattanaik S[2] and Mohanty R.C[2]

[1]*Department of Wildlife and Biodiversity Conservation, North Orissa University, Takatpur, Baripada, India*
[2]*Department of Botany, Utkal University, VaniVihar, Bhubaneswar, India*

Abstract

With the aim of restoring the disturbed sites on one of the premier tiger reserve of the country or the only biosphere reserve of the state, the study was planned to investigate the effect of disturbance on seed germination, early seedling growth and adaptation potential of *Cassia fistula* (L.), *Albizia lebbeck* (L.) and *Dalbergia sissoo* Roxb. in Similipal Biosphere Reserve (SBR), Odisha. The study sites of the biosphere reserve was categorised into disturbed, moderately disturbed and undisturbed stands on the basis of disturbance index. They differ from each other in forest floor light intensities along with other biophysical characters. High light intensity in the forest floor of Disturbed Stand (DS) than Undisturbed Stand (UDS) and Moderately Disturbed Stand (MDS) of the reserve was found suitable towards significantly higher rate of seed germination and seedling growth of the three investigated tree (over storey) species for all growth characteristics measured except S/R ratio. High soil nutrient content and low light intensity in the forest floor of undisturbed stand than disturbed and moderately disturbed stands of the reserve was not suitable for better seed germination and early stage of seedling growth. Moderate light intensity in the forest floor of moderately disturbed stand produced seed germination and seedling growth to come in the second order. All growth characteristic parameters of *Albizia lebbeck* seedlings had almost the highest values among species in all the forest stands of the reserve. Though *Cassia fistula* seedlings had lower growth characteristics than others, but its low S/R ratio in all the forest stands evidenced its higher adaptation potential.

Keywords: Disturbance, Forest stands, Similipal, Seedling growth, Seed germination, Light intensity, Forest floor

Introduction

The forest covers of Similipal Biosphere Reserve (SBR) in Mayurbhanj district of Odisha have faced serious problems over the years in extraction of over storey plants (trees) for timber, fuel wood and other energy requirements [1,2]. Identification of trees extracted for such practices and their rehabilitation within the depleting forest stands are urgently needed. Studies on ecological adaptation and growth characteristics of such species are therefore important for conservation, management and restoration of disturbed forest stands of SBR.

Plant growth and development is controlled by various environmental conditions [3-5] of which light is one of the most important factor [6]. In this regard studies stated that light availability on the forest floor is the primary limiting factor for seedling growth [7,8]. The major part of the forest floor receives only 1-2% photosynthetically active radiation [9]. Spatial variation in light availability in different stands of forest covers leads to variation in phytosociological characters of plants and physico-chemical parameters of soil [10,11]. Thus for successful seedling establishment it is crucial to study the seedling growth with respect to different growth parameters of seedlings. In addition species-specific responses to availability of light on the forest floor are important towards initial growth of seedlings [12,13]. The need for light during early stage of seedling growth is essential because seeds with smaller nutrient content are not able to sustain development for long periods in the shade [14].

In tropical forests it appears that seedling establishment and seedling growth face a more heterogeneous situation as compared to other forests. To study some of the mechanisms of seedling establishment and resource capture in tropical seasonal forests, it is helpful to investigate the growth responses of seedlings to different forest stands along with forest floor light content. Comparative quantitative information on seedling growth in terms of seedling height, shoot biomass, root biomass, total biomass and S/R ratio may also help to explain species co-existence and their adaptability to different forest stands [4,5,15]. Many authors confirmed the importance of light intensity on seedling growth and survival. Mayer and Mayber [16] and Osunkoya et al. [17] found that many trees at the seedling stage reduced their growth (e.g. root size and weight) with reduced light intensity. Moreover, Buylla and Ramous [18] indicated that reduced light was one of the factors that led to mortality of young plants. Similarly, Rao [19] studied the effect of forest floor light availability on the seedling growth of five important tree species in Tarai region of Uttaranchal and found that growth of all species was strongly affected by canopy gap position and availability of light on the forest floor.

Dalbergia sissoo, *Albizia lebbeck* and *Cassia fistula* belongs to the family, Fabaceae, Mimosaceae and Caesalpiniaceae, respectively are some of the timber yielding and ethno-medicinal tree species of Similipal that propagate by seeds. Different parts of *Cassia fistula* and *Albizia*

***Corresponding author:** Mishra R.K, Department of Wildlife and Biodiversity Conservation, North Orissa University, Takatpur, Baripada, India, E-mail: rabikumishra@gmail.com

lebbeck have medicinal values and were used by the tribal inhabited inside SBR to cure from different diseases [20,21]. Their woods were used for furniture, decorations, fine carving, etc. [2,22]. The woods of *Dalbergia sissoo* is economically very important and used in making high quality furniture, cabinet wood and general utility requirements [22]. Study on phytosociological analysis of woody perennials of SBR indicated that density of these species in comparison to others is too low [1]. Considering the ethno-medicinal values, use of woods of these species for various purposes and low density of these species inside the reserve, their introduction in afforestation programmes and planting trials is essential, which needs knowledge concerning about the optimum conditions for their germination and growth in different forest stands. Thus in the present study seed germination and seedling growth of these three indigenous tree species of Similipal was examined in different forest stands and light intensities on the forest floor.

Materials and Methods

Study site description

The study was carried out in Similipal Biosphere Reserve (SBR) located in Mayurbhanj district of Odisha (21°28′- 22° 08′ N latitude and 86°04′ - 86°37′ E longitude). The climate of the reserve is influenced by a monsoon pattern of rainfall. Maximum rainfall was occurred from mid-June to October and accounting for 75-80% of annual rainfall. Summer and winter were relatively dry generally with <15cm monthly rainfall (Figure 1). Summer was not unbearable, as the maximum temperature never exceeds 40°C. Winter was severe and the temperature comes down to 30°C with frosts in valleys. Spring was very pleasant. Because of luxuriant vegetation cover and a network of perennial streams, Similipal is relatively moist throughout the year. Minimum humidity of Similipal at 0600 hrs ranged from 40 to 70% and maximum at 1800 hrs ranged from 73 to 90% (Figure 1). The whole forest cover of Similipal was divided into three forest stands i.e. disturbed, undisturbed and moderately disturbed on the basis of disturbance regime.

Experimental design

Mature seeds of *Dalbergia sissoo* Roxb., *Albizia lebbeck* (L.) and *Cassia fistula* (L.) were collected from mother plants within the study area during end parts of winter and early parts of summer i.e. from February to March, 2013 and their germination potential was studied under three forest stands of SBR. For germination trial under each forest stand, 5 replicates of 100 seeds each were placed on the forest floor after removing all naturally fallen seeds from the ground in an area measuring 2 m × 2 m (3 forest stands × 3 species × 5 replicates of 100 seeds each). The seed plots were roofed with fine nylon net to avoid natural seed deposition until the seeds germinated. After emergence of seedlings nylon net was replaced with PVC pipes just to demarcate the seed plots for subsequent observations.

Seed germination and measurement of seedling growth

Observation on seed germination was made at an interval of 10 days for a total period of three months. Seed germination was recorded at each forest stand when epicotyls were visible above ground. After germination seedlings of each species were allowed to grow for one year. Species specific growth response (shoot length, root length and S/R ratio) and biomass (Shoot dry weight, root dry weight and total dry weight) of one year old seedlings of each species were measured in three different forest stands of SBR.

For extension growth five seedlings of each species from seed plots of three forest stands were randomly selected and shoot length was measured using centimeter scale. Means with standard deviation were calculated for each data set. Statistical test such as Least Significant Difference (LSD) was calculated at $p \leq 0.05$ level of significance to determine the overall influence of the forest stands on germination, seedling survival, and extension growth (shoot length, root length, shoot dry weight, root dry weight, total dry weight and S/R ratio) of *Dalbergia sissoo*, *Albizia lebbeck* and *Cassia fistula*. One way ANOVA (Analysis of Variance) was calculated to determine the degree of variation of growth parameters within the species among forest stands and the level of significance was also determined.

Measurement of bio-physical parameters

Each of the forest stands of SBR was sampled using square quadrats. A total of 120 quadrats of 20 m × 20 m for tree (≥ 30 cm cbh) species, 480 quadrats of 5 m × 5 m for shrubs and 600 quadrats of 1 m × 1 m for herbs were placed randomly. Plant species were collected, brought to the laboratory and identified with the help of Flora of Orissa [23]. Canopy area of tree species of each forest stand was calculated following Larson and Zaman [24]. Tree, shrub and herb density of each forest stand were calculated following Misra [25] and disturbance index was measured following Mohanty et al. [26].

Five number of soil samples were collected from each forest stand during February to April, 2013 at a depth of 0-15 cm. The moisture percentage was determined as per Misra [25] and the soil pH was measured using a digital pH meter (Systronics, 802) in a 1:5 soil water suspension. Soil organic carbon [27], available phosphorus [28] and nitrate nitrogen [29] was measured spectrophotometrically. The available potassium was measured by flame photometer method and the total nitrogen content by Kjeldahl digestion method [30]. Light intensity on the forest floor in each of the forest stand was measured by using Quantum Sensor (Hansatech, QSRED, UK).

Results

Biophysical condition of study sites

The three forest stands of SBR differ considerably in soil characters, tree density, canopy area (%) and light intensity (μmol/m²/Sec.) on the forest floor (Table 1). Tree density (Plants/ha) and canopy area (%) in undisturbed forest stand was comparatively higher than the disturbed and moderately disturbed forest stands of the reserve. However, a reverse trend was marked in case of shrub and herb density. Light intensity on the forest floor is one of the most important factor

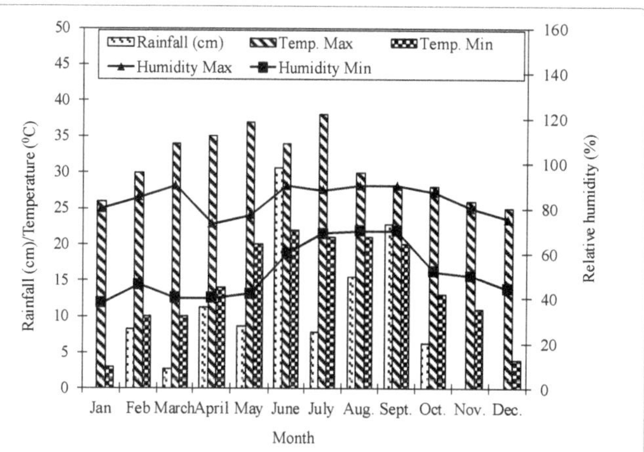

Figure 1: Climatic feature of Similipal biosphere reserve.

Character	DS	MDS	UDS
Disturbance Index (%)	>30	15-30	<10
Soil moisture (%)	11.74 ± 1.62	24.53 ± 2.81	34.16 ± 4.56
pH	6.26 ± 0.14	5.83 ± 0.33	5.69 ± 0.18
Organic Carbon (%)	0.96 ± 0.04	1.09 ± 0.07	1.25 ± 0.06
NO_3^--N (kg ha^{-1})	14.26 ± 3.27	17.26 ± 3.39	22.68 ± 5.56
Total Nitrogen (%)	0.16 ± 0.014	0.25 ± 0.02	0.38 ± 0.037
Available -P (kg ha^{-1})	19.03 ± 4.72	18.86 ± 3.29	18.51 ± 3.76
Available -K (kg ha^{-1})	205.33 ± 31.55	197.61 ± 31.27	171.65 ± 19.84
Tree density (Plants ha^{-1})	655 ± 23.53	750 ± 21.56	920 ± 31.41
Shrub density (individuals ha^{-1})	1680 ± 96.77	1560 ± 77.31	1040 ± 40.66
Herb density (individuals m^{-2})	28 ± 4.71	16 ± 3.12	5.0 ± 1.08
Tree crown area (%)	23.16 ± 5.79	31.42 ± 5.96	45.42 ± 4.87
Light intensity (µmol/m^2/Sec.)	344.91 ± 33.74	158.16 ± 22.84	59.36 ± 9.13

Table 1: Comparison of biophysical characters at disturbed (DS), moderately disturbed (MDS) and undisturbed (UDS) forest stands of Similipal Biosphere Reserve (SBR).

towards seedling growth differ among the forest stands. Maximum and minimum light intensity on the forest floor was measured at disturbed and undisturbed forest stands of the reserve, respectively (Table 1). Soil moisture in the studied forest stands varied between 11.74 ± 1.62% and 34.16 ± 4.56%. In all the forest stands the soil was slightly acidic (pH value ranging between 5.69 ± 0.18 and 6.26 ± 0.14) in nature. Average soil organic carbon ranged from 0.96 ± 0.04% to 1.25 ± 0.06% with a maximum at undisturbed forest stand and minimum at disturbed forest stand of the reserve. The same trend was marked in NO_3^- -N and total nitrogen content, while the trend in case of available phosphorus and available potassium was reverse (Table1).

Seed germination

The seed germination percentage of three indigenous tree species viz. *Albizia lebbeck, Cassia fistula* and *Dalbergia sissoo* were tested in three different forest stands of Similipal. Irrespective of forest stands maximum percentage of seed germination was exhibited by *Albizia lebbeck* followed by *Dalbergia sissoo* and *Cassia fistula*. The percentage of seed germination among the species was significantly differed from each other (Figure 2). Irrespective of species maximum percentage of seed germination was marked in the disturbed forest stand than undisturbed and moderately disturbed forest stands of the reserve and the variation was significant (Figure 3). Comparative analysis of seed germination of individual species among forest stands showed significant variation which may be due to variation in light intensity on the forest floor (Figure 4). As evidenced from the soil chemical parameters, high soil nutrient content in undisturbed and moderately disturbed forest stands of the reserve than disturbed forest stand may not be suitable for maximum seed germination of these tree species. But, high light intensity on the forest floor of disturbed forest stand than undisturbed and moderately disturbed forest stand may be the cause towards high seed germination.

Seedling growth

Statistical analysis of the obtained data on stem height, root length, shoot dry weight, root dry weight, total dry weight and S/R ratio revealed that there was significant difference in seedling growth among the tree species (Table 2). Irrespective of forest stands *Albizia lebbeck* seedlings had the highest values for stem height (15.1 ± 2.735 cm), root length (8.6 ± 1.831 cm), shoot dry weight (1.553 ± 0.231 g), root dry weight (0.635 ± 0.185 g), total dry weight (2.189 ± 0.395 g) and S/R ratio (1.793 ± 0.197) among the species and varied significantly than *Cassia fistula* and *Dalbergia sissoo* (Table 2).

Across forest stands seedlings of the investigated species had showed maximum growth in terms of shoot height, root length, shoot dry weight, root dry weight and total dry weight at the disturbed forest stand than undisturbed and moderately disturbed forest stands of the reserve. However, the shoot to root ratio was inverse i.e. maximum S/R ratio was observed at undisturbed forest stand and minimum at disturbed forest stand. The variation of each growth parameter of seedlings among the forest stands was significantly different (Table 3).

Seedling growth of each tree species was different with respect to

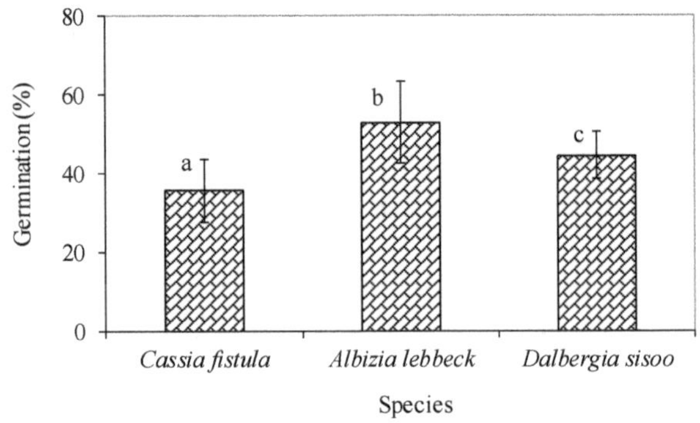

Figure 2: Seed germination percentage of three selected over-storey tree species of Similipal biosphere reserve. Bars represent SD of means. Different small letter above bars signify they significantly differ from each other at p ≤ 0.05

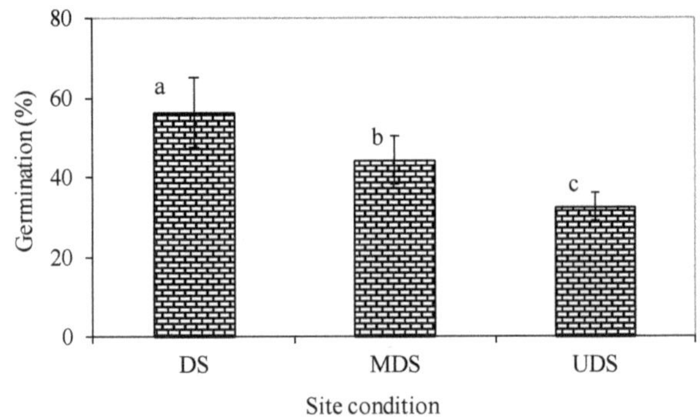

Figure 3: Effect of forest stand condition on seed germination percentage of three selected over-storey (tree) species of Similipal biosphere reserve. Bars represent SD of means. Different small letter above bars signify they significantly differ from each other at p ≤ 0.001.

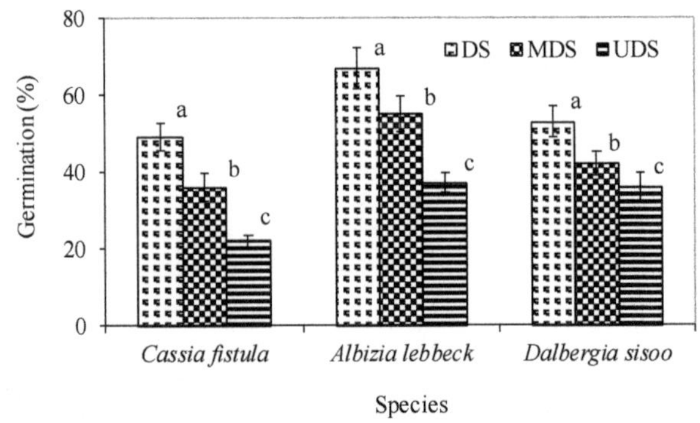

Figure 4: Comparative account of variation in seed germination percentage of individual tree species in different forest stands (disturbed, moderately disturbed and undisturbed) of Similipal biosphere reserve. Bars represent SD of means. Different small letter above bars signify they significantly differ from each other at p ≤ 0.05.

Character	*Cassia fistula*	*Albizia lebbeck*	*Dalbergia sisoo*	L.S.D.*
Stem height(cm)	8.4 ± 1.321	15.1 ± 2.735	10.7 ± 1.497	at 0.05=0.316 at 0.01=0.422 at 0.001=0.553
Root length (cm)	6.3 ± 1.29	8.6 ± 1.831	7.3 ± 1.582	at 0.05=1.833 at 0.01=2.448 at 0.001=3.203
Shoot dry weight (g)	0.393 ± 0.062	1.553 ± 0.231	0.464 ± 0.069	at 0.05=0.135 at 0.01=0.180 at 0.001=0.236
Root dry weight (g)	0.19 ± 0.025	0.635 ± 0.185	0.26 ± 0.053	at 0.05=0.143 at 0.01=0.191 at 0.001=0.25
Total dry weight (g)	0.583 ± 0.062	2.189 ± 0.395	0.724 ± 0.056	at 0.05=0.445 at 0.01=0.594 at 0.001=0.792
S/R ratio	1.355 ± 0.168	1.793 ± 0.197	1.537 ± 0.282	at 0.05=0.163 at 0.01=0.218 at 0.001=0.285

*L.S.D.: Least Significant Difference

Table 2: Mean values (mean ± SD) of stem height, root length, shoot dry weight, root dry weight , total dry weight and S/R ratio of *Cassia fistula, Albizia lebbeck* and *Dalbergia sissoo* seedlings of SBR.

Character	DS	MDS	UDS	L.S.D.*
Stem height(cm)	14.407 ± 2.396	11.6 ± 2.281	8.233 ± 1.217	0.05=2.594 0.01=3.317 0.001=4.34
Root length(cm)	10.333 ± 2.145	7.0 ± 0.829	4.933 ± 0.533	0.05=1.001 0.01=1.339 0.001=1.753
Shoot dry weight (g)	1.263 ± 0.384	0.737 ± 0.151	0.41 ± 0.031	0.05=0.438 0.01=0.586 0.001=0.766
Root dry weight (g)	0.562 ± 0.065	0.31 ± 0.043	0.213 ± 0.023	0.05=0.166 0.01=0.222 0.001=0.291
Total dry weight (g)	1.825 ± 0.346	1.047 ± 0.471	0.623 ± 0.129	0.05=0.598 0.01=0.8 0.001=1.047
S/R ratio	1.365 ± 0.228	1.643 ± 0.269	1.679 ± 0.253	0.05=0.184 0.01=0.246 0.001=0.322

*L.S.D.: Least Significant Difference

Table 3: Effect of forest stands (DS, MDS and UDS) on seedling growth characteristics (mean ± SD) of three selected tree species of Similipal.

different forest stand conditions. Comparison of growth parameters (stem height, root length, shoot dry weight, root dry weight, total dry weight and S/R ratio) among species and forest stands were in the order *Albizia lebbeck* > *Dalbergia sissoo* > *Cassia fistula* and DS > MDS > UDS, respectively. Low light intensity on the forest floor of undisturbed and moderately disturbed forest stands than disturbed forest stand may be the cause towards reduced growth (Table 1).

Species wise comparative analysis of each growth parameter between forest stands showed gradual decrease in shoot length, root length, shoot dry weight, root dry weight and total dry weight from disturbed forest stand to undisturbed forest stand through moderately disturbed forest stand (Table 4). However, a different trend was observed for S/R ratio i.e. S/R ratio was increased from disturbed forest stand to undisturbed forest stand through moderately disturbed forest stand in case of *Cassia fistula* and *Albizia lebbeck*. But no specific trend was marked in case of *Dalbergia sissoo* (Table 4). The difference in each growth parameter within species between forest stands was statistically significant.

Discussion

Light is one of the most important environmental variables responsible for synchronization of germination with conditions suitable for seedling establishment [8,31]. In the present study significant differences in seed germination and seedling growth of the three indigenous tree species is the outcome of the difference in bio-physical condition of the forest stands of the reserve (Table 1). The work fulfilled its primary aim by getting different results of seed germination and seedling growth at different forest stands. As reported by Hartmann and Kester [32] and Sacande and Some [33] seeds usually germinate at low light intensities. However, the present work supported Bahuguna and Pyare [34], Montgomery and Chazdon [35] and Makana and Thomas [11] indicated the importance of high light intensity for tropical over-story species during early stage of growth, since it enhances germination. High and moderate light intensity on the forest floor of disturbed and moderately disturbed forest stands of the reserve resulted enhanced seedling growth of the three investigated species. This result concur with that of Osunkoya et al. [17] who noticed that twelve forest trees under experiment showed reduced growth with decreasing light intensity. Montgomery and Chazdon [35] also reported the best growth of seedlings of *Dipteryx panamensis* and *Brosimum alicastrum* under high light intensity, compared to low light intensity. Similar results were obtained for different forest species like *Terminalia superba* and *Entandrophragma utile* [36]. The present work

Character	*Cassia fistula*				*Albizia lebbeck*				*Dalbergia sissoo*			
	DS	MDS	UDS	F-value	DS	MDS	UDS	F-value	DS	MDS	UDS	F-value
Height (cm)	9.5 ± 1.471	8.2 ± 1.316	7.5 ± 1.254	46.22***	20.9 ± 1.601	14.7 ± 1.316	9.8 ±0.608	65.89***	12.8 ± 0.671	11.9 ± 1.155	7.4 ± 1.380	46.47***
Root length(cm)	7.8 ± 1.504	6.3±1.255	4.5 ± 0.674	66.95***	12.7 ± 1.632	7.9 ± 1.474	5.3 ± 0.847	89.34***	10.5 ± 1.632	6.8 ± 0.867	4.6 ± 0.619	34.64***
Shoot dry weight (g)	0.63 ± 0.081	0.37 ± 0.051	0.18 ± 0.047	71.40***	2.41 ± 0.086	1.43 ± 0.088	0.82 ± 0.063	45.41***	0.75 ± 0.061	0.41 ± 0.055	0.23 ± 0.061	48.7***
Root dry weight (g)	0.28 ± 0.032	0.17 ± 0.041	0.12 ± 0.025	33.57***	0.96 ± 0.12	0.53 ± 0.061	0.38 ± 0.048	68.21***	0.41 ± 0.061	0.23 ± 0.041	0.14 ± 0.032	61.5***
Total dry weight (g)	0.91 ± 0.084	0.54 ± 0.051	0.30 ± 0.041	52.34***	3.37 ± 0.826	1.96 ± 0.473	1.2 ± 0.41	87.9***	1.16 ± 0.591	0.64 ± 0.066	0.37 ± 0.054	31.7***
S/R ratio	1.218 ± 0.041	1.304 ± 0.089	1.541 ± 0.144	16.25**	1.651 ± 0.096	1.866 ±0.119	1.867 ± 0.271	8.75*	1.226 ± 0.138	1.757 ± 0.099	1.629 ± 0.241	15.88**

***Significant at P ≤ 0.001; **Significant at P ≤ 0.01; *Significant at P ≤ 0.05

Table 4: Comparative mean values of growth characteristics of *Cassia fistula, Albizia lebbeck* and *Dalbergia sissoo* under different forest stands of Similipal.

also supported that of Rincon and Huante [37], concerning the effect on root length and root dry weight, and Ashton [38] concerning the effect on total biomass.

Seedling growth in terms of stem height (cm), root length (cm), shoot dry weight (g), root dry weight (g) and total dry weight (g) of the species studied was greater in disturbed and moderately disturbed forest stands than undisturbed forest stand of the reserve [8,39]. It was evident from the study that significantly declined seedling growth in undisturbed forest stand may be due to direct consequence of low light intensity and high percentage of canopy area. High percentage of canopy area and increased shading might increase seedling mortality indirectly by exacerbating fungal attack through increased humidity or by providing concealment for rodents, mollusks and land crabs that predate on seedlings [40].

Study on S/R ratio of the investigated species with respect to different forest stands of Similipal indicated that allocation of more photosynthate to shoot than root as evidenced S/R ratio was >1 for each species (Table 2) in each forest stand (Table 3). Allocation of more photosynthate to shoot resulting into more biomass to shoot is common in stable forests [41,42]. The high S/R ratio of seedlings (in all the tested species) during the initial stage of establishment in undisturbed forest stand than disturbed and moderately disturbed forest stand of the reserve could be at high priority to capture light rather than minerals. Higher root length and root mass of all the studied species in disturbed and moderately disturbed forest stand than undisturbed forest stand indicating the adaptation of these species to drier or disturbed forest stands of the reserve [42].

Conclusion

This study was focused on the germination and seedling growth of three indigenous species viz. *Albizia lebbeck*, *Cassia fistula* and *Dalbergia sissoo* in different forest stands of Similipal Biosphere Reserve (SBR). The results of the study have an important implication for the restoration of degraded forest stands inside the reserve. Seeds of the studied species germinate well and the growth characteristics of seedlings in terms of shoot length, root length, shoot dry weight, root dry weight and total dry weight were more in disturbed and moderately disturbed forest stands of the reserve than the undisturbed forest stand. Individual species followed their own adaptation in different forest stands depending upon the biophysical conditions. Test for S/R ratio indicated that all the three studied species showed more adaptability in the disturbed and moderately disturbed forest stands than undisturbed forest stand. This suggests that these species could potentially serve as frame work species for restoration of degraded forest stands of Similipal. Out of the three studied species, *Dalbergia sissoo* produces high quality timber having economic value. In addition to maintaining the forest structure, this species also provides economic returns to the Forest Managers, if the forest stands of SBR managed properly.

Acknowledgement

Authors are thankful to the Department of Council and Scientific Industrial Research (CSIR), Govt. of India for providing financial support and Field Director Similipal biosphere reserve for providing permission to carry out the research work inside the biosphere reserve.

References

1. Mishra RK, Upadhyay VP, Nayak PK, Pattanaik S, Mohanty RC (2012) Composition and stand structure of tropical moist deciduous forest of Similipal biosphere reserve, Orissa, India. Forest ecosystems-more than just trees. InTech Publisher, Croatia.
2. Sahoo S, Davidar P (2013) Effect of harvesting pressure on plant diversity and vegetation structure of Sal forests of Similipal tiger reserve, Odisha. Tropical Ecology 54: 97-107.
3. Sharma G, Nautiyal BP, Nautiyal AR (2009) Seedling emergence and survival of Cinnamomum tamala under varying micro-habitat conditions: conservation implications. Tropical Ecology 50: 201-209.
4. Tyagi JV, Rakesh K, Srivastava SL and Singh RD (2011) Effect of micro-environmental factors on natural germination of Sal (Shorea robusta). Journal of Forestry Research 22: 543-550.
5. Behera SK, Mishra AK, Sahu N, Kumar A, Singh N, et al. (2012) The study of microclimate in response to different plant community association in tropical moist deciduous forest from northern India. Biodiversity Conservation 21: 1159-1176.
6. Deb P, Sundriyal RC (2008) Tree regeneration and seedling survival patterns in old-growth lowland tropical rainforest in Nandapha national park, north-east India. Forest Ecology and Managemen 255: 3995-4006.
7. Burslem DFRP, Grubb PJ, Turner IM (1995) Responses to nutrient addition among shade tolerant tree seedlings of lowland tropical rain forest in Singapore. Journal of Ecology 83: 113-122.
8. Aud FF, Ferraz IDK (2012) Seed size influence on germination responses to light and temperature of seven pioneer tree species from the Central Amazon. Anais da Academia Brasileria de Ciencias 84: 759-766.
9. Chazdon RL, Pearcy RW, Lee W, Fetcher N (1996) Photosynthetic responses of tropical forest plants to contrasting light environments. Tropical Forest Plant Ecophysiology 5-55.
10. Dalling JW, Hubbell SP (2002) Seed size, growth rate and gap micro-site conditions as determinants of recruitment success for pioneer species. Journal of Ecology 90: 557-568.
11. Makana JR, Thomas SC (2005) Effects of light gap and litter removal on the seedling performance of six African timber species. Biotropica 37: 227-237.

12. Kitajima K (1996) Ecophysiology tropical forest tree seedlings. Tropical Forest Plant Ecophysiology 559-596.

13. Kennard DK, Gould K, Putz FE, Fredericksen TS, Morales F (2002) Effect of disturbance intensity on regeneration mechanisms in a tropical dry forest. Forest Ecology and Management 162: 197-208.

14. Souza RP, Valio IFM (2001) Seed size, seed germination and seedling survival of Brazilian tropical tree species differing in successional status. Biotropica 33: 447-457.

15. Davis-Colley RJ, Payne GW, Van Elswijk M (2000) Microclimate gradients across a forest edge. New Zealand Journal of Ecology 24: 111-122.

16. Mayer AM, Mayber A (1982) The germination of seeds. The Macmillan Company, USA.

17. Osunkoya O, Ash JE, Hopkins MS, Graham AW (1994) Influence of seed size and seedling ecological attributes on shade tolerance of rain forest tree species in Northern Queensland. Journal of Ecology 82: 149-163.

18. Buylla-Alvarez ER, Martinez-Ramous M (1992) Demography and allometry of a neo tropical pioneer tree: an evaluation of the climax-pioneer paradigm of tropical forests. Journal of Ecology 80: 275-290.

19. Rao PB (2003) Effect of shade on seedling growth and five important tree species in Tarai region of Uttaranchal. Bulletin of the National Institute of Ecology 15: 161-170.

20. Thatoi HN, Panda SK, Rath SK, Dutta SK (2008) Anatomical activity and ethnomedicinal uses of some medicinal plants from Similipal biosphere reserve, Orissa. Asian Journal of Plant Sciences 7: 260-267.

21. Jena M, Sahoo S, Sahu RK (2011) Some ethnomedicinal plants for the treatment of common health problems in Mayurbhanj district, Orissa, India. New York Science Journal 4: 87-92.

22. Mathur A, Joshi H (2013) Ethnobotanical studies of the Tarai region of Kumaun, Uttarakhand, India. Ethnobotany Research and Applications 11: 175-203.

23. Saxena HO, Brahmam M (1996) The flora of Orissa. Vol. I-IV. Regional Research Laboratory (CSIR), Bhubaneswar and Orissa Forest Development Corporation Ltd., Bhubaneswar.

24. Larson BC and Zaman MN (1986) Thinning guidelines for teak (Tectona grandis). Malaysian Forester 48: 288-297.

25. Misra R (1968) Ecology work book. Oxford and IBH Publishing Co., New Delhi.

26. Mohanty RC, Mishra RK, Bal S, Upadhyay VP (2005) Plant diversity assessment of Shorea robusta dominated forest stands of Similipal biosphere reserve. Indian Botanical Society 84: 21-29.

27. Islam KR, Weil RR (1998) A rapid microwave digestion method for colorimetric measurement of soil organic carbon. Soil Science and Plant Analysis 29: 2269-2284.

28. Olsen SR, Cole CV, Watnabe FS, Dean LA (1954) Estimation of available phosphorus in soils by extraction with sodium bicarbonate. USDA Circular 939: 1-19.

29. Kalra YP, Maynard DG (1991) Methods manual for forest soil and plant analysis. Information Report Nor-X-319. North Forestry Center, Canada.

30. Bremner JM, Mulvaney CS (1982) Method of Soil Analysis-Part-2: Chemical and Microbiological Methods. Agronomy Monograph 9.2. American Society for Agronomy and Soil Sciences.

31. Probert RJ (2000) The role of temperature in the regulation of seed dormancy and germination. 261-292 in M Fenner editor. The ecology of regeneration of plant communities. CABH Publishing, Wallingford.

32. Hartmann KT, Kester DE (1975) Plant propagation, principles and practices. Prentice Hall. Inc. Englewood Cliffs, New Jersey.

33. Sacande M, Some LM (1992) Influnce de la lumiere et de la temperature sur la germination des grains de neufespeces forstiere srecoltesakburkin faso. Center rational de Semences Forester 44: 319-345.

34. Bahuguna VK, Pyare LAL (1992) Standardization of nursery techniques of Celtisaustralis Linn. under Dehra Dun climatic conditions. Van Vigyan 30: 5-9.

35. Montgomery RA, Chazdon RL (2002) Light gradient partitioning by tropical tree seedlings in the absence of canopy gaps. Oecologia 131: 165-174.

36. Veenendaal EN, Swaine MD, Agyeman VK, Blay D, Abebress IK, et al. (1996) Differences in plant and soil water relations in and around a forest gap in West Africa during the dry season may influence seedling establishment and survival. Journal of Ecology 84: 83-90.

37. Rincon E, Huante P (1993) Growth responses of tropical deciduous tree seedlings to contrasting light conditions. Trees 7: 202-207.

38. Ashton MS (1995) Seedling growth of Shorea species in the simulated light environment of a rain forest. Forest Ecology and Management 72: 1-12.

39. Castro-Martin G, Tigabu M, Gonzalez-Rivas B, Oden PC (2011) Germination requirements and seedling establishment of four dry species from Nicaragua. Tropical Ecology 52: 1-11.

40. Sherman PM (2002) Effect of land crabs on seedling densities and distribution in a mainland neo tropical rain forest. Journal of Tropical Ecology 18: 67-89.

41. Fenner M (1985) Seed ecology. Chapman and Hall, London.

42. Bhat DM, Murali KS, Shastri CM, Ravindranath NH (2000) Resource allocation to roots and shoots in tree species during seedling stage and their survival strategies. CES Technical Report No. 82: 1-16.

13

The Effects of Flooding on Forest Floristics and Physical Structure in the Amazon: Results from Two Permanent Plots

Randall W Myster*

Biology Department, Oklahoma State University, Oklahoma City, USA

Abstract

I investigated how flooding affects floristics and physical structure of forests in the Amazon at the Area de Conservacion Regional Comunal de Tamshiyacu-Tahuayo, Loreto Provincelquitos, Peru and at the Yasuni Experimental Station, Yasuni National Park, Ecuador. Iset up and sampled 1 ha permanent plots next to a black-water river (igapó forest) in Peru and next to a white-water river (várzea forest) in Ecuador. I found (1)16 families in the 1 haigapó plot with Fabaceae the most common family, (2) várzeastems conformed to a reverse J size pattern for stems less than 40 cm dbh but had more large stems, total stems were within other várzea forest ranges with a slightly larger average dbh, trees were clumped at a low level with 45% canopy closure and while the basal area was also within other várzeaforest ranges, above-ground biomass was lower, (3) igapóstems conformed to the reverse J size pattern, total stems were lower than other igapóforest ranges with a slightly larger average dbh, trees were clumped at a higher degree than the várzea forest with 12% canopy closure while the basal area and above-ground biomass was less than both other igapó samplings and the várzea study plot, and (4) flooding produced reduced basal area in igapó, and smaller stems, stem densities and above-ground biomass for both flooded forests. I conclude that both study plots show a reduction of tree stem density and structure (basal area, above-ground biomass) with flooding, which reduces even more as months under water increase. More sampling in these forests is needed, however, before a conclusion about which aspect of the flooding regime –e.g., water quality, flooding duration or frequency – is most important in determining different aspects of forest structure. Permanent plot studies in the Amazon, like this one, provide much needed data for intelligent management decisions and the development of sustainability techniques.

Keywords: igapó; Tamshiyacu-Tahuayo;T ipitini; várzea; Yasuni

Introduction

The Amazon is the mightiest river in the world, having a discharge of fresh water 4–5 times greater than that of the next mightiest river, the Congo. The Amazon is also the second longest river in the world, originating in the foothills of the Andean Mountains of South America and running east into the Atlantic Ocean. It drains many smaller rivers along the way creating a huge watershed—the Amazon basin—which is generally located below 100 m a.s.l. Associated with this watershed is the largest continuous rainforest in the world, located in the equatorial regions of Brazil, Columbia, Ecuador, Bolivia and Peru. Besides the Andes and the Atlantic Ocean, the watershed is bounded to the north by the Guiana crystalline shield and to the south by the Brazilian crystalline shield [1] marked at their edges by cataracts in the rivers.

The Amazonian rainforest encompasses over 6,000,000 square kilometers [2] and is the most productive [3] and diverse terrestrial ecosystem on earth (containing more than 10% of its species) [1]. Not surprisingly this rainforest influences the entire world's weather patterns and climate [4] and may even control how much rainfall it itself receives [1]. Perhaps most importantly for the human future the Amazonian rainforest interacts intimately with the Earths carbon (C) cycle acting both as a carbon "sink", by taking in large amounts of CO_2 through photosynthesis, but also as a carbon "source" as, for example, when its plants decay or burn. Critically, this rainforest will continue in the future to both contribute to, and suffer from, the effects of global warming.

The majority of the Amazonian rainforest is unflooded (generally referred to as *terra firme*), located in areas lower than 100 m elevation and sharing much structural similarity with unflooded rainforests throughout the rest of the Neotropics [5-7]. Within that broad classification are types of *terra firme* which differ largely in soil characteristics (e.g., *terra firme* proper on clay or loam soils, white sand forests on soils with large amounts of quartz, palm or swamp forests often on standing water) [8].

The low relief of the Amazon basin leads to extensive flooding which varies in frequency, duration, depth and local spatial variation [9], explaining much of the tree distribution, composition, abundance, and association [10]. Most of the flooded water is nutrient rich "white" water from the Andes, which creates forests generally called várzea, and the rest is "black/clear" water which is nutrient poor forest runoff and creates forests generally called igapó [10] or a mixture of the two [9]. The resulting flooded forests cover at least 120,000 square kilometers [11] and have been shown to have a unique biology and ecology [5]. Further, because the Amazon and its tributaries are very dynamic – often changing their routes within a time span of a few decades [1,10] – it very well may be that forests that are unflooded today were flooded in the past and vice versa. This flooding dynamic then, along with differences in, at least, biota and soil characteristics [10,12], creates complex and diverse forest associations throughout the Amazon basin [9].

Therefore in order to better understand how Amazonian rainforests

***Corresponding author:** Randall W Myster, Biology Department, Oklahoma State University, Oklahoma City, USA, E-mail: rwmyster@gmail.com

are structured, and to collect data urgently needed in the Neotropics for sound, sustainable management efforts, Iexpand on past sampling of Amazon flooded forests [13-16] by setting up and sampling a 1 ha plot in an Amazon igapó forest and a1 ha plot in an Amazon várzea forest. I then use that data to compile floristics in the igapó forest, and to compute several forest structure parameters – tree stem size variation, mean, maximum and total, dispersion pattern and degree of clumping, canopy closure, basal area, above-ground biomass – in both forests, in order to address these questions:

(1) What species, genera, and families (floristics) are found in this medium-flooded igapó forest?

(2) What is the physical structure of this igapó forestand the less-flooded várzeaforest?

(3) Which aspects of that structure are similar to other samplings of these types of forests, and which are not?

(4) Putting together this sampling and others can we begin to say how those similarities and differences relate to water quality (white vs. black) and the numbers of months of flooding/ maximum water depth, within each forest type?

(5) Are there patterns in how these parameters vary among permanent plot samplings of other Amazonigapó, várzea and *terra firme* forests, which suggest an effect of flooding on forest structure?

Methods

Study sites

The first study site is the Area de Conservacion Regional Comunal de Tamshiyacu-Tahuayo (ACRCTT: www.perujungle.com) [9,15,16] located in Loreto Province, 80 miles southeast of Iquitos, Peru (~2° S, 75° W) with an elevation of ~100 m a.s.l. The reserve is part of one of the largest protected areas in the Amazon, containing wet lowland tropical rainforest [17] of high diversity [3,18]. It is comprised of low, seasonally inundated river basins of the upper Amazon and named for two of the major white-water rivers (the Tahuayo and the Tamshiyacu) which form boundaries to the north and west, creating large fringing floodplains [10]. The substrate of these forests is composed of alluvial and fluvial Holocene sediments from the eastern slopes of the Andes. Annual precipitation ranges from 2.4 – 3.0 m per year, and the average temperature is relatively steady at 26°C. Within the ACRCTT are areas of black/clear water runoff which createigapó forests of differing frequency, duration, and maximum water column height, where the rainy season is between November and April [5]. Common tree species include *Calycophyllum spruceanum*, *Ceiba samauma*, *Inga* spp., *Cedrela odorata*, *Copaifera reticulata*, *Phytelephas macrocarpa* with under-story palms such as *Guazuma rosea*, and *Piptadenia pteroclada* [3,15,19,20].

The second study site is the Yasuni Research Station (YRS: 0°41' S, 76°24' W), operated by the Pontificia Universidad Catolica of Ecuador and located within the Yasuni National park of eastern Ecuador [21-25]. Most of the YRS is *terra firme* forest which has been classified as lowland tropical rainforest [17]. The mean annual rainfall is 3081 mm with the wettest months April to May and October to November. August is the driest monthand the mean monthly temperature varies between 22°C and 35°C. Soils in the National park have been described as clayey, low in most cations but rich in aluminium and iron, whereas soils at the station in *terra firme* forest are acidic and rich in exchangeable bases with a texture dominated by silt [8]. The park has low topographic variation with a mean elevation of approximately 200 m above sea level. The station is the site of a long-term 50 ha vegetation plot in *terra firme* forest, maintained by the Smithsonian Tropical Research Institute [26], parts of which have been sampled [27-30]. Also found at YRS is várzea floodplain forest – located next to the nutrient rich whitewater Tiputini River - which is inundated a few weeks between the months of October and April to a maximum depth of 3 m.

Plot set-up and sampling

In May 2011, my field assistant and I set up a 1 ha plot in an igapó forest at the ACRCTT, which is underwater 3-4 months every year, and tagged and measured the diameter at breast height (dbh) of all trees at least 10 cm dbh in 10 m × 10 m continuous subplots. The dbh measurement was taken at the nearest lower point where the stem was cylindrical and for buttressed trees it was taken above the buttresses. Plots of this size have been used to study flooded forests in the Amazon for decades [31]. In June 2013, the tagged trees were identified to species, or to genus in a few cases, using Romoleroux et al. [32] and Gentry [33] as taxonomic sources. We also consulted the Universidad Nacional de la Amazonia Peruana herbarium and the web site of the Missouri Botanical Garden <www.mobot.org>.

In May 2010, my field assistants and I set up a 1 ha plot next to the Tipitini River and a few hundred meters from the 50 ha *terra firme* plot [26]. We tagged, identified, and measured the diameter at breast height (dbh) of all trees at least 10 cm dbh in 10 m×10 m continuous subplots. The trees were identified using the same protocol as the plot at ACRCTT. This data is archived at the Luquillo Experimental Forest as LTERDBAS#172 as part of the LTER program funded by the US National Science Foundation. One may visit their website (http://luq.lternet.edu) for further details.

Data analysis

From the igapó data set I first compiled Floristic tables of family, genus and species. Then for both the igapó data set and the várzea data set I generated (1) the total number of stems in the 1 ha plot, the mean and maximum among those stems, and the total number of stems divided into four size classes: 10<20 dbh, 20<30 dbh, 30<40 dbh and ≥ 40 dbh, (2) the stem dispersion pattern (random, uniform, clumped) computed by comparing plot data to Poisson and negative binomial distributions using Chi-square analysis and, if clumped, greens index was also computed to access degree of clumping [34,35], (3) canopy closure using the formula in Buchholz et al. [36] for tropical treeswith the resulting percentage of the 1 ha plot area closed,(4) total basal area as the sum of the basal areas of all individual stems (Πr^2; where r=the dbh of the individual stem/2) and (5) above-ground biomass(AGB) using the formula in Nascimento and Laurance [37] suggested for tropical trees of these stem sizes. A few years ago, other plots were also set up at ACRCTT in the same black-water flooded forest as the 1 ha plot and those results [9,15,16] create replication.

Results

There were a total of 16 families found in the 1 ha igapóplot (Table 1). Fabaceae was by far the most common family which also had the most genera and the most species. The families Moraceae, Annonaceae and Sapotaceae were also common, but there were 4 families with only one stem. The number of species was greater than or equal to the number of genera for every family. Dividing the stems by size class showed that most families have a monotonic decline in stem number as stems get thicker (Table 2). This was not true, however, of the families Elaeocarpaceae, Boraginaceae and Sapotaceae. The most

Family	total number of stems	No.of Genera	No., Of Species
Fabaceae	(62)	55	
Moraceae	(22)	44	
Annonaceae	(19)	23	
Sapotaceae	(19)	2	2
Chrisobalanaceae	(15)	33	
Rubiaceae	(9)	11	
Elaeocarpaceae	(5)	1	1
Malvaceae	(4)	2	2
Aquifoliaceae	(2)	1	1
Burseraceae	(2)	1	1
Combretaceae	(2)	1	1
Lamiaceae	(2)	1	1
Anacardiaceae	(1)	1	1
Boraginaceae	(1)	1	1
Rhizophoraceae	(1)	1	1
Salicaeae	(1)	1	1

Table 1: Each family sampled in the 1 ha igapóplot sorted in decreasing order by total number of stems (in parenthesis) followed by the total number of genera and the total number of species.

Family	10 < 19 cm	20 < 29 cm	30 < 39 cm	40 cm or greater
Fabaceae	62	28	2383	
Moraceae	22	11	722	
Annonaceae	19	10	72	0
Sapotaceae	19	7	453	
Chrisobalanaceae	15	10	50	0
Rubiaceae	9	5	40	0
Elaeocarpaceae	5	2	30	0
Malvaceae	4	22	0	0
Aquifoliaceae	2	2	0	0
Burseraceae	2	11	0	0
Combretaceae	2	11	0	0
Lamiaceae	2	2	0	0
Anacardiaceae	1	1	0	0
Boraginaceae	1	1	0	0
Rhizophoraceae	1	100	0	0
Salicaeae	1	100	0	0

Table 2: The family and total stems for each family from table 1 divided into size classes based on diameter at breast height (dbh) measured in whole cm.

common species were *Campsiandraangustifolia*, *Crudiaglaberrima*, and *Pseudolmedialaevigata* (Table 3).

The igapóstudy plot conformed to the reverse J stem size distribution pattern for all stems with a smaller proportion in the largest stem size class than the várzea study plot (Table 4). The igapó forest did, however, have the largest individual tree of 91 cm dbh, as well as a smaller average tree stem diameter. The total number of tree stems in the 1 ha plot – the stem density – was lower than the ranges sampled in other igapóforests which may have flooded less. The average dbh was slightly greater than the only other sample which reported it (in Brazil). It should be noted, however, that there were only half as many other samples for this forest compared to várzea. The trees were again clumped, but at a higher degree than the várzea forest. Canopy closure was only approx 12%, about half of the várzea forest. Basal area was also much less than both the other igapó sampling (in Brazil) and the várzea study plot. Above-ground biomass followed the same trend as basal area (Table 4).

The várzea study plot conformed to the reverse J stem size distribution pattern for stems less than 40 cm dbh, but had many more stems in the largest size class than expected included one of 80 cm dbh (Table 5). The total number of tree stems in the 1 ha plot – the stem density – was within the ranges sampled in other várzea forests, and close to the middle of the ranges. The average dbh was slightly greater than the only other sample which reported it (in Brazil). The trees were clumped, but at a low level (Table 5). Canopy closer was close to 45% and the basal area was again within the ranges of other várzea samplings, but at the lower end of that variation. Above-ground biomass was also lower than another várzea forest, sampled in Brazil to the east.

When comparing the two study plots which both received flooding, to unflooded samplings, we see a small reduction in average stem diameter (Table 6). In addition, there were more stems in the unflooded plots. Whereas igapó basal area was low, várzea basal area was in the middle of the range sampled in the three *terra firme* plots, one of which was located close to the várzea study plot (see papers by Valencia et. al.). Finally above-ground biomass was larger in *terra firme* compared to either flooded forest. The pie graphs reveal a "saddle" stem size distribution pattern for várzea (Figure 1a) and a more reverse J distribution pattern for igapó (Figure 1b).

Discussion

There was less than half the number of families in the 1 ha igapóplot compared to the 1 havárzeaplot (16 vs. 40: author, unpub. data). This could have been the result of poorer water quality or more flooding duration and maximum depth in the 1 ha igapóplot, or some other factor.There were nine familiesin common (out of 16), however. But no common genus or species were found between the two plots.

Family	Genus	species	number of stems
Annonaceae	Annona	montana Mac	7
Chrysobalanaceae	Couepia	subcordata Benth	5
Fabaceae	Crudia	glaberrima Mac	11
Fabaceae	Campsiandra	angustifolia Spruce	13
Moraceae	Pseudolmedia	laevigata Trecul	9
Rubiaceae	Genipa	spruceana Steyeermark	4
Sapotaceae	Sarcaulus	brasiliensis Eymaoppii	5

Table 3: All species in the 1 ha igapó plot with at least 4 stems, sorted by family, genus and species.

Parameter	Ecuador (at least 6)	Brazil1	Brazil2	Brazil3
Stems (dbh):				
10 < 19	84	--	--	--
20 < 29	58	--	--	--
30 < 39	17	--	--	--
40 or greater	8	--	--	--
mean	22.3	20.7	--	--
maximum	91	--	--	--
total	167	683	220-546	222
Dispersion:				
spatial pattern	clumped	-	-	-
green's index	0.17	-	-	-
Canopy:		-	-	-
Closure (m^2)	1231.22	-	-	-
% ha closed	12.3122	-	-	-
Basal area (m^2)	6.52	31.4	-	-
AG biomass (Mg)	202	387.8	-	-

[1]Haugaasen and Peres 2006, [2]Ferreira 1997, [3]Campbell et. al. 1986

Table 4: Structural parameters for all trees at least 10 cm dbh sampled in Amazon igapóflooded forests, normalized per ha. The number of months underwater (when known) is indicated in parentheses for each plot.

Parameter	Ecuador (1-2)	Brazil1	Brazil(6)2	Brazil3	Ecuador4	Brazil5	Brazil(1-5)6
Stems (dbh):							
10<19	366	--	--	--	--	--	--
20<29	87	--	--	--	--	--	--
30<39	39	--	--	--	--	--	--
40 or greater	81	--	--	--	--	--	--
mean	24.2	22.1	--	--	--	--	--
maximum	80	--	--	--	--	--	--
total	573	515.3	560-745	423-612	417	420-777	370-466
Dispersion:							
spatial pattern	clumped	--	--	--	--	--	--
green's index	0.02	--	--	--	--	--	--
Canopy:		--	--	--	--	--	--
Closure (m2)	4478.04	--	--	--	--	--	--
% ha closed	44.7804	--	--	--	--	--	--
Basal area (m2)	26.356	29.6	17-45	-	35.5	25-27	31-48
AG biomass (Mg)	292	417.1	--	--	--	--	--

[1]Haugaasen and Peres 2006, [2]Worbes et. al. 1992, [3]Ferreira 1997, [4]Balslev et. al. 1987, [5]Campbell et. al. 1992, [6]Wittmann et. al. 2004b

Table 5: Structural parameters for all trees at least 10 cm dbh,sampled in Amazon várzeaflooded forests and normalized per ha. The number of months underwater, when known, is indicated in parentheses for each plot. Ranges and means are given when there are multiple samplings. The first listed plot is the study plot.

Parameter	igapóplotvárzea	plotTerra firme	plot1Terra firme	plot2Terra firme	plot3
Stems (dbh):					
10<19		84	366	--	--
20<29	58	87	--	--	--
30<39	17	39	--	--	--
40 or greater	8	81	--	--	--
mean	22.3	24.2	--	21.7	--
maximum	91	80	--	--	--
total		167	573	604 – 725	605.3734
Dispersion:					
spatial pattern	clumped	clumped	--	--	--
green's index	0.17	0.02	--	--	--
Canopy:					
Closure (m^2)	1231.22	4478.04	--	--	--
per ha (%)	12.3122	44.7804	--	--	--
Basal area (m^2)	6.52	26.356	22.2 – 31.2	32.6	22.2
AG biomass (Mg)		202292	--	457.8	--

[1]Valencia et. al. 2004abc, [2]Haugaasen and Peres 2006, [3] Korning et. al. 1990

Table 6: Structural parameters for all trees at least 10 cm dbh sampled in the study igapóflooded forest, in the study várzea flooded forest, and in three *terra firme* forests. All data are expressed as per hectare.

The study plot in igapó forest was lower in stem density, basal area and AGB compared to other igapó samplings which may have been flooded for less time. For this study plot the flooding may have been so serve that there was few "threshold" sizes that stems could obtain to escape the flooding, and so the reduction in stems number with increasing stem size (reverse J) was maintained. While the average stem size was similar between the two study flooded forests and a *terrafirme* sampling, more stems were lost as flooding duration increased with proportionally more medium-sized stems gained. The amount of flooding in várzea was not enough to reduce basal area, but it was enough in igapó. Basal area shows the influence of the large individuals because a decrease in stem numbers is offset with larger stems.

The study plot in várzeaforest compared well to other várzea samplings in all regards except for low ABG. This study plot had more large stems than the reverse J distribution found in the 50 ha plot [27] and in *terra firme* forests within the Yasuni National Park [38] and at other Amazon sites [39]. This suggests that while smaller stems die from flooding, a stem may survive to a large size if it can reach a certain "threshold" size and take advantage of the resources that the dead stems are not now using. Consequently, the study plot in várzea forest had a large basal area for stems at least 40 cm in diameter in base diameter. The canopy opened up with the stem loss but the low amount of clumping suggests the dominance of flooding, over other factors such as biological interactions, in determining stem survival and growth. I also found in this plot (author unpub.data) that the seven most common families were also among the top ten families found in multiple samplings in the nearby 50 ha plot [27-29] but at the genus and species taxonomic level, similarities with the 50 ha plot samplings disappear except for the genera Cecropia, Lachornea, Inga, Zygia, Eschweilera and Virola and the species *Iriartea deltoidea* and *Coccoloba densifrons*. Also because the várzea forest loses families, genera and species proportionally more than it loses stems compared to *terra firme* forest, fisher's α was lower (author unpub. data).

I also found in ACRCTT forests that differed in flooding duration (unflooded [dry], flooded by black water 1-5 months per year [wet], flooded by black water 6 months per year [very wet]: [author unpub.

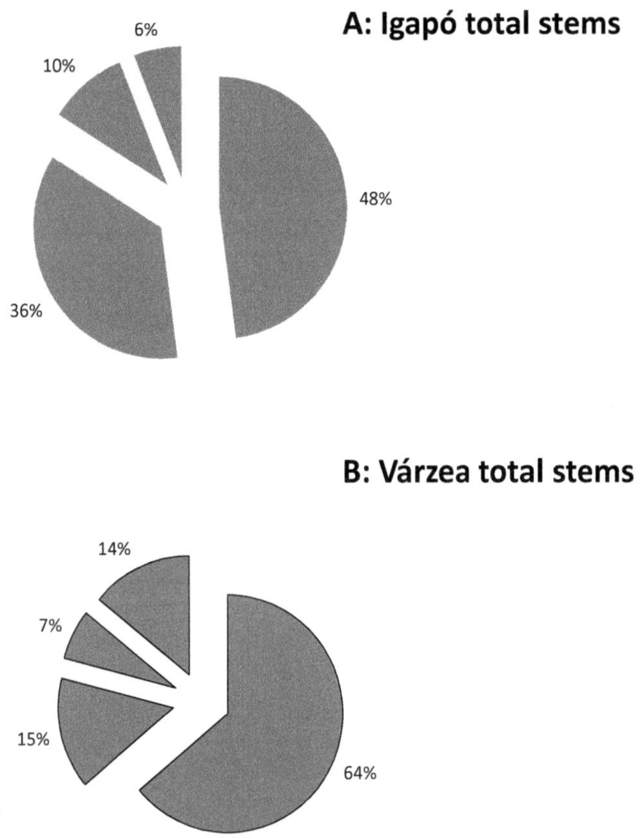

Figure 1: Pie graph of total stems **(A)** in the 1 ha igapóflooded forest divided by stem size class: 10<19 cm (48%), 20<29cm (36%), 30<39cm (10%) and 40 cm or greater (6%), **(B)** in the 1 ha várzeaflooded forest divided by stem size class: 10 < 19 cm (64%), 20 < 29 cm (15%), 30 < 39 cm (7%) and 40 cm or greater (14%).

data] [15,16] tree stems and canopy coverage declined as flooding increased, more so than reductions due to tree fall, trees were clumped only in the gaps for wet forest, and there were smaller stems in gaps compared to all adjacent forests. Consequently flooding was a greater stressor on these forests than tree fall where Amazonian forests may have gradients (flooding) and disturbances (tree fall) which overlap in their traditional roles, presenting plants with similar cues. Common species existed between wet forests and their gaps and between wet and very wet gaps, and tree richness was maximum in dry forest and minimum in very wet gaps. Finally there was a significant effect of degree of tree fall gap formation on canopy average height, canopy maximum height, basal area, density, above-ground biomass, turnover, and alpha diversity, and a significant effect of flooding on species richness, genera richness, density, turnover and alpha diversity. Moreover there were fewer but larger trees, and more production in the forest plots compared with the gap plots; and the very wet plots had fewer trees, species, and genera compared with the other forests. The greatest amount of turnover was also found in the very wet forest with the wet forest had the greatest richness and alpha diversity. Results supported a "mass effects" hypothesis where species from both the unflooded and most flooded forests and their gaps have overlapping ranges in the less flooded forest and gaps, causing continuous immigration which boosts diversity[15,16] [author unpub. data].

All four samplings of the 50 ha plot at YRS had more stems than either of the study plots, also true in the 15 other unflooded forests (all stems at least 10 cm dbh) sampled in Yasuni National Park [38]. Total basal area, however, is comparable among the flooded forests and the 50 ha plot, because the flooded forests make up in size what it losses in stem numbers. The density of tree stems and their size distributions in the study plots compared well with other Western Amazon flooded forests as well [9,40,41]. Several structural parameters conformed to the flooding gradient, decreasing in complexity as flooding increased, which may be due in part to root burial by sedimentation and oxygen deficiency in flooded forests [10]. The loss of tree stem density with flooding [13] may be explained by the loss of tree stems due to the action of moving water or the physical damage due to the weight of debris (falling branches as reviewed in Myster [42]. Clumping was less than that found in larger forest openings recovering from agriculture [35] which may have contained more perching opportunities or bird to land and disperse seeds.

In terms of forest structure, these forests do lose stems from flooding but that loss is not proportionally similar across all size classes. Flooded forests maintain a greater number of larger trees than unflooded forests and so their stem distribution is more of a "saddle" than a monotonic decline in numbers with increasing size, as seen in the basal area, leading to fewer trees, genera, and species as flooding increased [16]. Indeed flooding tends to eliminate both vertical and horizontal heterogeneity affecting, for example, the availability of commonly logged tree species and animal populations. It must be remembered, however, that those studies and this one were only "snapshots" of forest structure and that in order to completely understand forest structure, longer term sampling with larger plots is needed. Such studies will show that the underlying process of these forests, as for all plant communities, is plant-plant replacement [43]. These permanent plot studies, and others like them in the Amazon, provide baseline data on forest dynamics and fluctuations of forest structure. This knowledge will enable conservationists to develop sound management techniques for these forests in order to better utilize them as societal and human needs arise in the future. Sustainability of these flooded systems in the Amazon is critical for the lives of the local peoples that live there but also for the rest of us.

Conclusion

The most obvious conclusion of these two samplings, and others done in the same forest types, is that flooding reduces forest structure. This can be seen in total stem density, basal area and ABG. Flooding can, however, merely change structure, as seen in stem size distribution pattern. Such results from the study plots beg the question: What aspects of the flooding regime [9] are most important and for what aspects of structure? To date researchers have pointed to differences in water nutritional quality vs. flooding duration with its correlated maximum water height. With the only two samplings done here, it is difficult to tease apart these differences. The comparison with other forests of the same water type helps to a degree but much more sampling and computation in the Amazon is needed. There may be other ways to look at flooding which warrant further study, such as the effects of flooding frequency and sedimentation. Only then may we be able to discover the causes of Amazon flooded forest structure.

Acknowledgements

I thank Alfredo Reyes, Dr. Paul Beaver and the staff of Amazonian expeditions for their help in executing the research in Peru, and Milton Zambrano Reyes, AnellioLoor and the staff at the Yasuni Research station for their help in executing the research in Ecuador. I also thank Dr. Juan Celedonio Ruiz Macedo of the Universidad Nacional de la Amazonia Peruana (UNAP) herbarium, and UNAP students Christian Renan Silva Arirama and Josias Tello Huanaquiri, for taxonomic assistance.

References

1. Pires JM, Prance GT, TE Lovejoy (1985) The vegetation types of the Brazilian Amazon. Amazonia Pergamon press, Oxford, Pp: 109-145.

2. Walter H (1973) Vegetation of the earth and the ecological systems of the geo-biosphere. springer-Verlag, Berlin.

3. Daly DG, Prance GT, Hammond, HD Campbell (1989) Brazillian Amazon. Floristic inventory of tropical countries New York Botanical Garden, Bronx, NY, Pp: 401-426.

4. Keller M, Alencar A, Asner GP, Braswell B, Bustamante M, et al. (2004) Ecological research in the large-scale biosphere atmosphere experiment in Amazonia: early results. Ecological Applications, 14: S3-S16.

5. Kalliola R, Jukka S, Puhakka M, Rajasilta M (1991) New site formation and colonizing vegetation in primary succession on the western Amazon floodplains. Journal of Ecology 79: 877-901.

6. Everham EM III, MysterRW, Vandergenachte E (1996) Effects of Light, Moisture, Temperature, and Litter on the Regeneration of Five Tree Species in the Tropical Montane Wet Forest of Puerto Rico. American Journal of Botany, 83:1063-1068.

7. Pitman NC, Terborgh JW, Silman MR, Nunez PV, Neill DA, et al. (2001) Dominance and distribution of tree species in upper Amazonian *terra firme* forests. Ecology, 82:2101-2117.

8. Tuomisto H, Poulson AD, Ruokolainen K, Moran RC, Quintata C, et al. (2003) Linking floristic patterns with soil heterogeneity and satellite imagery in Ecuadorian Amazonia. Ecological Applications, 13:352-371.

9. Myster RW (2009) Plant communities of Western Amazonia. The Botanical Review, 75:271-291.

10. Junk WJ, Holm-NielsenL B, Nielsenl C, Balslev H (1989) Flood tolerance and tree distribution in central Amazonian flood plains. Tropical forests: Botanical dynamics speciation and diversity, Academic press, NY, Pp: 47-64

11. Parolin P, De Simone O, Hasse K, Waldhoff D, Rottenberger S, et al. (2004) Central Amazonian Floodplain Forests: Tree adaptations in a pulsing system. The Botanical Review, 70:357-380.

12. Honorio EN (2006) Floristic relationships of the tree flora of Jenaro Herrera an unusual area of the Peruvian Amazon. M.S.thesis University of Edinburgh, Edinburgh, UK.

13. Balslev H, Luuteyn J, Ollgaard B, Holm-Nielsen LB (1987) Composition and structure of adjacent unflooded and floodplain forest in Amazonian Ecuador. Opera Bot. 92:37-57.

14. Worbes MH, Klinge JD, Martinez C (1992) On the dynamics floristic subdivision and geographical distribution of varzea forests in Central Amazonia. Journal of Vegetation Science, 3:553-564.

15. Myster RW (2007) Interactive effects of flooding and forest gap formation on tree composition and abundance in the Peruvian Amazon. Folia Geobotanical, 42:1-9.

16. Myster RW (2010) Flooding duration and tree fall gap interactive effects on plant community structure richness and alpha diversity in the Peruvian Amazon. Ecotropic, 16:43-49.

17. Holdridge LR (1967) Life zone Ecology. Tropical Science Center, San Jose, CR.

18. Daly DG, Mitchell JD, Lentz DL (2000) Lowland vegetation of tropical South America, Imperfect balance: landscape transformation in the Pre-Columbia Americas, Columbia University Press, NY, pp: 391-453

19. Prance GT (1979) Notes on the vegetation of Amazonia III The terminology of Amazonian forest types subject to inundation. Brittonia, 31:26-38.

20. Puhakka ME, Kalliola R, Rajasita M, Salo J (1992) River types site evolution and successional vegetation patterns in Peruvian Amazonia. Journal of Biogeography,19:651 -665.

21. Svenning JC (1999) Microhabitat specialization in a species-rich palm community in Amazonian Ecuador. Journal of Ecology, 87:55-65.

22. Duivenvoorder JF, Balslev H, Cavelier J, Grandez C, TuomistoH, et al. (2001) Evaluaciondo recursos vegetales no maderables en la Amazonia noroccidental. The Netherlands Centre for Geo-ecological Research (ICG), The Netherlands.

23. Myster RW, Santacruz PG (2005) Una comparación de campo de insectos de suelomorar de Amazonas: Tierra firme y bosques de tierras inundadas vs. espacios abiertos en el Parque Nacional Yasuní Ecuador.Ponti. Revista Universidad Catolia, Ecuador, 76:111-124.

24. Metz MR, Comita LS, Chen YY, Norden N, Condit R, et al. (2008) Temporal and spatial variability in seedling dynamics: a cross-site comparison in four low land tropical forests. Journal of Tropical Ecology, 24: 9–18.

25. Myster RW (2012a) Spatial heterogeneity of light and soil water along a *terra firme* transect in Amazonian Ecuador: effects on tree seedling survivorship growth and allocation. Canadian Journal of Forest Research, 42:1-4.

26. Losos EC, Leigh EC (2004) Forest diversity and dynamism: findings from a network of large-scale tropical forests plots. University of Chicago, Press Chicago Il.

27. Valencia R, Condit R, Romoleroux K, Foster RB, Munoz GV, et al. (2004a) Tree species diversity and distribution in a forest plot at Yasuni National park Amazonian Ecuador. Forest diversity and dynamism: findings from a network of large-scale tropical forests plots. University of Chicago Press, Chicago Il. Pp. 107-118.

28. Valencia R, Condit R, Foster RB, Romoleroux K, Munoz GV, et al. (2004b) Yasuni forest dynamics plot Ecuador. Forest diversity and dynamism: findings from a network of large-scale tropical forests plots. University of Chicago Press, Chicago Il, Pp. 609-620.

29. Valencia R, Foster RB, Villa G, Condit R, Svenning J, et al. (2004c) Tree species distributions and local habitat variation in the Amazon: large forest plot in eastern Ecuador. Journal of Ecology, 92:214-229.

30. Valencia R, Condit R, Muller-Lamdau HC, Hernandez C, Navarrete H (2009) Dissecting biomass dynamics in a large Amazonian forest plot. Journal of Tropical Ecology, 25:473-482.

31. Junk WJ, Piedade MTF, Parolin P, Wittman F , Schongart J (2010). Amazonian floodplain forests: ecophysiology biodiversity and sustainable management. Ecological Studies, Springer-Verlag, Berlin.

32. Romoleroux K, Foster R, Valencia R, Condit R, Balslev H, et al. (1997) Especies lenosas (dap => 1 cm) encontradas en dos hectareas de un bosque de laAmazonia ecuatoriana. Estudios SobreDiversidad y Ecologia dePlantas. Pontificia Universidad Catolica del Ecuador Quito Ecuador, Pp: 189-215.

33. Gentry A (1993) A field guide to woody plants of northwest South America (Colombia Ecuador Peru).Conservation International, Washington, DC.

34. Ludwig JA, Reynolds JF (1988) Statistical ecology. John Wiley & sons, NY.

35. Myster RW, Pickett STA (1992) Effects of palatability and dispersal mode on spatial patterns of tree seedlings in old fields. Bulletin of the Torrey Botanical Club, 119:145-151.

36. Buchholz T, Tennigkeit T, Weinreich A (2004). *Maesopsis eminii* – a challenging timber tree species in Uganda – a production model for commercial forestry and smallholders. Proceedings of the international union of forestry research organizations (IUFRO) conference on the economics and management of high productivity plantations Lugo, Spain, Pp:15.

37. Nascimento HEM, Laurance WF (2001) Total above ground biomass in central mAmazonian rainforests: a landscape-scale study. Forest Ecology and Management,157:1-11.

38. Pitman NCA, Terborgh JW, Silman MR, Nunez P, Neill DA, et al. (2002) A comparison of tree species diversity in two upper Amazonian forests. Ecology 83:3210–3224.

39. Korning J, Thomson K, Ollgaard B (1990) Composition and structure of a species rich mAmazonian rain forest obtained by two different sample methods. Nordic Journal of Botany 11:103-110.

40. Ferreria LV (1997) Is there a difference between the white water floodplain forests (várzea) and black water floodplain forests (igapó) in relation to number of species and density? Revista de Brazilian Ecology 1:60-62.

41. Pitman NCA, Terborgh J, Silman MR, Percy NV (1999). Tree species distributions in an upper Amazonian forest. Ecology 80:2651-2661.

42. Myster RW (1993) Tree invasion and establishment in old fields at Hutcheson Memorial Forest. The Botanical Review, 59:251-272.

43. Myster RW (2012b) Plants replacing plants: the future of community modeling and research. The Botanical Review 78:2-9.

14

Aboveground Biomass Dynamics in the Amazonian Rainforest under Influence of Reduction in Rainfall

Vicente de P. R. da Silva[1]*, Glayson F. B. das Chagas[1], Rafaela S. R. Almeida[2], Vijay P. Singh[3] and Vanessa de A. Dantas[1]

[1]*Federal University of Campina Grande/Center of Technology and Natural Resources/Academic Unity of Atmospheric Sciences, Av. Aprígio Veloso, 882, Bodocongó, 58109 970, Campina Grande, PB, Brazil*

[2]*State University of Paraíba, R. Domitila Cabral de Castro, 148-572, Bodocongó, 58429-570 Campina Grande, PB, Brazil*

[3]*Department of Biological and Agricultural Engineering & Department of Civil & Environmental Engineering, Texas A&M University, College Station, TX 77843-2117, USA*

Abstract

Exchange of biomass between the ecosystem and the atmosphere plays an important role in regional and global carbon cycles that have a major impact on biodiversity. This study evaluated the effect of reduction in rainfall on the aboveground biomass in an Amazonian rainforest. Data for this study were obtained from the "Long-term drought impact on water and carbon dioxide fluxes in Amazonian Tropical Rainforest Experiment (ESECAFLOR)" which was a subproject of Large Scale Biosphere Atmosphere Experiment in Amazon forest (LBA), carried out in *terra firma* rainforest in Caxiuanã National Forest, Pará, Brazil. The experimental design entailed two experimental sites each with one hectare of natural forest: control TFE (a simulated soil drought or 'throughfall exclusion' experiment) under normal conditions of climate and treatment TFE with rainfall exclusion of about 50%. The tree growth parameters employed in the study were based on monthly data from the experimental period from January 2005 to May 2009. Results indicated that a decrease in rainfall significantly affected the tree growth parameters, resulting in a decrease of biomass (21.1 t ha^{-1} $year^{-1}$) and basal area (1.04 m^2 ha^{-1} $year^{-1}$). The Amazonian rainforest may become increasingly vulnerable to higher background tree mortality rates in response to drought events, such as El Niño.

Keyword: Basal area, Biomass increment, Caxiuanã reserve, Tree mortality

Introduction

Recent analyses suggest that carbon dynamics of the Amazonian tropical forests vary interannually in response to rainfall and temperature anomalies [1]. Using simulated soil-moisture droughts or TFE experiments (throughfall exclusion experiment), effects of severe droughts as a consequence of climate change, have been investigated in the eastern Amazonian rainforest. These investigations have analyzed tropical forests which have been subjected to increasingly severe drought episodes through the El-Niño/Southern Oscillation (ENSO). Davidson et al [2] studied the effect of an experimental drought on soil emissions of carbon dioxide, methane, nitrous oxide, and nitric oxide in Tapajós forest, Brazil. Using the same design as for the Caxiuanã forest, Fisher et al [1] analyzed the response of an eastern Amazonian rainforest to drought stress using data from the TFE experiments. Both investigations were carried out as part of LBA, using two 100 m x100 m plots, one control plot and one treatment 'TFE' plot.

The aboveground biomass in tropical forests is mainly contained in trees. The tree biomass is a function of wood volume, obtained from the tree diameter and height, architecture and wood density. It can be quantified by a direct or an indirect method, where the biomass is quantified using a mathematical model. An allometric model can be site specific when elaborated to a particular ecosystem or can be general when used at different sites. Nogueira et al [3] presented biomass equations developed from trees directly weighed in open forest on fertile soils in the southern Amazon and allometric equations for bole-volume estimates of trees in both dense and open forests. These equations were used to improve the commonly used biomass models based on large-scale wood-volume inventories carried out in the Amazonian forest. However, estimates of biomass storage are discordant when the same method is applied or when estimates from allometric equations are compared with the biomass obtained from large-scale wood-volume inventories [4]. Also, there are difficulties in accurately measuring the total or even merchantable tree height of tropical trees. Therefore, the diameter at breast height (DBH) has become the most important variable for allometric equations [5]. Also, the diameter increment measurements have been used to examine the dynamics of natural forests [6].

The tropical forests in Amazon and elsewhere are subjected to increasingly severe droughts as a consequence of climate change which seems to cause severe reductions in rainfall and other ecological functions. However, reduced rainfall on forest gives rise to major uncertainties as regards the future climate and their consequences are still not well understood. On the other hand, the fundamental mechanisms underlying tree survival and mortality during droughts remain less than well understood. Mortality of canopy trees is an important process in forest dynamics, and can be sudden, with no relationship to past events or the culmination of a long decline [7]. Some studies have included an extensive tree mortality of Austrocedrus chilensis during the El-Niño droughts in Argentina [8], and die-off of multiple pine species during the 1950s drought in the southwestern USA [9].

Using throughfall exclusion experiments in the Amazonian

***Corresponding author:** Vicente de P. R. da Silva, Federal University of Campina Grande/Center of Technology and Natural Resources/Academic Unity of Atmospheric Sciences, Av. Aprígio Veloso, 882, Bodocongó, 58109 970, Campina Grande, PB, Brazil, E-mail: vicente@dca.ufcg.edu.br

rainforest, Fisher at al [1] and Costa et al [10] investigated forest transpiration by the sap flow method in the Caxiuanã National Forest. Fisher et al [1] concluded that the forest was not able to withstand a 50% reduction in rainfall over 1-2 years without impacting the canopy gas exchange, while Costa et al [10] found a decrease of 68% in the mean transpiration of *E. Coriacea* from the control plot to the rainfall exclusion plot. Rainforests are forests characterized by high rainfall. However, the link between reduced rainfall and the aboveground biomass dynamics is poorly understood. Thus, it is important to know how the forest responds to drier soil conditions than those concurrently experienced in the control plot. The objective of this study, therefore, was to establish a throughfall exclusion experiment and analyze the biomass dynamics in the Amazonian rainforest under the influence of large reduction in rainfall, by excluding about 50% of rainwater from the soil for a long period (mimicking an extreme El-Niño).

Materials and Methods

Experimental site

This study was carried out at Ferreira Penna Scientific Station (FPSS) in the Caxiuanã National Forest (CNF) in Pará State (Latitude: 1°42'30"S, Longitude: 51°31'45"W and Altitude: 62 m above sea level), Brazil. The FPSS is located in the 300 000 hectare Caxiuanã National Forest, about 400 km to the west of Pará capital city of Belém, Brazil. The region has a well-preserved forest with a canopy of 35 m high. The Amazon is covered predominantly by moist dense tropical forest, but with several other vegetation types, including savannas, montane forests, open forests, floodplain forests, grasslands, swamps, bamboos, and palm forests [10]. Figure 1 shows the localization of FPSS in Pará State.

The predominant trees species in the landscape are *Eschweilera carinata* (Matamata), *Voucapoua americana* (Acapu), *Protium paraense* (Latilla), *Dinizia excelsa* (angelim-vermelho), *Marmaroxylon racemosum* (Marblewood), *Couratari guianensis* (Mahot), *Buchenavia capitata* (Yellow Sanders), *Swartzia racemosa* (pitaíca), and *Dipteryx odorata* (Sarrapia). The forest is a lowland *terra firme* rainforest. The mean annual temperature is 25.78°C and the mean annual rainfall is 2272 mm, with a dry season when only 555 mm of rainfall occurs on average [1]. The wet season is from January to June, while the dry season is from July to December. Most soils are yellow Oxisols (Brazilian classification Latossolo), but there are large differences in texture. The water table has been observed at a soil depth of 10 m during the wet season and the site elevation is 15 m above the river level [1].

ESECAFLOR experiment

The data used in this study were obtained from the "Long-term drought impact on water and carbon dioxide fluxes in Amazonian Tropical Rainforest Experiment" (ESECAFLOR) which was a subproject of Large Scale Biosphere Atmosphere Experiment in Amazon forest (LBA). The ESECAFLOR experiment compared water and carbon dioxide fluxes over the tropical rainforest ecosystem with those of an experimental plot in which rainfall was artificially excluded (about 50%) to simulate a severe drought in eastern Amazon which was associated with the El-Niño years. A "throughfall" exclusion (TFE) experiment was conducted at the experimental site where rainfall was excluded from the soil over a 100 m x 100 m plot (treatment) and the data were compared to those concurrently experienced in another 100 m x 100 m plot (control). Further details about the "TFE" experiment in eastern Amazon are provided in Fisher et al [1]. Our study analyzed observations of biomass dynamics from January 2005 to May 2009, and the observations were not replicated.

Description of data

Field measurements were performed on Plot A (control TFE) and Plot B (treatment TFE) in which rainfall was artificially excluded to simulate a drought as previously mentioned. Based on at least one stem with a diameter at the breast height (DBH) > 5 cm, the number of trees were 532 and 502 for control TFE and treatment TFE, respectively. The most predominant tree species in control TFE were *Escheweleira odorata* (Siam Weed), *Lecythis lanceolata* (sapucaia-mirim), *Licania octandra* (Caraipé), *Pouteria macrophylla* (abiu), *Swartzia racemosa* (Pacapeuá), *Rinoria guianensis* (acariquarana) and *Vouacapoua Americana* (Bruinhart). On the other hand, the tree species in the treatment TFE included Eschweilera coriacea (matamata), Manilkara huberi (Macaranduba), *Swartzia polyphylla* (Paracutaca) and *Tetragastris panamensis* (Amesclao).

The forest tree species diversity varied from 150 to 160 trees per hectare and the individual density ranged from 450 to 550 trees ha^{-1}. Each plot of 100 m x 100 m (1 ha) was divided into 100 subplots each 10 m x 10 m including different species. Metal dendrometer bands were fixed to the trunk of each selected tree for measuring changes in stem diameter through the return spring displacement. Displacement measurements were taken by a digital caliper with a precision of 0.01 mm. Since DBH measurements are important for assessing the biomass dynamics of tropical forest growth, the basal area was obtained by summing cross-sections of trees for a range of three classes of DBH.

As no allometric equations have been developed for old-growth Atlantic forest sites encompassing a range of tree diameters suitable for this study, we applied the allometric equations developed by Higuchi et al [11]. The tree aboveground biomass through experimental period was grouped into three diameter classes, defined as the largest (DBH ≥ 20 cm), the smallest (5 ≤ DBH < 20 cm) and unique class (DBH > 5 cm), as follows:

$$\ln P = -0.151 + 2.170 \ln DBH, \text{ for } DBH \geq 20 \text{ cm} \quad (1)$$

$$\ln P = -1.754 + 2.665 \ln DBH, \text{ for } 5 \;\; DBH < 20 \text{ cm} \quad (2)$$

$$\ln P = -1.497 + 2.548 \ln DBH, \text{ for } DBH > 5 \text{ cm} \quad (3)$$

where, DBH is the diameter at the breast height (cm), and P is the total fresh weight (kg). These equations were derived by four statistical models from the data set with 315 trees with DBH greater than 5 cm on a site covered by a typical dense *terra firme* in Central Amazon. Higuchi et al [11] found that the logarithmic model using a single independent variable (DBH) produced results as consistent and precise as those with two variables (DBH and total height), with the differences between observed and estimated biomass being below 5%. In order to evaluate the forest structure and biomass variation under rainfall exclusion

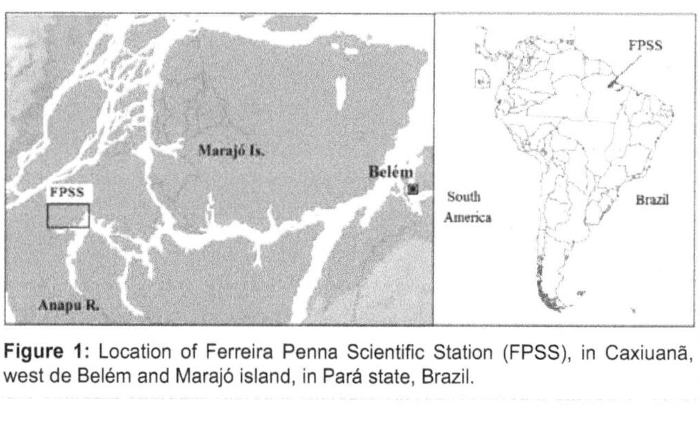

Figure 1: Location of Ferreira Penna Scientific Station (FPSS), in Caxiuanã, west de Belém and Marajó island, in Pará state, Brazil.

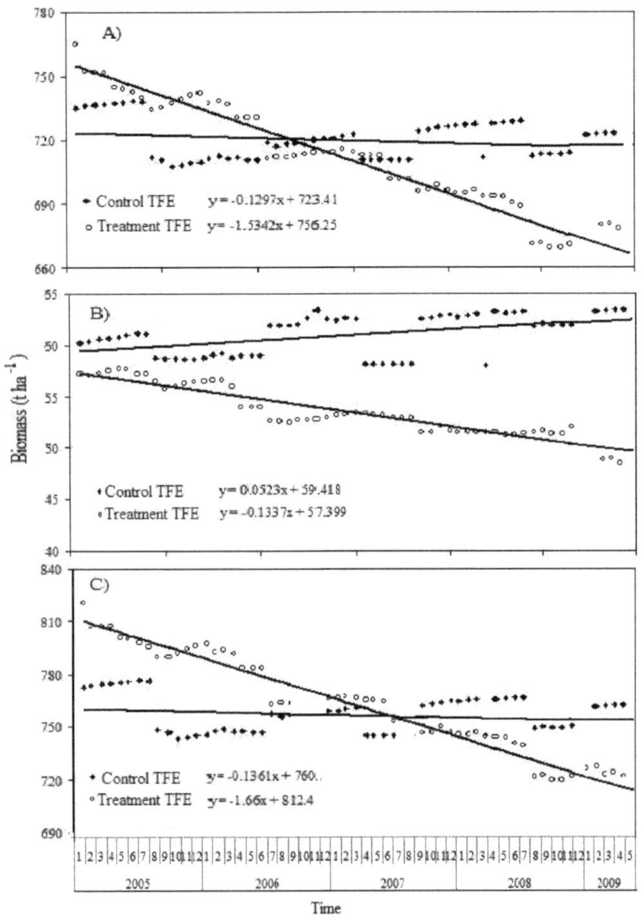

Figure 2: Variability in the aboveground biomass during the period 2005-2009 for three classes: (i) largest (DBH ≥ 20 cm), (ii) smallest (5 ≤ DBH < 20 cm), and (iii) unique class (DBH > 5 cm).

in our study, the biomass data were collected for the two plots from January 2005 to May 2009. The allometric equations relating diameter at the breast height to stem were derived from destructive sampling. The tree DBH and diameter measurements were made monthly on living trees in both plots.

Results

Biomass data and basal area

There was an appreciable difference in the aboveground biomass between control and treatment plots for all tree diameter classes throughout the study period, as shown in Figure 2, because rainfall was artificially excluded to simulate a severe drought which is associated with the El-Niño years. The aboveground biomass on control TFE did not practically vary during the experimental period, except for the smallest class with an increase of only 2.7 t ha^{-1}, while the other classes showed a little sign of decreasing trend. The biomass values declined from 755 to 670 t ha^{-1} for the largest class, from 57 to 52 t ha^{-1} for the smallest class and from 815 to 715 for the unique class.

While the forest biomass for the smallest class in control TFE increased with values ranging from 58 to 63 t ha^{-1}, the aboveground biomass in treatment TFE decreased from 57 to 52 t ha^{-1}. This result is particularly important, because the role of environmental variables that control the distribution and abundance of biomass in tropical lowland forests has been a key property of ecosystems. In addition, forests can play an important role, since they can supply biomass residues that may constitute an important source of energy. Baker et al [12] discussed that uncertainty in biomass estimates is one of the greatest limitations of models of carbon flux in tropical forests.

The average basal area in control TFE was 32.3±0.45 (range 32.9 - 31.9 m^2 ha^{-1}); while in treatment TFE it was 30.8±1.41 (range 32.8 - 28.7 m^2 ha^{-1}). The highest difference between basal areas occurred at the end of the experimental period; the whole difference was up to 9%. There was a significant decrease in the basal area of 1.04 m^2 ha^{-1} year^{-1} in treatment TFE and almost no change in control TFE with an increase of only 0.11 m^2 ha^{-1} year^{-1}, because rainfall did not practically vary from 2006-2009 (Table 1). However, in 2005, the biomass was lower than the average from 2005-2008 (755.7 t ha^{-1}), because rainfall was 50.3% lower than the long term mean (2272 mm). This rainfall decrease in 2005 was influenced by a strong El-Niño weather pattern producing droughts in the region and had amplified the annual rate of tree mortality in tropical forests of Brazil. It is important to highlight the fact that the aboveground biomass data in 2009 was not for the entire year (January-May).

Trees mortality

The losses of biomass through mortality under normal climate and drought events are shown in Figure 3. From the tree death data record (2005-2009), the loss of aboveground biomass through mortality was of 5.3 t ha^{-1} year^{-1} in control TFE and 21.1 t ha^{-1} year^{-1} in treatment TFE. Obviously, the average density (± standard deviation) of dead trees on treatment TFE was higher (54 ± 20.4 t ha^{-1}) than in control TFE (25 7.1 t ha^{-1}) which can be attributed to a drought and heat stress.

	Biomass (t ha^{-1})				
	2005	2006	2007	2008	2009
	Control TFE				
Average	765	750	752	763	750
Standard deviation	4.9	14.6	7.5	5.9	0.6
	Treatment TFE				
Average	800	782	763	745	721
Standard deviation	13.9	9.1	5.8	2.9	1.3
	Rainfall (mm)				
Annual total	1143.0	2062.4	2079.8	2130.2	-
Standard deviation	123.7	137.7	122.7	168.5	-

Table 1: Annual rainfall, average, and standard deviation of the aboveground biomass for each year of the experimental period.

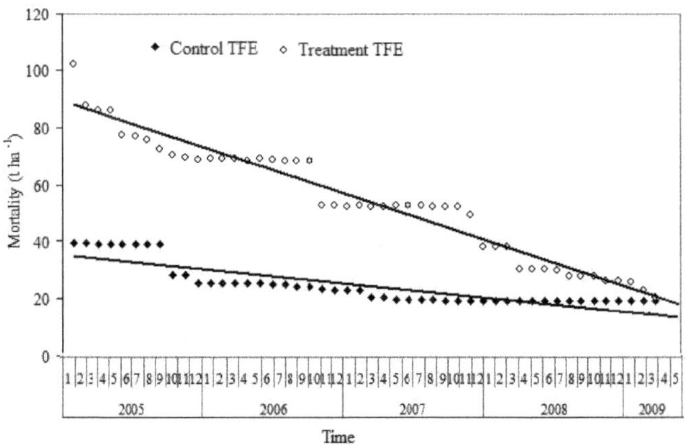

Figure 3: Temporal pattern of the loss of the aboveground biomass through mortality in control TFE and treatment TFE.

The frequency of tree death by year in control TFE and treatment TFE from 2005 to 2009 is shown in Figure 4. The frequency of tree deaths was averaged for a 1-year period for each plot. The temporal variation in the frequency of tree deaths was associated with droughts. In 2009, however, the tree deaths in the plot with rainfall exclusion were lower than those with normal climate, possibly due to the short record period (January to May) which was not representative for the entire year. Apart from that, due to rainfall exclusion, the frequency of trees that died was higher in treatment TFE than in control TFE, except in 2007, as a consequence of insects and disease activity on the plot under normal climate.

The mortality tree rate in treatment TFE was 3.7% year^{-1} (ranging from 10 to 21 trees), whereas it was 2.6% year^{-1} (raging from 19 to 39 trees) in control TFE. Such values are similar to those obtained by Hu and Wang [13] who reported an annual tree mortality rate of 1.03±0.38% for South Carolina Piedmont. On the other hand, Laarmann et al [14] found an average annual tree mortality rate of 1.3% based on the initial stem numbers when analyzing forest naturalness and tree mortality patterns in Estonia. The biomass date was negatively correlated with the frequency of tree death in control TFE and treatment TFE (Figure 5). The decrease in the aboveground biomass in treatment TFE was explicated up to 97% by tree deaths as a consequence of reduced rainfall from January 2005 to May 2009. However, the tree mortality was not significantly correlated with biomass in control TFE, indicating that other factors, such as forest insects and disease activity, can be dominant to reduce the biomass [10]. The coefficient of determination of 0.9748 between biomass and tree mortality in treatment TFE is statistically significant at 1% level by the Student-t-test. However, there is not a statistically significant relationship between the variables for control TFE ($r^2 = 0.2481$).

Discussion

The seasonal pattern of tree density for three diameter classes in treatment TFE was essentially in decline, because rainfall was artificially excluded to simulate a severe drought in the forest. This result is particularly important, because extensive drought conditions experienced in eastern Amazon coincide with the El Niño/Southern Oscillation (ENSO) events. In addition, continued increase in the concentration of carbon dioxide in the atmosphere due to anthropogenic emissions is predicted to lead to significant changes in climate, especially in tropical rainforests, like the Amazon region. Analyzing the biomass and net primary productivity of forests in central Himalaya, India, Lodhiyal and Lodhiyal [15] observed that the basal area and biomass of trees increased with increase in the forest age,

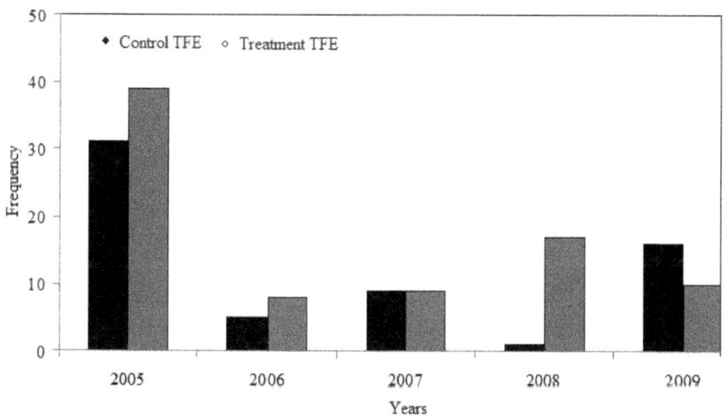

Figure 4: Frequency of tree death in control TFE and treatment TFE in Caxiuanã, Pará, Brazil.

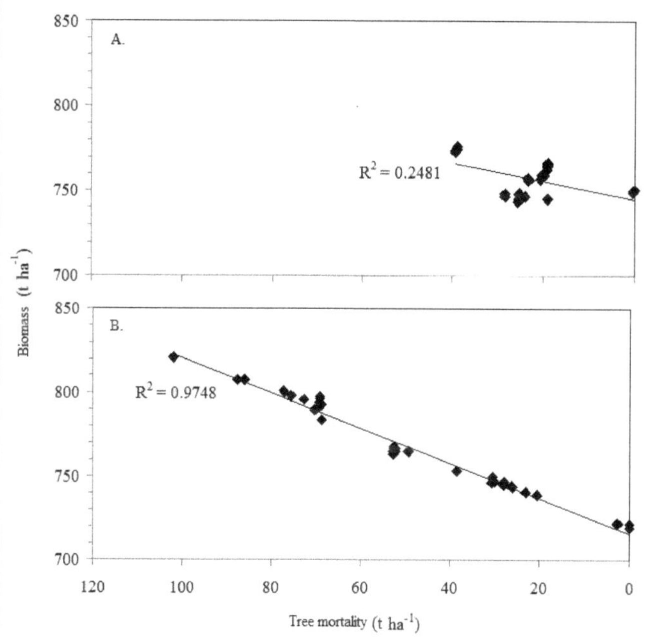

Figure 5: Relationship between mortality and the aboveground biomass in control TFE (A) and treatment TFE (B).

whereas the herb biomass significantly decreased with increasing forest age. They also found that the total vegetation biomass ranged from 52.5 (5 years) to 118.1 t ha^{-1} (15 years).

The basal area decreased substantially and linearly in treatment TFE as a consequence of reduced soil water content produced by rainfall exclusion, while for the plot under normal conditions of climate the basal area was approximately constant (around 32.5 m^2 ha^{-1}) during the experimental period (Figure 6). Comparison of basal areas for two plots revealed no changes through the first year of experimental period, with the basal area in treatment TFE being higher than in control TFE. This difference was less than 1%, which can be explained by the fact that at the beginning of experiment the effect of drought in forest was not felt. However, the effect of drought impact became evident after 2005 because the basal was reduced by 12.5% at the end of experimental period in May 2009. Marín et al [16] found basal areas of 15.62 and 23.13 m^2 ha^{-1}, respectively, in deciduous and forest gallery forest types in a tropical dry forest reserve in Nicaragua. They also observed that in the deciduous forest small stems contributed to more than half of the basal area, whereas in the gallery forest large stems (>70 cm DBH) contributed to almost half the basal area.

A significant decrease in the basal area of 4.6 m^2 ha^{-1} during the experimental period in treatment TFE was likely due to the effect of severe droughts in forest. Marín et al [16] also found a significant decrease in the basal area of 1.2% year^{-1} in the deciduous forest and no change in the gallery forest when the stand dynamics and basal area change were analyzed in a tropical dry forest reserve in Nicaragua. Analysis of the intra-annual variability of data indicated that the basal area decreased from July to August because of the reduction in rainfall during the dry season. The values of biomass and basal area were similar to the findings of Baker et al [12] who found the biomass and basal area of 846 t ha^{-1} and 38.9 m^2 ha^{-1}, respectively, for *terra firme* rainforest forest in the Amazon. Although episodic mortality occurs in the absence of climate change, the decrease in biomass of 17.1 t ha^{-1} in the plot under normal climate during the period of record was considered

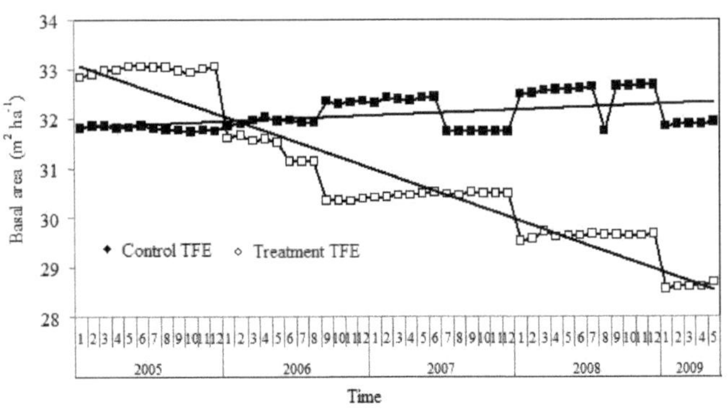

Figure 6: Change in the basal area from 2005-2009 in control TFE and treatment TFE during the experimental period.

abrupt. Analyzing a total of 2493 trees that died during the study period from 2001 to 2007 in Estonia, Laarmann et al [14] reported an average annual tree mortality rate of 1.3%, while grey alder (Alnus incana (L.) Moench) experienced the highest mortality rate (4.3%) but pine and spruce the lowest (0.9%). On the other hand, Stephens and Gill [17] found the cumulative tree mortality raging from 2.7 to 3.6% and the annual rate of tree mortality of 0.162% $year^{-1}$ in north-western Mexico.

The extensive increase in tree mortality in both plots in 2005 was linked to anomalously warm sea surface temperatures in the North Atlantic which produced a hot and severe drought across the Amazon basin. The tree mortality rate in treatment TFE was high throughout the experimental period with the subsequent aboveground biomass loss, indicating the vulnerability of Amazonian rainforests to moisture stress. The percentage of trees that died from 2005 to 2009 in response to drought was 17% while in control TFE the percentage of trees that died was of 12%. Effectively, treatment TFE had a total of 86 trees that died through the period of record from 2005 to 2009, while control TFE had 62 dead trees. Therefore, more trees died in years that had below normal annual rainfall.

Another important finding is that dead trees were more frequent in the smallest class than in tress with DBH > 20 cm. Similar results were obtained by Guarín and Taylor [18] when analyzing the effects of drought for tree mortality in mixed conifer forests in Yosemite National Park, USA. Guarín and Taylor [18] also investigated the influence of drought on recent patterns of tree mortality in forests. They observed that the frequency of tree death dates was negatively correlated with annual and seasonal Palmer Drought Severity Index (PDSI) and the April snowpack depth, and more trees died in years with below normal PDSI and snowpack. Analyzing the sap flow data from a throughfall exclusion experiment in eastern Amazonian rain forest, Fisher et al [1] showed large dry season declines in transpiration, with tree water use restricted to 20% of that in the control plot at the peak of both dry seasons. These results suggest that the forest is not able to withstand a 50% reduction in rainfall over 1-2 years without impacting canopy gas exchange, and are in contrast with the results of Nepstad et al [19] for their TFE experiment, located in the Tapajo`s national forest.

The frequency of tree death was associated with below normal moisture conditions over the experimental period. On the other hand, a higher coefficient of determination between tree mortality and biomass in treatment TFE indicated that mortality was higher for higher tree-size diversity during severe drought episodes. However, the relationship between biomass and mortality was less clear for treatment TFE with a low coefficient of determination, because the tree mortality commonly involved multiple, interacting factors, ranging from particular sequences of climate stress and insect pests. Although the mortality rate was not significantly correlated with biomass in the plot under normal climate, the loss of aboveground biomass through mortality was decreasing for both plots. However, the mortality rate in control TFE was nonrandom with respect to tree size classes and species. Ganey and Vojta [20] found proportions of trees dying at the greatest rate in the largest size classes, particularly in mixed-conifer forest, where the mortality in the largest size class exceeded 22% from 2002 to 2007. These results indicate that the Amazonian rainforest is not resilient to severe droughts, and that treatments to increase the resilience of forest biodiversity to climate change may be appropriate. A drought can also reduce tree growth, increase tree mortality particularly on forest edges as well as increase leaf shedding. In this context, the Amazonian rainforest droughts have been strongly related to El Niño events [21]. However, declining rainfall over the Amazon region is likely to be impacting other anthropogenic forcing factors, such as deforestation and fires [22].

Conclusions

Experimental TFE, removing an estimated 50% of rainfall, caused soil drying and a resultant decrease in the basal area of 12.5% at the end of experimental period in May 2009 compared with the control plot. The tree mortality rate in treatment TFE was high throughout the experimental period with the subsequent aboveground biomass loss, indicating the vulnerability of Amazonian rainforests to moisture stress. The percentage of trees that died from 2005 to 2009 in response to drought was 17% while in control TFE the percentage of trees that died was of 12%. A significant environmental stress due to a long-term drought can have a profound impact on the vegetation dynamics in the Amazon region. The greater decrease in the aboveground biomass on treatment plots can be explained by reduced rainfall. The forest biomass is negatively correlated with tree mortality, especially on the rainfall exclusion plot, because mortality occurs in all size classes in both plots, with most mortality occurring in the subplot of higher tree-size diversity. The relationship between the aboveground biomass and mortality is less clear in normal climate with a low coefficient of determination. There was more mortality in 2005 than in 2006 to 2009 because of severe drought across the Amazon basin produced by the stronger El Niño in that year.

References

1. Fisher RA, Williams M, Costa ACL, Malhi Y, Costa RF (2007) The response of an Eastern Amazonian rain forest to drought stress: results and modelling analyses from a throughfall exclusion experiment. Glob Change Biol 13: 1-18.

2. Davidson EA, Ishida FY, Nepstad DC (2004) Effects of an experimental drought on soil emissions of carbon dioxide, methane, nitrous oxide, and nitric oxide in a moist tropical forest. Glob Change Biol 10: 718-730.

3. Nogueira EM, Fearnside PM, Nelson BW, Barbosa RI, Keizer EWH (2008) Estimates of forest biomass in the Brazilian Amazon: New allometric equations and adjustments to biomass from wood-volume inventories. Forest Ecol Manag 256: 1853-1867.

4. Malhi Y, Roberts JT, Betts RA, Killeen TJ, Li W (2008) Climate change, deforestation, and the fate of the Amazon. Science 319: 169-172.

5. Silva RP, Santos J, Tribuzy ES, Chambers JQ, Nakamura S (2002) Diameter increment and growth patterns for individual tree growing in Central Amazon, Brazil. Forest Ecol Manag 166: 295-301.

6. Conner WH, Day Jr, JW (1992) Diameter growth of Rhodium distichum (L) Rich and Nyssa aquatica L from 1979 to 1985 in four Louisiana swamp stands. Am Midl Nat 127: 290-299.

7. Antos JA, Parish R, Nigh GD (2008) Growth patterns prior to mortality of mature

Abies lasiocarpa in old-growth subalpine forests of southern British Columbia. Forest Ecol Manag 255: 1568-1574.

8. Villalba R, Veblen TT (1998) Influences of large-scale climatic variability on episodic tree mortality in northern Patagonia. Ecology 79: 2624-2640.

9. Swetnam TW, Betancourt JL (1998) Mesoscale disturbance and ecological response to decadal climatic variability in the American southwest. J Climate 11: 3128–3147.

10. Costa RF, Silva VPR, Ruivo MLP, Meir P, Costa ACL (2007) Transpiração em espécie de grande porte na Floresta Nacional de Caxiuanã, Pará. Agriambi 11: 180-189.

11. Higuchi N, Santos J, Ribeiro RJ, Minette LJ, Biot Y (1998) Biomassa da parte aérea da vegetação da Floresta Tropical Úmida de Terra-Firme da Amazônia Brasileira. Acta Amazon 28: 153-166.

12. Baker TR, Phillips OL, Malhi Y, Almeida S, Arroyo L, et al. (2004) Variation in wood density determines spatial patterns in Amazonian forest biomass. Glob Change Biol 10: 545-562.

13. Hu H, Wang GG (2008) Changes in forest biomass carbon storage in the South Carolina Piedmont between 1936 and 2005. Forest Ecol Manag 255: 1400-1408.

14. Laarmann L, Korjus H, Sims A, Stanturf JA, Kiviste A (2009) Analysis of forest naturalness and tree mortality patterns in Estonia. Forest Ecol Manag 258S: S187–S195.

15. Lodhiyal N, Lodhiyal LS (2003) Aspects of nutrient cycling and nutrient use pattern of Brabar Shisham forests in central Himalaya, India. Forest Ecol Manag 176: 237-252.

16. Marín GC, Nygardb R, Rivasa BG, Oden PC (2005) Stand dynamics and basal area change in a tropical dry forest reserve in Nicarágua. Forest Ecol Manag 208: 63-75.

17. Stephens SL, Gill SJ (2005) Forest structure and mortality in an old-growth Jeffrey pine-mixed conifer forest in north-western Mexico. Forest Ecol. Manag 205: 15-28.

18. Guarín A, Taylor AH (2005) Drought triggered tree mortality in mixed conifer forests in Yosemite National Park, California, USA. Forest Ecol Manag 218: 229-244.

19. Nepstad DC, Moutinho P, Dias-Filho MB, Davidson E, Cardinot G, et al. (2002) The effects of partial through-fall exclusion on canopy processes, aboveground production, and biogeochemistry of an Amazon forest. J Geophys Res 107: 8085.

20. Ganey JL, Vojta SC (2011) Tree mortality in drought-stressed mixed-conifer and ponderosa pine forests, Arizona, USA. Forest Ecol Manag 261: 162-168.

21. Marengo J A (2004) Interdecadal variability and trends in rainfall in the Amazon basin. Theoretical And Applied Climatology 78: 79-96.

22. Cochrane MA, Laurance WF (2002) Fire as a large-scale edge effect in Amazonian forests. J Trop Ecol 18: 311-325.

ISSR Primer Selection for Genetic Variability Analyses with Jaborandi (*Pilocarpus microphyllus* Stapf ex Wardlew., Rutaceae)

Jefferson Almeida Rocha[*1,5], **Santelmo Vasconcelos**[4], **Fabrícia Meireles Meneses da Silva**[3], **Anne Jurkiewicz Melo**[1], **Maria Francilene Souza Silva**[2], **João Antônio Leal de Miranda**[2], **Ana Maria Benko-Iseppon**[4] **and Ivanilza Moreira de Andrade**[1]

[1]*Graduate Program Biotechnology (PPGBiotec), Campus of Parnaíba, Federal University of Piauí, Brazil.*

[2]*Graduate Program Biomedicine. Federal University of Piauí, Brazil.*

[3]*Anhydrous Brazil Extractions SA, Povoado Rosápolis s/n, Brazil.*

[4]*Laboratory of Plant Genetics and Biotechnology, Department of Genetics, Center of Biological Sciences, Federal University of Pernambuco, Brazil.*

[5]*Center for Natural Sciences, Federal University of Maranhão, Brazil.*

Abstract

Pilocarpus microphyllus Stapf ex Wardlew. (jaborandi) is native to the North and Northeast regions of Brazil. Intensive exploitation for the extraction of pilocarpine, a valuable alkaloid for the pharmaceutical industry, has killed off or brought about a loss of vigor of naturally occurring plant populations. As a result, the species is officially listed as an endangered species of the Brazilian flora. Genetic diversity is required for the adaptation of populations to environmental changes, and maintaining it is a central objective for biological conservation. ISSRs are dominant markers widely used in genetic diversity studies of endangered species, allowing for the identification of genotypes and cultivars, as well as helping in phylogenetic studies based on DNA fingerprinting. This study presents an ISSR primer selection for genetic structure analyses of natural populations and cultivated collections of *P. microphyllus*.

Keywords: Jaborandi; Pilocarpine; Selection of markers

Introduction

The Rutaceae family has a pantropical distribution and includes approximately 161 genera and 2,070 species [1]. In Brazil, 33 genera and 193 species have been recorded, and among these there are 14 species known of *Pilocarpus Vahl* (jaborandi), of which 11 are endemic to Brazil [2]. *Pilocarpus microphyllus Stapf* ex Wardlew. is native but not endemic, occurring in the north and northeast of Brazil, more specifically in eastern Pará, western and northern Maranhão and northern Piauí. It is a shrub growing in open forests, and frequent in the rocky outcrops known as "canga". In forest interiors, the species can be found in areas with high light incidence [3].

Pilocarpus microphyllus contains a significant amount of secondary metabolites with diverse pharmacological properties, and specially stimulants of the secretive and diuretic systems [4]. Among these, the alkaloid pilocarpine stands out, being known for stimulating salivation and transpiration and for its use in the composition of eye drops [4]. The intensive collection of wild *P. microphyllus* populations has caused widespread loss of populations and mortality or loss of vigor in those that remain, reducing the size of the plants and their leaves [5]. Consequently, the species was officially included in the list of the endangered species of the Brazilian native flora [6].

Successful adaptation of plant populations to environmental changes depends on their genetic diversity, and maintaining this diversity has a central role in biological conservation programs [7]. There are several types of molecular markers that can easily detect genetic diversity levels within natural populations and cultivated materials [8]. ISSR markers, developed by ZIETKIEWICZ et al. [9], are PCR based technique which is easy, quick, simple and economical. The advantages of ISSRS over other markers like AFLPs, RAPDs and more specific SSRs are reproducible due to their better stringency (high annealing temperature), require no gene sequence information and no prior genetic studies are required for these analyses [10,11]. ISSR markers have been successfully used for the assessment of genetic diversity and gene pool origin in Amw plants [12-16].

An important step prior to the analysis of the genetic diversity of a population is the development of appropriate markers in order to avoid those that either fail to amplify or generate only few fragments and low polymorphism levels. It is known that the nucleotide sequence of ISSR primers recognize sequences arbitrarily distributed in the genome and the preliminary stage of primer selection and testing is essential prior to using them to evaluate all the samples of the studied population.

The aim of this study was to select suitable ISSR marker applicable to the evaluation of natural populations and germplasm resources of *P. microphyllus*, in order to estimate the genetic diversity of such an endangered and economically important plant species [17].

Materials and Methods

For marker selection eight individuals of P. microphyllus were used, two from each sampled population in Piauí (Parnaíba and Luzilandia), Maranhão (Mata Roma) and Pará (Parauapebas) (Table 1 and Figure 1). Young leaves were collected from each individual and immediately stored in 2% NaCl-saturated CTAB buffer [18].

Genomic DNA was extracted from leaf fragments in 2% CTAB as described by Doyle and Doyle [17], with some modifications. Approximately 200 µg of young leaves were macerated with extraction buffer [2% CTAB, Tris-HCl 0.1 mM (pH 8.0), 20 mM EDTA (pH

***Corresponding author:** Jefferson Almeida Rocha, Graduate Program Biotechnology (PPGBiotec), Campus of Parnaíba, Federal University of Piauí, Brazil, E-mail: jeffersonkalel@hotmail.com

Population	Collection Site	Voucher	Coordinates
TL	Tabuleiros Litorâneos, Parnaíba, Piauí	Rocha, J. A. 258	3°6'S, 41°47'W
LU	Assentamento Cutias, Luzilândia, Piauí	Rocha, J. A. 255	3°41'S, 42°28'W
MR	Barra dos Macacos, Mata Roma, Maranhão	Silva, F.M.M. 257	3°31'S, 43°6'W
PA	Flona Carajás, Parauapebas, Pará	Silva, F.M.M. 253	6°2'S, 50°12'W

Table 1: Identification and location of sampled populations of *Pilocarpus microphyllus*.

8.0), 1.4 M NaCl and 2% β-mercaptoethanol]. After the maceration, the material was transferred to 1.5 mL polypropylene microtubes and incubated in a water bath at 60°C for 20 min. After cooling the extract, 800 μL of a chloroform-isoamyl alcohol (24:1, v/v) solution were added and the mix was shaken for one hour. The homogenized contents of the tubes were centrifuged for 10 min at 13,000 rpm. The resulting supernatant (~500 μL) was transferred to a new tube, and 300 μL of isopropanol (0.6 volumes) were added, being carefully mixed by inversion and stored overnight in a freezer (–20°C). Then, the tubes were centrifuged at 13,000 rpm for 5 min, concentrating the precipitated DNA. The resulting pellet was washed with 1 mL of 70% ethanol and centrifuged for 5 min at 13,000 rpm (this process was repeated three times) and finally it was left to dry overnight at room temperature. The DNA obtained was re-suspended in 100 μL in TE buffer [10 mM Tris-HC1 (pH 8.0) and 0.1 mM EDTA] for 48 h or until the pellet was homogenized in the solution.

A total of 48 primers [19,20] were tested in eight individuals of P. *microphyllus* ('TL1', 'TL2', MR1', 'MR2', 'PA1', 'PA2', 'LU1', 'LU2'). The PCR procedures were conducted in a final volume of 10 μL containing: enzyme buffer 1×, 1 mM dNTPs, 2.5 mM $MgCl_2$, 50 μM primer, 0.8 U of Taq polymerase enzyme and 25 ng of genomic DNA [19]. The PCR reactions were carried out in a Techne TC-412 thermocycler, with a program consisting of initial denaturing for 4 min at 94°C, followed by 35 cycles of (1) denaturing for 30 s at 94°C, (2) annealing of the primer for 45 s at 50.4-52.0°C (depending on the primer used; Table 2) and extension for 2 min at 72°C; the final extension occurred for 7 min at 72°C. The products of each reaction were separated in a 1.8% agarose gel electrophoresis in 0.5x TBE buffer, stained with ethidium bromide (1.5 μL/100 mL of gel), under 100 V and 159 mA for 3.5h, and then photographed under ultraviolet light.

For the analysis of the amplified fragments, a binary matrix was constructed by the visual criteria of presence (1) and absence (0) of every molecular weight observed with their respective primers. The same weight (1) was given for all accounted fragments, independent of their fluorescence intensity on the gel. Indistinct fragments, with poor visualization, were counted as missing data, remaining neutral in the data analysis.

The discriminating power of each primer was determined by means of three parameters: (1) PIC (polymorphism information content); (2) MI (marker index): and (3) RP (resolving power). The PIC value for each amplified fragment was calculated as $PICi = 2\ fi\ (1 - fi)$, PICi being the individual PIC of marker i, fi the frequency of fragments present in the marker per accession and 1 – fi the frequency of absent fragments [21]. Then, the mean PIC values were obtained for each primer. The MI was obtained using the following formula: $MI = PIC \times n \times np/(np + nm)$, n being the average number of fragments per primer, np the number of polymorphic fragments and nm the number of monomorphic fragments [22]. The RP of each primer was calculated as follows: $RP = \Sigma\ Ib$, where Ib is the level of information of each fragment. The Ib can be represented as an interval of 0-1 by the following formula: $Ib = 1 - (2 \times |0.5 - fi|)$ [23]. Additionally, the correlation between the indices was tested using the Pearson coefficient with aid of the software BioEstat 5.0 [24].

The genetic distance between accessions was estimated using the Dice coefficient [25] using the program DARWIN 5.0.158. Subsequently, the dissimilarity matrices obtained were used to generate a phenogram in the same program, using the neighbor-joining algorithm (30.000 bootstrap replications).

Results and Discussion

Of the 48 primers tested, 34 were informative, generating 336 bands or fragments, ranging from 100-1000 bp in size, of which 226 (67%) were polymorphic (Table 2). The total number of amplified fragments (TNF) in each ISSR reaction ranged from five (UBC 847, UBC 887, UBC 888 and UBC 890) to 19 (UBC 825), with an average of 9.8 bands per primer (Table 2). The polymorphic fragments were categorized according to their respective intervals: unique fragments ($0 < x \leq 0.2$), rare fragments ($0.2 < x \leq 0.4$), frequent fragments ($0.4 < x \leq 0.6$), common fragments ($0.6 < x \leq 0.8$) and shared fragments ($0.8 < x < 1.0$) [20] (Table 2 and Figure 2).

Four genotypes (TL1, TL2, LU1 and LU2) presented unique fragments (data not shown), which were amplified with the primers

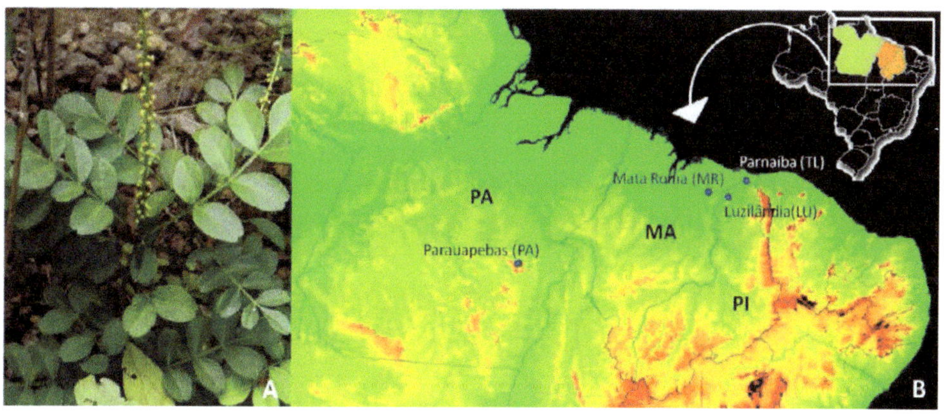

Figure 1: *Pilocarpus microphyllus* in its natural habitat (A). Collection site of the sampled populations of *P. microphyllus* (B) (program DIVA-GIS 7.5; HIJMANS et al. 2001).

Primer UBC [repetitions][a]	Ta(°C)[b]	TNF[c]	NPF[d]	NMF[e]	fp(%)[f]	NUF[g]	NRF[h]	NFF[i]	NCF[j]	NSF[K]	MPICL	MIM	RPN
807 [$(AG)_8$T]	50.4	14	13	1	92	0	4	2	6	1	0.388	3.56	8.78
809 [$(AG)_8$G]	52.8	8	5	3	62	0	0	2	3	0	0.300	1.85	4.25
810 [$(GA)_8$T]	50.4	16	11	5	68	0	1	2	7	1	0.280	1.90	6.57
811 [$(GA)_8$C]	52.8	8	3	5	37	0	1	1	1	0	0.160	0.59	2.00
812 [$(GA)_8$A]	50.4	12	7	5	58	0	0	1	1	5	0.164	0.94	2.75
813 [$(CT)_8$T]	50.4	11	10	1	90	1	1	2	2	4	0.338	3.04	5.66
814 [$(CT)_8$A]	50.4	7	5	2	71	0	1	2	1	1	0.309	2.18	3.66
815 [$(CT)_8$G]	52.0	6	6	0	100	0	2	2	1	1	0.408	4.03	3.71
816 [$(CA)_8$T]	50.4	16	11	5	68	0	1	1	5	4	0.244	1.66	5.42
818 [$(CA)_8$G]	52.8	7	4	3	57	0	1	0	0	3	0.182	1.03	1.66
821 [$(GT)_8$T]	50.4	12	10	2	83	0	5	2	3	0	0.388	3.19	7.75
824 [$(TC)_8$G]	52.0	14	13	1	92	1	2	0	9	1	0.401	3.68	9.42
825 [$(AC)_8$T]	50.4	19	10	9	52	0	1	1	5	3	0.179	0.93	4.75
826 [$(AC)_8$C]	52.8	6	5	1	83	0	0	2	3	0	0.385	3.17	4.00
827 [$(AC)_8$G]	52.8	7	2	5	28	0	0	1	1	0	0.125	0.35	1.50
828 [$(TG)_8$A]	52.8	12	5	7	41	0	1	1	0	3	0.136	0.56	2.28
830 [$(TG)_8$G]	52.8	11	11	0	100	0	0	2	9	0	0.423	4.18	7.25
831 [$(AT)_8$YA]	50.4	9	6	3	66	0	1	0	3	2	0.236	1.55	3.00
843 [$(CT)_8$RA]	52.0	8	8	0	100	1	6	0	1	0	0.425	4.20	5.25
847 [$(CA)_8$RC]	54.0	5	4	1	80	1	2	0	1	0	0.337	2.66	3.00
853 [$(TC)_8$RT]	52.0	14	14	0	100	2	3	3	3	3	0.367	3.63	7.42
855 [$(AC)_8$YT]	52.0	11	1	10	9	0	0	1	0	0	0.045	0.04	1.00
856 [$(AC)_8$YA]	52.0	17	9	8	52	0	1	3	3	2	0.211	1.10	5.42
858 [$(TG)_8$RT]	52.0	9	3	6	33	0	1	1	0	1	0.131	0.43	2.00
860 [$(TG)_8$RA]	52.0	9	2	7	22	0	1	0	0	1	0.065	0.14	0.75
873 [$(GACA)_4$]	52.0	12	11	1	91	0	1	2	8	0	0.380	3.44	7.00
884 [HBH$(AG)_7$]	52.0	8	2	6	25	0	0	1	2	0	0.109	0.27	1.50
885 [BHB$(GA)_7$]	52.0	10	6	4	60	0	1	1	1	3	0.209	1.24	3.25
886 [VDV$(CT)_7$]	52.0	8	8	0	100	0	0	2	6	0	0.428	4.23	5.14
887 [DVD$(TC)_7$]	52.0	5	4	1	80	0	2	1	1	0	0.362	2.86	3.00
888 [BDB$(CA)_7$]	52.0	5	3	2	60	0	0	0	3	0	0.262	1.55	2.00
889 [DBD$(AC)_7$]	52.0	7	3	4	42	0	0	2	1	0	0.174	0.73	2.25
890 [VHV$(GT)_7$]	52.0	5	4	1	80	0	0	1	2	1	0.310	2.45	2.28
891 [HVH$(GT)_7$]	52.0	8	7	1	87	0	0	0	5	2	0.316	2.73	3.42
MEAN	-	9.8	6.6	3.2	67	0.17	1.17	1.23	2.85	1.23	0.270	2.06	4.09
TOTAL	-	336	226	110	67	6	40	42	97	42	-	-	-

Table 2. Primers tested for generating polymorphic ISSR markers among *Pilocarpus microphyllus* genotypes, showing the respective levels of polymorphism and information generated by each primer.

[a]Degenerate bases used: Y (C or T); [b]annealing temperature; [c]total number of fragments; [d]number of polymorphic fragments; [e]number of monomorphic fragments; [f]frequency of polymorphic fragments; [g]number of unique fragments; [h]number of rare fragments; [i]number of frequent fragments; [j] number of common fragments; [k]number of shared fragments; [l] mean PIC; [m]marker index, [n] resolving power.

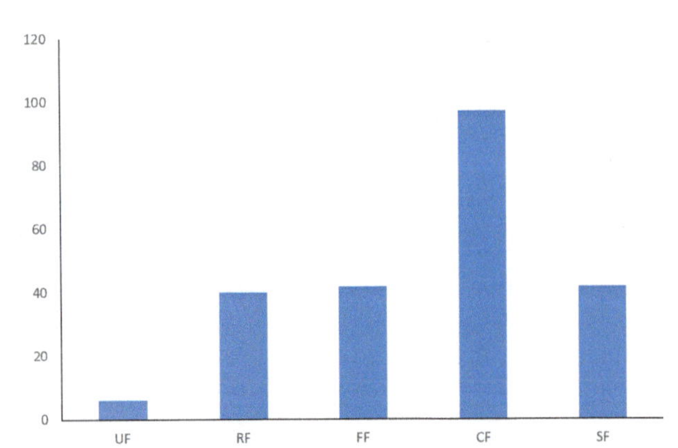

Figure 2: Distribution of polymorphic fragments for the tested ISSR primers in *Pilocarpus microphyllus*. Unique fragments (UF), rare fragments (RF), frequent fragments (FF), common fragments (CF) and shared fragments (SF).

UBC-813, UBC-824, UBC-843, UBC-847 and UBC-853. The primer UBC-853 generated the highest number of unique fragments (2 out of 6; Table 2). A total of 40 rare fragments were amplified, comprising 11.9% of the total number of amplified fragments (TNF) and 17.6% of the polymorphic fragments, with an average of 1.17 per primer. The highest number of rare fragments (six) was observed for the primer UBC-843 (Table 2). The number of frequent fragments was 42 (12.5% of the TNF and 18.5% of the recorded polymorphism), with an average of 1.23 fragments per primer, and a higher value (three) for the primers UBC-853 and UBC-856 (Table 2). Ninety-seven common fragments were amplified, comprising 28.8% of the TNF and 42.9% of the number of polymorphic fragments (NPF), with an average of 2.85 fragments per primer, and the maximum of nine fragments generated by the primers UBC-824 and UBC-830 (Table 2). The shared fragments were visualized 42 times, showing the same values observed for the frequent fragments. In turn, the primer UBC-812 generated the highest number of shared fragments (five; Table 2).

The mean PIC values varied between 0.045 (UBC-855) and 0.428 (UBC-886), with an average of 0.270 (Table 2). The average MI was 2.06, varying between 0.04 (UBC-855) and 4.23 (UBC-886) (Table 2). A strong positive correlation was observed between PIC and MI values (r=0.979; p<0.0001). The average RP was 4.09 and values ranged from 1.00 (UBC-855) to 9.42 (UBC-824). There was also a strong positive correlation between RP and PIC (r=0.724; p<0.0001), similar to that observed between MI and RP (r=0.722; p<0.0001).

Of the 48 primers tested in P. *microphyllus*, 26 were polymorphic. Thus, more than 50% generated polymorphic fragments, eight were completely monomorphic (UBC-811, UBC-827, UBC-828, UBC-855, UBC-858, UBC-860, UBC-884 and UBC-889), and 14 showed no amplification (UBC-820, UBC-829, UBC-832, UBC-833, UBC-837, UBC-838, UBC-839, UBC-846, UBC-849, UBC-852, UBC-859, UBC-878, UBC-882 and UBC-883).

The primers UBC-807, UBC-810, UBC-812, UBC-813, UBC-816, UBC-821, UBC-824, UBC-825, UBC-830, UBC-853, UBC-856, UBC-873, and UBC-885 were the ones that showed the highest number of amplified fragments, with at least ten fragments and a minimum proportion of 50% of polymorphism. The primer UBC-873 was the best one, generating 12 bands, 91% of which were polymorphic, corroborating results obtained by other authors for other plant species [26-29].

The number of fragments by ISSR primer may vary considerably, depending on the species. Some authors [3,30-33], observed a total number of bands per primer lower than 10 (about five) with the primers UBC-807, UBC-810, UBC-812, UBC-813, and UBC-856, which were relatively uninformative in their respective analyses. However, other authors, such as NAN et al. [27,34-37] reported an abundant generation of bands for these same primers, showing an average of 20 fragments per primer. This discussion shows the importance of knowing the profile of bands of a certain primer before the beginning of a study of the genetic variability of a given species.

The PIC value is another criterion that should be taken into account in the selection of primers. Average PIC values similar to the obtained for P. *microphyllus* (0.270) have been described for several other species (e.g. [29,38,39]). The RP and MI can also be used to infer differences between genotypes. Therefore, the average values of these parameters (4.09 and 2.06, respectively) are also indexes of discrimination that have been widely used to measure the information content generated by molecular markers in various plant species, such as sesame (*Sesamum indicum* L.) [25] castor (*Ricinus communis* L.) [20,40], potato (*Solanum tuberosum* L.) [23], physic nut (*Jatropha curcas* L.) [41] and barley (*Hordeum vulgare* L.) [22].

Of the 48 ISSR tested primers, 34 are suitable for studies of genetic diversity in P. microphyllus, especially the following thirteen: UBC-807, UBC-810, UBC-812, UBC-813, UBC-816, UBC-821, UBC-824, UBC-825, UBC-830, UBC-853, UBC-856, UBC-873, and UBC-885. This cited primers evidenced the highest levels of polymorphism among the selected individuals, with an average amplification of 14 fragments per primer, of which 11 were polymorphic, i.e. 78.5% of polymorphism (Figure 3).

Figure 3 shows the phenetic tree dissimilarity for the eight accessions. The high number of replications used (30.000), produced basal branches with high support values (98,90,89,80 and 63). Also observed the formation of three groups being the first (TL-1, LU-1 and LU-2), formed by natural accessions present in the state of Piaui, and TL-1, cultivated individual also present in Piauí. The second group (MR-1 and MR-2) that are present in natural accessions Maranhão,

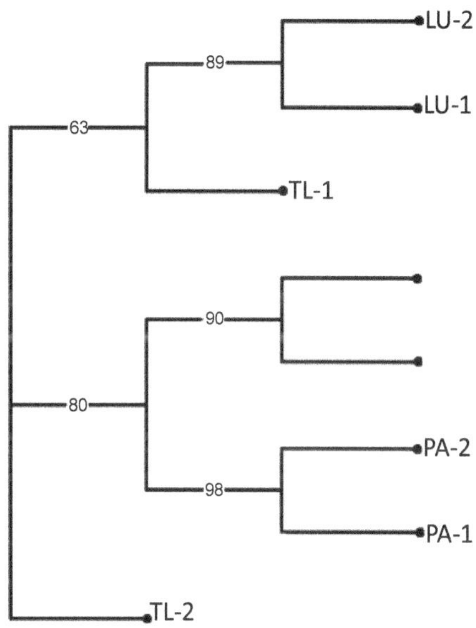

Figure 3: Phenogram Constructed of simple coefficients ISSR data , showing the structure of eight accessions of *P. microphylus*

and the third group (PA-1 and PA-2) individuals collected in Pará. At the same time there is the proximity the second and third groups, tending to form one group. And the TL-2 access, appears as an external individual to groups formed, may be explained by the TL accesses, are grown in other states and not natural populations.

Conclusion

According to the indices analyzed, such as PIC, MI, and RP, the 34 primers indicated herein are informative for genetic diversity analysis in the studied species, and can be used in future studies aimed at the diversity characterization and conservation of natural populations. As there was a significant positive correlation among all the tested indices, any of the three can be taken into account in discriminating more informative polymorphic markers for future studies of genetic diversity among P. *microphyllus* natural populations.

Acknowledgements

The authors thank Anidro do Brasil Extrações S.A. for logistical support. We also thank the Brazilian agencies CAPES (Coordination for the Improvement of Higher Education Personnel) and CNPq (Brazilian Council for Research and Development) for fellowships and financial support. We are also grateful to Dr. Simon Mayo, Royal Botanic Gardens, Kew, for reviewing the abstract.

References

1. Stevens PF (2001) Angiosperm Phylogeny Website. Version 13.
2. Pirani JR, Groppo M, Rutaceae (2013) In: Lista de Espécies da Flora do Brasil. Jardim Botânico do Rio de Janeiro.
3. Almeida CMA, Lima SEM, Lima GSA, Brito JZ, Donato VMTS, et al. (2009) Caracterização molecular de cultivares de cana-de-açúcar utilizando marcadores ISSR. Ciência e Agrotecnologia 33: 1771-1776.
4. PINHEIRO CUB (2002) Extrativismo, cultivo e privatização do jaborandi (Pilocarpus microphyllus Stapf ex Holm.; Rutaceae) no Maranhão, Brasil. Acta Botânica Brasílica 16: 141-150.
5. Taveira FSN, Andrade EHA, Lima WN, Maia JGS (2003) Seasonal variation in the essential oil of Pilocarpus microphyllus Stapf. Anais da Academia Brasileira de Ciências 75: 27-31.

6. IBAMA (2008) Instituto Brasileiro do Meio Ambiente e dos Recursos Naturais Renováveis. Lista oficial de espécies Brasileiras ameaçadas de extinção. Instrução normativa MMA nº6.
7. Gepts P (2006) Plant genetic resources conservation and utilization: the accomplishments and future of a societal insurance policy. Crop Science 46: 2278-2292.
8. Nybom H (2004) Comparison of different nuclear DNA markers for estimating intraspecific genetic diversity in plants. Molecular Ecology 13: 1143-1195.
9. Zietkiewicz E, Rafalski A, Labuda D (1994) Genome fingerprinting by simple sequence repeat (SSR)-anchored polymerase chain reaction amplification. Genomics 20: 176-183.
10. Archak S, Gaikwad AB, Gautam D, Rao EVVB, Swamy KRM (2003) DNA fingerprinting of Indian cashew (Anacardium occidentale L.) varieties using RAPD and ISSR techniques. Euphytica 230: 397- 404.
11. Thimmappaiah S, Shobha, WG, Melwyn, GS (2009) Assessment of genetic diversity in cashew germplasm using RAPD and ISSR markers. Scientia Horticulturae,v. 120: 411-417.
12. Galvan MZ, Bornet B, Balatti PA, Branchard M (2003) Inter simple sequence repeat (ISSR) markers as a tool for the assessment of both genetic diversity and gene pool origin in common bean (Phaseolus vulgaris L.). Euphytica 132: 297-301.
13. Yao H, Zhao Y, Chen DF, Chen JK, Zhou TS (2008) ISSR primer screening and preliminary evaluation of genetic diversity in wild populations of Glycyrrhiza uralensis. Biologia Plantarum 52: 117-120.
14. Dje Y, Tahi CG, Zoro BA, Baudoin JP, Bertin P (2010) Use of ISSR markers to assess genetic diversity of African edible seeded Citrullus lanatus landraces. Scientia Horticulturae 124: 159-164.
15. Cekic C, Battey NH, Wilkinson MJ (2001) The potential of ISSR-PCR primer-pair combinations for genetic linkage analysis using the seasonal flowering locus in Fragaria as a model. Theoretical and Applied Genetics 103: 540-546.
16. Noor Cemellia NA, Thohirah Lee A, Abdullah NAP (2012) Genetic relationships and diversity of Jatropha curcas accessions in Malaysia. African Journal of Biotechnology 11: 3048-3054.
17. Doyle JJ, Doyle JL (1987) Isolation of plant DNA from fresh tissue. Focus 12: 13-15. 1987.
18. Rogstad SH (1992) Saturated NaCl-CTAB solution as a means of field preservation of leaves for DNA analyses. Taxon 41: 701-708.
19. Bornet B, Branchard M (2001) Nonanchored inter simple sequence repeat ISSR markers: reproducible and specific tools for genome fingerprinting. Plant Molecular Biology Reporter 19: 209-215.
20. Vasconcelos Jr SS (2001) Análise da diversidade genética entre genótipos de Mamona (Ricinus communis – Euphorbiaceae) Através de Fingerprinting de DNA com marcadores AFLP e ISSR. 2011. Master Thesis – Universidade Federal do Pernambuco, Recife 61.
21. Roldán-Ruiz I, Dendauw J, Van Bockstaele E, Depicker A, De loose M (2000) AFLP markers reveal high polymorphic rates in ryegrasses (Lolium spp.). Molecular Breeding 6: 125-134.
22. Varshney RK, Chabane K, Hendre PS, Aggarwal RK, Graner A (2007) Comparative assessment of EST-SSR, EST-SNP and AFLP markers for evaluation of genetic diversity and conservation of genetic resources using wild, cultivated and elite barleys. Plant Science 173: 638-649.
23. Prevost A, Wilkinson MJ (1999) A new system of comparing PCR primers applied to ISSR fingerprinting of potato cultivars. Theoretical and Applied Genetics 98: 107-112.
24. Ayres M, Ayres JM, Ayres DL, Santos AS (2007) Bioestat 5.0: aplicações estatísticas nas áreas das ciências biológicas e médicas. 5 ed., Belém, Para. Sociedade Civil Mamirauá.
25. Laurentin H, Karlovsky P (2007) AFLP fingerprinting of sesame (Sesamum indicum L.) cultivars: identification, genetic relationship and comparison of AFLP informativeness parameters. Genetic Resources and Crop Evolution 54: 1437-1446.
26. Ruas PM, Ruas CF, Rampim L, Carvalho VP, Ruas EA, et al. (2003) Genetic relationship in Coffea species and parentage determination of interspecific hybrids using ISSR (Inter Simple Sequence Repeat) markers. Genetics and Molecular Biology 26: 319-327.
27. Pharmawati M, Yan G, Finnegan PM (2005) Molecular variation and fingerprinting of Leucadendron cultivars (Proteaceae) by ISSR markers. Annals of Botany 95: 1163-1170.
28. Oliveira LO, Venturini BA, Rossi AAB, Hastenreiter SS (2010) Clonal diversity and conservation genetics of the medicinal plant Carapichea ipecacuanha (Rubiaceae). Genetics and Molecular Biology 33: 86-93.
29. Saleh B (2011) Efficiency of RAPD and ISSR markers in assessing genetic variation in Arthocnemum macrostachyum (Chenopodiaceae). Brazilian Archives of Biology and Technology 54: 859-866.
30. Xiao LQ, GE XJ, Gong X, Hao G, Zheng SX (2004) ISSR Variation in the endemic and endangered plant Cycas guizhouensis (Cycadaceae). Annals of Botany 94: 133-138.
31. Velu D, Ponnuvel KM, Muthulakshmi M, Sinha RK, Qadri SM (2008) Analysis of genetic relationship in mutant silkworm strains of Bombyx mori using inter simple sequence repeat (ISSR) markers. Journal of Genetics and Genomics 35: 291-297.
32. Pradeep AR, Jingade AH, Singh CK, Awasthi AK, Kumar V, et al. (2011) Genetic analysis of scattered populations of the Indian eri silkworm, Samia cynthia ricini Donovan: differentiation of subpopulations. Genetics and Molecular Biology 34: 502-510.
33. Kareem VK, Rajasekharan PE, Ravish BS, Mini S, Sane A, et al. (2012) Analysis of genetic diversity in Acorus calamus populations in south and north east India using ISSR markers. Biochemical Systematics and Ecology 40: 156-161.
34. Nan P, Shi S, Peng S, Tian C, Zhong Y (2003) Genetic diversity in Primula obconica (Primulaceae) from central and south-west China as revealed by ISSR markers. Annals of Botany 91: 329-333.
35. Souza VQ, Pereira AS, Kopp MM, Coimbra JLM, Carvalho FIF, et al. (2005) Dissimilaridade genética em mutantes de aveia tolerantes e sensíveis a ácidos orgânicos. Bragantia 64: 569-575.
36. Bandeira AJ, Deimling LI, Georg-Kraemer JE (2010) Variabilidade genética do jaborandi (Pilocarpus pennatifolius Lemaire; Rutaceae) em populações naturais da região noroeste do Rio Grande do Sul. Revista de Iniciação Científica da ULBRA 1: 43-52.
37. Huang W, Zhao XY, Zhao X, Zhao H, Wang S, et al. (2011) A combined approach using ISSR and ITS analysis for the characterization of Artemisia halodendron from Horqin sandy land, northern China. Biochemical Systematics and Ecology 39: 346-351.
38. Kalpana D, Choi SH, Choi TK, Senthil K, Lee YS (2011) Assessment of genetic diversity among varieties of mulberry using RAPD and ISSR fingerprinting. Scientia Horticulturae 134: 79-87.
39. Verma S, Rana TS (2011) Genetic diversity within and among the wild populations of Murraya koenigii (L.) Spreng., as revealed by ISSR analysis. Biochemical Systematics and Ecology 39: 139-144.
40. Pecina-Quintero V, Anaya-López JL, Núnez-Colín CA, Zamarripa-Colmenero A, Montes-García N, et al. (2013) Assessing the genetic diversity of castor bean from Chiapas, México using SSR and AFLP markers. Industrial Crops and Products 41: 134-143.
41. Tatikonda L, Wani SP, Kannan S, Beerelli N, Sreedevi TK, et al. (2009) AFLP-based molecular characterization of an elite germplasm collection of Jatropha curcas L., a biofuel plant. Plant Science 176: 505-513.

Sensible and Latent Heat Storage Fluxes within the Canopy Air-Space in the Amazon Rainforest

Vicente de P. R. da Silva[1]*, Rafaela S. R. Almeida[2], Vanessa de A. Dantas[1], Antonio C. L. da Costa[3], Vijay P. Singh[4] and Glayson F. B. das Chagas[1]

[1]*Federal University of Campina Grande/Center of Technology and Natural Resources/Academic Unity of Atmospheric Sciences, Av. Aprígio Veloso, 882, Bodocongó, 58109 970, Campina Grande, PB, Brazil*
[2]*State University of Paraíba, R. Domitila Cabral de Castro, 148-572, Bodocongó, 58429-570 Campina Grande, PB, Brazil*
[3]*Federal University of Pará, Av. Augusto Corrêa 1, Guamá, CEP 66075-110, Belém, PA, Brazil*
[4]*Department of Biological and Agricultural Engineering & Department of Civil & Environmental Engineering, Texas A&M University, College Station, TX 77843-2117, USA*

Abstract

Ecosystem–atmosphere exchanges of energy display considerable variability over a range of temporal and spatial scales. This study evaluated the seasonal and annual variations in sensible and latent heat storage fluxes within the canopy air-space in the Amazon rainforest. The data for this study were obtained from the "Long-term drought impact on water and carbon dioxide fluxes in Amazonian Tropical Rainforest Experiment (ESECAFLOR)" which is a subproject of Large Scale Biosphere Atmosphere Experiment in Amazon forest (LBA), carried out in *terra firma* rainforest in Caxiuanã National Forest, Pará, Brazil. The data sets also included observations to obtain the sensible and latent heat storage fluxes within forest collected throughout the 1-year period. The vapor pressure and air temperatures were obtained for each 8 m interval from the surface to 32 m. Results indicated that the cumulative sensible heat storage flux in the Amazon rainforest canopy was 167.9 Wm^{-2} in 2008 and the average daily magnitude was always low for the same period. The latent heat storage flux (ranging from -32.7 to -10 Wm^{-2}) was more influenced by rainfall producing high humidity.

Keywords: Energy fluxes; Energy balance; Air temperature gradient

Introduction

The amount of water intercepted by the Amazon forest canopy during rains is responsible for the replacement of considerable amount of water vapour in the atmosphere, contributing to the local water balance. The biosphere-atmosphere interaction processes in the Amazonian region have been investigated over the last years in order to evaluate their importance for the local and global climate [1]. The large extent of vegetation covering the Amazon forest produces the highest source of energy to the atmosphere from the continental origin. Due the significance of the Amazon forest to biodiversity and global carbon cycle, several studies have focused on the National Forest Reserve Caxiuanã in the framework of the LBA (Large Scale Biosphere Atmosphere Experiment in Amazonia) project [1-4].

Although changes in the biomass of Amazonian region forest represent an important component of the global carbon cycle, the biomass of these forest remains poorly quantified [5]. Other important studies aimed at quantifying the forest biomass in the Brazilian Amazon include those of Brown et al. [6], Sales et al. [7], Fearnside et al. [8], Nogueira et al [9], Alves et al. [10], among others. Researches on the effect of drought episodes on heat storage fluxes within the canopy air-space in the Amazon rainforest have just started. Although changes in the biomass of Amazon region forest represent an important component of the global carbon cycle, the biomass of these forest remains poorly quantified [5]. Using through fall exclusion experiments in the Amazon rainforest, Fisher et al. [11] and Costa et al. [12] investigated forest transpiration by the sap flow method in Caxiuanã National Forest. Fisher et al. [13] concluded that the forest was not able to withstand a 50% reduction in rainfall over 1-2 years without impacting the canopy gas exchange, while Costa et al. [12] found a decrease of 68% in mean transpiration of *E. Coriacea* from control plot to rainfall exclusion plot. The objective of this study was to analyze the surface energy balance, particularly the sensible and latent heat storage fluxes within the canopy air-space Amazon rainforest.

Materials and Methods

Experimental site

This study was carried out at Ferreira Penna Scientific Station (FPSS) in the Caxiuanã National Forest (CNF) in Pará State (Latitude: 1°42'30"S, Longitude: 51°31'45"W and Altitude: 62 m above sea level). The ECFPn is located in the 300000 hectare Caxiuanã National Forest, about 400 km to the west of Pará capital city of Belém, Brazil (Figure 1). The region has a well-preserved forest with a canopy of 35 m high. The Amazon is covered predominantly by moist dense tropical forest, but with several other vegetation types, including savannas, montane forests, open forests, floodplain forests, grasslands, swamps, bamboos, and palm forests.

The predominant trees species in the landscape are *Eschweilera coriacea* (White Matá-matá), *Voucapoua americana* (Acapu), *Rotium pallidum* (White Pitch), *Dinizia excelsa* (angelim-vermelho), *Marmaroxylon racemosum* (angelimrajado), *Couratari guianensis*

***Corresponding author:** Vicente de P. R. da Silva, Federal University of Campina Grande/Center of Technology and Natural Resources/Academic Unity of Atmospheric Sciences, Av. Aprígio Veloso, 882, Bodocongó, 58109 970, Campina Grande, PB, Brazil; E-mail: vicente@dca.ufcg.edu.br

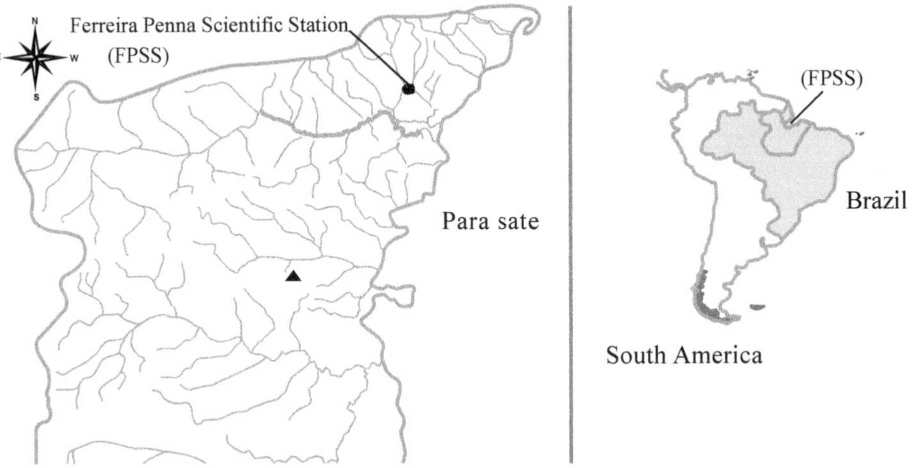

Figure 1: Localization of Ferreira Penna Scientific Station (FPSS) in the Caxiuanã National Forest (CNF) in Pará State (Latitude: 1°42'30"S, Longitude: 51°31'45"W and Altitude: 62 m above sea level).

(tauari), *Bucheniavia grandis* (tanimbuca), *Swartzia racemosa* (pitaíca) and *Dipteryx odorata* (cumaru). The forest is a lowland *terra firme* rainforest. The mean annual temperature is 25.78°C and the mean annual rainfall is 2272 mm, with a dry season when only 555 mm of rainfall occurs on average [13]. The wet season is from January to June while the dry season is from July to December. Most soils are yellow Oxisols (Brazilian classification Latossolo), but there are large differences in texture. The water table has been observed at a soil depth of 10 m during the wet season and the site elevation is 15 m above the river level [13].

Our study analyzed observations of 1-year (2008) of canopy heat storage fluxes. This period is enough to characterize the energy fluxes during rainy and dry seasons in the study area.

Data descriptions

The most predominant tree species in the experimental plot were *Escheweleira, Licania octandra, Lecythis, Pouteria decorticans, Swartzia racemosa, Rinoria guianensis* and *Vouacapoua americana*. On the other hand, the tree species in Plot B included *Escheweleira coriacea, Manilkara bidentata, Swartzia racemosa e Tetragastris panamensis*. The forest tree species diversity varied from 150 to 160 trees per hectare and the individual density ranged from 450 to 550 trees ha^{-1}. The net radiation was measured with a net radiometer (CNR1, Kipp and Zonen, Delft, the Netherlands) and the vapor pressure was calculated based on wet and dry bulb temperatures for each 8 m interval from the surface to 32 m. All data were collected using a data logger (CS10X, Campbell Scientific) for every 10 s and averaged over each 30 min period. The data sets also included observations to obtain the canopy heat energy flux within forest collected throughout the 1-year period. Energy storage flux can be a significant component of the sub-diurnal surface energy budget in a tall forest, owing to the large volumes of air and biomass in the canopy.

Canopy heat storage flux

The large height of native forest with a large volume of air and biomass in the canopy covering water supply catchments in the Amazon basin means that the change in canopy heat storage flux through the change in moisture content of the canopy air play an important role in the energy balance of the ecosystem. The sensible heat storage flux in the canopy air-space (ΔS_h) was obtained following McCaughey [14] as:

$$\Delta S_h = \int_0^{z_r} \rho c_p \frac{\partial T}{\partial t} dz \cong \rho c_p \sum_{i=1}^{n} \left(\frac{\Delta T_a}{\Delta t} \Delta z_i \right) \quad (1)$$

where ρ is the air density, c_p is the specific heat, z_r is the height of 32 m, T is the air temperature (°K) below r_r, and T_a is a representative layer-average of T in each layer (°K). Since ρ and c_p are approximately constant within the canopy layer, the ratio $\Delta T_a/\Delta T$ can be calculated as an average of measurements at several levels in the canopy [15]. Similarly, the latent heat storage flux in canopy air-space (ΔS_w) was calculated as [14]:

$$\Delta S_w = \int_0^{z_r} \rho L_e \frac{\partial e_i}{\partial t} dz \cong \rho L_e \sum_{i=1}^{n} \left(\frac{\Delta e_i}{\Delta t} \Delta z_i \right) \quad (2)$$

where L_e is the latent heat of vaporization and *e* is the vapour pressure, calculated from Ta and relative humidity measurements. Both ΔS_h and ΔS_h were calculated in four levels up to 32 m.

Results and Discussion

Climate data

Temporal patterns of meteorological variables throughout the 1-year of experimental period in 2008 are shown in Figure 2. The total rainfall of 2230 mm was close to the long-term mean of 2272 mm [11]. The amount of low precipitation from July to November coincided with a decrease in relative humidity, but differed markedly in air temperature and wind speed. Higher air temperature and lower relative humidity through the dry season were mainly due to the small amount of rain as well as the increase in global solar radiation and consequently in the net radiation. Therefore, the climate did not show any remarkable differences between other years and the experimental period in 2008. The global solar radiation had a similar seasonal course to that of the net radiation with maximum and minimum in dry and wet seasons, respectively.

Canopy heat storage flux

Sensible heat flux: The sensible heat storage flux values were generally low through the experimental period ranging from 4.94 to -2.22 Wm^{-2} (Figure 3). Although the canopy heat storage was important for diurnal and seasonal time scales, their seasonal integration accounted for only a small portion of the net radiation. The average canopy heat storage flux within the canopy air-space over the experimental period in 2008 was 0.46 Wm^{-2}. The daily total heat storage flux resulted in the annual total canopy heat storage flux of 167 Wm^{-2} $year^{-1}$.

Analyzing seasonal variations of energy and water vapour fluxes above a tropical seasonal rainforest in China, Junxia et al. [16] also found low values of the heat storage flux ranging between 0.9 and 0.8% of the net radiation. The maximum and minimum values in the sensible heat storage flux occurred in the dry and rainy seasons, respectively, because the differences of temperature within the canopy are mostly driven by differences in radiation penetration to the ground, governed by canopy density.

The sensible heat storage flux was eventually an important sink or source of energy, with a typical daily maximum value of about 43 Wm^{-2} in dry season, but because the term usually changed signs between daylight and nighttime hours, the average daily magnitude was always low through the whole experimental period. Similar results were obtained by Wilson and Baldocchi [17] who analyzed the components

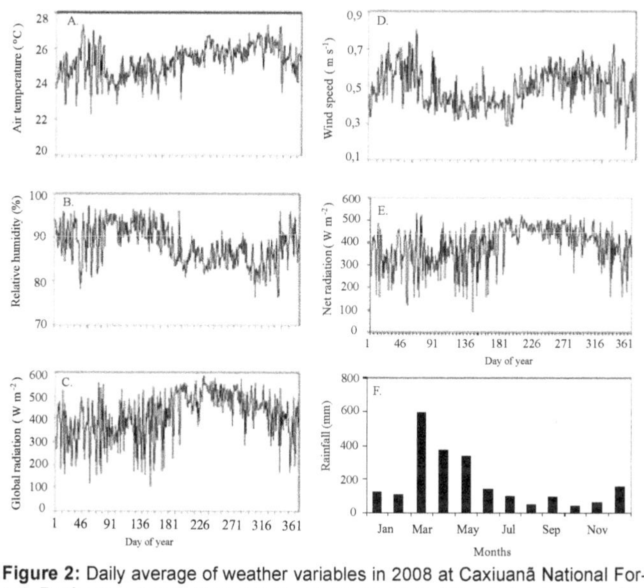

Figure 2: Daily average of weather variables in 2008 at Caxiuanã National Forest, Pará state, Brazil. Data include air temperature (A), relative humidity (B), global radiation (C), wind speed (D), net radiation (E) and rainfall (F).

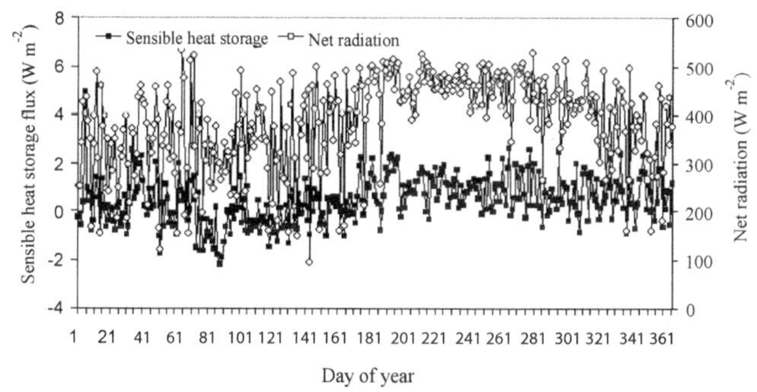

Figure 3: Seasonal sensible heat storage flux and net radiation during the experimental period in 2008 at Caxiuanã, PA, Brazil.

of surface energy balance over a broadleaf deciduous forest in North America. Results obtained in this study, however, were not consistent with that reported Moore and Fisch [18] and Silberstein et al. [19]. They found maximum absolute hourly canopy storage fluxes of around 80 Wm^{-2} for forests in the Amazonian tropical and Western Australia. This difference is probably because a good estimate of heat storage requires a large number of temperature and humidity measurements throughout the canopy air space, as well as temperatures for each of the biomass components (leaves, twigs, fruit, branches, stems and litter) [19]. The sensible heat storage flux in the surface-air space cannot be neglected on an hourly basis. However, it represents a little proportion of the net radiation on a daily basis. Similar results were obtained by Sánchez et al. [20] for hourly sensible heat storage flux using radiometric temperature observations for estimating energy balance fluxes above a boreal forest. The cumulative sensible heat storage flux in the wet season was 6% of that in the dry season. The lower value of sensible heat storage flux in the canopy air in January-June than that in June-December was a consequence of high rainfall and low vapour pressure deficit in the wet season of that year. Wang et al. [21] also found similar values who analyzed seasonal variations in energy and water fluxes in a pine forest in Finland.

Seasonal variation of sensible heat storage flux and net radiation in dry and wet seasons are shown in Figure 4 which indicates a close link between canopy heat storage and net radiation throughout the wet season in which the sensible heat storage flux and net radiation were on average -0.08 and 340 Wm^{-2}, respectively (Figure 4A). However, results of this study also indicate that these variables do not exhibit a similar performance throughout the dry season (July to November). For this season, the sensible heat storage flux and net radiation were on average 0.83 and 430.7 Wm^{-2}, respectively (Figure 4B). As expected, the net radiation was greater in the dry season than in the wet season due to the increase in the atmospheric demand and this effect was estimated to have reduced the sensible heat storage flux to 9.4%. The mean values of the net radiation in the wet season decreased 21% when compared to the dry season. This difference in the net radiation between seasons was the main source of variability in the sensible heat storage flux. Since the net radiation is mainly affected by global solar radiation, the seasonal change in cloud cover was the main controlling factor of the net radiation.

The transition from dry to wet season coincided with a marked reduction in the incoming radiation. The seasonal net radiation values presented here are greater than those reported by Rocha et al. [22] for the Amazon region, Santarém, PA. This difference between the mean values of net radiation can be attributed to the variability in cloud cover in the region. The net radiation was the most important source of energy, with typical daily maximum values for both seasons between 520-540 Wm^{-2}, while the daily minimum values varied from 120.8 to 263.4 Wm^{-2}. Similar results were obtained by von et al. [23] who analyzed the scale variability of atmospheric surface layer fluxes of energy and carbon over a tropical rainforest in southwest Amazon. During the experimental period in 2008 the proportion of energy storage within the canopy was, on average, only 1.23% of the net radiation. The very low consumption of net radiation during the day is attributed to the widely open canopy and high wind speed regime.

The sensible heat storage flux in 2008 was 167.93 Wm^{-2}, which equals 14.5 $MJm^{-2}year^{-1}$. Similar values were obtained by Oliphant et al. [24] who analyzed heat storage and energy balance fluxes for a temperate deciduous forest in south-central Indiana, USA. They found that the annual storage heat balance for the three complete years showed a small but consistent deficit averaging about 16.18 $MJm^{-2}year^{-1}$. This study indicates that the cumulative values of energy storage within the canopy air-space for wet and dry periods were 76.1 and 91.8 Wm^{-2}, respectively. This means that less net radiation was available for turbulent energy dissipation within the canopy during the wet season.

Latent heat flux: Figure 5 shows the seasonal latent heat storage flux within the canopy air-space in 2008, indicating that the seasonal course of latent heat storage followed a pattern inverse of the net radiation during the whole year. The storage flux had a maximum absolute value of 32.7 Wm^{-2}, while the net radiation had a lowest value of 171.2 Wm^{-2} in the wet season (day of year 166). On the other hand, the net radiation had a maximum value of 526.7 Wm^{-2} and the latent heat storage flux had a lowest value of 12.9 Wm^{-2} in the dry season (day of year 280). Therefore, the latent heat storage flux within canopy is most likely due to rainfall amount summing to 1652 mm during the wet season.

The mean values of air temperature and relative humidity in 2008 was 25°C and 88%, respectively. On the other hand, the cumulative rainfall during the dry season was 477.8 mm. The latent heat storage flux was influenced by the low atmospheric demand but mainly by the high amount of water vapour in the canopy air-space through the wet season. The net radiation followed the temporal variation in global solar radiation but became more irregular with depth in the forest because of the highly variable penetration of beam radiation in space and time. The solar radiation pattern into the forest canopy is a fundamental process of energy exchange which regulates the availability of water for evaporation. Although solar elevation continues to rise each day until the summer solstice, the amount of net radiation and its variability affect water storage within the canopy forest.

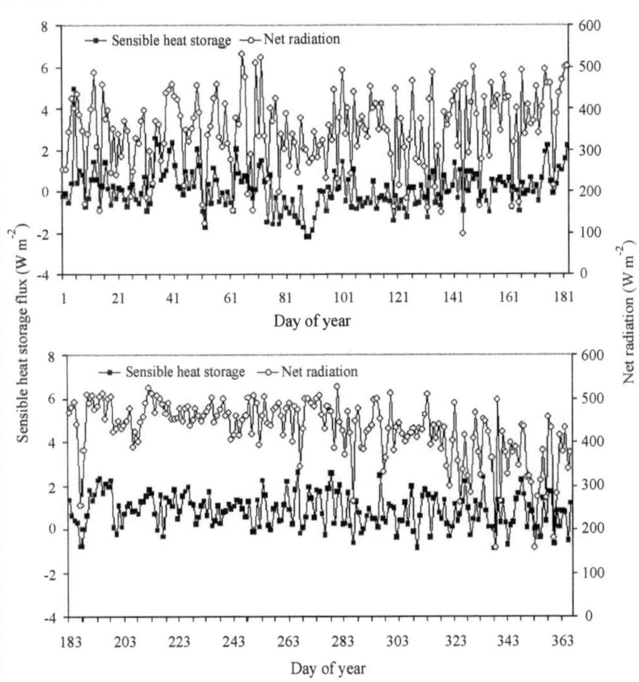

Figure 4: Comparison between mean daily patterns of sensible heat storage flux and net radiation in wet (A) and dry seasons (B) in Caxiuanã, Pará, Brazil.

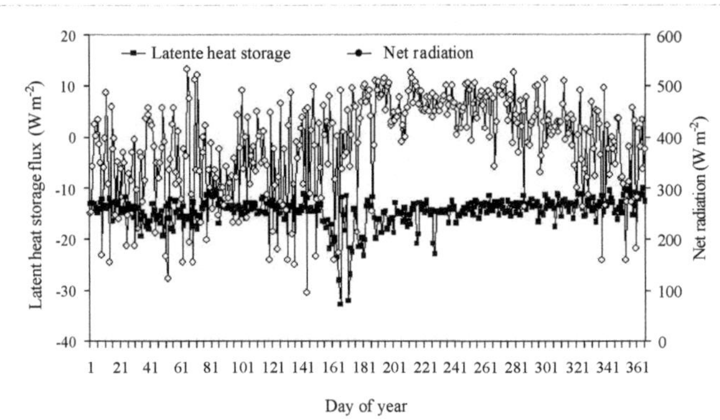

Figure 5: Seasonal latent heat storage flux and net radiation during the experimental period in 2008 at Caxiuanã, PA, Brazil.

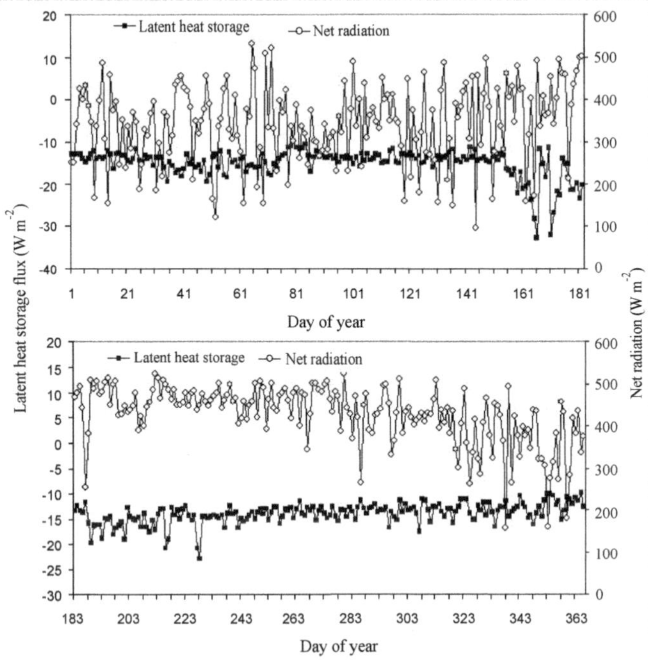

Figure 6: Seasonal latent heat storage flux and net radiation in Caxiuanã, PA, in 2008.

Despite a reduction in the net radiation from the dry season to the wet season of around 20%, on average, the latent heat storage flux had similar values, 4.4% of the net radiation for the wet season and 3.3% for the dry season (Figure 6). Although the maximum values of latent heat storage flux were nearly the same during both seasons, the minimum values declined by 30% from the wet to the dry season. The average annual latent heat storage flux was nearly 15 Wm^{-2} for both periods, suggesting ample water vapour storage within the canopy air during the whole year. The daily total latent heat storage flux resulted in the seasonal total latent heat storage of 2750.80 and 2581.19 Wm^{-2} for wet and dry seasons within canopy air-space, respectively. The net radiation, consumed by the latent heat storage within the forest canopy, varied from 6.1 to 27% during the experimental period. The net radiation was greater in the dry season than the wet season, but this did not seem to have increased the latent heat storage.

Despite the net radiation pattern into forest canopy being a fundamental process of energy exchange which regulates the availability of water for evaporation, the difference in the net radiation among months was only a minor source of variability in the latent heat storage, while the high annual rainfall interaction with forest canopy was responsible for transport of water within forest.

Conclusions

Although sensible heat storage flux is important on a diurnal time scale, its annual value integrated within the canopy air space is rarely more than 1% of the net radiation. The sensible heat storage flux in the surface-air space can be neglected on a daily basis, but not on an hourly basis. However, the latent heat storage flux represents a considerable proportion of the net radiation on both hourly and daily bases. The latent heat storage flux within canopy is most likely equal to the amount of water vapour in canopy air-space, summing to 1652 mm during the wet season. The difference in the net radiation among months, because of the highly variable penetration of beam radiation, is only a minor source of variability of latent heat storage, while the high annual rainfall interaction with forest canopy is responsible for the high quantity of water transported within forest. Despite a reduction of around 20% in the net radiation from the dry season to the wet season, maximum values of latent heat storage flux are nearly the same for both seasons. However, their minimum values are reduced by 30% from the wet season to the dry season.

References

1. Gonçalves FLT, Martins JA, Silva Dias MA (2008) Shape parameter analysis using cloud spectra and gamma functions in the numerical modeling RAMS during LBA Project at Amazonian region, Brazil. Atmos Res 89: 1-11.

2. Behling H, Hooghiemstra H (2000) Holocene Amazon rainforest–savanna dynamics and climatic implications: high-resolution pollen record from Laguna Loma Linda in eastern Colômbia. J Quaternary Sci 15: 687-695.

3. Iwata H, Malhi Y, von Randowd C (2005) Gap-filling measurements of carbon dioxide storage in tropical rainforest canopy airspace. Agr Forest Meteorol 132: 305-314.

4. Sotta ED, Veldkamp E, Guimaraes BR, Paixão RK, Ruivo MLP, et al. (2006) Landscape and climatic controls on spatial and temporal variation in soil CO2 efflux in an Eastern Amazonian Rainforest, Caxiuanã, Brazil. Forest Ecol Manag 237: 57-64.

5. Keller M, Palace M, Hurt G (2001) Biomass estimation in the Tapajós National Forest, Brazil examination of sampling and allometric uncertainties. Forest Ecol Manag 154: 371-382.

6. Brown FI, Martinelli LA, Thomas WW, Moreira MZ, Ferreira CAC, et al. (1995) Uncertainty in the biomass of Amazonian forests: An example from Rondônia, Brazil. Forest Ecol Manag 75: 175-189.

7. Sales MH, Souza Jr CM, Kyriakidis PC, Roberts DA, Vidal E (2007) Improving spatial distribution estimation of forest biomass with geostatistics: A case study for Rondônia, Brazil. Ecol Model 205: 221-230.

8. Fearnside PM, Barbosa RI, Graça PMLA (2007) Burning of secondary forest in Amazonia: Biomass, burning efficiency and charcoal formation during land preparation for agriculture in Apiaú, Roraima, Brazil. Forest Ecol Manag 242: 678-687.

9. Nogueira EM, Fearnside PM, Nelson BW, Barbosa RI, Keizer EWH (2008) Estimates of forest biomass in the Brazilian Amazon: New allometric equations and adjustments to biomass from wood-volume inventories. Forest Ecol Manag 256: 1853-1867.

10. Alves LF, Vieira SA, Scaranello MA, Camargo PB, Santos FAM, et al. (2010) Forest structure and live aboveground biomass variation along an elevational gradient of tropical Atlantic moist forest (Brazil). Forest Ecol Manag 260: 679-691.

11. Fisher RA, Williams M, Vale RL, Costa ACL, Meir P (2006) Evidence from Amazonian forests is consistent with isohydric control of leaf water potential. Plant Cell Environ 29: 151-165.

12. Costa RF, Silva VPR, Ruivo MLP, Meir P, Costa ACL, et al. (2007) Transpiração em espécie de grande porte na Floresta Nacional de Caxiuanã. Pará Agriambi 11: 180-189.

13. Fisher RA, Williams M, Costa AL, Malhi Y, Costa RF, et al. (2007) The response of an Eastern Amazonian rain forest to drought stress: results and modelling analyses from a throughfall exclusion experiment. Glob Change Biol 13: 2361-2378.

14. Mccaughey JH (1985) Energy balance storage terms in a nature mixed forest at petawawa, Ontario – a case study. Bound Lay Meteorol 31: 89-101.

15. Vogt KA, Grove M, Asbjornsen H, Maxwell KB, Vogt DJ, et al.(2002) Linking Ecological and Social Scales for Natural Resource Management. Cambridge University Press 1-45.

16. Junxia D, Yiping Z, Guirui Y, Shuangju Z, Qinghai S (2007) Interannual and seasonal variations of energy and water vapor fluxes above a tropical seasonal rain forest in Xishuangbanna, SW China. Acta Ecologica Sinica 27: 3099-3109.

17. Wilson KB, Baldocchi DD (2000) Seasonal and interannual variability of energy fluxes over a broadleaved temperate deciduous forest in North America. Agr Forest Meteorol 100: 1-18.

18. Moore CJ, Fisch G (1986) Estimating heat-storage in amazonian tropical forest. Agr Forest Meteorol 38: 147-168.

19. Silberstein R, Held A, Hatton T, Viney N, Sivapalan M (2001) Energy balance of a natural jarrah (Eucalyptus marginata) forest in Western Australia: measurements during the spring and summer. Agr Forest Meteorol 109: 79-104.

20. Sánchez JM, Caselles V, Niclòs R, Coll C, Kustas WP (2009) Estimating energy balance fluxes above a boreal forest from radiometric temperature observations. Agr Forest Meteorol 149: 1037-1049.

21. Wang KY, Kellomaki S, Zha T, Peltola H (2004) Seasonal variation in energy and water fluxes in a pine forest: an analysis based on eddy covariance and an integrated model. Ecol Modell 179: 259-279.

22. Rocha HR, Goulden ML, Miller SD, Menton MC, Pinto LDVO, et al. (2004) Seasonality of water and heat fluxes over a tropical Forest in Eastern Amazonia. Ecol Appl 14: S22-S32.

23. Von Randow C, Sa LDA, Gannabathula PSSD, Manzi AO, Arlino PRA, et al. (2002) Scale variability of atmospheric surface layer fluxes of energy and carbon over a tropical rain forest in southwest Amazonia. I. Diurnal conditions. J Geophys Res 107: 1-12.

24. Oliphant AJ, Grimmond CSB, Zutter HN, Schmid HP, Su HB, et al. (2004) Heat storage and energy balance fluxes for a temperate deciduous forest. Agr Forest Meteorol 126: 185-201.

17

Isozymatic Characterization of Accessions of *Arundinaria falcata* (Nees)

Chandrakant Tiwari* and Meena Bakshi

Plant Physiology Discipline, Botany Division, Forest Research Institute, Dehradun, India

Abstract

Bamboos are arborescent grasses belonging to the family *Poaceae*. To study the genetics of bamboo specially, one has to face enormous difficulties such as rare flowering and we cannot make hybridizations as easily as like other plant taxa. Unlike other tropical bamboos, studies on hill bamboos are scarce. *Arundinaria falcata*, an important hill bamboo, acts as soil stabilizer in hilly terrain. This study assessed the genetic diversity of 10 accessions of *A. falcata* collected from different localities of Garhwal Himalayas (India) and established at Hill bamboo Germplasm at Khirsu (India), using isozyme marker with four enzyme system (peroxidase, esterase, malate dehydrogenase and malic enzyme). Isozymatic analyses were performed with polyacrylamide gels (one system), bands were scored as binary data. Cluster analyses were conducted, using Jaccard´s similarity coefficient and UPGMA method. Very high degree of similarity was reported i.e. 63- 94% among different accessions. Dendrogram revealed two major clusters with three (A8- A10) and seven (A1-A7) accessions respectively. The results obtained inferred low genetic diversity in the species and urgent need of the *in situ* conservation of the natural genetic resources of the *A. falcata* species.

Keywords: Genetic diversity; dendrogram; Polyacrylamide; Jaccard's coefficient

Introduction

Genetic conservation programmes are directed towards the long-term preservation of genetic resources either *in situ* or *ex situ* so that the potential for continuing evolution or improvement could be sustained. *In situ* conservation includes the organization and/or servicing of natural supplies where species are permitted to stay in maximum environments with the lowest of management. On the other hand, *ex situ* conservation includes the use of botanic landscapes, field farms, seeds shops and gene financial banks and germplasm.

The characterization of germplasm is required to maintain identity and purity for proper conservation and management. Various breeding strategies of plants are based on the knowledge of germplasm. Germplasm characterization is thus an important link between the conservation and utilization of plant genetic resources for improvement.

Arundinaria falcata vs. Gol Ringal, is socio- economically and ecologically, an important member of hill bamboo, present in Uttarakhand Garhwal Himalayas (India). Like other bamboo species, Ringal (hill bamboo) also has erratic and long flowering cycles. *Arundinaria falcata* flowers irregularly at times, gregarious over large areas, while few culms may be found in flowers almost every year. After seeding the clumps died [1].

Due to over usage of genetic resources and heavy extraction of the material from natural forests, the species are depleting at an alarming rate, gregarious flowering in turn intensifying the depletion of Ringal resources since flowering result into death of entire clumps following seeding. If a single regeneration event coincides with unsuitable conditions, the entire population might lose its chance to reproduce. Heavy forest degradation and agricultural encroachment, forest fires, human impact, heavy grazing pressures, etc. narrows the genetic base and eventually degradation of the gene pool of these species.

Isozymes are commonly used as biochemical markers as detectably different enzymes, which catalyze the same reaction [2]. Enzymatic analyses are added tools for detecting diversity [3]. The relationship between observed phenotypes and unobserved genotypes is simpler and better understood for electrophoretic evidence. Allozymes are the biochemical consequence of the substitution, deletion, or addition of amino acids in the polypeptides that comprise the enzymes and they can be distinguished if these changes affect their electrophoretic migration [4]. Isozyme electrophoresis is used to describe population structure, breeding structure and gene flow; to know species boundaries, and to document adaptive differences in allozymes; and to investigate phylogenetic relationships, rates of evolution, origin of polyploid plants and ploidy levels [5].

The present work was therefore, undertaken to assess variability among accessions of *A. falcata* growing in a Germplasm at Khirsu (Pauri) through morphological and genetic tools with the objective i.e. to study the genetic polymorphism in four hill bamboo species on the basis of morphological traits.

Materials and Methods

Sample collection

The young leaves of selected accessions were collected from Ringal Germplasm, Khirsu, tagged properly in polybags and stored in ice bucket till they were brought to Plant Physiology Laboratory (FRI, Dehradun). The leaves were stored at -20°C in freezer (vest frost DFS 345) till use.

Extraction of enzymes

Two grams of young leaves were macerated to powder with

***Corresponding authors:** Chandrakant Tiwari, Plant Physiology Discipline, Botany Division, Forest Research Institute, Dehradun, India
E-mail: tiwari.chandrakant08@gmail.com

Weimin Xi, Department of Biological and Health Sciences,Texas A&M University
E-mail: weimin.xi@tamuk.edu

liquid Nitrogen in a mortar – pestle and then 0.1 gm PVP and 5 ml of extraction buffer was added. The extraction buffer consisted of 1 M Sucrose, 0.2 M Tris and 0.056 M β - Marcaptoethanol. The volume was made up to 100 ml with distilled water and pH was adjusted at 8.5. The macerates were centrifuged at 1500 rpm for 20 minutes at 4°C to collect the supernatants.

Casting of polyacrylamide gels

Separating gel: Two slab gel plates were assembled in a casting tub using 1.5 mm thick spacers. Required amount of stock solutions were mixed thoroughly excepting freshly prepared Ammonium Per Sulphate (APS) and TEMED (N'N'N'N' Tetra Methyl Ethylene Diamine), which were mixed just prior of pouring the mixture into the slab sandwich. The mixture was pipetted into the glass up to a level about 4 cm from the top. The gel solution was allowed to polymerize at room temperature.

Staking gel: Measured amount of stock solutions were mixed for stacking gel. The staking gel mixture was gently poured into the glass slab over the polymerized separating gel. A comb was inserted into the glass slab to make wells in the staking gel for loading the sample materials before it polymerization. The staking gel was allowed to polymerize for half an hour.

Electrophoresis of the enzyme extracts: After the complete polymerization of the staking gel, the comb was removed from the staking gel. Both the upper and lower tanks of the electrophoretic apparatus were filled with the Tris- Glycine (pH 8.3) running buffer and the lower surface of the separating gel should remain immersed in the running buffer.

25 µl of the enzyme extracts of the different accessions were loaded in the wells of the staking gel. A constant current of 30 mA was applied across 1.5 mm thick gel. Initially the current applied was 20 mA, but increased to 30 mA when the dye front entered into the separating gel. The electrophoresis was performed in a cold chamber having a constant temperature of 4° C. With the completion of electrophoresis, dye approached the end of the gel, the gel was removed and stained as per staining protocol of different enzymes.

Staining of the gel

Peroxidase (PRX, EC 1.11.1.7): For the staining of peroxidase isozymes, staining solution was prepared with 100 ml of redistilled water and 1 ml solution of acetone with 0.1 gm of the Orthodiacidine. Immediately after putting the gel into the above solution, 0.4 ml of H_2O_2 was added over the gel. The gel was agitated gently till brown colored bands were visible.

Esterase (EST, EC 1.1.1.1): D' α- Napthyl Ester, α- Napthol + Salt, α- Napthol + Fast Blue RR, Colour Precipitate.

Gel was kept in 50 ml of Phosphate A Buffer; pH 7.0 (Add 3.12 gm of $NaH_2PO_4.2H_2O$ and 2.8 4gm of Na_2HPO_4 in 1000 ml of distilled water). Further, 1 ml of α- Napthyl Acetate and 20 mg of Fast Blue RR stain was added in the gel. The gel was incubated at room temperature until bands appear. The staining solution was replaced with fixative quickly.

Malate dehydrogenase (MDH, EC 1.1.1.37): Gel was kept in 50 ml of Tris Buffer; pH 8.5. Further, 70 mg of L- malic acid, mono-Na salt, 0.5 ml of 1% NAD solution, 0.5 ml of 0.6% PMS solution (6 mg/ml) and 1 ml of 0.6% MTT solution (6 mg/ml) (10 mg/ml)and 20 mg of Fast Blue RR stain was added in the gel. The gel was incubated at room temperature until bands appear. The staining solution was replaced with fixative quickly. The gel was incubated at room temperature until bands appear. The staining solution was replaced with fixative quickly.

Malic enzyme (ME, EC 1.1.1.40): Gel was kept in 50 ml of Tris Buffer; pH 8.0. Further, 100 mg of L- malic acid, mono-Na salt, 1.0 ml of 1% NADP solution (10 mg/ml), 0.5 ml of 10% $MgCl_2$ solution (100 mg/ml), 0.4 ml of 0.6% PMS solution (6 mg/ml) and 1.5 ml of 0.6% MTT solution (6 mg/ml) was added as a staining solution in the above gel. The gel was incubated at room temperature until bands appear. The staining solution was replaced with fixative quickly. The gel was incubated at room temperature until bands appear. The staining solution was replaced with fixative quickly.

Fixation and storage of the gel

After appearance of the bands, the gel was transferred to a fixative solution that contained 100 ml of 50% methanol, 20 ml of 10% acetic acid and 40 ml of distilled water. The gel was stored at 4°C in refrigerator.

Statistical analysis of isozyme profile data

Each accession was considered a taxonomical operational unit (OTU) and isozyme bands were analysed as binary characteristics. The presence (coded as 1)/ absence (coded as 0) data matrix was prepared by scoring the gel. Relationships among the taxa were inferred with UPGMA clustering from Jaccard's similarity matrix of the bands. Data were analysed using the SAHN (Sequential Agglomerative Hierarchical and Nested) module of NTSYS-PC 2.11 (Numerical Taxonomy and Multivariate Analysis System) software program and dendrogram with bootstrap values was constructed by Darwin (version 5.0) software program [6].

Results

A total of 31 bands were recovered for esterases (EST), in 10 accessions with six loci viz. EST-1, EST-2, EST-3, EST-4, EST-5 and EST-6. Esterase loci (E1-E6) were numbered in sequence from the anode according to their decrease in negative charge. The locus EST-4 was monomorphic for all OTU's whereas EST- 5 was present in only accession A8. Locus EST-3 was present in three accessions viz. A2, A8 and A9.

With Peroxidase (PRX) resolution, a total of twenty two bands were resolved with maximum (three) loci viz. PRX-1, PRX-2 and PRX-3 for accessions A3 and A5. Loci PRX-1 and PRX-2 were monomorphic for all accessions. Locus PRX- 3 was found to be polymorphic and represented by the accessions A3 and A5.

Malate Dehydrogenase (MDH) enzyme system was monomorphic for all the accessions showing single locus for each. Overall, 10 bands were recovered in all the accessions.

Malic enzyme was represented with thirty seven bands with four loci viz. ME- 1, ME- 2, ME- 3 and ME- 4. Locus ME-2 was found to be polymorphic whereas remaining three loci i.e. ME-1, ME-2 and ME-3 were monomorphic for all accessions of *A. falcata*.

A total of 100 bands with fourteen putative loci in the four different enzyme assays were resolved with sufficient consistency and clarity. The number of polymorphic loci was 7. The polymorphism ranged from 25 - 83.33%. Maximum polymorphism (83.33%) was achieved through Esterase followed by Peroxidase (33.33%). and minimum for Malic enzyme (25%). MDH revealed only monomorphic bands (Table 1). Overall 50% polymorphism was depicted by ten accessions of *A. falcata* with four enzyme systems (Table 1).

Cluster analysis

Data scored for 10 accessions of *A. falcate* (Table 2) with four enzyme systems, were used to generate similarity coefficients. The genetic relatedness among the accessions revealed by unweighted pair group methods with arithmetic mean (UPGMA) cluster analysis is presented in dendrogram as below (Figure 1).

Cluster analysis was performed using isozyme data and it produced stable and consistent patterns. At 73% similarity level, the dendrogram revealed two major clusters; First cluster consisted of seven accessions (A1 to A7) representing different localities. The second cluster consisted of only three accessions viz. A8, A9 and A10 (Figure 2A-2D) with overall similarity of 82%, which was further sub divided into two sub clusters at overall similarity of 88%, consisting of A10 and A9 as one sub cluster and A8 as second sub cluster. Similarity between different accessions was laid between 63 to 94%. Maximum of 94% similarity was found between accessions A4- A5 and A4- A6 (Table 3).

Discussion

Isozymes electrophoresis is a powerful tool for population genetics [7] and the Sodium Dodecyl Sulphate-Native Poly Acrylamide Gel Electrophoresis technique is particularly considered as a reliable way because storage proteins are largely independent of environmental fluctuations [8,9]. The International Union for Protection of New Varieties of Plants (UPOV) has harmonized and adopted test guidelines and procedures for the use of isozyme electrophoresis as a characteristic for establishing uniqueness of plants [10]. Due to its simplicity and validity for describing genetic structure of groups of plants [11] resolving systematic relationships and inter and intra specific studies [12].

Isozyme profiling of four enzyme systems viz. Esterases, Peroxidases, Malate dehydrogenase and Malic enzyme were exploited to find out the diversity within and among accessions of Ringal species. The isozyme profiling revealed significant diversity among and within accessions of Ringal species. Isozyme analysis of limited selection of bamboos from five genera was reported by [13].

In present study, Esterase and Peroxidase were found to be more polymorphic than other two systems i.e. Malate dehydrogenase and Malic enzyme. The Esterase fingerprints were the most distinguishing of the 4 enzymes analysed and indicated that the species constitute a relatively heterogenous group. Esterase is useful as a diagnostic tool for cultivar identification in view of the extensive polymorphism for this enzyme [14]. Esterase banding patterns in different species had shown variable loci (isoforms) with high polymorphism [14] among all four enzyme system. In *A. falcata*, total six isoforms were resolved on the gel whereas. The Esterase (EST), isozyme pattern recorded a total of 31 bands produced from six loci viz. EST-1, EST-2, EST-3, EST-4, EST-5 and EST-6. Esterase loci (E1-E6) were numbered in sequence from the anode according to their decrease in negative charge. Accessions A8 and A9 showed maximum of five loci. The locus EST-4 was monomorphic for all OUT's whereas EST- 5 was present in only accession A8. EST-3 was present in three accessions viz. A2, A8 and A9. A total of 7 isoforms were observed for this particular enzyme. Esterase has also been studied in different plants such as *Nicotiana* [15], *Musa* [14], *Allium* [16], *Pinus* [17] and Cotton [18]. Along with esterase, peroxidase was also appreciably polymorphic with maximum number of isoforms i.e. three. Various numbers of loci from one locus (*Citrullus*) to thirteen loci (maize) was reported [19].

A biochemical study was done using esterase and Peroxidase and phenolics to study the different banding patterns in genera *Arundinaria, Pseudosasa, Semiarundinaria, Shibataea, Sinobambusa*and *Yushania* [20]. Phylogenetic study among 16 species of eight genera were conducted by Li Shengfeng [21] by using peroxidase and esteraseand concluded that isozymes are suitable markers to distinguish among species and genera.

The mitochondrial MDH isozyme (Malate Dehydrogenase, MDH) pattern were more conservative and uniform with one band (monomorphic) for all species studied hence, less differentiating. Overall, 10 bands were recovered in all the accessions. In the study on *Lythrumsalicaria* [22], eight bands were reported.

Malic enzyme was represented with thirty seven bands with four loci viz. ME- 1, ME- 2, ME- 3 and ME- 4. ME-2 was found to be polymorphic whereas remaining three loci were monomorphic for all accessions of *A. falcata*. In *Pinushalepensis* [23], this enzyme is expressed by one locus only.

Malic enzyme showed four to five isoforms but Malate dehydrogenase was totally failed to induce polymorphism among accessions. Similar banding pattern of MDH was reported by Siddiquee et al. [24].

Results of isozyme analysis revealed very high percentage of similarity due to less discriminatory power of the isozyme marker [25]. Cluster analysis revealed very close proximity among the accessions of individual species, since dendrogram of each individual species inferred grouping of accessions into two major clusters with very high similarity percent in marvel grass [25].

Results and findings of the study revealed high percentage of similarity or low variability which is the reflectance of deterioration of natural resources of *A. falcata*. Therefore, there is a need to develop *ex situ* as well as *in situ* approach for conservation and enrichment of genetic resources of the species and unscientific harvesting should be kept under control.

Acknowledgment

Our profound thanks to Forest Research Institute (FRI), Dehradun, India for providing the facilities. Authors are also thankful to the staff of Plant Physiology Discipline, Forest Research Institute (FRI), Dehradun, India for providing necessary assistance.

Locus name	Total no. of loci	No. of polymorphic loci	No. of monomorphic loci	Polymorphism (%)
EST	6	5	1	83.33
PRX	3	1	2	33.33
MDH	1	0	1	0
ME	4	1	3	25.0
Total	**14**	**7**	**7**	**50**

Table 1: Isozyme loci and the properties of resolved loci per accession of *A. falcate.*

Accession. No.	Place of Collection	Lattitude	Longitude	Altitude (m)
	Arundinaria falcata			
A1.	Hanumanchatti, Uttarkashi	30° 55.923 ' N	78° 23.951 ' E	2072.0
A2.	PithondiKhirsu. Pauri	30° 8.827' N	78° 46.472' E	1685.5
A3.	Maroda Village, Khirsu, Pauri	30° 8.827' N	78° 46.472' E	1685.5
A4.	Mandal I, Chamoli.	30° 27.593' N	79° 16.483' E	1499.5
A5.	Solan, H.P.	30° 54.238' N	77° 5.816' E	1446.0
A6.	Netwar, Uttarkashi	31° 3.789 ' N	78° 6.393 ' E	1524.0
A7.	Dewal, Chamoli	30° 3.665 ' N	79° 34.632 ' E	1488.9
A8.	Mandal II, Chamoli.	30° 27.593' N	79° 16.483' E	1499.5
A9.	Mandal III, Chamoli.	30° 27.593' N	79° 16.483' E	1499.5
A10.	Jarmola, Uttarkashi	30° 43.998 ' N	78° 26.394 ' E	1457.6

Table 2: Geographical details of accessions of *A. falcate.*

	A1	A2	A3	A4	A5	A6	A7	A8	A9	A10
A1	1.00									
A2	0.75	1.00								
A3	0.88	0.88	1.00							
A4	0.81	0.69	0.81	1.00						
A5	0.75	0.75	0.88	0.94	1.00					
A6	0.75	0.63	0.75	0.94	0.88	1.00				
A7	0.81	0.81	0.94	0.88	0.94	0.81	1.00			
A8	0.63	0.75	0.63	0.69	0.63	0.75	0.69	1.00		
A9	0.81	0.81	0.81	0.75	0.69	0.69	0.75	0.81	1.00	
A10	0.81	0.69	0.81	0.75	0.69	0.81	0.75	0.81	0.88	1.00

Table 3: Similarity matrix index showing relatedness among the accessions of *A. falcata* using Isozyme marker.

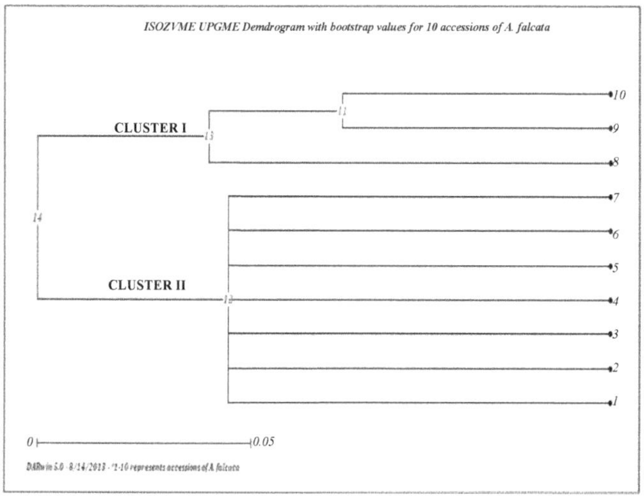

Figure 1: Illustration of UPGMA based dendrogram exhibiting genetic relationships among the accessions of *A. falcata* using Isozyme markers.

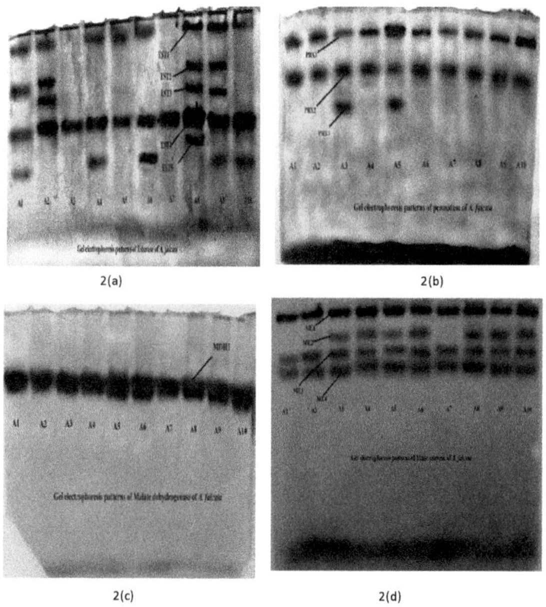

Figure 2A-D: Isozyme profile of accessions of *A. falcate*: represent isozyme profiles of esterase, peroxidase, malate dehydrogenase and malic enzyme respectively of 10 accessions of *A. falcata* (A1- A10)

References

1. Naithani HB, Pal M, Lepcha, STS (2003) Gregarious flowering of Thamnocalamus spathiflorus and Thamnocalamus falconeri, bamboos from Uttaranchal, India. Indian Forester 129: 517-526.
2. Hamrick JL, Godt MJW, Sherman-Broyles SL (1992). Factors influencing levels of genetic diversity in woody plant species. New Forest 6: 9-124.
3. Zeidler M (2000) Electrophoretic analysis of plant isozymes. Acta Univ Palacki Olomuc Fac Rerum Nat Biol 38: 7-10.
4. Gottlieb LD (1977) Electrophoretic evidence and plant systematics. Annals of Missouri Botanical Garden 64:161-180.
5. Murphy RW, Sites JW, Buth DG, Haufler CH (1996) Proteins: isozyme electrophoresis. In: Molecular Systematics, Hills DM, Moritz C, Mable BK (Ed), Sinauer Associates, Sunderland, Massachusetts USA. Pp: 51-121.
6. Sharifi G, Kouhsari SM, Ebrahimzadeh H, Khatamsaz M (2006) Isozyme analysis of seedling samples in some species of Hyoscamus from Iran. Pakistan Journal of Biological Sciences 9: 1685-1692.
7. Parker PG, Snow AA, Schug MD, Booton GC, Fuerst PA (1998) What molecules can tell us about populations: choosing and using a molecular marker. Ecology 79: 361-382.
8. Javid A, Ghafoor A, Anwar R (2004) Seed storage protein electrophoresis in groundnut for evaluating genetic diversity. Pakistan Journal of Botany 36: 87-96.
9. Iqbal SH, Ghafoor A, Ayub N (2005) Relationship between SDS- PAGE markers and Ascochyta blight in chick pea. Pakistan Journal of Botany 37: 87-96.
10. UPOV (1997) Adopted report of the Technical committee. Thirty third Session, Geneva, October 16 – 18, 1996.UPOV TC/33/11.
11. Sher AK, Habib A, Muhhamed S (2010) Conformation of Sunflower F1 hybrids using SDS- PAGE analysis. Africal Journal of Biotechnology 9: 4516-4520.
12. Karihaloo JM, Kaur M, Singh S (2002) Seed protein diversity in Solanum melogena L.and its wild and weedy relatives. Genetic Resources and Crop Evolution 49: 533-539.
13. Heng HP, Yeoh HH, Tan CKC, Rao AN (1996) Leaf isozyme polymorphisms in bamboo species. Journal of the Singapore National Academy of Science 22: 10-4.
14. Jerret RL, Litz RE (1986) Enzyme polymorphism in Musa acuminata Colla. Journal of Heredity 77: 183-188.
15. Bhatiya CR, Buiatti M, Smith HH (1967) Electrophoretic variation in proteins and enzymes of the tumor forming hybrid Nicotiana glauca X Nicotiana langdorfii and its parent species. American Journal of Botany 54: 1237-1241.
16. Abdul Wahab AH (1994) Acrylamide gel electrophoresis of several isozyme systems in 27 Allium species and their use in taxonomy. MARDI Res J 221: 1-13.
17. Gonzalez- Andres F, Pita JM, Ortiz JM (1999) Identificationof Iberian and Canarian species of the genus Pinuswith four isoenzyme systems. Biochemical Systematics and Ecology 27: 235-242.
18. Farooq S, Iqbal N, Zaidi AA (1999) Isozyme markers in cotton breeding 1.Standardization of different isozyme systems for identification of different cultivars of cotton (Gossypium hirsutum).Pakistan Journal of Botany 31: 5-20.
19. Brewbaker JL, Nagai C, Liu EH (1985) Genetic polymorphism of 13 maize peroxidase. Journal of Heredity 76: 159-167.

20. Chou CH, Sheen SS, Hwang YH (1986) A biochemical aspect of phylogenetic study of Bambusaceae in Taiwan II. The genera Arundinaria, Pseudosasa, Semiarundinaria, Shibataea, Sinobambusaand Yushania. Bot Bull. Academia Sinica, 27: 177-131.

21. Li SF (1989) Application of isozymes in Arundinarieae classification. Journal of Bamboo Research 8: 13-21.

22. Strefeler MS, Darmo E, Becker RL Katovich EJ (1996) Isozyme characterization of genetic diversity in Minnesota populations of purple loosestrife, Lythrumsalicaria (Lythraceae). American Journal of Botany 83: 265-273.

23. Loukas M, Vergini Y, Krimbas CB (1983) Isozyme variation and heterozygosity in Pinus halepensis L. Biochem Genet 21: 497-509.

24. Siddiquee S, Tan SG, Yusof UK (2010) Isozyme analysis and relationships among three species in Malaysian Trichoderma isolates. J Microbiol Biotechnol 20: 1266-1275.

25. Saxena R, Chandra A (2010) Isozyme, ISSR and RAPD profiling of genotypes in marvel grass (Dichanthium annulatum). J Environ Biol 31: 883-890.

Changes and Status of Mangrove Habitat in Ganges Delta: Case Study in Indian Part of Sundarbans

Pramanik MK*

Research Scholar, Centre for International Politics, Organization and Disarmament (CIPOD), Jawaharlal Nehru University, NewDelhi-110067, India

Abstract

This paper quantifies the changes and present status of mangrove forest in Indian part of Sundarban from 1975 to 2014 using Landsat MSS (1975), TM (1990), ETM (2002) and OLM (2014) satellite imageries. The study used two image processing techniques: Maximum Likelihood Classification for the Land use and land cover analysis and NDVI for the vegetation characteristics and their temporal changes. The research found that the area of mangrove gradually decreases from 203752 hector (44%) to 132723 hector (31 %) and the barren land increases from 15078 hector (2.86 %) to 37247 hector (7.12%) due to natural (sea level rise, salinization etc.) and anthropogenic (livelihood collection and shrimp farming etc.) disturbances and continuous land reclamation. Other land use categories like agriculture, water body, and sand deposition have approximately remained constant. The NDVI values were changed significantly only in 1990 due to the landward migration and defragmentation of mangrove forest. However, the paper signifies that the forest cover is constantly evolving due to deforestation, aggradation, erosion and forest rehabilitation programs in the Indian Sundarban.

Keywords: Mangrove habitat; Sundarban biosphere reserve; Ganga-Brahmaputra delta; Mangrove change; Mangrove vulnerability

Introduction

Mangrove forests are one of the most significant biodiversity of intertidal zones along the subtropical and tropical coastal regions of the world. These ecosystems take the role of attenuating wave energy protect the shoreline and adjacent dependent ecosystem [1-2]. The mangrove ecosystem also plays a significant role in the biogeochemical cycle, coastal protection and also has an economic significance of the region through different activities like fishing, aquaculture, and farming. However, mangrove dependent livelihood such as a collection of fuelwood, forest fire, flood, erosion and other human activities have resulted in the reduction of the mangrove forest in Sundarban [3-6].

The Indian part of Sundarban mangrove is a part of the world's largest delta of Bengal estuarine region. The delta has formed in the estuarine area of Ganga-Brahmaputra and Meghna River. The region is also named as Region of the largest halophytic formation along the coastline. It was also declared a "biosphere reserve"(1989) and "world heritage site"(1987) by United Nations Educational and Scientific Cooperation [7-9] and International Union for Conservation of Nature [10] respectively where 2125 sq. Km area occupied by mangroves across 56 islands. The aerial extent of the Indian part of Sundarban is about 4262 sq. Km where having 106 deltaic islands within 54 has human habitation. The total region characterized by most of the tidal creeks, innumerable rivers, and small channels. The biodiversity has been significant as an abandoned due to the presence of endangered plant species and animals of which 37 are mangrove species in 87 plant species.

In the last century, the erosion, flood probabilities increased heavily due to the increasing rate of siltation of river beds. Also, present continuous rises of sea level are now being a magnificent threat for existing vulnerable or fragile ecosystem [11-17]. So, the present study analyzes an overview of the distribution and changes of mangroves in Sundarban delta by using Remote Sensing and Geographical Information System for the conservation planning and strategies [18,19].

Study Area

The present study area situated on the eastern coast of India in the southern part of West Bengal in between 21°13′- 22°40′ north latitude and 88°05′–89°06′ east longitude (Figure 1). It has located about 100 km southeast of Kolkata in South 24 Pargana district. Most of this district predominated by flood and cyclone prone area. The tides of the study area are semi-diurnal, and floor varies from 0.9 m to 2.11 m above sea level in different regions during different season, mainly it reaching a maximum during monsoon, and minimum during the winter [29]. This tidal level influenced by the direct connection with the sea at the Hugli river mouth and also through the Ganga-Brahmaputra estuaries [20,21].

Materials and Methods

Data sources

Four multispectral satellite images (Landsat) that covers whole Indian parts of Sundarban Biosphere Reserve were downloaded from the freely available USGS (United States Geological Service) GLOVIS. These orthorectified images (with universal transverse Mercator projection and world geodetic system 84 datum) are Landsat Multispectral Scanner (MSS) data (Columns 1953, Rows 1459, Dated 5 Dec 1973), Landsat Thematic Mapper (TM) data (Columns 3708, Rows 2754, Dated 12 Dec 1990), Enhanced Thematic Mapper plus (ETM+) data (Columns 7809, Rows 5833, Dated 27 Nov 2002) and Operational land imager (OLI) data (Columns 3710, Rows 2771, Dated

***Corresponding author:** Malay Kumar Pramanik, Research Scholar, Centre for International Politics, Organization and Disarmament (CIPOD), Jawaharlal Nehru University, New Delhi, India, E-mail: malaygeo@gmail.com; malay65ise@jnu.ac.in

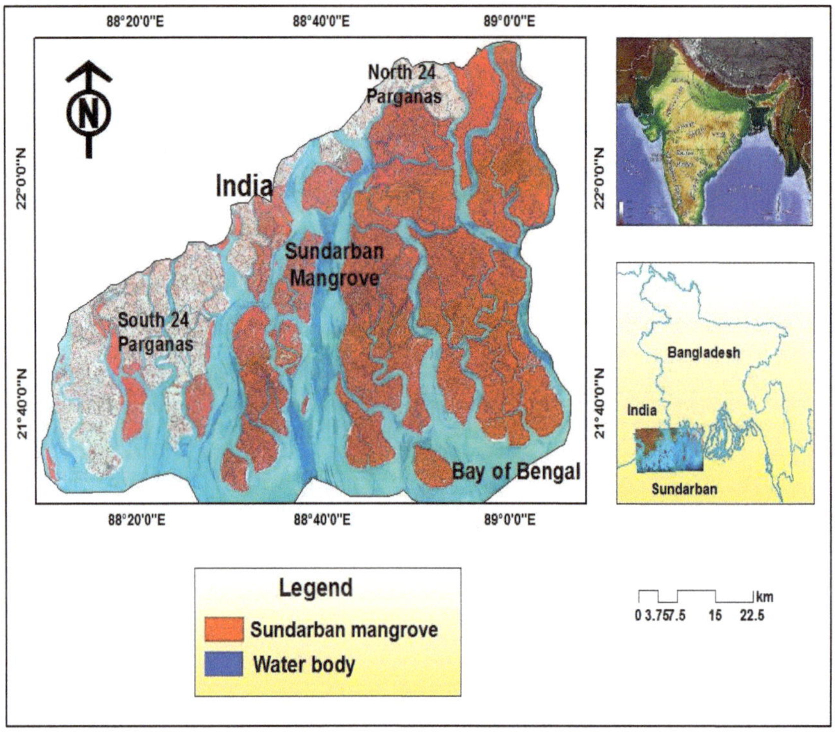

Figure 1: Map of the study area.

23 Jan 2014). These images are possibly collected at same season and less likely to have had miss classification error during spectral analysis of different LULC categories. Topographic maps (NF 45/11, NF 45/12, NF 45/7 and NF 45/8) are used that has been downloaded from Texas library at scale of 1: 250000.

Identification of mangrove boundary

Boundaries of the study area have extracted from four images as well as Toposheet by manual digitizing along the near boundary of mangrove forest.

Image processing

The whole image processing procedure was processed ERDAS IMAGINE 13.0 software. Individual band data stacking to prepare false color composites (FCCs) was done for all bands of MSS, band 1-5 and 7 of TM, band 1-5 and 7, 8 of ETM+, BAND 1-7 of OLI data then haze reduced of all images. After layer stack extract the study area by masking using Arc GIS 10.

Image classification and accuracy assessment

For supervised classification follow a parametric decision rule of each images , i.e., classification techniques adopted by Maximum Likelihood Classification method because this classification algorithms produces consistently a better results for the mangrove habitat zonation [22]. However, five land use and land cover classes were taken for image classification based on Google Earth visit and locational experience of the study area. The classes are mangrove, water body, sand deposition, agricultural land and barren land. After classification, accuracy assessment has performed by error matrices of all images. All accuracy has different to each other.

Mapping of vegetation attributes

Spectral vegetation attributes used for the analysis of vegetation density and canopy structure. NDVI is the most important of all vegetation indices. The basic principle is that the rates of the difference between NIR (4th band) and Red (3rd band) of an image that can help to know the green plants. The green characteristics of plants influenced by topography as well as regional deltaic slope [3]. So, the NDVI calculated by the following equation in Arc GIS.

$$NDVI=NIR-RED/NIR+RED$$

NDVI maps have generated from the NDVI ratios of four LANDSAT images used for the present study. Generally, Biomass contribution of mangrove leaves (chlorophyll portion) reflect more in middle infrared region and NDVI value ranges from -1to+1 where the 0 indicate the bare earth but higher negative value denotes water content and the value beyond 0.5 represents green vegetation. However, the study areas NDVI are 0.7 to 0.9 for the dense mangrove.

Change of land use and land cover

The change of land utilization and land cover analyzed by overlaying GIS techniques (Figure 2) adopted within ERDAS IMAGINE to explore the different temporal land cover change at approximately ten years interval. Using GIS database these data are tabulated and quantifying their temporal transformation.

Results

Land utilization and land cover change through classified image

The present study of mangrove forest area has been classified into five categories of Land use and land cover area as mangrove, agricultural

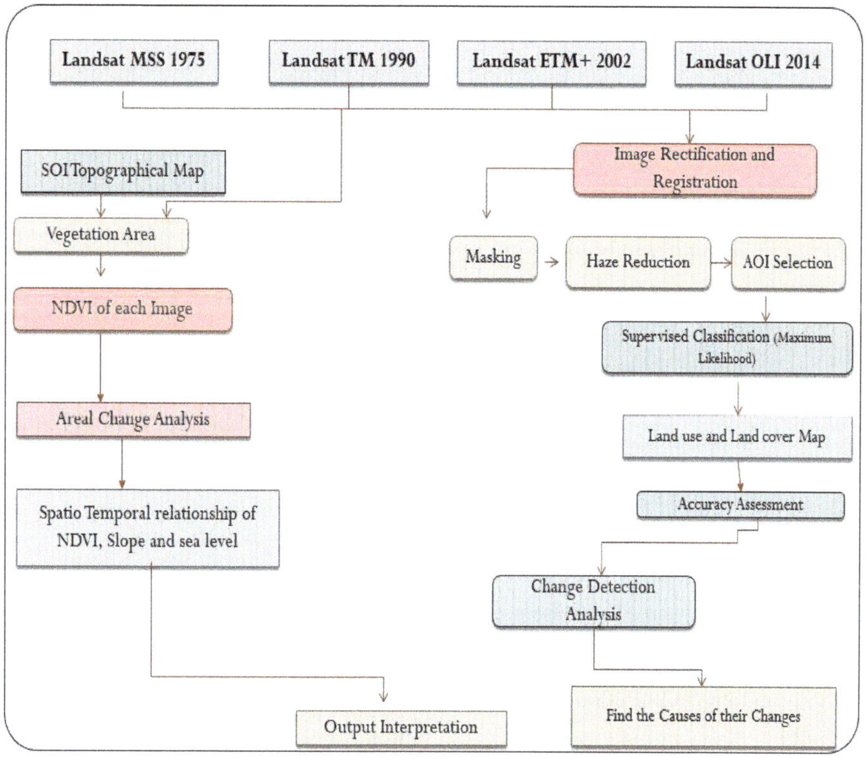

Figure 2: Framework of methodology followed in the present study.

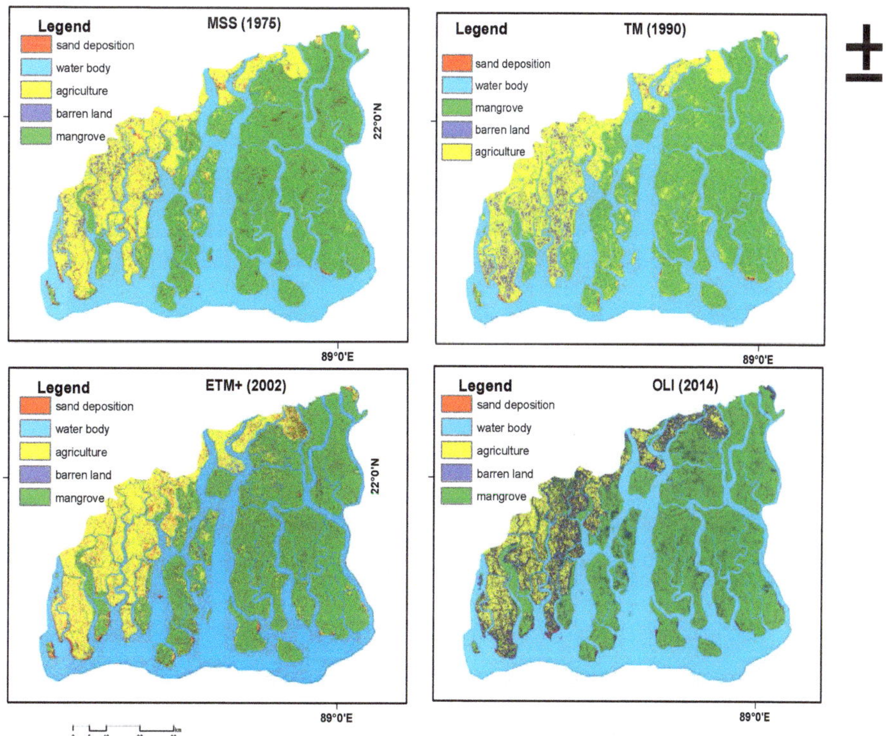

Figure 3: Land use and land cover of the study area in different time period of 1975, 1990, 2002, and 2014.

land, Barren land, sand deposit and water body (Figure 3) of all satellite images. These estimated areas under different land use categories of the study area have shown in figure 2 and 3. In 1975, areas of mangrove and agriculture land were approximately 203752 hector (44%) and 94764 hector (18%), respectively. These mangroves were found in the remote areas of human habitation, partly along the river banks and mouths. The area of mangroves is spatially a continuous zone along the coastline of the study area. The agriculture area is less due to dense vegetal cover but most of the agriculture areas occupied in sparse mangrove areas. Barren land occupied the less area approximately 15078 hector (2.86%) along the River within the study area.

In 1990, 2002 and 2014, the area of mangrove gradually decrease is about 182345 hector (40%), 156711 hector (35%) and 132723 hector (31%) respectively (Figure 4). Mangrove area decreases probably due to forest clearing and continuous land reclamation during this period [3,20]. Agricultural lands are remains constant, but barren land drastically increase from 2002 (32355 Hector) to 2014 (69596 Hector) is about 7.12% (37241 hectors).This drastic change of barren land occurred due to the error of miss classification (kappa coefficient value 81.26) and may be of image that collected in different time period for the not availability of cloud-free images in same season. So, during this field studies several agricultural locations and open mangrove were found in corresponding to the barren land category. Most of the farmland has found in the northwestern part of the study area, and the conversion of mangrove to barren land and agriculture to barren land is a maximum of this part also.

Change detection in vegetation indices

NDVI was used for the analysis of land use change to delineate the mangrove dynamics from 1975 to 2014. Four NDVI maps are used for visual interpretation to understand the density of mangrove and their fragmentation due to sea level change. The change observed higher along the river and coastal margins due to high rates of bank erosion. The value of NDVI gradually decreased from 1975 to 2014 due to positive changes in human population and gradual decline of open mangrove, as well as conversion of agricultural lands, barren land and built-up areas [23,24]. However, positive changes of NDVI found only limited areas like Lothian, Chulkati island, other protected forests and vegetal riverbanks. However, higher positive changes found in the north-western part of the study area due to the initiative of man-made vegetation. Moreover, landward migration of mangrove forest and defragmentation of mangrove were also handling the declining NDVI values.

In 1973, the NDVI values were more or less same but the NDVI of 1990, 2002 and 2014 (Figure 5) differ significantly from the previous one. In 1990, NDVI valued higher in north-western part and lowered in the mangrove areas due to defragmentation of dense mangrove. However, in the year of 2002 south-eastern part of the study area covers higher values remaining part holds their medium NDVI values. Moreover, recent NDVI (2014) values not differ significantly from 2002.

However, the pattern and condition of healthy upper layer mangrove characteristics is different in different classification periods. Therefore, more healthy areas in 1973s are different from the least healthy areas of 2014s correlates to the areas of 'Top dying disease' of Sundari (Heritiera fomes) [25]. As mentioned above, the lack of multiple set of images for each period, the different image acquisition seasons of different era and the variation in the tidal inundation degree of different satellite images restrict comparison of absolute values of canopy closure layers. However, the canopy closure layer has been reduced in different time period of 1973, 1990, 2002, and 2014 (Figure 6). The positive value of NDVI gradually decreases due to the increasing flooded areas and barren lands in last 40 years. The value of NDVI over 0.40 indicates the dense mangrove cover area that significantly changes over the periods where 18.91% of area in 1973 changes to 8.7% in 2002, and finally lost this healthy vegetation in 2014 due to intense cyclonic effect of 'Sidr'. The dense mangrove areas in 1973 are equally distributed all over the area of Sundarbans but in 1990s areas of dense mangrove only found in the south-eastern part of Bangladesh Sundarbans and in 2002s areas are found in the middle portion of the Sundarbans (Table 1). The degraded land, barren lands are gradually increased over different time period where 7.3% area of 1976s converted to 60.08% area in 2014 (Figure 3). This may signify that the mangrove dominated total Sundarbans are totally influenced by climate change and their related effects over the periods.

Discussion

Sundarbans ecosystem has located in climatic hotspot and also has highly susceptibility sea level rise [4-6]. Except sea level fluctuation, the region faces many difficulties like Monsoonal high floods, cyclones, storm surge and high drainage density, and Stalinization. Moreover, the sea level change, salinization, and other related events are the major

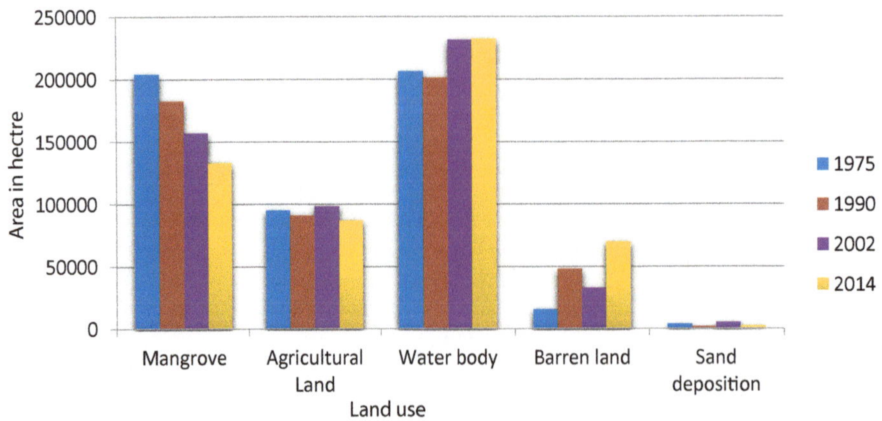

Figure 4: Amount of land under different land use and land cover classes in 1975, 1990, 2002, and 2014.

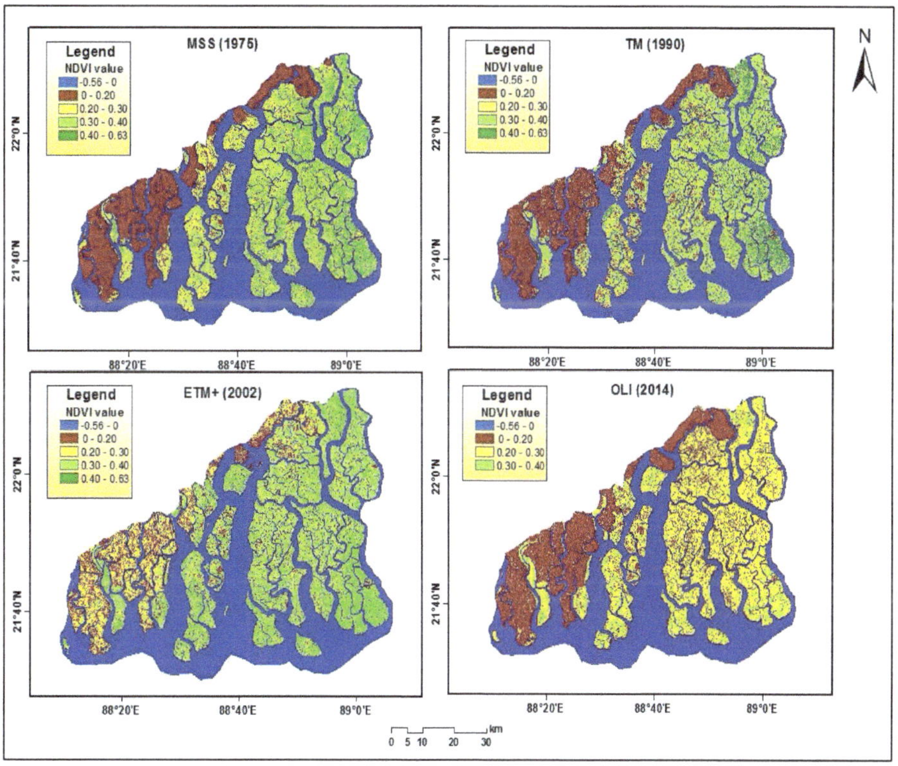

Figure 5: Normalized difference Vegetation Index (NDVI) in different time period of 1973, 1990, 2002, and 2014.

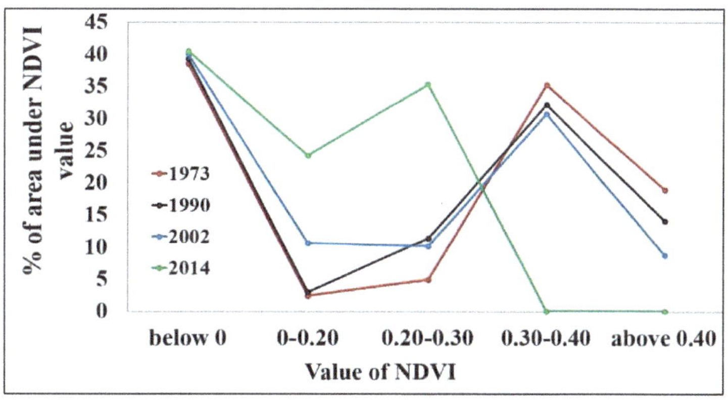

Figure 6: Area (%) under different NDVI value of Sundarbans in 1973, 1990, 2002, and 2014.

Value of NDVI	Area in %				Remarks
	1976	1990	2002	2014	
Below 0	38.49	39.26	39.95	40.52	Water bodies, flooded areas
0-0.20	2.4	3.0	10.6	24.20	Soil, degraded land, barren land, bare rock
0.20-0.30	4.9	11.4	10.1	35.28	
0.30-0.40	35.3	32.2	30.65	0	Less dense mangrove areas
Above 0.40	18.91	14.14	8.7	0	Very dense mangrove areas

Table 1 Area under different Normalized Difference Vegetation Index (NDVI) value in Sundarbans.

factors that may influence the changes and dynamics of mangrove ecosystems [26,27]. In respect to sustainable management of natural resource in coastal areas closely related to the association and density of mangrove community [28-30]. These characteristics primarily associated with slope, soil, habitat stratigraphy as well as salinity regimes [31]. that can alter the mangrove systems. The increasing sea level and salinity regimes may affect some edaphic changes, general changes of soil and salinity, tidally dominated mud flat and groundwater

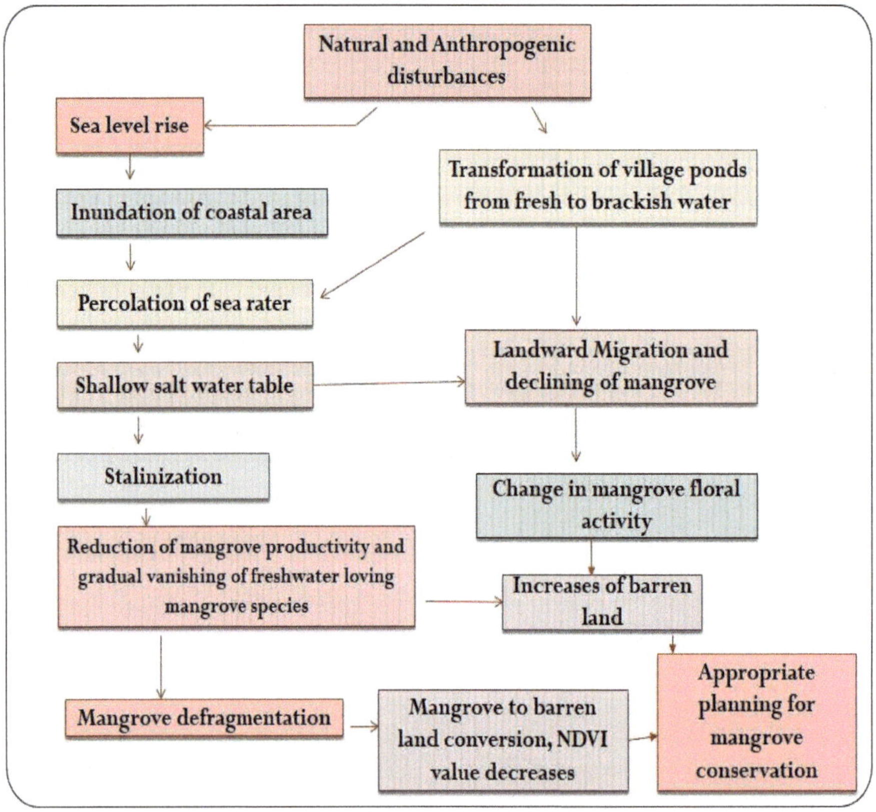

Figure 7: Impacts of SLR on mangroves via ground water system (Source: Modified from Avijit Mitra, 2013).

fluctuation and their quality. The slope of this region also influences the dynamics and adjustment of mangrove forests along the coastal areas. The dynamics and change mangrove association depends on:

1. The inundation of high tidal areas.
2. Ground water fluctuation.
3. Increasing inundation frequency.

In the view of the sustainable mangrove management, Sundarban has overcome many difficulties like cyclones, inaccessible terrain condition, and high flood events. However, harmful human activities for their traditional livelihood generation like, timber collection, farming, and honey collection are not conducive to their sustainability [3]. In the last few decades experienced that the rural peoples had destroyed the open mangrove covers without any management initiative [32,33]. However, the existing mangroves are not able to protect against the cyclone like Aila (25 may 2009) and havoc tsunami waves, which caused of landslide, flooding, tree uprooting, bank erosion and loss of human lives and property lead to the defragmentation and destruction of mangrove habitat in the intertidal region [34-37]. The northern part of Indian border and the adjacent portion has experienced maximum growth of shrimp farming that influenced the higher rate of deforestation activities (Figure 7). According to Wikramanayake et al. report, unsustainable shrimp farms leads to the destruction of mangrove habitat in Sundarban areas [38]. Islam suggests that the ecosystem management system is inadequate, and the highest numbers of people are economically dependent on the mangrove ecosystem [26].

Conclusion

Management issue of the intertidal mangrove ecosystem is a major problem due to the climatic complexity and present human interventions. The academician, planner, and researcher can follow a comprehensive way to incorporating qualitative and quantitative information for the decision-making processes and management initiative. As demonstrated, the present study of integrated remote sensing and GIS technique provide some important information about mangrove dynamics, changes, existing status but the optimum possibility for success cannot achieve [39,19]. Multispectral satellite images can be used potentially to measure forest cover and mangrove forest cover change if carefully image processing methods has adopted. The completely deforested and fragmented area can be simply detected using the maximum likelihood algorithm. Also subpixel classification algorithm and NDVI differencing method can be used for the differences of forest cover of the Indian part of Sundarbans.

The potential changes and dynamics vary from one place to another location along the coastline. Multiple environmental stresses likely make the most vulnerability [40,41,27]. The low-lying coastal areas and higher population with greater mangrove density are the most vulnerable because of they are lower social, technological and financial adaptation [5]. An effort, in particular, the low-lying coastal areas need an adoptive option to their proper management. Thus, the adaptive options of mangrove sustainability are mostly dependent on their locational characteristics and long-term thinking as well as economic activities.

References

1. Brinkman RM, Massel SR., Ridd PV and Furukawam K (1997) Surface wave attenuation in mangrove forests. Proceedings of 13th Australasian Coastal and Ocean Engineering Conference 2: 941-949.
2. Quartel S, Augustinus A, Tri P, Augustinus PG, Tri HN (2007) Wave attenuation in coastal mangroves in the Red River Delt, Vietnam. Journal of Asian Earth Sciences 29P: 576-584.
3. UNESCO (2002b) Ecotourism and Sustainable development in Biosphere Reserves: Experiences and Prospects, WorkshopSummary Report, UNESCO, Paris.
4. Ardil ER, Wolff M (2009) Land use and land cover change affecting habitat distribution in the Segara Anakan lagoon, Java, Indonesia. Regional Environmental Change 9: 235-243.
5. Ellison JC, Stodart DR (1991) Mangrove Ecosystem Collapse During Predicted Sea-Level Rise: Holocene Analogues and implications. Journal of Coastal Research: 151-165.
6. Kebede AS, Nicholls RJ, Hanson S, Mokrech M (2010) Impacts of Climate Change and Sea-Level Rise: A Preliminary Case Study of Mombasa, Kenya. Journal of Coastal Research. 8-19.
7. Babo NR., Froehlich JW (1998) Community-Based Mangrove Rehabilitation: A Lesson Learned from East Sinjai, South Sulawesi, Indonesia. Indonesia: The World Bank.
8. UNESCO (2003) Five trans boundary biosphere reserves in Europe, Biosphere Reserves Technical Notes, UNESCO, Paris.
9. Emch M, Peterson M (2008) Mangrove Forest Cover Change in the Bangladesh Sundarbans from 1989-2000: A Remote Sensing Approach. Geocarto International: 5-12.
10. Kathiresan K (2000) A review of studies of Pichavaram Mangroves, Southeast coast of India. Hydrographia 430: 185-205.
11. Field C (1995) Impacts of expected climate change on mangroves. Hydrobiologia 295: 75-81.
12. Gilman E, Ellison J ,Coleman R (2007a) Assessment of mangrove response to projected relative sea-level rise and recent historical reconstruction of shoreline position. Environ. Monit. Assess. 124: 112-134.
13. Ding H, Nunes PA (2013) Modeling the links between biodiversity, ecosystem services and human wellbeing in the context of climate change, Results from an econometric analysis of the European forest ecosystems, Ecological Economics 97: 60-73.
14. Ellison J (2000) How South Pacific mangroves may respond to predicted climate change and sea level rise. In Gillespie A and Burns W, Climate Change in the South Pacific: Impacts and Responses in Australia, New Zealand and Small Islands States. Dordrecht, Netherlands: Kluwer Academic Publishers: pp 289-301.
15. Church J, Hunter J, McInnes K, White N (2004) Sea-level rise and the frequency of extreme events around the Australian coastline. In: Coast to Coast '04 - Conference Proceedings Australia's National Coastal Conference, Hobart, 1-8.
16. Cohen MC, Souza Filho PW, Lara RL, Behling H, Angulo R (2005) A model of Holocene mangrove development and relative sea-level changes on the Bragança Peninsula (northern Brazil). Wetlands Ecology and Management 13: 433-443.
17. Lua CY, Gua W, Daid AH, Weib HY (2012). Assessing habitat suitability based on geographic information system (GIS) and fuzzy: A case study of Schisandra sphenanthera Rehd. et Wils in Qinling Mountains, China. Ecological Modelling 242: 105-115.
18. UNESCO (2002a) Biosphere Reserves: special places for people and nature, UNESCO, Paris.
19. Rajkumar M, Prasad MNV, Freitas H, Ae N (2009) Biotechnological Applications of Serpentine soil Bacteria for phytoremediation of trace elements. Critical Review of Biotechnology 29: 120-130.
20. Kebede AS, Nicholls RJ (2012) Exposure and vulnerability to climate extremes: population and asset exposure to coastal flooding in Dar es Salaam, Tanzania. Regional Environmental Change **12**: 81-94.
21. Store R, Jokimaki J (2003) A GIS-based multi-scale approach to habitat suitability modeling. Ecological Modelling 169: 1-15.
22. Berger U, Rivera Monroy VH, Doyle TW, Dahdouh-guebas F, Duke NC et al. (2008) Advances and limitations of individual-based models to analyze and predict dynamics of mangrove forest. A Review. Aquatic Botany 89: 260-274.
23. Datta D, Chattopadhyay RN, Guhan P (2012) Community based mangrove management: A review on status and sustainability. Journal of Environmental Management 107: 84-95.
24. Gilman EL, Ellison L, Duke NC, Field C (2008) Threats to mangroves from climate change and adaptation option, A review: Aquatic Botany 89: 237-250.
25. Jagtap TG, Nagle VL (2007) Response and Adaptability of Mangrove Habitats from the Indian Subcontinent to Changing Climate. Royal Swedish Academy of Sciences: 328-334.
26. Dev Roy AK, Alam K, Gow J (2011) A review of the role of property rights and forest policies in the management of the Sundarbans Mangrove Forest in Bangladesh. Forest Policy and Economics 15: 46-53.
27. Cahoon D H (2006) High-resolution global assessment of mangrove responses to sea-level rise: A review in E Gilman, Proceedings of the Symposium on Mangrove Responses to Relative Sea Level Rise and Other Climate Change Effects, 13 July 2006, Catchments to Coast, Society of Wetland Scientists 27th International Conference, Cairns Convention Centre, Cairns, Australia: Western Pacific Regional Fishery Management Council, Honolulu, HI, USA, 9-17.
28. Biswas SR, Mallik AU, Choudhury JK, Nishat A (2009) A unified framework for the restoration of Southeast Asian mangroves-bridging ecology, society and economics. Wetlands Ecol Manage 365-383.
29. Church J, White N (2006) A 20th century acceleration in global sea-level rise. Geophys Res Lett 33: L01602.
30. Datta D, Chattopadhyay RN, Guhan P (2012) Community based mangrove management: A review on status and sustainability. Journal of Environmental Management 107: 84-95.
31. Uddin S, Shah M A, Khanom S, Nesha M K (2013) Climate change impacts on the Sundarbans mangrove ecosystem services and dependent livelihoods in Bangladesh. Asian Journal of Conservation Biology: 152-156.
32. Dev Roy AK, Alam K, Gow J (2011) A review of the role of property rights and forest policies in the management of the Sundarbans Mangrove Forest in Bangladesh. Forest Policy and Economics 15: 46-53.
33. Islam MS (2011) Biodiversity and livelihoods: A case study in Sundarbans Reserve Forest, World Heritage and Ramsar Site (Bangladesh). Austria: A Master thesis submitted of the requirements for the degree of Master of Science (M.Sc.) in Management of Protected Areas at the University of Klagenfurt.
34. Haq SA (2010) Impact of climate change on Sundarbans, the largest mangrove forest: ways forward", 18th Commonwealth Forestry Conference 2010, Edinburgh.
35. Gilman E, Ellison J, Sauni J, Tuaumu S (2007b) Trends in surface elevations of American Samoa mangroves. Wetl Ecol Manag 15: 391-404.
36. IUCN (1989) the impact of climatic change and sea level rise on ecosystems. Report for the Commonwealth Secretariat, London.
37. Gilman E, Ellison J, Jungblat V, VanLavieren H, Wilson L et al. (2006) Adapting to Pacific Island mangrove responses to sea level rise and other climate change effects. Climate Res 32: 161-176.
38. Semeniuk V (1993) The Mangrove systems of Western Australia-Presidential address. Journal Royal Society of Western Australia 76: 99-122.
39. UNESCO (2002a) Biosphere Reserves: special places for people and nature, UNESCO, Paris.
40. Li Z, Saito Y, Mao L, Tomura T, Li Z, Song B (2012) Mid-Holocene mangrove succession and its response to sea-level change in the upper Mekong River delta, Cambodia. Quaternary Research, 386-399.
41. Limaye RB, Kumaran, KP (2012) Mangrove vegetation responses to Holocene climate change along Konkan coast of south western India. Quarternary International 263: 114-128.
42. Islam MS (2011) Biodiversity and livelihoods: A case study in Sundarbans Reserve Forest, World Heritage and Ramsar Site (Bangladesh). Austria: A Master thesis submitted of the requirements for the degree of Master of Science (M.Sc.) in Management of Protected Areas at the University of Klagenfurt.

19

Fuelwood Consumption in Takoli Gad Watershed of TehriGarhwal in Garhwal Himalaya, India

Suraj Kumar and Munesh Kumar*

Department of Forestry and Natural Resources, HNB Garhwal University, Srinagar Garhwal Uttarakhand, India

Abstract

The aim of the present study was to understand fuelwood consumption pattern of the villager at different altitudes in Takoli Gad Watershed. Three different altitudes i.e., 500-800 m (lower altitude) 800-1100 m (middle altitude), 1100-1500 m (upper altitude) were selected. At each altitude three villages were selected i.e., Dangchaura, Pali, Sauru (500-800 m), Takoli Tolu Amoli (800-1100 m) and Maikhandi Riscoti Kwili (1100-1500 m). Each selected village was further categorized based on family size into small, medium and large to observe impact of family size on fuelwood consumption.

The results indicate that fuelwood consumption among family size (irrespective villages) ranged from 430.70 to 643.61 kg/capita/year (lower altitude), 486.66 to 689.90 kg/capita/year (middle altitude), 406.57 to 675.25 kg/capita/year (upper altitude). The consumption of fuelwood increased with increasing altitudes as 541.8 (500-800 m), 554.97 (800-1100 m) and 557.71 kg/capita/year (1100-1500 m). However, fuelwood consumption with family size (irrespective altitudes), showed reduced with increasing family size from small > medium > large as 669.58 (small family), 543.35 (medium family) and 441.88 (large family). In the study, it has been observed that villagers collected fuelwood from forest and agro forestry field where agroforestry contributed 25-30% fuelwood consumption for villager's need, which directly helps in reducing pressure on forest.

Keywords: Fuelwood; Consumption; Forest; Agroforestry; Per capita; Household

Introduction

In many developing countries the consumption of biomass as fuel has been identified as one of the most significant cause of forest degradation. According to Academy of Science (1980), fuelwood scarcity is most acute in the countries of Indian sub-continent alone requires about 274 million m^3 fuelwood annually [1]. Fuelwood demand in India has increased from 60 million tons in 1991 to 158.45 million tons in 2000 [2].

The use of wood as a means of generating heat energy, particularly for cooking and heating is still very much relevant in the developing countries, most especially those of sub-Saharan Africa [3]. Reports such as Arnborg [4] had earlier shown that as much as half of the timber harvested in the world is used as fuelwood for cooking and heating, most of which are used in the developing countries [5], where recent estimates showed that fuelwood accounts for about 90% of the timber harvested [6]. It is also estimated that over 90% of the people of Africa depend on either fire wood or charcoal for cooking and heating. Apart from the use of biomass, especially wood, for cooking and heating, the inhabitants of these countries have depended, for livelihood and sustenance, on their indigenous plants resources for centuries. They depend on them for food security and a host of everyday products, from medicines to fibers [7].

The increases in the use of fuelwood have partly contributed to the depletion of forests [8] and the implications of this on the environment, biodiversity stability and by extension on the wellbeing of human beings are well documented in literature. Owing to the increasing consumption of woody biomass as fuel in the world, it is imperative to be in tune with other traditional uses to which accepted fuelwood species are put by the cross section of users.

The major determinants of the energy needs and consumption in rural India are income distribution preferences, private discount rates and micro financing [9]. The energy use pattern in rural India is changing, with uptake of clean energy, but traditional fuels including fuelwood, crop residue and cow dung still constitute the main source of household cooking energy due to inadequate and unreliable supply of clean energy [10].

The fuelwood demand in the country ranges from 96 to 157 million tons annually, including a rural demand of 80-128 million tones, thus raising the consumption level to 148-242 kg per capita [11]. However, the per capita annual consumption of dry wood in various parts of the Himalayas is reported to be much higher, ranging between 500 and 1200 kg [12,13]. A survey of use pattern in the region revealed that non-commercial energy formed 98.59% of the total household energy demand. Thus, the commercial energy component formed only 1.41% of the total, comprising of kerosene and electricity [14]. Thus, it is evident that most of this demand is met from the adjoining forests almost uninterrupted [15,16] and biomass extraction is the major reason for such depletions [17].

In India, 80% of the population live in villages and wood is the main source of fuel for these villages, which is continuously depleting fuelwood resources and therefore, the rural poor are being forced to use some time other forms of energy such as crop residues and dung [18]. People of the Himalaya had traditional rights to collect limited amount

***Corresponding author:** Munesh K, Department of Forestry and Natural Resources, HNB Garhwal University, Srinagar Garhwal Uttarakhand, India
E-mail: muneshmzu@yahoo.com

of fuelwood from forested areas. However, new environmental law by Government agencies have restricted fuelwood collection without providing the alternative source of energy, thus local people are facing problem [19]. Although fuelwood collection is still going on which may affect the existing forest resources in the Himalaya in coming periods.

In Indian mountain, fuelwood is the only main source of energy for villages, for cooking purposes. In India, about 70% of energy requirement is taken by the fuelwood, collected from the forests and nearby sites and about 50 million tons of wood are removed every year [20].

In the mountains of Garhwal Himalaya, generally ordinary people, due to their socio-economic conditions people fulfill their energy demand almost entirely from the forests. Alternative fuels such as animal dung and crop residues may also be utilized.

In the Garhwal Himalayan region of present study, many species have been identified by the villagers, which have been used for fuelwood and other purposes. The trees identified in the forest areas are; *Quercus leucotrichophora, Mallotus philippensis, Terminalia tomentosa, Anogeissus latifolia,* and in agroforestry fields; *Grewia oppositifolia, Quercus leucotrichophora, Adina cordifolia, Celtis australis, Ficus palmata, F. auriculata, F. glomerata, Melia azedarach, Bauhinia variegata, Boehmeria rugulosa, Terminalia bellerica, Butea monosperma, Toona ciliata, Woodfordia fruticosa, Rhus parviflora* etc. In the villages of Garhwal Himalaya, it has been noticed that usually villagers collected fuelwood from adjacent existed traditional agroforestry field and forest.

The aim of the present study was to understand the available fuelwood collection sources to the villager's needs, impact of family size on fuelwood consumption pattern and impact of altitude on fuelwood consumption pattern. Therefore, following objectives were selected for the study are; (i) Fuelwood collection sources to the villages need (ii) Impact of family size on fuelwood consumption pattern and impact of altitudes on fuelwood consumption pattern.

Materials and Methods

Location of the study area

The present study was carried out in 'Taki Gad Watershed' in a part of Tehri District of Uttarakhand State. Uttarakhand state extend over an area of 53000 km^2 from 28°43' to 31°27' N latitude and from 77° 34' to 81° 21'E longitude. The state is drained by eight major river systems and is thus divided into eight major catchment areas. These are the Yamuna, the Ganga 'A', the Ganga 'B', the Bhagirathi, the Alaknanda, the Ramganga, the Koshi and the Kali. The river Alaknanda represents the second biggest river system of Uttarakhand after the Kali river system. The study is carried out in the Takoli Gad watershed which is the part of Alaknanda catchment area. The total human population of villages in Takoli Gad Watershed is given in Table 1 as per census 2011.

Geographically the catchment "Takoli Gad Watershed" is between the 30° 14' to 30° 23' N latitude and 78° 37' to 78°46' longitudes in the Survey of India toposheet No. 53 J/11, 53 J/12 and 53 J/15 with an area of about 131.43 km^2. It comes under jurisdiction of district Tehri Garhwal, Uttarakhand. The area falls in inner Garhwal lesser Himalaya and is characterized by gentle and mature topography. The Takoli Gad originates from the Eastern slope of the Chandrabadni Peak (2278 meter) and joins the AlaknandaRiver at JuyalGarh (605 meter). Jakhand and Dagar Gad are the two main sub streams/tributaries of the Takoli Gad watershed.

Methodology adopted

In the 'Takoli Gad watershed' agriculture is practiced between 500 to 2000 m elevations where villagers are intestinally managed trees in their agriculture field for their daily needs such as fodder, timber, minor forest products including fuelwood. In the present study as per suitability of the study sites three different altitudinal zones viz., 500-800 m, 800-1100 m and 1100-1500 m above mean sea level (a.m.s.l) were selected for the assessment of fuelwood consumption. In each altitude three representative villages were selected i.e., 500-800 m (Dangchaura, Pali, Sauru), 800-1100 m (Takoli Tolu Amoli) and 1100-1500 m (Maikhandi Riscoti Kwili) (Figure 1) to conduct detailed study on fuelwood consumption functioning and each village were further categorized on the basis of family size i.e., small, medium and large. In each village for family size three representatives household were selected for data collection. Basic household information like number of family members, livestock holding, source and requirements of fodder for livestock, source and requirement for fuelwood, distance from the forest, irrigation water, road and other infrastructure, socio-economic status and educational level has been collected through questionnaire during the survey in the study area. The fuelwood consumption of selected was calculated based on collected data. The questionnaire survey covered 81 households in entire study area at different altitudes. The quantity of fuelwood consumption in the sampled villages was estimated over a period of 24 hour by using a weight survey method [21]. Quantification of fuelwood was done using collected data. To quantify the fuelwood consumption initially a wood lot was weighted and left in the kitchen of the household and the household was requested to burn wood only from given bundle. After 24 hrs the household was visited again and the actual fuelwood consumption was measure. Per capita per day consumption of fuelwood was calculated on the basis of total wood consumed by the family, divided by the total number of family members, as followed by Gupta et al, [22], as; F=TFc/ TFm Where; F=per day per capita fuelwood consumption, TFc=total fuelwood consumption by the sample household, and TFm =total number of members in the family.

Altitude (m amsl)	Village Name	House holds	Population			Literacy/illiteracy	
			Male	Female	Total	Literate	Illiterate
500-800	Dangchaura	53	116	110	226	146	80
	Pali	18	29	35	64	45	19
	Sauru	93	166	209	375	243	132
800-1100	Takoli	84	136	163	299	192	107
	Tolu*	-	-	-	-	-	-
	Amoli	52	93	96	189	128	61
1100-1500	Maikhandi	137	291	328	619	420	199
	Riscoti*	-	-	-	-	-	-
	Kwili	52	79	94	173	131	42

Table 1: Population and literacy pattern in selected villages of Takoli Gad Watershed (Source: census, 2011) *The villages not mentioned in census 2011.

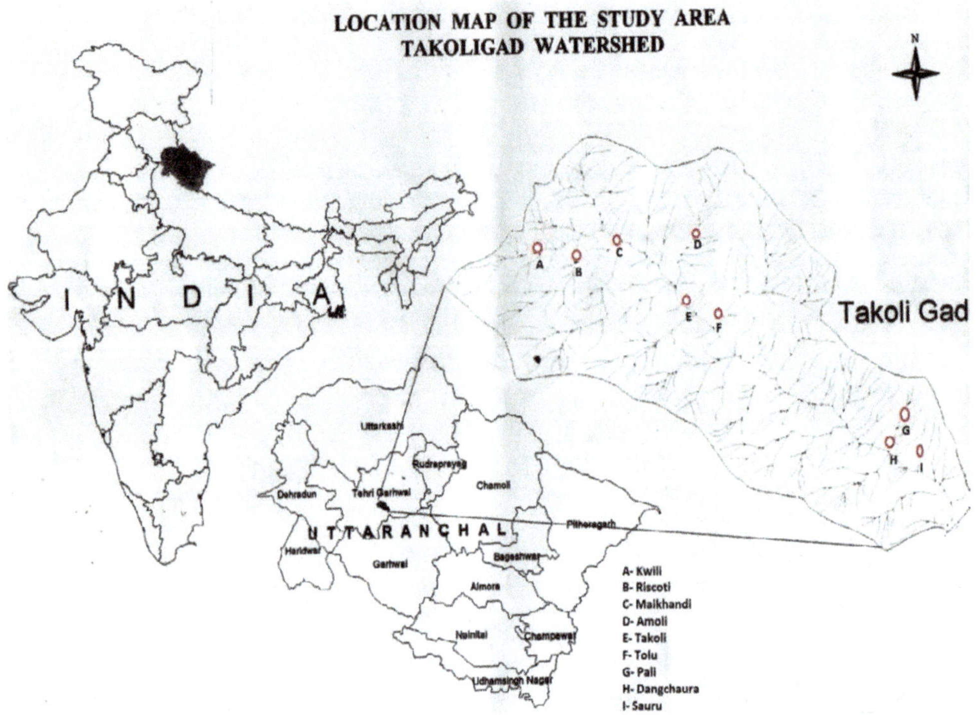

Figure 1: Location map of the study area.

Altitude (m amsl)	Village	Forest			Agroforestry		
		Small	Medium	Large	Small	Medium	Large
500-800	Dangchaura	496.40	408.80	350.40	146.00	109.50	105.85
	Pali	456.25	339.45	313.90	135.05	102.20	91.25
	Sauru	536.55	383.25	332.15	160.60	113.15	98.55
800-1100	Takoli	496.40	448.95	354.05	146.00	135.05	105.85
	Tolu	454.80	339.45	346.75	164.25	102.20	102.20
	Amoli	339.45	321.20	350.40	168.80	113.15	105.85
1100-1500	Maikhandi	627.80	514.65	324.85	186.15	135.05	94.90
	Riscoti	394.20	383.25	427.05	116.80	113.15	124.10
	Kwili	554.80	485.45	124.83	146.00	142.35	124.10

Table 2: Fuel consumption (kg/capita/year) from forest and agroforestry in different villages of Takoli Gad.

Results

Fuelwood consumption (kg/capita/year)

Forest: Table 2 revealed fuelwood consumption in different altitude. At lower altitude (500-800 m), in village Dangchaura, fuelwood consumption in small family was observed the highest (496.40 kg/capita/year), whereas, the lowest value of fuelwood consumption was reported in large family (350.40 kg/capita/year). The value of fuelwood consumption in the medium family was 408.80 kg/capita/year. In village Pali fuelwood consumption in the small family was 456.25 kg/capita/year, in medium family 339.45 kg/capita/year and in large family reported 313.90 kg/capita/year. In village Sauru fuelwood consumption in small family was the highest (536.55 kg/capita/year), followed by medium family (383.25 kg/capita/year) and in large family consumption the lowest (332.15 kg/capita/year (Table 2).

At medium altitude (800-1100 m), in village Takoli, fuelwood consumption reported maximum (496.40 kg/year) in small family followed by medium family (448.95 kg/capita/year) and in large family (354.05 kg/capita/year). In village Toluthe consumption in small family was again reported highest (454.80 kg/capita/year) and lowest in medium family (339.45 kg/capita/year). However, in large family the fuelwood consumption was 346.75 kg/capita/year. In village Amoli, fuelwood consumption in small family was maximum (339.45 kg/capita/year) and minimum in medium family (321.20 kg/capita/year). The consumption of fuelwood in large family was reported 350.40 kg/capita/year (Table 2).

At high altitude (1100-1500 m), in village Maikhandi, fuelwood consumption in small family was maximum (627.80 kg/capita/year) followed by 514.65 kg/capita/year (medium family) and in large family (324.85 kg/capita/year). In village Riscoti, the consumption in small family observed 394.20 kg/capita/year, in medium family 383.25 kg/capita/year and in large family 427.05 kg/capita/year. In village Kwili, fuelwood consumption was highest (1248.3 kg/capita/year) in large family size followed by small family (554.80 kg/capita/year) and medium family (485.45 kg/capita/year) (Table 2).

Agroforestry: Similar as forest the consumption of fuelwood was also calculated from agroforestry field of Takoli Gad Watershed, which is shown in Table 2. At lower altitudes (500-800), in village Dangchaura, fuelwood consumption in small family was reported highest (146 kg/capita/year), followed by medium family (109.50 kg/capita/year) and large family (105.85 kg/capita/year). In village Pali fuelwood consumption in small family was highest (135.05 kg/capita/year), than in medium family (102.20 kg/capita/year) and in large family (91.25 kg/capita/year). In village Sauru, fuelwood consumption in small family was 160.60 kg/capita/yearfollowed by medium family (113.15 kg/capita/year) and in large family (98.55 kg/capita/year) (Table 2).

At medium altitude (800-1100 m), in village Takoli, fuelwood consumption in small family was 146 kg/capita/year which reported highest and it reduced in medium family and large family with the values of 135.05, 105.85 kg/capita/year respectively. In village Tolu fuelwood consumption in the small family was 164.25 kg/capita/year, however, the fuelwood consumption value in medium family and large family was reported same (102.20 kg/capita/year) for both. In village Amoli, fuelwood consumption in the small family was 168.80 kg/capita/year, while in medium family and large family the values were of 113.15 kg/capita/year and 105.85 kg/capita/year respectively (Table 2).

At the high altitude (1100-1500 m), in village Maikhandi fuelwood consumption in the small family was maximum (186.15 kg/capita/year) reduced in medium family (135.05 kg/capita/year) and in large family (94.90 kg/capita/year). In village Riscoti fuelwood consumption in small family was 116.80 kg/capita/year, followed by medium family (113.15 kg/capita/year) and in large family (124.10 kg/capita/year). In village Kwili fuelwood consumption in small family were 146.00 kg/capita/year, in medium family 1423.50 kg/capita/year and in large family 124.10 kg/capita/year (Table 2).

Fuelwood consumption (kg/household/year)

Forest: Fuelwood consumption of villages of Takoli Gad watershed in different altitude on household basis is shown in Table 3. In village Dangchaura, fuelwood consumption was reported highest in small family size (1489.20 kg/household/year) followed by medium family (1226.40 kg/household/year) and in large family size (1058.50 kg/household/year). In village Pali fuelwood consumption in small family was highest (1368.75 kg/household/year), followed by medium family (1022.00 kg/household/year) and large family (949 kg/household/year). In village Sauru, fuelwood consumption in small family was 1609.65 kg/household/year, followed by medium family (1153.40 kg/household/year) and in the large family (1003.75 kg/household/year) (Table 3).

At the medium altitude (800-1100 m), in village Takoli, fuelwood consumption in small family was 1489.20 kg/household/year, which was reported highest followed by medium family (1346.85 kg/household/year) and in large family (1062.15 kg/household/year). In village Tolu, fuelwood consumption in small family was maximum (1671.70 kg/household/year) than in medium family (1022.00 kg/household/year) and in large family (1040.25 kg/household/year). In village Amoli, fuelwood consumption in the small family was 1022.00 kg/household/year, in medium family 970.92 kg/household/year and in large family 1058.50 kg/household/year (Table 3).

At high altitude (1100-1500 m), in village Maikhandi the highest (1834.40 kg/household/year) value of fuelwood consumption was reported in the small family, which reduced in medium family (1547.60 kg/household/year) and large family (981.85 kg/household/year). In village Riscoti, fuelwood consumption values in small family was 1186.25 kg/household /year, however, the values in medium and large family size were 1153.40 kg/household/year and 1281.15 kg/household/year respectively. In village Kwili, fuelwood consumption reduced in order of small family, medium family and large family with the observed values of 1671.70, 1423.50 and 1248.30 kg/ household/year respectively (Table 3).

Agroforestry: Similar as forest the consumption of fuelwood was also calculated household basis from agroforestry field. At lower altitude (500-800 m), in village Dangchaura, fuelwood consumption in the small, medium and large families reported as 445.30, 328.50 and 525.60 kg/household/year respectively. In village Pali, fuelwood consumption in the small family was 408.80 kg/household/year, followed by medium family (306.6 kg/household/year) and in large family (281.05 kg/household/year). In village Sauru, fuelwood consumption in small family was 481.80 kg/household/year, which reduced medium family (346.75 kg/household/year) and in large family (299.30 kg/household/year) (Table 4).

At the medium altitude (800-1100 m), in village Takoli, fuelwood consumption in the small, medium and large families were reported as 445.30, 405.15 and 317.55 kg/household/year respectively. In village Tolu, fuelwood consumption in small family was 500.05 kg/household/year, in medium family 306.60 kg/household/year and in large family 310.25 kg/household/year. In village Amoli, fuelwood consumption in the small family was 354.05 kg/household/year, however, in medium and large family the values were 346.75 kg/household/year and 317.55 kg/household/year respectively (Table 3).

At the high altitude (1100-1500 m), the selected villages were Maikhandi, Riscoti and Kwili. In village Maikhandi, fuelwood consumption was in order of 565.75, 405.15 and 292 kg/household/year in small, medium and large families respectively. In village Riscoti, fuelwood consumption in the small family was 354.05 kg/household/

Altitude (m amsl)	Village	forest			agroforestry		
		Small	Medium	Large	Small	Medium	Large
500-800	Dangchaura	1489.20	1226.40	1058.50	445.30	328.50	525.60
	Pali	1368.75	1022.00	949.00	408.80	306.60	281.05
	Sauru	1609.65	1153.40	1003.75	481.80	346.75	299.30
800-1100	Takoli	1489.20	1346.85	1062.15	445.30	405.15	317.55
	Tolu	1671.70	1022.00	1040.25	500.05	306.60	310.25
	Amoli	1022.00	970.92	1058.50	354.05	346.75	317.55
1100-1500	Maikhandi	1834.40	1547.60	981.85	565.75	405.15	292.00
	Riscoti	1186.25	1153.40	281.15	354.05	346.75	379.60
	Kwili	1671.70	1423.50	1248.30	445.30	427.05	372.30

Table 3: Fuelwood consumption (kg/household/year) from forest of different villages of Takoli Gad.

Altitude (m amsl)	Village	Small (2-4)	Medium (4-6)	Large (6-9)
500-800	Dangchaura	642.40 (22.72)	518.30 (21.12)	456.25 (23.20)
	Pali	591.30 (22.83)	441.65 (23.12)	405.15 (20.00)
	Sauru	697.15 (23.03)	696.40 (16.24)	430.70 (22.88)
	Total	**643.61(22.86)**	**552.11(20.16)**	**430.70(22.02)**
800-1100	Takoli	642.40(22.72)	584.00(23.12)	559.90(18.90)
	Tolu	619.05(26.53)	441.65(23.14)	448.95(22.76)
	Amoli	808.25(20.88)	434.35(26.05)	456.25(23.200
	Total	**689.90(23.37)**	**486.66(20.77)**	**488.36(21.62)**
1100-1500	Maikhandi	813.95(22.86)	649.70(20.78)	419.75(22.60)
	Riscoti	511.00(22.85)	496.40(22.79)	551.15(22.51)
	Kwili	700.80(20.83)	627.80(22.67)	248.83(49.87)
		675.25(22.18)	**591.30(22.08)**	**406.57(31.66)**

Table 4: Total Fuelwood consumption (kg/capita/year) from forest and agroforestry in different villages of Takoli Gad watershed (values in the parenthesis of small, medium and large show range of family size; however values in parenthesis of each villages is percent contribution of agroforestry of total fuelwood consumption.

Author(s)	Name of the region	Tribe/ sub- region	kg/capita/year
Maikhuri (1991)	Arunachal Pradesh	Nishis, Karbi, Kacharis, Chakma	949-5511.50
Maikhuri and Gangwar (1991)	Meghalaya	Garos, Khasis, Mikirs, Nepalese	146-912.50
Bhatt and Sachan (2004b)		Jaintia, Khasi, Garos	879.65-2620.70
Sundriyal et al. (1994)	Sikkim	Sikim Himalaya	1277.50
Bhatt and Sachan (2004a)	Uttarakhand	Garhwal Himalaya	390.55-1022.00
Bhatt *et al.* (1994)		Garhwal Himalaya,Tehri	408.80-890.60
Bartwal, 1987		Garhwal Himalaya	560
Dhanaiet *al.*2014		Garhwal Himalaya	646.05-1091.35
Kumar and Sharma(2009)		Garhwal Himalaya	1215.45-1803.00
Sagar *et al.* 1985		Dehradun	540
Srivastava (1981)		National average	600
Rajwar and Kumar (2009)		Nanda Devi Biosphere Reserve	1032.90 to 2157.15
Awasthi *et al.* (2003)		Garhwal Himalaya	1095
Reddy (1981), Hedge (1984)		South India	693.50 to 803.00
Mehatet *al.* (1987)	Nepal	Nepal Himalaya	448.95
Donovan (1981)	Asian countries.	South-east Asian countries.	620.50
Wijesinghe, (1984)		South-east Asian countries	985.50
Morton, (2007)	West Africa	Mali, West Africa	574, 380, and 291
Vishvakarma *et al.* (1998)	Himachal Pradesh	Kullu valley, H.P	1569.50
Rai and Chakrabarti (2001)		Hill forest urban areas, hill forest urban areas, plane forest areas, hill non-forest rural areas	441.65-5515.15
Rawatet *al.* (2009)		Khoksar, Jahlma, Hinsa, Kuthar	332.15-1095
Present study	Garhwal Himalaya (Uttarakhand)	Takoli Gad (Tehri Garhwal): 500-800	430.70-643.61
Present study		Takoli Gad (Tehri Garhwal):800-1100	486.66-689.90
Present study		Takoli Gad (Tehri Garhwal):1100-1500	406.57-675.25

Table 5: Fuelwood consumption (kg/capita/year) pattern in the Himalayan and other regions.

year, which reduce in medium family (346.75 kg/household/year) and again the value reported higher in large family (379.60 kg/household/year). In village Kwili, fuelwood consumption in the small family was 445.30 kg/household/year, followed by medium family (427.05 kg/household/year) and large family (372.30 kg/household/year) (Table 3).

Discussion

Fuelwood in the Himalayan regions is considered one of the important sources of energy, where majority of the villagers collect fuelwood from different sources such as forest, agro forestry, pasture land, community land etc. The present study focused on fuelwood collection by the villagers from forest and agroforestry at different altitudes. Among the altitudes, fuelwood consumption in lower altitude ranged from 430.70 to 643.61, middle altitudes 486.66 to 689.90 and upper altitudes 406.57 to 675.25. Many workers carried out study on fuelwood consumption in different parts of the world, country and Garhwal Himalayan region reported consumption values comparatively higher, lower and closed to present study.

Maikhuri [18] carried out fuelwood consumption study on different tribes of Arunachal Pradesh based on family size and recorded fuelwood consumption values from 2409.00-5511.5 kg/capita/year (Nishis), 1314-2774 kg/capita/year (Karbi), 949-1679 kg/capita/year (Kacharis) and 1095-2409 kg/capita/year (Chakma) tribes, where these values lowest to highest ranges from 949-5511.5 for Kacharis and Nishis respectively (Table 5). Bhatt and Sachan [23] Carried out a fuelwood consumption study in Garhwal Himalaya, ranges from 390.55-1022.00 kg/capita/year. Maikhuri and Gangwar [24] conducted fuelwood study in different tribes of Meghalaya and reported fuelwood consumption values between 328.50-912.50 kg/capita/year (Garos), 292.00-547.50 kg/capita/year (Khasis) 219-365 kg/capita/year (Mikirs), 146-365 kg/capita/year (Nepalese), where the lowest to highest consumption values ranges from 146-912 kg/capita/year for Nepalese tribes Garos (Table 5). Bhatt and Sachan [25] also carried out a study in the tribes of Meghalaya and reported fuelwood consumption values of 879.65-2051.30, 1573.15-2657.20 1365.10-2620.70 kg/capita/year for Jaintia, Khasi and Garos tribes respectively. The lowest and highest values ranged from 879.65-2620.70 for Jaintia and Garos respectively. Bhatt et al. [26] reported fuelwood consumption values between (408.80-890.60 kg/capita/year) for Western Himalaya are within the range of present study. However, the value reported (1095 kg /capita/year) by Awasthi et al. [27] in Garhwal Himalaya is quite higher than the present study. The fuelwood consumption of Raath area in Garhwal Himalaya reported 560 kg/capita/year [28]. Pandey and Singh [29] reported average fuel consumption (Table 5) value of 543.85 kg/capita /year for Central Himalaya.

Kumar and Sharma [30] have conducted a study in Garhwal Himalaya (Table 5). at three different altitudes (tropical region, sub-tropical region and temperate region) and reported fuelwood consumption values maximum (1803 kg/capita/year) in tropical region followed by 1492.85 kg/capita/year (temperate region) and 1215.45 kg/capita/year (sub-tropical region). Although the trend of fuelwood consumption for different purposes (cooking, heating, boiling of water and other purposes) was increased with increasing altitude, but higher values of fuelwood consumption in tropical region could be because of extra fuelwood consumption used for protection of crops from wild animals. In a similar study Sagar et al., [31], reported fuelwood consumption 540 kg/capita/year in rural areas of District Dehradun, however, the values reported (national average) of fuelwood consumption by Srivastava [32] was 600 kg/capita/year.

Rajwar and Kumar [33] conducted a study for fuelwood consumption in the villages Lata and Dunagiri in Nanda Devi Biosphere Reserve where fuelwood consumption was ranged between 1032.95 to 2157.15 kg/capita/year. Between the villages, higher values were recorded in village Dunagiri due to its location at higher elevation and its cooler condition than the village Lata. In other studies [34,35] of fuelwood consumption values were reported between 693.50 to 803.00 kg/capita/year for southern region of India (Table 5). Mehat et al. [36] reported fuelwood consumption value (448.95 kg/capita/year) for Nepal Himalaya. The fuelwood consumption for South and South East Asian countries reported as 620.50 kg/capita/year [37] and 985.50 kg/capita/year [38]. Jaiswal et al. [39] reported in their study that the fuelwood consumption was 657 kg/capita/year and at household level reported 5183 kg/household/year. The household level consumption is quite similar to the consumption level in low altitude areas of Uttarakhand reported as 4745 kg/capita/year [40]. In the study of Bhattacharya and Joshi [41], in the Eastern Himalayan area of North Bengal, the per household fuelwood consumption in average five-member household was found to be 4015-4745 kg/household/year. In rural areas of Garhwal Himalaya presented the per capita fuelwood consumption for cooking ranging from 412.45 to 664.30 kg/capita/year depending upon the altitudinal difference of the region [30].

A study carried out by [20] in the similar range of watershed of present study reported fuelwood consumption values ranged from 646.05 to 1091.35 kg/capita/year, these reported values are higher than present study. The reason for higher consumption values could be because of the study conducted at higher altitude up to 2000 m whereas; in the present study the sampling for fuelwood consumption was taken only up to 1500 m. It is evident from the several studies carried in Himalayan regions that increasing altitude leads more consumption of fuelwood because of severe cold with altitude. Therefore, the consumption reported lower

It is evident from our study region that villagers fulfill their major fuelwood demand from agroforestry field and forest. Although the major source of fuelwood consumption was taken from forest area which is almost 70% of total fuelwood consumption and rest 30% consumption was taken from agroforestry field. Jaiswal [39] suggested that major source of fuelwood was forest and local people try to meet fuelwood demand from agriculture field, pasture land, roadside plantations and open fallow land etc. Jaiswal [39] also suggested that nearly 87% households fulfill their fuelwood requirement completely from forest. Mahapatra and Mitchell [42] have suggested 50% household dependence in Orissa. In North-West Bengal, around 84% of the rural households collect fuelwood directly from the State forests [41].

The results of various workers have been given in Table 5, indicated that the fuelwood consumption varies with site to site, variation in altitudinal gradients. The higher consumption fuelwood reported in north-eastern regions which might be because of high consumption of meat and for cooking them need extra sources of energy including other daily routine works. Similarly, in Himalayan region consumption comparatively increased with increasing attitudes, where villagers consume more energy for space heating due to severe cold particularly in winter seasons and need more time for cooking because of low temperature which provides more consumption of fuelwood.

Figure 2, indicates the fuelwood consumption (irrespective family size) increased with increasing altitudes i.e., 541.8 (500-800 m), 554.97 (800-1100 m) and 557.71 kg/capita/year (1100-1500 m). Figure 3, indicates the fuelwood consumption (irrespective altitudes) reduced with increasing family size as small > medium > large i.e., 669.58 (small family), 543.35 (medium family) and 441.88 (large family).

The distance travelled by villagers for fuelwood collection is shown in Table 6. The distance of forest was the highest (5 km) at upper altitude (800-1100 m) followed by middle altitude (3.16 km) and lower altitude (1.43 km). However, the distance travelled by villagers for fuelwood collection from agroforestry again reported the highest (0.56 km) at upper altitude followed by middle altitude (0.33 km) and lower altitude (0.3 km). Although distance travelled by the villagers for the collection of forest produce especially fuelwood would not affect the consumption requirement but affect other activities of villager where they can contribute more time for other activities.

Conclusion and Suggestions

The study in the Takoli Gad Watershed of Garhwal Himalaya is

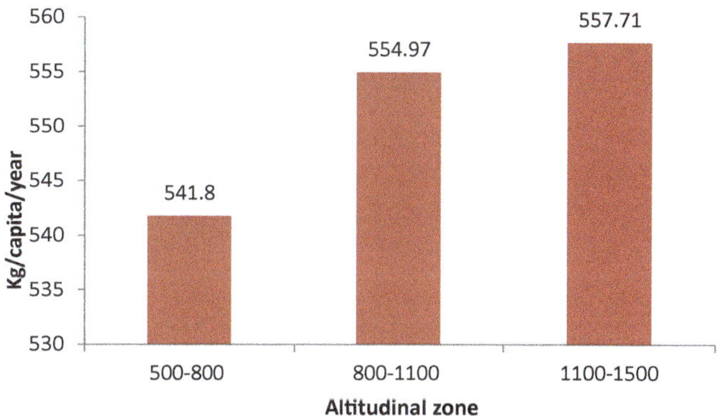

Figure 2: Fuelwood consumption along altitudinal gradient (irrespective family size) in Takoli Gad Watershed.

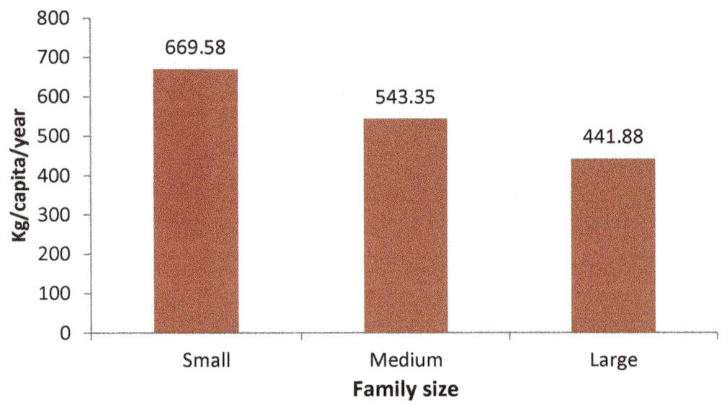

Figure 3: Fuelwood consumption pattern of different family size (irrespective altitude) in Takoli Gad Watershed.

Altitude	Distance from forest (km)	Distance from agroforestry (km)
500-800 amsl	1.43 ± 0.40	0.3 ± 0.1
800-1100 amsl	3.16 ± 0.76	0.33 ± 0.21
1100-1500amsl	5 ± 1	0.56 ± 0.15

Table 6: Distance travelled (km) by the villagers for fuelwood collection in Takoli Gad watershed.

focused on consumption of fuelwood from forest and agro forestry by the villagers in different altitudes. It has been noticed that the proportion of fuelwood consumption of agroforestry compared to total fuelwood was 25-30%. This consumption of fuelwood from agroforestry reduces direct pressure on forest resources. Therefore introduction of more existed traditional tree species in agroforestry systems can reduce pressure on forest and will enhance health of our degraded/degrading forests. But before introducing trees in agroforestry systems the suitable screening should be done with agricultural crops for its least adverse effect i.e., shading, allelopathic effect, etc. Thus the system can run for longer period.

References

1. FAO (2001) State of world forest FAO report. Rome, Italy: Food and Agriculture Organization 181.
2. Purohit AN, Dhar U (1997) Himalayan tree diversity - an update. Proc. Indian National Science Academy 63: 187-209.
3. Erakhrumen AA (2008) Wood Biomass as a Major Source of Energy in Sub-Sahara African Region: Implications for Sustained Research and Education in Agroforestry Technologies. In: Onyekwelu, J.C., Adekunle,V.A.J. and Oke, D.O. (eds.). Research for Development in Forestry, Forest Products and Natural Resources Management. Proceedings of the FirstBiannual Conference of the Forests and Forest Products Society of Nigeriaheld at the Federal University of Technology, Akure, Ondo State, Nigeria. 205-211.
4. Arnborg T (1984) Agroforestry. A Paper presented at the Seminar of Faculty of Agriculture, University of Sokoto, Sokoto 14.
5. IEA (2002) Electricity in India. International Energy Agency Paris: OECD.
6. FAO (2007) State of the World's Forests. Food and Agriculture Organization of the United Nations, Rome. 1-144.
7. Leakey, RRB and Izac, AMN (1996) Linkages between Domestication and Commercialization of non-timber Forest Products: Implications for Agroforestry, Eds. Leakey R.R.B. et al. Domestication and Commercialization of non-timber Forest products in Agroforestry Systems. Proceedings of an International Conference held in Nairobi, Kenya. Food and Agriculture Organization of the United Nations, Rome.
8. FAO (2002) Forests and Forestry in Small Island Developing States by Wilkie ML, Eckelmann, CM, Laverdiere M and Mathias A. Forest Management Working Paper No. FM 22. Food and Agriculture of the United Nations, Rome, Italy.
9. Eckholm E (1975) The Other Energy Crisis: Firewood. World watch Paper 1. World watch Institute.
10. Das D, Srinivasan R (2012) Income levels and transition of cooking fuel among rural poor in India. Energy Science and Technology 4: 85-91.
11. Bhattacharya B, Nanda SK (1992) Building fuel wood demand supply scenario. Journal of Rural Development 11: 773-787.
12. Singh V (1989) Energetic of agroecosystem and its relation to forest ecosystem in the Central Himalaya. Environmental Conservation 11: 45-53.
13. Metz JJ (1990) Conservation practices at upper elevation village of west Nepal. Mountain Research and Development 10: 7-15.
14. Kumar P (2005) Natural Resource Policy: Some Related Issues. In: Dewan ML, Bahadur J (Eds.), Uttaranchal Vision and Action Programme. Concept Publishing Company, New Delhi.
15. Shah SL (1982) Ecological degradation and future of agriculture in the Himalaya. Indian Journal of Agriculture Economics 37: 1-22.
16. Khoshoo TN (1987) Strategies for meeting the fire wood needs in the hills. In: Dhar TN, Sharma PN (Eds.), Himalayan Energy System. Nainital Gyanodaya Prakashan 11-19.
17. Singh JS, Singh SP, Ram J (1998) Fodder and fuelwood resources of central Himalayan. Problems and Solutions. Report Submitted for Study Group on Fuel and Fodder, Planning Commission, Government of India, New Delhi.
18. Maikhuri RK (1991) Fuel wood consumption pattern of different tribal communities living in North-East India. Bioresources Technology 35: 291-6.
19. Negi AK, Todaria NP (1993) Fuel wood evaluation of some Himalayan trees and Shrubs. Energy 18: 799-801.
20. Dhanai R, Negi RS, Parmar MK, Singh S (2014) Fuel wood & Fodder Consumption Pattern in Uttarakhand Himalayan Watershed. International Journal of Environmental Biology 4: 35-40.
21. Mitchell R (1979) An analysis of Indian agro ecosystem Interprint New Delhi, India.
22. Gupta RK, Pathania MS, Gupta T (1997)A study on fuel consumption pattern in Himachal Pradesh: a case study. Range Management & Agroforestry 18: 181-8.
23. Bhatt BP, Sachan MS (2004) Firewood consumption along an altitudinal gradient in mountain villages of India. Biomass and Bioenergy 27: 69-75.
24. Maikhuri RK, Gangwar AK (1991) Fuel wood use by different tribal and non-tribal communities in North East India. Natural Resources Forum 15: 162-165.
25. Bhatt BP, Sachan MS (2004) Fire wood consumption pattern of different tribal communities in Northeast India. Energy Policy 32: 1-6.
26. Bhatt BP, Negi AK, Todaria NP (1994) Fuel wood consumption pattern at different altitudes in Garhwal Himalaya. Energy 19: 465-8.
27. Awasthi A, Uniyal SK, Rawat GS, Rajvanshi A (2003) Forest resource availability and its use by the migratory villages of Uttarkashi, Garhwal Himalaya, India. Forest Ecology and Management 174: 13-24.
28. Bartwal PS (1987) Fuel wood consumption pattern in rural areas case study. Journal of Tropical Forestry 3: 136-41.
29. Pandey U, Singh JS (1984) Energy-flow relationship between agriculture and forest ecosystems in Central Himalaya. Environmental Conservation 11: 45-53.
30. Kumar M, Sharma CM (2009) Fuel wood consumption pattern at different altitudes in rural areas of Garhwal Himalaya. Biomass and Bioenergy 33: 1413-1418.
31. Sagar SR, Chandola LP, Ansari MV (1985)Pilot survey of fuel consumption in rural areas. Indian Forester 111: 305-17.
32. Srivastava BP (1981) High density short relation forestry for mitigating the energy crisis in India. Indian Forester 107: 767-770.
33. Rajwar GS, Kumar M (2011) Fuel wood consumption in two tribal villages of the Nanda Devi Biosphere Reserve of the Indian Himalaya and strategies for fuel wood sustainability Environ Dev Sustain 13: 727-741.
34. Reddy AKN (1981) An Indian village agricultural ecosystem case study of Ungra village. Part II. Discussion. Biomass 1: 77-88.

35. Hedge MS (1984) Fuel problem in villages: challenges and opportunities. Bulletin of Science, 8-13.

36. Mehat TBS, Grigflin DM, Shephered KP (1987) Human impacts onsome forests of the middle hills of Nepal. Part 4: a detailed study in South-east SindhuPalanchock and Northeast KabherePalanchock. Mountain Research and Development 7: 114-34.

37. Donovan DG (1981) Fuel wood: how much do we need? Han over, NH: Institute of Current World Affairs.

38. Des Wijesinghe LCA (1984) A sample study of biomass fuel consumption in Sri Lanka households. Biomass 5: 261-82.

39. Jaiswal A, Bhattacharya P (2013) Fuel wood Dependence around Protected Areas: A Case of Suhelwa Wildlife Sanctuary, Uttar Pradesh. J. Hum Ecol 42: 177-186.

40. Sati VP, Song C (2012) Estimation of forest biomass flow in the Mountain Mainland of the Uttarakhand Himalaya. International Journal of Forest, Soil and Erosion 2: 1-7.

41. Bhattacharya P, Joshi B (2001) Public Forests, Fuel wood Collection and Migration: A Case Study in North-West Bengal. Field Document No. 60.Bangkok: FAO.

42. Mahapatra AK, Mitchell CP (1999) Biofuel consumption, deforestation, and farm level tree growing rural India. Biomass and Bioenergy 17: 291-303.

Woody Species Diversity of Traditional Agroforestry Practices in Gununo Watershed in Wolayitta Zone, Ethiopia

Bajigo A[1]* and Tadesse M[2]

[1]*Jigjiga University College of Dryland Agriculture (Natural Resource Management Department), Jigjiga, Ethiopia*
[2]*Mekelle University College of Dryland Agriculture and Natural Resources, Mekele, Ethiopia*

Abstract

In Gununo watershed at Wolayitta zone, Ethiopia, to determine the variation of woody species structure and composition among agroforestry practices and along elevation gradients in the watershed, this study was conducted. The elevation gradient was stratified in to three transects (upper, middle and lower); and three dominant agroforestry practices (homegarden, parkland and woodlot) were used as treatments. Along each transect line, three agroforestry practices with three replications, a total of 27 sampling points, a complete enumeration of woody species in homegarden which has the average are of 900 m2. Whereas, 50 m × 100 m sampling quadrates in parklands, and 10 m × 10 m sample quadrates for woodlot were used. Number of individuals per plot, DBH, height, crown diameter, and plot area were measured and recorded. The structure and composition of woody species through important value index (IVI), basal area, canopy cover, and diversity indexes determine were determined. A total of 32 woody species belonging to 19 families were recorded in the three agroforestry practices. Fabaceae family 28%, (9 species) was the dominant family of the woody species recorded followed by Euphorbiaceae 13%, (4 species). From these species, 69% (22 species) were indigenous and 31% (10 species) were exotic. From the indigenous species, two were endemic to Ethiopia *(Erythrina brucei* and *Millettia ferruginea*). The largest indigenous species, and highest species diversity (20) were recorded in homegardens followed by parklands (11)). Middle elevation was exhibited higher indigenous species, and highest species richness (19): followed by upper (15) and lower elevation. The Shannon, Simpson and evenness diversity indexes were highest in homegardens. Comparing to parkland, homegarden diversity indexes were highest in upper, middle and lower elevations in decreasing order. Attention should be given to the existing agroforestry practices, and the practices must be promoted to lower elevation to enhance biodiversity conservation in agroforestry land use system.

Keywords: Diversity; Endemic; Homegarden; Indigenous; Parkland; Woodlot

Introduction

Ethiopian highlands are typically characterized by high population density, small land holdings and associated land degradation [1]. This aggravates unsustainable land management and drastically affects the contribution of agriculture [2]. This is also true in Southern Ethiopia particularly to Gununo area: with high population density. Therefore, the watershed is prone to land degradation and resource depletions such as soil erosion, flooding, loss of biodiversity and associated impacts on household economy [1,3,4]. In relation to this, the local people have been using different strategies to combat such degradations. One of the strategies is a traditional agroforestry practices [5]. Traditional agroforestry practices with it woody species diversity, have been serving different protective and productive roles in Gununo watershed.

Woody vegetation is characterized by its plant morphological characters (structure) or the recognized plant species (composition). Structure denote the vertical arrangement of different plant form and its associated canopy layers whereas composition refers to the number of species and its associated diversity in a system [6]. The structure of tropical agroforestry looks a forest like appearance (complex multilayer structure) and their plant diversity is rich i.e. comparable to that of natural forest [7]. This is due to selective and repeated planting and management of useful woody species from a spontaneous regeneration [7,8,9].

Therefore, because of their high plant diversity, structural complexity, and management (possibly that favor for useful woody species) agroforestry practices are very essential in conservation and management of tropical forest landscapes [8], and species richness at regional and global scales [10]. Homegardens and agroforests are a valuable example for this [8,11]. Structure and composition of tropical agroforestry is reported to be rich by many authors [7,9,12].

Many studies mainly based on floristic composition indicated that pattern of vegetation and floristic diversity is generally governed by climate, elevation, moisture availability, nutrient and topographic variables. Yitebitu, [13] has reported as elevation gradient have effect on the structure and composition of woody species. Nuberg [14] has also found as the species diversity tend to decrease with altitude. Further, the farm characteristics such as farm size, shape, species adaptability, nature of cropping pattern, and management variation also affect the structure and composition in agroforestry [7,15]. Therefore, this study was conducted with the objective to determine the variation of woody species structure and composition among agroforestry practices and along elevation gradients in the watershed.

***Corresponding author:** Bajigo A, Jigjiga University College of Dryland Agriculture (Natural Resource Management Department), Jigjiga, Ethiopia
E-mail: bajigoaklilu05@gmail.com

Materials and Methods

Study area description

Wolayitta zone is in southern nations, nationalities and peoples (SNNP) region of Ethiopia with a total land area of 4537.5 sq kms is located between 6°4′N to 7° 1′N latitudes and 37°4′E to 38°2′E longitudes. It is located 22 km from Sodo town and about 330 km from Addis Ababa (Figure 1). The watershed has an area of about 544 hectare with three rural districts namely: Demba Zamine (middle elevation), Doge Hunchucho (lower elevation) and Chew kare (upper elevation).

Gununo watershed has plain lands, plateaus, hills and rugged mountains topography with an altitude ranging from 1937 to 2100 meter above sea level [16,17]. The mean annual temperature of the site is 19.2°C and the mean annual rainfall is 1335 mm with bimodal rain pattern [17].

Soil of the watershed is Eutric Nitosol according to FAO/UN classification system [4]. Soil erosion in watershed is severe due to conversion of natural forests to other land uses. The study area has high population pressure i.e. around 450 person per km^2 [17], and an average land holding is about 0.25 ha per household and drive farmers to cultivate slope lands [16]. Agroforestry is one of the major land uses at the area. Different species (tree crops and woody species) such as *Enset ventricosum Musa accuminata, Moringa oleifera* and *Brassica oleracea* serve as primary food source while *Croton macrostachyus* and different Acacia species are the dominant trees in the degraded natural forest of Wolayitta [17].

Sampling procedures

Since the area is with steep slopes and undulating topographic, the watershed was classified in to three transects. Consequently, upper (2006 to 2040), middle (1972 to 2006) and lower zone (1937 to 1971) meters above sea level (m.a.s.l.) ware considered. At each zone, the middle point was chosen for horizontal transect walk i.e. at 2023, 1989 and 1954 m.a.s.l. for upper, middle and lower transect respectively. These transect lines thatch all the three districts that the watershed contains.

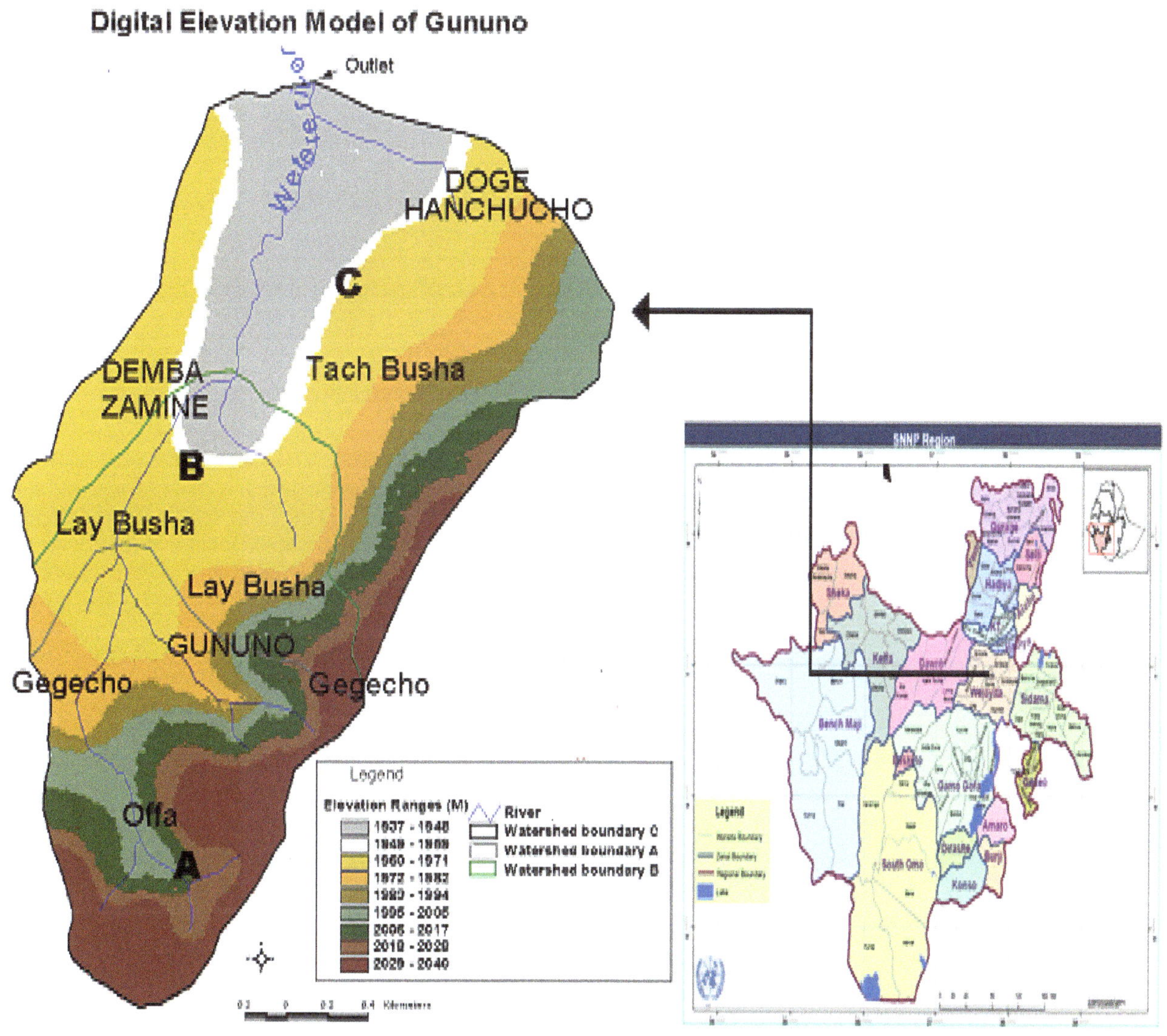

Figure 1: Map of Wolayitta Zone and Gununo watershed, Southern Ethiopia (adopted from Mowo et al., 2011).

The distance of each transect line was measured from the delineated watershed map and sampling points were distributed proportionally. Accordingly, at each horizontal transect line the first sampling point was randomly selected i.e. some distance away from initial standing point. Then the next sampling point (quadrate) was allocated systematically at every two kilometer by using GPS. Therefore, along each transect line, three agroforestry practices (homegarden, parkland and woodlots) with three replications, a total of 27 sampling points were used. These sampling plots per land use/agroforestry practices were found sufficient according to the plot number-species accumulation curve done after data collection. No quadrates were laid in homegarden; rather a complete enumeration of woody species with average area of 900 m^2 following Motuma [18]. In parklands vegetation data was collected from 50 m × 100 m sample size quadrates [19], and for woodlot 10 m × 10 m sample quadrates were taken [20].

All the woody species in each sample plot greater than or equal to five centimeter DBH was taken [18]. Height less than 1.5 m was considered as seedlings while height between 1.5 and 3 m was taken as sapling and height greater than 3 m was taken as tree [20]. The average height of woody species was used for homegardens (average of upper and middle story). Woody species very close to the sampling points were taken for species richness estimation [21]. At every sampling point, number of individuals per plot counted, DBH was measured by caliper, height by hypsometer, and crown diameter by measuring tape, were measured recorded.

Plant species determination was done with the help of local communities, and identification of the scientific names of species was carried out using books of Edwards [22], and Woldemichael [23,24].

Data analysis

The structure and composition of woody species were determined through quantitative analysis. For the determination of species structure, parameters such as height, important value index (IVI), which includes relative frequency, relative density, and relative dominance, were calculated following Leul [25], and Newton [26]. Basal area, frequency, density and canopy cover were also measured. To determine the species composition, diversity indexes (Shannon diversity index, Simpson diversity index, and evenness) were computed following Newton [26].

Crown area was calculated following (Crown area= π (0.5 * average crown diameter)^2). The Shannon diversity index was calculated as:

$$H' = \sum_{i=1}^{s} (P_i)(\ln P_i)$$

Where: H′=index of species diversity, S=number of species, pi=proportion of total sample belonging to the ith species, ln=natural log. Simpson diversity index was also calculated as:

$$SI = \sum_{i=1}^{s} P_i^2 ,$$

Where: SI = Simpson's index of species diversity, S = number of species, pi = proportion of total sample belonging to the ith species. Evenness (E') was calculated as Begon [27]:

$$E = \frac{H'}{\ln S},$$

where H'=is the Shannon diversity index, S=is the number of species in a particular elevation contour.

Two way ANOVA was carried out at P<0.05 with the help of (SPSS versions 16), agroforestry practices and elevation were considered to analyze the structure and composition of woody species. Least significance difference (LSD) test was used to separate the means.

Results

Dominant agroforestry practices in the watershed

The common agroforestry practices in the watershed are homegardens, parklands, woodlots and boundary plantation. However, there is also alley cropping and soil erosion control with woody perennials in some households. The homegarden (HG) of the study site is *Enset-coffee* based. Woody species such as *Persea americana*, *Croton macrostachyus*, and *Cordia africana* are mainly dominating the upper story while *Enset ventricosum*, *Coffee arabica*, *Rhamnus prinoides* and *Musa accuminata* dominate the middle story. Whereas, different vegetables, root and tuber crops and pulses such as *Brassica integrifolia*, *Capsicum frutescens*, *Solanum tuberosum*, *Ipomoea batatas*, *Dioscorea alata*, *Colocasia esculenta*, *Phaseolus vulgaris*, and *Pisum sativum* dominate the lower strata/on ground.

The main crop component in homegardens and parklands are *Dioscorea alata*, *Colocasia esculenta*, *Solanum tuberosum*, *Ipomoea batatas*, *Phaseolus vulgaris*, *Manihot esculenta* (*kassava*), *Brassica oleracea* (cabage), *Enset ventricosum*, *Musa accuminata*, *Pisum sativum*, *Coffee arabica* and *Zea mays*. The most important agricultural crops such as *Eragrostis tef*, *Enset ventricosum*, *Ipomoea batatas*, *Solanum tuberosum*, and *Phaseolus vulgaris*, and plantation crop such as *Coffee arabica*. These crops were integrated together in the same land management unit, except *Eragrostis tef* and *Phaseolus vulgaris* which are poorly integrated in simultaneous agroforestry practices.

Among the tree component of traditional agroforestry practices, *Cordia africana*, *Acacia sieberiana*, *Maesa lanceolata*, *Croton macrostachyus*, *Ficus* species (in the upper and middle slope) and *Moringa oleifera* (in the lower slope) are the most chosen woody species in homegardens and parklands for multiple products and services.

Structure and composition

A total of 32 woody species belonging to 19 families were recorded in the three agroforestry practices. The family Fabaceae was 28%, (9 species) which was the dominant family of the woody species recorded in the study area. Following this, Euphorbiaceae was 13%, (4 species) which was the commonly observed family among woody species. However, many of the families were represented by single species. From this species, 69% (22 species) were indigenous and 31% (10 species) were exotic (Table 1). From the indigenous species, two were endemic to Ethiopia (*Erythrina brucei* and *Millettia ferruginea).* The largest indigenous species were recorded in homegardens; followed by parklands while woodlots were absent in indigenous species.

In terms of woody species distribution across agroforestry practices, 54% (14 species) of woody species existed only in homegardens, 19% (five species) occurred only in parklands whereas 4% (one species) occurred only in woodlot. Again, 23% (6 species) were found in both homegardens and parklands, whereas no species were found in common to all agroforestry practices (Table 2). Consequently, homegardens were highest in species diversity (20); followed by parklands (11) and woodlots (only one).

Seven woody species (27%) existed in common among the three elevations. Six unique woody species (23%) were recorded in upper contour. Again, five unique woody species (19%) were recorded in middle elevation, whereas one species was uniquely recorded only in the lower elevation. Common woody species in upper/middle and

Woody species	Vernacular name	Family name	Origin
Asteracantha longifolia nees.*	Xuxuwa	Acanthaceae	E
Acacia senegal (L.) Willd.*	Gaammo Gaadiya	Fabaceae	I
Acacia sieberiana DC.	Gara	Fabaceae	I
Albizia gummifera (J.F. Gmel.) C.A. Sm.	Shuwa/ Chata	Fabaceae	I
Annona senegalensis Pers.	Eta	Annonaceae	I
Brucea antidysenterica J.F. Mill.	Shulshudhya	Simaroubaceae	E
Carica papaya L.	Papaya	Caricaceae	E
Casimiroa edulis Llave & Lex.*	Kasmiree	Rutaceae	E
Citrus sinensis L.	Birtukaniyaa	Rutaceae	E
Coffea arabica L.	Tukkiyaa	Rubiaceae	I
Cordia africana Lam.	Moqotta	Boraginaceae	I
Croton macrostachyus Del.	Anka	Euphorbiaceae	I
Entada abyssinica Steud. ex A. Rich.*	Kontre	Fabaceae	I
Erythrina abyssinica Lam.Ex.Dc.	Korra/ Quarra	Fabaceae	I
Erythrina brucei schweinf.	Korch /Bortwa	Fabaceae	I(en.)
Eucalyptus camaldulensis Dehnh.	Zafiya	Myrtaceae	E
Euphorbia candelabrum Kotschy	Akirsa	Euphorbiaceae	I
Euphorbia tirucalli L.	Maaxxuwa	Euphorbiaceae	I
Faidherbia albida (Del.) A. Chev.	Odoruwa	Fabaceae	I
Ficus sycomorus L.	Wolaa	Moraceae	I
Ficus thonningii Blume*	Etta	Moraceae	I
Maesa lanceolata Forssk.*	Gergecuwa/ Gergecho	Myrsinaceae	I
Mangifera indica L.	Manguwa	Anacardiaceae	E
Manihot esculenta Crantz	Cassava /Mita Boyiya	Euphorbiaceae	E
Millettia ferruginea (Hochst.) Baker	Zage/ Zagiya	Fabaceae	I (en.)
Moringa oleifera Lam.	Moringa	Moringaceae	E
	Mayluwa	Fabaceae	I
Olea africana Mill.	Wogara	Cuspidate	I
Persea americana Mill.	Avocadosiya	Lauraceae	E
Podocarpus falcatus (Thunb.)* R.Br. ex Mirb.	Ziga	Podocarpaceae	I
Prunus africana (Hook.f.) Kalkm.	Garbia	Rosaceae	I
Rhamnus prinoides L'Hérit.	Geeshuwa	Rhamnaceae	I

Note: The letter I is for indigenous, E for exotic whereas en. is for endemic species. Similarly T is for tree and S is for shrub. * Sign is for woody species that exist very adjacent to the sample plot and used for species richness estimation in the area

Table 1: Woody species in traditional agroforestry practices of Gununo watershed.

Elevation	Agroforestry practices	Simpson	Shannon	Evenness
Upper	Homegarden	0.88	2.23	0.61
	Parkland	0.8	1.63	0.5
	Woodlot	0	0	0
Middle	Homegarden	0.84	2.16	0.52
	Parkland	0.81	1.56	0.5
	Woodlot	0	0	0
Lower	Homegarden	0.83	1.67	0.48
	Parkland	0.8	1.54	0.5
	Woodlot	0	0	0

The highest relative frequency exhibited in homegarden was for *Coffea arabica, Persea*

Table 2: Tree species diversity in agroforestry practices of Gununo Watershed.

middle/lower were two (8%), and five (19%) respectively. No common woody species were encountered to the upper and lower elevation. The upper and middle elevations have fewer common species than the middle and lower contours. Consequently, the middle elevation was highest in species richness (19): followed by upper (15) and lower elevation (13).

The Shannon, Simpson and evenness diversity index were highest in homegardens (average H'=2.02, Simpson index=2.55 and evenness=0.54), parklands (average H'=1.58, Simpson index=0.8 and evenness=0.5) and woodlots (average H´=0, Simpson index 0 and evenness=0) in decreasing order. Since, the woodlots is composed of single species, the diversity index was relatively zero. Comparing homegarden and parkland, diversity indices were highest in homegarden particularly to upper, middle and lower elevation in decreasing order (Table 2).

americana and *Cordia africana* while the highest relative density was recorded for *Coffea arabica, Cordia africana* and *Croton macrostachyus*. The relative dominance was highest for *Coffea arabica, Euphorbia candelabrum* and *Rhamnus prinoides*. The important species according to important value index was *Coffea arabica, Euphorbia candelabrum* and *Coffea africana* in deceasing order (Table 3).

The highest relative frequency was exhibited for *Cordia africana, Euphorbia tirucalli* and *Moringa oleifera* in parkland while the highest relative density was recorded for *Ficus sycomorus* and *Millettia ferruginea*. In addition, the highest relative dominance was recorded for *Coffea arabica, Faidherbia albida* and *Moringa oleifera* in parkland. Further, the important species in parkland according to important value index was *Cordia africana, Moringa oleifera* and *Euphorbia tirucalli* in deceasing order (Table 4).

and Homegarden (7.15 m). There was significant difference in density and crown cover among agroforestry practices. Hence, woodlots were the highest in density and crown cover (Table 5).

Significant difference in density/ha across elevation gradients was observed between middle and lower elevations. Consequently, the highest density/ha was in the middle elevation (Table 6).

Discussions

Structure and composition of woody species

Homegarden, parkland and woodlot are the most commonly used traditional agroforestry practices in Gunono watershed. The traditional agroforestry practices of the watershed are among the commonly known agroforestry practices in Southern Ethiopia [28,29] and throughout tropics [9]. Consequently, within these traditional agroforestry practices different woody species were dominating the upper and middle story. The dominancy of fruit and timber trees in the upper story is related to farmers' tree species preference. Consequently, the woody species, plantation crops and agricultural crops are chosen due to their shade, soil fertility improvement, less competitive effect to middle and lower story, and income generation such as timber production. For example *Enset* is chosen because of its drought resistance, good in erosion control, soil fertility enhancement and staple food source in the area. Fruit tree (*Persea americana*) and timber tree (*Cordia africana*) were also reported to mainly dominate the shade canopy layer in the traditional coffee agroforests of South Ethiopian [13] while species of *Enset* and coffee was also reported to dominate the middle story in Dryland regions. In addition, different vegetables, root and tuber crops and pulses found in lower strata/on the ground in the study area. This is in line with [30].

Species name	RD	RF	RDOM	IVI
Coffea Arabica	92.02056	18	22.25151	132.2721
Euphorbia candelabrum	0.057623	2	21.96966	24.02728
Cordia Africana	2.542204	10	10.55404	23.09625
Persea americana	0.606166	12	3.281196	15.88736
Rhamnus prinoides	0.576232	4	10.98483	15.56106
Croton macrostachyus	1.091808	6	7.708653	14.80046
Annona senegalensis	0.276591	6	5.858576	12.13517
Mangifera indica	0.712433	10	0.976438	11.68887
Brucea antidysenterica	0.293355	6	2.662989	8.956343
Erythrina brucei	0.115246	2	6.590898	8.706144
Manihot esculenta	0.834549	6	0.631329	7.465878
Erythrina abyssinica	0.057623	2	2.196966	4.254589
Acacia sieberiana	0.230493	4	0	4.230493
Euphorbia tirucalli	0.288116	2	1.098483	3.386599
Citrus sinensis	0.038415	2	1.220537	3.258952
Carica papaya	0.115246	2	0.732322	2.847568
Prunus africana	0.057623	2	0.732322	2.789945
Myluwa (local name)	0.057623	2	0.366161	2.423784
Olea africana	0.028812	2	0.183081	2.211892

Note: RD means relative density, RF means relative frequency, RDOM means relative dominance, IVI means important value index

Table 3: Importance value index of woody species for homegarden in Gununo Watershed.

Species	RD	RF	RDo	IVI
Cordia Africana	44.74034	41.66708	37.45729	123.8647
Moringa oleifera	15.97869	8.333417	20.50582	44.81793
Euphorbia tirucalli	18.64181	16.66683	2.734109	38.04275
Faidherbia albida	5.99201	8.333417	23.58169	37.90712
Maesa lanceolata	6.657789	8.333417	5.468218	20.45942
Ficus sycomorus	3.994674	8.333417	6.151745	18.47984
Millettia ferruginea	3.994674	8.333417	4.101163	16.42925

The highest woody species length was observed on woodlot (19m) followed by parkland (10.3m)

Table 4: Important value index of woody species in parkland agroforestry practice in Gununo Watershed.

AFP	Den/ha	Cc/ha (%)
Homegarden	1,125.23 ± 334.6 b	15.81 ± 3.06 b
Parkland	8.79 ± 1.3 c	0.74 ± 0.14 c
Woodlot	28,600.13 ± 1772 a	94.35 ± 4.09 a
P value	P<0.0001	P<0.0001

Note: Mean value is shown for agroforestry practices and values next to ± indicate the standard error of the mean. ns refers to not significant. Den/ha and Cc/ha means density per hectare and crown cover per hectare, respectively. These values were the transformed into log scale for comparison before ANOVA

Table 5: Woody species structural parameter within agroforestry practices in Gununo Watershed.

Elevation	Den/ha	Cc/ha (%)
Upper	10,277.98 ± 5072.3 ab	32.41 ± 14
Middle	11,012.54 ± 5383.547a	37.93 ± 15.2
Lower	8,443.64 ± 3697.6 b	40.55 ± 14.9
P value	P=0.027	Ns

Note: Mean value is shown along elevation level and values next to ± indicate the standard error of the mean. ns refers to not significant. Den/ha and Cc/ha means density per hectare and crown cover per hectare, respectively. These values were transformed into log scale for comparison before ANOVA

Table 6: Woody species structural parameter along elevation gradient in Gununo Watershed.

The majority of woody species in the watershed was represented by the family of Fabaceae and Euphorbiaceae. These families were also dominant in the south-eastern rift valley escarpment of Ethiopia [31], Southern Tigray [25] and other tropical region such as SW china [25]. Among these family woody species such as *Erythrina brucei* and *Millettia ferruginea* are under red list of endemic trees of Ethiopia [32]. The existence of these species in traditional agroforestry practices showing that it has given the advantage of conserving these species.

Variation in structure and composition of woody species among agroforestry practices in the study area is attributed to the complex nature of homegarden; which is related to the diverse products expected from it. Whereas, simple structure of woodlot is due to that it is composed of single species. Misana [33] also found species diversity and richness differences among agroforestry practices. In relation to this, higher diversity indexes in homegarden (followed by parkland and woodlot) is attributed to difference in distribution of number of individuals and species richness as a result of variation in woody species efficiency and farmers species preference. Consequently, higher species diversity and richness of homegarden than parkland and woodlot is consistence with the claim of high tropical homegarden complexity, species richness and diversity than other agroforestry practices in both vertical and horizontal arrangement [9,12]. In addition, highest even distribution of woodlot than parkland and homegarden is related to the dominant of single species in woodlot i.e. *Eucalyptus* tree.

The average Shannon index for tree diversity (H′=2.02) and evenness (1.61) in homegarden of the study area was higher than previous *Enset-Coffee* agroforestry systems of the Southern Ethiopia (H′=1.41). However, species richness in homegarden of the study area was lower than Sidama agroforestry (120), Southern Central Highlands of Ethiopia (64) [18] and fruit tree diversity (104) in homegarden of Western Amhara of Ethiopia [34]. In addition, evenness of homegarden of the study area was lower than fruit tree diversity (0.69) in homegarden of Western Amhara of Ethiopia [34].

Parkland agroforestry of the study area had lower Shannon index as compared to the parklands in the region of semi-arid West Africa (H´=1.737) [19], Northern Togo (H´=4.53) [35] and Queens County, New York (H´=2.17 and Simpson=0.162) [36]. The species richness in parkland of the study area was observed to be lower than Northern Togo (21) [35] and Queens County, New York (13 to 19) [36]. Consequently, these variations of species diversity among different areas could be associated with difference in species structure and composition of woody species i.e. due to variation in topographic variables, moisture availability, nutrient availability [14]. In addition, species adaptability in different area, farm characteristics such as farm size, shape and nature of cropping pattern [15,7] and age of agroforestry practices can also determine the species diversity in this agroforestry practice.

Additionally, lower species diversity in the agroforestry practices of the study area as compared to reports of Southern Ethiopia could be probably related to higher population growth, land shortage and deforestation. Although, homegardens are appreciable in conserving the species diversity of the area and to reduce deforestation, population growth was reported to be the main cause for land degradation in Gununo area [4].

High species diversity and richness in middle/upper elevation can also be attributed to population pressure i.e. expansion of settlement from lower to higher elevations. High population growth and pressure in the lower elevations is mentioned to be the cause for deforestation and land scarcity (focus group discussion with elders). Gautam [37] has also found species diversity and richness was decreased in the lower elevation than upper. Variation in structure and composition of woody species along elevation is linked to environmental variability such as elevation/topographic effect [13], soil type/depth, species adaptability and management variation etc. [15].

Variation of IVI for various woody species among agroforestry practice is related to difference in species preference by farmers, difference in growth performance of species, number/existence of individual/dominance per plot, basal area and agroforestry plot area. For example high relative frequency of woody species could be linked with high species preference by farmers i.e. those frequently existed woody species are considered to be fairly fast growing, shade tree, tolerable for different managements and provide different uses like firewood, timber, etc.

The greater tree height in woodlot is linked to fast growing Eucalyptus species dominating woodlots in the study area. The high planting density increases competition for light among individuals and facilitates taller height growth in woodlots [6]. High dominance/density of *Eucalyptus* is due to high market access and value of eucalyptus in lower elevations. The market demand of Eucalyptus species, is coupled with land scarcity in the lower elevation to force farmers to plant fast growing and high market value Eucalyptus species.

Highest woody species density per hectare in the middle elevation is linked with species richness of the elevation and more land cover. Also, high crown area in the lower elevation is attributed to the dominancy of Eucalyptus woodlot in lower elevation with closer spacing (about 0.5 m spacing).

Conclusions

Homegarden, parkland and woodlot are the main agroforestry practices in the land use systems of Gununo watershed, Wolayitta, Ethiopia. The natural vegetation in the watershed is removed due to high population growth and associated deforestation for crop production. Land degradation is still widespread in the watershed despite some efforts made for tree planting and agroforestry development activities are promising. The agroforestry practices in the area holding and conserving more woody species and saved the endemic and endangered species.

The structure and composition of woody species vary among agroforestry practices: homegardens had shown higher complexity in structure and species diversity, while woodlots being monoculture, mostly *Eucalypus* and *Gravellia* species. The effect of elevation gradient on some of the studied parameters was not significant. However, woody species density/ha and species richness found to be highest in middle elevation.

Recommendations

Based on the result and discussion, the three dominant traditional agroforestry practices in the study area should be continued with professional advice to encourage it as a means of biodiversity conservation and sustainable land use system. Agroforestry practices with high species diversity, such as homegarden practices must be promoted to lower elevation where species diversity is very limited due to conventional agriculture system and tree plantation with single species as woodlots.

References

1. Amede T, Belackew T, Geta E (2001) Reversing the Degradation of Arable Land in the Ethiopian Highlands. Managing Africa Soils. No. 23.
2. Berry L, Olson J, Campbell D (2003) Assessing the extent, cost and impact of land degradation at the national level: findings and lessons learned from seven pilot case studies. Commissioned by Commissioned by Global Mechanism with support from the World Bank.
3. Beyene S (2011) Characterization of Soils along a Toposequence in Gununo Area, Southern Ethiopia. Journal of Science and Development, 1: 31-41.
4. Tegene B (1992) Farmers' perceptions of erosion hazards and attitudes towards soil conservation in Gununo, Wolaita, Southern Ethiopia. Etiopian journal of development research, 14.
5. Ethiopia's Climate-Resilient Green Economy Green economy (CRGE) strategy. Federal Democratic Republic of Ethiopia II. Addis Ababa, September 2011.
6. Maarel EVD. (edt) (2005) Vegetation Ecology. Blackwell Science Ltd UK. 411pp.
7. Kumar BM, Nair PKR (2004) The enigma of tropical homegardens. Agroforestry Systems, 61: 135–152.
8. Götz S, FonsecaABG, Harvey CA, Gascon C, Heraldo L, et al. (eds) (2004) Agroforestry and Biodiversity Conservation in Tropical Landscapes. Island Press. USA, pp. 537.
9. Nair PKR (1993) An introduction to agroforestry. Kluwer Academic publisher with cooperation ICRAF, Dordrecht, London. pp. 489.
10. Edouard GB (2001) Network on agroforestry and soil conservation: secretariat of the convention to combat desertification. June 2001
11. Kang BT, Akinnifesib FK (2000) Agroforestry as alternative land-use production systems for the tropics. Natural Resources Forum, 24: 137-151.
12. Young A (1989) Agroforestry for soil conservation. International Council for Research in Agroforestry. C A B International, p. 318.
13. Moges Y (2009) The impact of overstorey trees on sustainable coffee (*Coffea arabica* L.) Production in southern Ethiopia. Ph.D Dissertation, Horizonte Bd. 25, Der Andere Verlag, Tönning, Lübeck and Marburg.
14. Ian N, Brendan G, Rowan R (2009) Agroforestry for natural resource management. Csiro Publishing. Australia. p.360.
15. Rebecca LB, Anne JL, Robert KP (2007) Species Richness: Small Scale. Encyclopedia of life sciences. John Wiley & Sons, Ltd.
16. Mowo J, Adimassu Z, Masuki K, Lyamchai C, Tanui J, et al. (2011) The

Importance of Local Traditional Institutions in the Management of Natural Resources in the Highlands of Eastern Africa. Working Paper No 134. Nairobi: World Agroforestry Centre.

17. WZFEDD (2008) Zonal Basic Socio-Ecnomic & Demographic Information. Wolaita Soddo.

18. Tolera M, Asfaw Z, Lemenih M, Karltun E (2008) Woody species diversity in a changing landscape in the south-central highlands of Ethiopia. Agriculture, Ecosystems and Environment, 128: 52-58.

19. Albert N (2005) Agroforestry Parkland Species Diversity: Uses and Management in Semi-Arid West Africa (Burkina Faso). PhD thesis Wageningen University, Wageningen.

20. Senbeta F, Teketay D, Naslund B (2002) Native woody species regeneration in exotic tree plantations at Munessa-Shashemene Forest, southern Ethiopia. New Forests, 24: 131-145.

21. Negash M (2002) Socio economic aspects of farmer's eucalypt planting practice in the Enset coffee based agroforestry system of Sidama, Ethiopia: The case of Awassa and Shebedino district M.Sc. thesis.

22. Sue E, Tadese M, Inga H (Eds) (1995) Flora of Ethiopia and Eritrea, Addis Ababa, Etiopia, Uppsala, Sweden, 2.

23. Kelecha W (1987) A glossary of Ethiopian plant names by woldemichael kelecha. Fourth edition revised and enlarged. January 1987. Addis ababa, Ethiopia. p. 339.

24. Kelecha W (1977) A glossary of Ethiopian plant names. Second edition revised. March 1977. Addis ababa, Ethiopia. p.469.

25. Woldemichael LK, Bekele T, Nemomissa S (2010) Vegetation Composition in Hugumbirda-Gratkhassu National Forest Priority Area, South Tigray. *CNCS:* Mekelle University, 2: 27-48.

26. Newton AC (2007) Forest ecology and conservation: a handbook of techniques. Oxford University Press, New York, p. 132.

27. Begon M, Townsend CR, Harper JL (2006) Ecology: from individuals to ecosystems, 4th ed. Blackwell, Malden, USA.

28. Asfaw Z (2003) Tree species diversity, top soil condition and arbusecular mycorrhizal association in the Sidama traditional agroforestry land use, southern Ethiopia. Doctoral Thesis, Swedish University of Agricultural Science.

29. Asfaw Z, Nigatu Z (1995) Home gardens in Ethiopia: Characteristics and plant diversity. SINET: Ethiop. J. Sci. 18: 235-266.

30. Asfaw Z, Agren, G (2007) Farmers' local knowledge and topsoil properties of agroforestry practices in Sidama, Southern Ethiopia. Agroforest Syst, 71: 35-48.

31. Negash M, Yirdaw E, Luukkanen O (2011) Potential of indigenous multistrata agroforests for maintaining native floristic diversity in the south-eastern Rift valley escarpment, Ethiopia. Agroforest syst, 85: 9-28.

32. Luis VJ, Ensermu K, Sebsebe D (2005) The Red List of Endemic Trees & Shrubs of Ethiopia and Eritrea. Fauna & Flora International. Cambridge, UK, p. 28.

33. Salome BM, Majule AE, Herbert VL (2003) Linkage between changes in land use, biodiversity and land degradation on the slopes of Mount Kilimanjaro, Tanzania. Project working paper 38. Nairobi, Kenya: international Livestock research institute.

34. Tiruneh FM (2008) Fruit tree species in the wild and in homegarden agroforestry: species composition, diversity and utilization in western Amhara region, Ethiopia. Department for Forest and Soil Sciences. Institute of Forest Ecology. PhD dissertation theses, Vienna. September, 2008.

35. Fousseni F, Samake G, Yu ZC, Xiu HZ, Kperkouma W, et al. (2011) Evaluation of agroforestry species in potential fallows of areas gazetted as protected areas in North-Togo. African Journal of Agricultural Research, 6: 2828-2834.

36. Carsten WG (2000) The Floristic Composition and Community Structure of the Forest Park Woodland, Queens County, New York. Urban Habitats, 4.

37. Gautam R., Sthapit B.R. and Shrestha PK (eds) (2006) Home Gardens in Nepal: Proceeding of a workshop on "Enhancing the contribution of home garden to on-farm management of plant genetic resources and to improve the livelihoods of Nepalese farmers: Lessons learned and policy implications", 6-7 August 2004, Pokhara, Nepal. Li-Bird, Bioversity International and SDC.

Comparative Study of Epidemiological and Anthropological Aspects of Diabetes and Hypertension in Cameroon

Tsabang N[1,*], Fongnzossie E[2], Donfack D[3], Yedjou CG[4], Tchounwou PB[4], Minkande JZ[5], Nouedou C[5], Van PD[6] and Sonwa[7]

[1]*Center for Research on Medicinal Plants and Traditional Medicine, Institute of Medical Research and Medicinal Plants Studies, Ministry of Scientific Research and Innovation, Yaounde, Cameroon*
[2]*University of Douala, Cameroon*
[3]*Faculty of Education, Department of Specialized Education Sciences, Yaounde, Cameroon*
[4]*Cellomics and Toxicogenomics Research Laboratory, NIH-RCMI Center for Environmental Health, Jackson State University, Jackson, Mississippi, USA*
[5]*General Hospital of Yaounde, Cameroon and Faculty of Medicine and Biomedical Sciences, Cameroon*
[6]*Faculty of Science Engineering, Universitteit Ghent, Coupure Links 653B-9000 Ghent, Belgium*
[7]*Center for International Forestry Research (CIFOR), Yaounde, Cameroon*

Abstract

The traditional medicine in Africa in general and specifically in Cameroon does not manage diabetes and arterial hypertension very well. Yet, these pathologies are becoming more prevalent among the populations that need adequate knowledge to fight against them. Therefore the present study was designed to determine the knowledge, attitudes and practices of indigenous people regarding diabetes and hypertension control, and to assess the epidemiological aspects of these diseases in order to reinforce their health education and promote a better health care through traditional medicine.

To achieve this objective, 1,131 households including 70 traditional healers, 114 diabetics, 167 hypertensive patients, 30 hypertensive patients-diabetics and other Cameroonians were questioned on their ethnomedical knowledge of diabetes and arterial hypertension. Fifty-eight randomly distributed tribes were taking in account. The elucidation of anthropological and epidemiological aspects of diabetes and hypertension improved the beliefs of indigenous people and facilitated the modernization of diabetes and hypertension comprehension that remained focused on the elucidation of diseases' causes and complications, as well as on the behaviors that could help translate biomedical terms into locally meaningful metaphors.

Keywords: Diabetes; Arterial hypertension; Indigenous people knowledge; Traditional health care improvement

Introduction

High blood pressure and high glycaemia are universally accepted as two of the most important risk factors in the development of cardiovascular disease and they still represent a serious public health problem worldwide [1]. Diabetes and hypertension have been for a long time considered to be more prevalent in developed countries. Nevertheless, 62% of people with diabetes are undiagnosed in Africa and 86% of known diabetics live in low and middle income countries [2,3]. Particularly in Cameroon, where epidemiological studies have shown that 10% of the population is diabetic, 90% of them are obese at the beginning of the disease. Among the Cameroonian patients, these pathologies are the foremost cause of extremity amputation and acquired renal diseases. They are also associated with an increased risk for ischemic heart disease and infections [4] and contribute to higher risk for depression [5]. In Cameroon, the prevalence of hypertension adjusted to the age is 16.6% in men and 12.6% in women in urban population [6]. While in Teaching Hospital of Brazzaville, Congo Democratic Republic, arterial hypertension constituted 5.1% among 825 admissions. Patients were divided in 4 groups: Chronic hypertension (n=4), pre-eclampsia presumed on chronic hypertension (n=4), pregnancy hypertension only (n=15), pre-eclampsia (n=19). Primary gravidity (38.1%) and family history of hypertension (40.4%) were the most frequent risk factors. Obesity, gemality, previous pre-eclampsia represented respectively 14.3%, 9.5% and 4.8% [7]. Arterial hypertension represents the cardio-vascular affection most frequent among the black population in Africa [8]. In this study the anthropological and the epidemiological aspects of diabetes and hypertension were determined and were addressed for promoting the indigenous people' knowledge reinforcement and their better health care through traditional medicine.

Cultural factors in the prevalence of diabetes and hypertension in Cameroon are mainly attributed to diet, although other relevant cultural characteristics include lifestyles, attitudes about disease prevention, and practices related to personal autonomy and giving up advice. In the African continent the cultural diversity is favorable to the use of alternative modes of health care including familial self-medication, traditional healers' treatments and market procurement of medicines. Hence, the following research questions were developed: Are the hypertensive patients and/or diabetics able to confront the anthropological aspects of these dangerous diseases with their epidemiological manifestations to have more useful understanding for high-quality of early treatments? In other words how to promote successful treatments of diabetics and/or hypertensive patients in the households and at the traditional healers? The search for potential

***Corresponding author:** Tsabang N, Center for Research on Medicinal Plants and Traditional Medicine, Institute of Medical Research and Medicinal Plants Studies, Ministry of Scientific Research and Innovation, Yaounde, Cameroon
E-mail: tsabang2001@yahoo.fr, tsabang@hotmail.com

answers to these research questions is the key determinant for the set-up of the methodology.

Methodology

Condition of selection of patients

Patients admitted in this study have been diagnosed at least once in medical centers and confirmed by presenting a medical book or a bulletin of medical exam that certified their diabetic and/or hypertensive state at the moment of their selection.

Anthropological aspects

The survey of the beliefs of indigenous Cameroonians regarding diabetes and arterial hypertension diseases was carried out on 1131 households that included 114 diabetics, 167 hypertensive patients, and 30 hypertensive patients with diabetes. These patients respectively represented 10.08%; 14.76% and 2.65% of respondents, and all together constituted 27.50% of the total interviewees. Fifty-eight socio-cultural groups were taking in account, in a random distribution in Cameroon. These respondents provided information on their cultural and ethno medical knowledge on diabetes and arterial hypertension. The following questions were addressed to each of them: Do you know about diabetes and arterial hypertension diseases? If yes, how do you diagnose them? What do you think about these pathologies? Are you a diabetic and/or hypertensive patient? Is there someone in the family who is a diabetic and/or hypertensive patient? Are there myths, tales or legends around these diseases?

Representativeness of recorded respondents

From a sample of 1131 households that included 328 diabetics and hypertensive patients, data on the beliefs of Cameroonians were collected on diabetes and hypertension. Fifty percent of respondents were males and 50% were females. From the first nine hundred and twenty-three respondents all the data were recorded. Therefore from the 208 remaining respondents, no new data were registered. So the number of 1131 interviewees was representative of Cameroon population. The answers of the following questions permitted to establish the representativeness of recorded types of patients using statistical analysis. Are there significant differences between the recorded types of diabetics, of hypertensive patients and of hypertensive patients-diabetics when compared two by two? In the other words: Do the recorded types of patients (diabetics, hypertensive patients and diabetics-hypertensive patients) present the same prevalence in Cameroon' population, if in the same condition the survey is repeated?

Statistical analysis

Study participants were categorized in different subgroups including: type 1 diabetes, type 2 diabetes, essential hypertension and secondary hypertension, diabetes with essential hypertension, diabetes with secondary hypertension, males, and females. From the data generated from these groups, statistical analyses were performed to compare the disease prevalence rates between specific paired groups including type 1 diabetes and type 2 diabetes; essential hypertension and secondary hypertension; diabetes-hypertension with essential and with secondary hypertension, and male and female patients. Statistical differences were assessed using standard protocol as described in "Statistical Methods in Biology" [9].

Results

Number of recorded patients

Three hundred and twenty-eight (328) patients were recorded and divided into groups: clinical followed up patients and not clinical followed up patients (Table 1). The first group used medicinal plants in familiar herbal remedies and the second did not use.

Patients	Diabetics		Hypertensive patients		Diabetic-hypertensive patients	
	NIDD	IDD	ETH	SHT	NIDD-SHT	NIDD- EHT
Clinically followed up patients						
Men	30	15	4	20	03	01
Women	40	21	6	36	04	02
Total	**70**	**36**	**10**	**56**	**07**	**03**
Not clinically followed up patients						
Men	05	3	15	5	6	2
Women	10	0	24	7	12	7
Total	15	3	89	12	18	9
Total per type of diseases	85	39	99	68	25	12
Total per diseases	124		167		37	

NIDD: Non Insulin-Dependent Diabetes; IDD: Insulin-Dependent Diabetes; EHT: Essential Hypertension; SHT: Secondary Hypertension.

Table 1: Distribution of diabetics and/or hypertensive patients recorded.

Epidemiological aspects

Prevalence comparison of type 1 diabetics (insulin-dependent diabetics) and type 2 diabetics (non- insulin-dependent diabetics): The number of insulin-dependent diabetics and non-insulin-dependent diabetics recorded was respectively 39 and 85. If the two types of diabetes have the same chance to be met during the survey, there is equal probability: p=q=½. The two types of diabetes form each a binomial distribution. It was about to compare the observed percentage for each type of diabetes to hypothetic value. According to the hypothesis zero (no difference between the two types of diabetes) the two types of diabetes have the same chance to appear during the survey. In this case they have the same prevalence in the population. The variance of this probability is

V=pq/n with n=124

V=1/2 × 1/2/124=0. 0020161. Standard error:=0.044901325.

The observed values were:

From IDD sample:

a/n ± =0. 6855 ± 0. 0416

a=85.

From NIDD sample:

a/n ± =0, 3145 ± 0, 0416

a=39

The different observed gap of hypothetic value is:

From NIDD sample: 0.6855–0, 5000=0.1855;

From IDD sample: 0.5000–0.3145=0.1855.

By definition, this gap is 2.6 times superior to standard error; we concluded that there was a significant difference between these 2 types of diabetes at 95%. IDD was dominant. Until what limits the variations of each disease can be put on the account of chance fluctuations? To answer this question, the confidence interval of diabetes was determined.

The normal variable with zero mean or the unit standard deviation was

d=a/n-p: $\sqrt{p(1-p)/n}$ =4. 131.

The confidence intervals for the true unknown proportion were: 80.869 and 89.131 for NIDD: and 34.869 and 43,131 for IDD at 95% coefficient of security. The probability so that the observed proportions of diabetic's types remained in these intervals, if the survey began again in the same conditions, was more increased. This sample was widely representative for the two types of diabetes in Cameroonian's population.

Prevalence' comparison of hypertensive patients with essential and secondary hypertensive patients: The similar analysis previously realized has permitted to demonstrate the non-existence of a significant difference between essential hypertension and secondary hypertension, because the observed gap of hypothetic value (0.0928) was < Standard error (0.03869 × 2.6).

Prevalence' comparison of diabetics with essential and diabetics with secondary hypertension: In the same condition the similar analysis previously realized has permitted to demonstrate the existence of a non-significant difference between diabetics with essential hypertension and diabetics with secondary hypertension. The observed gap of hypothetic value (0.1756) was < Standard error (0.038219 × 2.6).

Sex distribution of recorded patients

The similar analysis previously realized has permitted to demonstrate the nonexistence of a significant difference between type 2 male diabetics and type 2 female diabetics. The observed gap of hypothetic value (0.0882) was < Standard error (0.0542) × 2.6. Also between male secondary hypertensive patients and female secondary hypertensive patients there wasn't a significant difference. The observed gap of hypothetic value was (0.1323) < Standard error (0.060632 × 2.6). Nevertheless, there is a significant difference between male essential hypertensive patients and female essential hypertensive patients. The observed gap of hypothetic value (0.1969) was > Standard error (0.03869 × 2.6). The normal variable with zero mean or the unit standard deviation was d=a/n-p: =0.1972. The confidence intervals were: 29.802 and 31.1972 for essential male hypertensive patients and 68.802 and 69.1972 for female essential hypertensive patients at 95% coefficient of security (Table 2). The probability so that the observed proportions of male essential hypertensive patients remained in these intervals, when we began again the survey in the same conditions was more increased. This sample was widely representative for the proportions of men diabetics and women diabetics, in Cameroonian's population.

Anthropological aspects

According to the respondents, the anthropological aspects described below were raised from the beliefs of population. Indeed 62% of respondents considered diabetes and arterial hypertension as being two mystic diseases. They believed that the coma (Menoubia), the numbness (Aloute me kouc), the dermal clues (Emock Dem), the gangrenes (Ta-Nteu Nfeung), the retinopathy (Meneck Se'eung), the nephropathy (Letse kekan) and the cardiac complications like arteriosclerosis (Ate'eu ngan net) and the grave noose bleeding (Ta-Nteu Afifi), were linked to the suffering of patients' totems, killed by the traditional sporting-gun or victim of a grave fall in the invisible world. The above vernacular names in brackets were the complications' designations in Bamileke. At these state of diseases, the patients were usually considered as sorcerers and the complications they developed came from their expeditions in the invisible world. The patients could also suffer from severe attacks of witchcraft, the punishments of ancestors, the maledictions or the false gods. Then these false gods must be appeased at all costs. Likewise, the gangrenes of the foot and the benumbed inferior members came from the stampings of conspicuous or invisible poisons. The paralysis and the hemiplegia attacked bewitched patients.

The survey suggested that 10% of 70 traditional healers, living in the hinterland were illiterate; therefore they diagnosed diabetes by the color (clear or dim) of urine, the strong density of insects (flies, bees and ants) around it on the soil surface, and also by its sweet taste. It is not recommended to taste the urines because it is rich in toxic substances such as uric acid and by metabolic products. Only traditional healers are able to make a good traditional diagnosis and cure of victims. The healings were researched during the nocturne sessions of treatment of mystic, the ceremonies of healing bewitched patients, the cults to ancestors, the gift of foods, drinks, clothes, and/or money to family or clan members, and the sacrifices to idols.

The restitution of the present study, three years ago, has changed gradually the behavior of indigenous Cameroonians. Many traditional healers, especially those of hinterland produce less and less the incantations and the speculations that retarded the health care of diabetics and/or hypertensive patients in the hospital centers. The good understanding of diabetes and hypertension epidemiology and manifestations including progression and complications, can save the broken relationships between familial' members due to the consequences of the consultation of native doctors. Nevertheless, the diagnostic of traditional healers, for those who effusively believed them, reduced the stress of patients and enhanced their health care.

Manifestations and difficulties of health care

Far to claim a pejorative judgment for these indigenous considerations, diabetes and hypertension are insidious diseases (hidden beginning and slow evolution), with detection often late and with sudden release. There are chronic pathologies with generally long and expensive treatments. The monthly costs for treating each diabetic, hypertensive, and diabetic-hypertensive patient are respectively 22,500 Fcfa, 22,500 Fcfa and 45,000 Fcfa in the case of uncomplicated diabetes and hypertension. The cost of diet increases the financial expenses. Certain combinations of antidiabetic or antihypertensive treatments, composed from the following products: insulatard HM (14,525 Fcfa), Actrapid HM (14,510 Fcfa) like insulins, Hexen 50 (13,860 Fcfa), Lodoz (10,415 Fcfa) like hypotensive products and Glucophage 850 mg (9,190 Fcfa) representative of oral hypoglycemic products, are rare in rural or onerous for patients. Then the difficulties of access to the medicines can facilitate the appearance of the following severe complications: coma, numbness of members and muscle atrophy, on diabetics; hypertrophy of left ventricle, occlusion of heart artery (infratus) and of brain artery (cerebral softening), vascular-cerebral

Patients	Diabetics		Hypertensive Patients		Diabetics-hypertensive patients	
	NIDD	IDD	ETH	SHT	NIDD-SHT	NIDD- EHT
	Men					
	35	18	30	25	9	3
	Women					
	50	21	69	43	16	9
Total	**85**	**39**	**99**	**68**	**25**	**12**

NIDD: Non Insulin-Dependent Diabetes; IDD: Insulin-Dependent Diabetes; EHT: Essential Hypertension; SHT: Secondary Hypertension.

Table 2: Sex distribution of diabetics and/or hypertensive patients recorded.

accidents (AVC), on patients; gangrenes (local necrosis of a tissue), nephropathy (deficiency or renal insufficiency), on diabetics and/or hypertensive patients, rendering the treatment difficult and prohibitive. The patients are terminally sick and face access restriction to modern medicines. Therefore 76% of recorded patients live with families who are financially deprived, have limited resources, and rely on alternative modes of treatment over modern medicine. However, the 24% of patients who are financially sufficient also rely on medicinal plants for a total recovery because of the chronic character of the two diseases that makes the modern treatment very lengthy.

The ethnopharmacological preparations of the plants used were previously described by Tsabang [10]. Plants are potential sources for medicinal treatment because the rural population, mostly of the hinterland and far from urban zones well supplied with manufactured pharmaceutical products, have developed important experiences on the use of medicinal plants and natural products.

Discussions

There was a statistically significant difference ($p<0.05$) between the proportion of type 1 diabetics and that of type 2 diabetics. Hence, the two forms of diabetes did not affect the young population (0 to 30 years) and the adult population (more than 30 years) at the same prevalence rate. It has been reported that most diabetics eat much food and this favors obesity near 45 years [11]. Thirty percent of obese persons become type 2 diabetics [12]. In Cameroon, 90% of population is type 2 diabetic [13]. This is the confirmation of why in our sample; the number of type 2 diabetics was dominant. The comparison between essential hypertensive patients and secondary hypertensive patients revealed a non-significant difference in disease rate between these two types of hypertensive patients. Yet, essential hypertension (also called primary hypertension or idiopathic hypertension) is the most common type of hypertension, affecting 95% of hypertensive patients [14,15]. Prevalence of essential hypertension increases with age, and individuals with relatively high blood pressure at younger ages are at increased risk for the subsequent development of hypertension. The observed variations of each type of hypertension can be attributed to fluctuations in specific risk factors. Hypertension can increase the risk of cerebral, cardiac, and renal complications [16].

There was no significant difference between diabetics and/or hypertensive male (50%) and female (50%) patients. Previous works presented below show that this tendency is observed elsewhere in the study of the prevalence of hypertension and diabetes. The age-adjusted prevalence of hypertension was 34%, 25.4%, and 23.2% for men and 31%, 21%, and 21.6% for women among blacks, whites, and Mexican Americans, respectively. NHANES III reported the prevalence of hypertension was 12% for white men and 5% for white women aged 18-49 years. However, the age-related blood pressure rise for women exceeded that of men. The prevalence of hypertension was reported at 50% for white men and 55% for white women aged 70 years or older [17]. The prevalence of diabetes was 5.7% (95% confidence interval-CI: 4.8-6.8) overall, with a slight difference between the men (5.5%; 95% CI: 4.1-7.2) and women (5.9%; 95% CI: 4.6-7.6). The prevalence of diabetes among men was not significantly different from that among women (Owner's Risk: 1.01; 95% CI: 0.91-1.11) [18].

There is a significant difference between male essential hypertensive patients and female essential hypertensive patients, in favor of women who are fatter than men. The relation of hypertension with obesity is well established. The reduction of body mass index (BMI) to values less than 23 would prevent 46.7% of future hypertensive cases in men. Hypertension increases progressively with age in both genders, although the prevalence is higher for men than for women (45.2% vs 28.9%) in all age groups [1].

Diabetes and hypertension are the exceptionally difficult diseases to manage. For the individual, it requires a lifelong commitment to dietary change, physical exercise, and self-monitoring of a medication regime (oral or injectable products) and lastly dialysis. Diabetes and hypertension health care requires the work of a team of health providers-internal medicine or family medicine practitioners, diabetes and hypertension educators, nutritionists, podiatrists, and dentists-ideally in regular communication with each other [19].

Conclusion

This study reveals that diabetes and hypertension are increasingly prevalent in the population of Cameroon. It is also divulges that many factors including the anthropological considerations, the increasing number of diabetics and hypertensive patients, the chronic nature of diabetes and arterial hypertension, and the expensive costs of modern treatments, facilitate the recourse to traditional medicine. Because of their poor financial conditions, the patients living in affected families look for alternative modes of therapy instead of physicians' treatment using modern medicine [20]. Collaborations with physicians are needed for a better diagnostics and a good health care of patients. Nevertheless the mystic considerations are harmful for the familial' cohesion in Cameroon and retard the medical care of patients. However, they present some good cultural aspects that include stress reduction due to the self-confidence of patients, because of the appeasement of false gods, of bewitches and diverse aspects of witchcraft. For rapid diagnosis, we recommend patients to consult physicians first and to use medicinal plants in collaboration with physicians. We also recommend traditional healers to avoid the diagnosis of diabetes by tasting the urines rich in toxic substances like uric acid [21]. In general four domains (political, economic, biological/genetics, and socio-cultural) influence diabetes and hypertension management. Each of these domains should be reflected in program models and tribal policies for the control/prevention of diabetes and hypertension in Africa.

Acknowledgements

The authors acknowledge the households, the traditional healers, the diabetic and hypertensive patients and the nurses and physicians for their contribution during the ethnobotanical survey. The support of NIH-NIMHD RCMI Grant # G12MD007581 is also acknowledged.

References

1. Sulbara´n T, Silva E, Calmo´n G, Vegas A (2000) Epidemiologic aspects of arterial hypertension in Maracaibo, Venezuela Journal of Human Hypertension 14: S6-S9.
2. World Health Organization (2014) Global Health Estimates: Deaths by Cause, Age, Sex and Country, 2000-2012. Geneva, WHO.
3. International Diabetes Federation (2015) Calling the World to act on diabetes, Briefing on diabetes Post.
4. Narayan KM, Venkat (1997) Diabetes Mellitus in Native Americans: The Problem and Its Implications. Population Research and Policy Review 116: 169-192
5. Alan H (1981) Ethnicity and Medical Care. Cambridge, Mass.: Harvard University Press.
6. Mbanya JCN, Ngogang J, Salah JN, Minkoulou E, Balkau B (1997) Prevalence of NIDD and impaired glucose tolerance in a rural and an urban population in Cameroon, Diabetologia 40: 824-829.
7. Kimbally KG, Barassoumbi H, Buambo SF, Gombet T, Kibeke P (2007) Arterial hypertension: epidemiological aspects and risk factors on pregnant and delivered woman. Dakar Med 52: 148-52.

8. Bertrand E (1995) Hypertension of people of Black African origin Pradel Edition.

9. Norman TJ, Bailey MA (1992) Statistical methods in biology. Second edition, Cambridge University Press 37-38.

10. Tsabang N (2008) Study ethnobotanical plants antidiabetic virtues/or antihypertensive Cameroon. Thesis / PhD. University of Yaounde 1: 389.

11. OMS (1990) Série de rapports techniques n 797, Régime alimentaire, nutrition, prévention des maladies chroniques. Rapport d'un groupe d'étude de l'OMS, Organisation de la santé, Genève P: 79.

12. Mbanya JCN, Minkoulou EM, Salah JN, Balkau B (1998) Prevalence of hypertension in rural and urban Cameroon. International Journal of Epidemiology, Great Britain 27: 181-185.

13. Darnaud C (1959) Que sais-je, Le point de connaissances actuelles sur le diabète. Presse Universitaire de France P: 126.

14. Carretero OA, Oparil S (2000) "Essential hypertension Part I: definition and etiology". Circulation 101: 329-35.

15. Oparil S, Zaman MA, Calhoun DA (2003) "Pathogenesis of hypertension". Ann Intern & Med. 139: 761-776.

16. Messerli FH, Williams B, Ritz E (2007) Essential hypertension 370: 591-603.

17. Cornoni-Huntley J, La Croix AZ, Havlik RJ (1989) Race and sex differentials in the impact of hypertension in the United States; The National Health and Nutrition Examination Survey I Epidemiologic Follow-up Study. Arch Intern Med 149: 780-800.

18. Hilawe EH, Yatsuya H, Kawaguchi L, Aoyama A (2013) Differences by sex in the prevalence of diabetes mellitus, impaired fasting glycaemia and impaired glucose tolerance in sub-Saharan Africa: a systematic review and meta-analysis. Bulletin of the World Health Organization 91: 671-682.

19. Carolyn M, Smith-Morris (2004) Reducing Diabetes in Indian Country: Lessons from the Three Domains Influencing Pima Diabetes Human Organization 63.

20. Nkongmeneck BA, Tsabang N, Zafack L, Carlson TJ, Keita A (2000) Therapeutic remedy for diabetes and plants used by diabetics in Cameroon: Case of Yaounde and its environs. Med Journal And Pharm Afr Pp: 89-98.

21. Pamplona GR (1995) Croquez la vie. Editions Vie et Santé 60, avenue Emile Zola, 77192 Dammarie Lès lys Cedex France p: 190.

The Potential of P Loss from Livestock Farms to Watercourses in Wensleydale; A Risk Mapping Approach

Romero Tejero G*

Geography Department, University of Leeds, UK

Abstract

Hydrologic, chemical and physical factors controlling P export from 10 livestock farms located in Wensleydale were considered in order to estimate the risk of P loss to surface waters in a Non Nitrate Vulnerable Zone (non NVZ). Phosphorus source factors were obtained from a farm investigation questionnaire as well as from soil analysis results. Transport factors were examined using a GIS as a tool to implement a predictive model of delivery of P to watercourses. The classification of grassland into different risk categories was carried out using the Pennsylvanian P Index in a GIS. Finally, this risk assessment has yielded that non NVZ areas can pose a risk of breaching the Water Framework Directive (WFD) objectives. A wide range of mechanisms including voluntary schemes are needed in order to reach good ecological status of freshwaters by 2015 in the UK.

Keywords: NVZ; Phosphorus Index (PI); GIS; Water framework directive; Livestock farming

Introduction

Several published studies have demonstrated the impact of diffuse agricultural pollution on rivers ecological functions as a result of poor water quality [1-4].

The increase use of fertilisers in order to increase production efficiency in farming systems has been pointed out as the main factor that leads to nutrient emissions to freshwaters [5,6]. In this regard, agricultural stewardship alongside physicochemical soil properties, geomorphology or even weather, are other factors that may contribute to increasing nutrient water pollution [7,8].

Nitrogen (N) and Phosphorous (P) have been widely studied due to the important role they play on water eutrophication [9,10]. This excessive enrichment of water with N and P leads to algal blooms and hence, anoxic conditions of receiving waters that have proven to be harmful to aquatic life [9,11].

Nitrate Directive

In spite of the fact that P is a nutritional limiting factor for algal growth [12,13] that can substantially increase the risk of eutrophication in waterbodies, more attention was paid to N due to the EU Nitrate Directive. This makes sense if we focus on the quality of drinking water, as some human health problems such as methemoglobinemia are associated with high nitrate concentrations [6,14]. Furthermore, N is more likely to leach with rainwater moving through the soil [5] what brings about lower nitrate retention by soil and hence, a higher nitrate concentration in freshwaters. By contrast, high concentrations of P in freshwaters are difficult to find as phosphate anions are mostly retained by soil particles [15]. All these factors were mainly considered when MS including the UK, were required by the Nitrate Directive to identify nitrate pollution in freshwaters [16].

The designation of NVZs (Nitrate Vulnerable Zones) depends on nitrate concentration exceeds the threshold value of 11.3 mg N l^{-1} in freshwaters [14]. In England 68 of the territory has been designated as NVZ [17]. This implies a set of measures that farmers in NVZs are obliged to implement in order to minimise water pollution. In the case of England, farmers in NVZ are required to meet the Nitrate Pollution Prevention Regulations 2008 (Statutory Instrument 2008/2349). According to this regulation, all farmers must have a planning for N use where they can show the amount of N fertiliser that they need to apply. This should be based on N content of organic fertilisers, this is Farm Yard Manure (FYM), slurries, or both, as well as Soil Nitrogen Supply (SNS). Some of these requirements differ depending on the type of farm. For example pig farms are required to provide storage for organic manures for at least 6 months (1 October to 1 April inclusive) whilst dairy farms need 5 months (1 October to 1 March inclusive). This is fundamentally based on estimations of the nutrient content of organic manures, as Table 1 shows.

Furthermore, there are restrictions for spreading livestock manure as farmers cannot exceed the amount of 170 kg ha^{-1} of total N per year. There are also spreading restrictions, mostly in the winter period depending on the type of land. In the case of grassland, spreading of organic fertilisers is restricted between September and January depending on soil characteristics. Similarly, spreading mineral fertiliser is not permitted between the 15th Sep and 15th Jan. Further restrictions include for example, not spreading manure within 10 m. from surface waters as well as not spreading when soil is waterlogged, snow covered or frozen.

However, farms located in the remaining 32 of the English territory, this is non-NVZs, are not required to comply with these rules.

SSAFO Regulations

Nevertheless, storage for organic manures is regulated in all the territory by the Water Resources (Control of Pollution) (Silage, Slurry and Agricultural Fuel Oil) (England) Regulations [18]. The application of this regulation brings about that every livestock farm in England must provide enough slurry storage capacity for the maximum amount

***Corresponding authors:** Romero Tejero G, Geography Department, University of Leeds, UK, E-mail: gromerote@gmail.com

	Dry matter (%)	Total Nitrogen (Kg N/m³)	Available phosphate Kg P_2O_5/t
Cattle	6	2.6	0.6
Pig slurry	6	4.4	1.3

Table 1: Nutrient content of slurries depending on the type of farm [34].

of slurry "likely to be produced in any four month period". This rule aims at reducing the number of organic fertiliser applications in order to reduce the risk of nutrients loss to freshwaters, in cold seasons characterised by the low nutrient uptake rates by vegetation [19].

Nonetheless, there is a gap in this regulation as storage systems built for that purpose before 1991 are excluded. According to Figure 1, over 70 of slurry storage systems in areas not included in NVZ, date back more than 10 years and hence, many of them could have been constructed before 1991.

Water Framework Directive (WFD)

With the advent of the WFD in 2000, freshwaters in the UK including the non-NVZs, should meet good ecological and chemical status by 2015 [20,21]. This framework brought about The England Catchment Sensitive Farming Delivery Initiative (ECSFDI) in order to meet the WFD objectives. This Iniciative tries to raise awareness and deliver advice among farmers for reducing nutrient water pollution [14]. In this regard, the ECSFDI is a project where is up to farmers to get involved, unlike farms in NVZ where they are required to apply the Code of Good Agricultural Practice [17].

Moreover, some recent studies undertaken to determine the effectiveness of NVZ and ECSFDI have indicated the lack of evidence of significant water quality improvements in the UK [14,22]. In fact, Worral [22] questioned the effectiveness of measures that mainly aim at reducing fertiliser inputs in farms included in NVZ, whereas other factors such as connectivity of nutrient sources and hydrological flow paths can be determinant for nutrient pollution at a catchment scale. In this sense, the catchment-based approach introduced by the WFD is crucial to reduce diffuse water pollution, as farms in non-NVZ may pose a risk of failing good status objective.

Assessment Methods

As stated above, P has been pointed out by numerous authors [10,23-25], including the WFD, as a key nutrient for addressing water eutrophication derived from diffuse nutrient loss from agriculture. However, this is a hard task considering the complexity of the processes involved in the delivery of P to freshwaters [15].

Due to time constraints to meet WFD targets, consistent approaches should be implemented in order to assess the risk of nutrient pollution posed by agricultural systems. This information would be useful for identifying P emission hotspots and hence, increasing the effectiveness of actions to mitigate diffuse pollution [26].

Suitability of assessment methods will vary according to the target [27]. For example, the measurement method could provide an accurate assessment of nutrient delivery at a catchment scale as long as analytical equipment was available in representative sample sites [23]. Unfortunately, this is not always possible as monitoring is expensive and time consuming.

Moreover, the nutrient budgeting method has proven useful to estimate nutrient surpluses in soil considering inputs such as animal feed, internal cycling and outputs [28]. Domburg [29] demonstrated higher P surpluses were linked to livestock farms mainly due to fertiliser inputs. Similarly, Edwards [30] published that higher P surpluses were associated with grassland farming systems (Table 2). In this sense, other studies have yielded that a common practise among farmers is to apply fertiliser on the basis of N, not considering concentrations of P already present in the soil [31,32]. Thus, this information can contribute to optimising farm management, in particular nutrient management.

However, some additional considerations appear when the target is to assess the risk of P loss from grassland farming systems to freshwaters. Source factors are important in order to estimate nutrient availability such as P availability, yet delivery of P is conditioned by transport factors [27]. In fact, "transport factors are what transform potential P sources into actual P losses from a field or watershed" [33]. Hence, a risk assessment method is selected in this study in order to determine the potential for P loss from livestock farms in a non NVZ. The specific objectives of this research are to i) ascertain organic fertiliser management on livestock farms on a non NVZ area, ii) determine if there is a relationship between P application to land and soil P index, iii) produce a P risk map for the study area, and iv) ascertain the risk of failing the WFD objectives in non-NVZ area.

Materials and Methods

Study area

The current study was undertaken in the Upper Wensleydale catchment, within the Yorkshire Dales, in North Yorkshire (Figure 2). The most important urban areas located near the study area are Askrigg, Bainbridge and Hawes. All this area has a long tradition of livestock farming according to the Yorkshire Dales National Park Authority.

The main waterbodies included in the study area are River Ure, River Bain and Semerwater Lake. It should be highlighted that Semerwater Lake has many inlet streams as it is the largest glacial Lake within the Yorkshire Dales. Many of these inlet streams are found in a small valley called Raydale and cross most of the farm fields included in this research. The lowest point in the study area is adjacent to the River Ure at 205 m whilst the highest elevation rises to almost 600 m. Moreover, it was observed that the annual precipitation data from July 2011 to June 2012 presented a significant variation between gauge stations that are only 8 Km apart. In this sense, the gauge at Hawes (227 mAOD) shows a total annual precipitation of 1599 mm whilst the gauge at Moorland

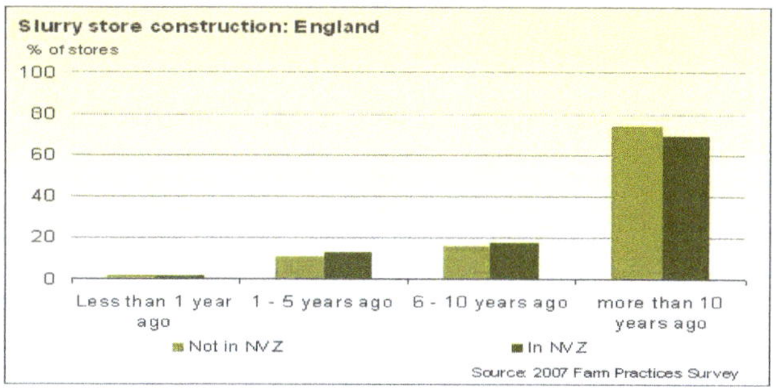

Figure 1: Slurry store construction in England [48].

Farm categories	Land area (%)	P surplus (%)
Dairy	19	29
Cropping	41	31
Pigs/poultry	1	17
Remaining	39	23

Table 2: Contribution of various farm categories to the annual total P surplus for the UK, expressed as a proportion of total agricultural land. [30].

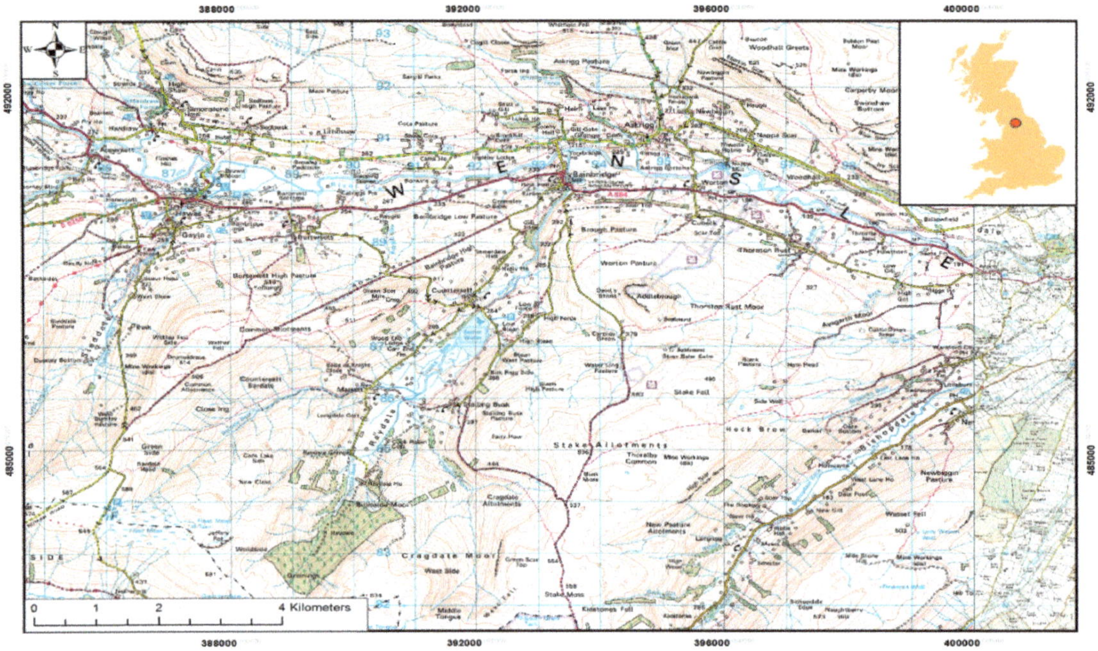

Figure 2: Map of the study area (digimap.edina.ac.uk).

Cottage (343 mAOD) was 2502.1 mm. This shows that it rains more at higher elevations. Furthermore it shows that the study area is a very humid region where soils are predominantly acidic.

Farms characterization

The types of livestock farms included in the study were mostly dairy and sheep farms (Table 3). Furthermore, different types of farms may bring about differences in the risk of P loss to watercourses. Information associated with livestock numbers, slurry storage systems and farm management was collected from a Farm Investigation Questionnaire (Appendix). Farm level data on the organic and mineral fertiliser rates to grass as well as timing of spreading was also obtained from the questionnaire. In fact, this data were different from the figures shown in The British Survey of Fertiliser Practice [17] where the five-year mean for grazed grass was 10 kg ha^{-1}. The N:P:K mass ratio of mineral fertiliser applied to land was used to obtain the application rate of P (kg ha^{-1}) considering that P fertiliser is present as P_2O_5 (56.4 oxygen and 43.6 elemental phosphorus). Similarly, the available P from organic fertiliser (manure/slurry) applied to the soil in some farm fields, was obtained from the nutrient analyses made by the ECSFDI. On the other hand, for those farms that an analysis report from ECSFDI was not available, data for P concentration in slurries/manure were obtained from the Fertiliser Manual (RB209) [34]. In this regard, no overall quantification of the accuracy of the calculations could be given.

These data were included in the GIS model to calculate the average application rate of P per field block.

The livestock farms included in this study extend to 2014 hectares of meadow, inbye pasture, woodland and moorland. The farm fields comprise two main land uses; rough grazing that includes moorland (70) and improved pasture (30).

Also relevant to this study is the livestock units per hectare (LU/Ha) as it is well known that animal trampling is a factor that increases the compaction of the soil and hence, brings about an increase of runoff volumes [35]. LU/Ha was estimated depending on the type of land use. Moorland is normally used for grazing sheep but not hardy cattle. In this respect, rough areas will have lower LU values compared with improved areas. The values are obtained in the GIS by dividing the correspondent number of livestock units by the area (in hectares), considering the length of time the stock are kept on the area. In this study, we considered that cattle are kept indoors between October and April. In addition to this, a table of ratios based on feed requirements of different livestock [36] was used for these estimations.

Soil sampling

The timing of the farm visits coincided with a period of very wet weather by the middle of July. Soil cores were taken from the top 7.5 cm. according to Rowell [37] recommendations for grassland. A composite soil sample for each study farm was made up of 3 individual cores taken at sites close to surface waters (less than 50 metres). This sampling aimed to obtain soil P concentration in areas that pose a higher risk due to its proximity to watercourses (Figure 3). The exception was Farm 1 as one more composite soil sample (1B) was taken from a field where dairy cattle were trampling on the river bank.

Farm Id	Sample Id	Type of livestock
1	1B	Dairy and sheep
1	1T	Dairy and sheep
2	2S	Dairy and sheep
3	3S	Dairy
4	4S	Beef, sheep and lamb
5	5S	Dairy and sheep
6	6S	Beef and sheep
7	7S	Dairy, sheep, lamb and beef
8	8S	Beef and sheep
9	9S	Dairy and sheep
10	10S	Dairy, sheep and lamb

Table 3: Identification of farms and composite soil samples included in the study.

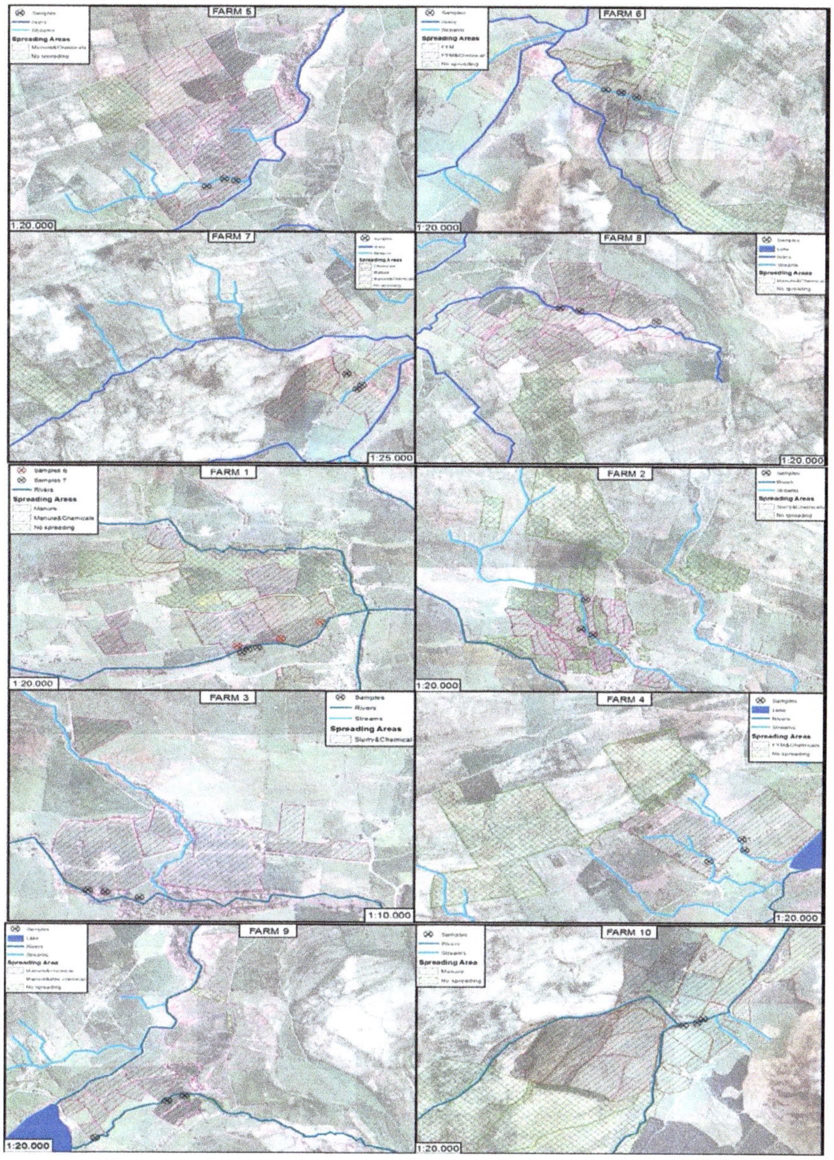

Figure 3: Soil sample sites for each study farm. The maps include the areas where fertiliser is applied.

Soil Analysis

In the field, soil texture was assessed by hand following a guide for mineral soils taken from Rowell [37]. On return to the laboratory, soil moisture content was determined gravimetrically after oven-drying soil samples overnight at 105°C. The loss-on-ignition method was used to estimate the soil organic carbon using the procedure presented in Ball [38].

Soil pH was measured by placing a glass electrode in a mixture of soil and deionised water

Furthermore, pH was also measured in a salt solution in order to release "reserve" acidity of colloid surfaces.

Olsen's P was determined according to the method presented in Rowell [37]. Soil samples were dried at 40°C for 48 hr. Subsequently, they were crushed in a porcelain mortar to break up large aggregates and sieved using a 0.2 mm stainless steel sieve. 3 reagent blanks were extracted in order to determine if phosphorus was present in the reagents. Due to high P concentrations in the soil extracts, the absorbance was greater than the top standard and therefore the amount of extract/sulphuric acid had to be reduced in order to match the standards.

Statistical Analysis

Pearson's correlation tests were undertaken in Minitab v16 (Minitab Inc., State College, PA, USA) and considered significant where $P<0.05$. These analyses were conducted to assess any linear relation between P concentration in farm soils and farm practices.

Risk mapping

All the research was built on a predictive model carried out using ArcGIS V.10. Risk areas for P losses from field blocks were mapped using the Pennsylvanian P Index as it is well documented (Table 4) [39].

Source factors: All source factors were estimated using the data listed above. Data on soil test P (Olsen Method) was obtained from

Part A- Screening tool					
	Evaluation Category				
Soil Test P	>200 mg P kg^{-1}		**If yes to either factor then proceed to Part B**		
Contributing Distance	<45 m.				
[a]Contributing Distance	>45 m. AND field artificially drained				
Part B – Source factors					
Soil test	Soil Test P (mg P kg^{-1})				
	Soil Test P Rating = 0.20* Soil Test P (mg P kg^{-1})				
Fertiliser P rate	Fertiliser (mg P kg^{-1})				
Manure P rate	Manure (mg P kg^{-1})				
P source application method	0.2 Placed or injected 5cm or more deep	0.4 Incorporated <1 week	0.6 Incorporated>1 week or not incorporated April – October	0.8 Incorporated>1 week or not incorporated Nov – March	1.0 Surface applied to frozen or snow covered soil
	Fertiliser Rating = Rate × Method				
Manure P availability	0.5 Treated manure/Biosolids		0.8 Dairy	1.0 Poultry/Pigs	
	Manure Rating = Rate × Method × Availability				
	Source Factor = Soil Test P Rating + Fertiliser Rating + Manure Rating				
	Part C – Transport factors				
Erosion	Soil Loss (tonnes ha^{-1})				
Runoff potential	0 Very low	2 Low	4 Medium	6 High	8 Very high
Sub-surface drainage	0 None		1 Some		2[b] Patterned
Contributing distance	0 >150 m.	2 150 to 100 m.	4 100 m. to 75 m.	6 45 to 75 m.	8 <45 m.
Transport Sum = Erosion + Runoff Potential + Sub-surface Drainage + Contributing Distance					
Modified Connectivity	0.7 Riparian buffer – applies to distance<45 m.		1.0 Grassed waterway or none	1.1 Direct connection – applies to distance>45 m.	
Transport Factor = Modified Connectivity × (Transport Sum/22)					
[a]Modification introduced from the Danish P Index [b]Or rapid permeability soil near a stream					
Phosphorus Index Value = 2 × Source Factor × Transport Factor					

Table 4: The Pennsylvania P Index [39].

the analysis results for each farm. The different fertiliser and manure application rates, depending on the farm, between April to October and November to March, brought about two different P risk estimations for each farm. This is PIV_S (Phosphorus Index Value_April to October) and PIV_W (Phosphorus Index Value_November to March).

Transport factors: Soil erosion (tonnes ha^{-1}) was calculated by the Revised Universal Soil Loss Equation (RUSLE) in a GIS in order to create a model to obtain the annual soil loss for each farm holding. The model was created according to Shiferaw [40] and is based on the RUSLE equation (1).

$$\mathbf{A = LS^{*}\ R^{*}\ K^{*}\ C^{*}\ P} \qquad (1)$$

Where:

- **A** is the annual soil loss (metric tons ha^{-1}yr^{-1});

- **LS** = slope length factor (dimensionless); this was derived from a 5 m. resolution DEM where local depressions (sinks) were previously removed. The equation (2) given by Morgan [41] was used to estimate the LS factor.

$$\text{LS} = \sqrt{\frac{l}{22}}\left(0.065 + 0.45\text{s} + 0.0065\text{s}^2\right) \qquad (2)$$

Where:

l = slope length in m.; it was obtained by using the ArcHydro tool “Flow Length”.

s = percent slope; it is the result of using the Spatial Analyst tool “Slope”.

- **R** is the rainfall runoff erosivity factor [MJ mm h^{-1} ha^{-1} yr^{-1}]; Due to the lack of rainfall intensity data as well as storm kinetic energy, R was derived from mean annual and monthly rainfall data obtained from the previously mentioned gauge stations (Table 5).

- **C** is land cover and management factor (dimensionless, ranging between 0 and 1); According to Shiferaw [40] the C-Value for grassland is 0.01.

- **P** is conservation practice factor (dimensionless, ranging between 0 and 1). In this study we considered P = 1.

The relationship between rainfall data and altitude was expressed in a linear equation in order to interpolating the rainfall values across the study area. A 5 m. resolution DEM was used along with the derived equation to obtain the mean annual rainfall (MAR) raster. Fournier Index (3) was obtained according to Arnoldus [42]:

$$\text{F} = \sum_{j=1}^{12} \frac{pj^2}{P} \qquad (3)$$

Where, F = Fournier Index;

p_j is the monthly rainfall for month;

P is the annual rainfall

F was linked to R through the empirical formula given by

Gauge Station	2011						2012					
	Jul	Aug	Sep	Oct	Nov	Dec	Jan	Feb	Mar	Apr	May	Jun
Hawes	118	158	141	170	78,8	238	154	84,8	23,8	148	97,6	187
MoorlandCottage	125	244	273	241	160	428	165	220	51,1	151	127	317

Table 5: Monthly rainfall data for the study area. (Environment Agency).

R = (4.17F – 152)/17.2 [42].

K factor was obtained by using the following equation [43]:

$K = 2.1 \times 10^{-6} \times M^{1.14}(12\text{-}OM) + 0.025\ (S\text{-}3) + 0.0325\ (P\text{-}2)$

Where;

K = soil erodibility factor in t.h/MJ.mm; this layer is based on data such as soil texture and percentage of organic matter, obtained from the analysis results.

K factor was obtained by using the following equation [43]:

M = (Percentage very fine sand + Percentage silt) x (100 – Percentage clay)

OM = Percentage of organic matter obtained from the soil samples.

S = Code according to the soil structure (very fine granular = 1, fine granular = 2, coarse granular = 3, lattice or massive = 4);

P = Code according to the permeability/drainage class (fast = 1, fast to moderately fast = 2, moderately fast= 3, moderately fast to slow = 4, slow = 5, very slow = 6). This information was obtained from Soilscapes viewer.

Table 6 shows the different textures found in farm soils (all of them histosols).

Runoff potential

The relationship between potential runoff and soil moisture has been demonstrated in different studies [44,45]. The Topographic Index model [46] was implemented in this study in order to establish different thresholds for runoff potential. Threshold values for classifying areas where saturation excess overland flow is likely to occur, were based on field work as well as the waterbodies layer derived from the 1:10,000 topographic map. These values were finally reclassified according to the Pennsylvania P index.

The Topographic Wetness Index (TWI) was calculated as follows [47]:

$Wi = \log (a / \tan ß)$ Where,

a is the upslope contributing area per unit contour length (cell size * flow accumulation) and tan ß is the local slope gradient for estimating a hydraulic gradient (ß is the slope converted in radians). In addition, flow accumulation layer was derived from a DEM where sinks were removed and null (No Data) cells were filled using focal function (e.g. 'mean').

It was assumed that wetter cells have more soil moisture and hence, more likely to induce saturation excess overland flow.

Furthermore, values were transformed to the same unit of measurement scale. This was done on a 0 to 1 scale using this generic formula (4).

("Oldgrid" – minimum value) / (maximum value – mininmum value) (4)

Finally values were reclassified according to Table 7.The values used for "sub-surface drainage" were based on the farm visits.

The contributing distance was calculated in GIS by appending the waterbodies raster layers (rivers, streams and lakes) in order to obtain a cost distance raster. In this regard, the flow direction raster was used as the input cost distance raster, as water always flows down a slope. Finally, distance values were reclassified according to Table 4.

Sub-surface drainage information was provided by the Yorkshire Dales National Park Authority.

The connectivity values were based on the analysis of high-resolution aerial images. Therefore, values were specific for each field block.

Results

Despite the fact that all the farms are not in a NVZ area, 90 accepted to receive advice from the ECSFDI according to the WFD objectives. However the farms management practices could substantially be improved according to the data collected during the farm visits.

Fertiliser intputs

Data for fertiliser inputs, this is mineral as well as organic fertiliser, were crucial to determine the risk of P loss when the P Index assessment was carried out (Table 8).

The procedure followed by farmers to estimate the manure application rates was basically to avoid "smothering" the grassland. Only 20 of the farmers assured that manure application rates varied depending on the season (applying higher rates in spring). All the farmers provided data for slurry application rates, unlike data for FYM where 40 of the farmers did not have figures for it. This posed a problem for the risk assessment of farms 1, 5, 6 and 10.

Information about soil analysis and nutrient content of manure deserve special attention. In this regard, 30 of the farmers assured to make soil analysis, normally every 3-4 years, whilst Farm 5 also measured the nutrient content of manure and the P concentration of soil. This information links to the mineral fertiliser mass ratio, as no P is applied to soil in Farm 5. Regarding the rest of the farms, estimations of mineral fertiliser application rates were made on the basis of N and varied widely among them.

Moreover, it should be highlighted that some of the farms received between 1 and 3 years ago, from the ECSFDI a Nutrient Management Plan. This report was built on a soil nutrient analysis as well as on standard figures from the RB209 Fertiliser Manual in order to advise farmers about the fertiliser application rates. However, the results for P concentration obtained from soil analysis in the laboratory for each farm, were generally high and even very high. In this regard, Table 8 shows that over 50 of the results are included in high and very high categories according to the P availability index used in Britain by ADAS. This means that soils are so enriched with P.

Pearson's correlation tests showed no significance relationship ($P < 0.05$) between P application rates and P Index. This could be due to the lack of accurate and complete data for the P application rates. In this respect, special attention deserves farm 7 as it obtained the highest P concentration value despite no mineral P was applied this last year

Sample Id	Textural class	M	OM (%)	S	P	K
1B	Sandy Loam	8100	6.14	3	3	**0.38**
1T	Sandy Loam	8100	3.64	3	3	**0.53**
2S	Sandy Loam	8100	4.77	3	3	**0.47**
3S	Silty Clay Loam	4900	6.14	2	5	**0.27**
4S	Clay Loam	4900	6.09	3	4	**0.26**
5S	Silt Loam	8100	6.30	2	4	**0.38**
6S	Loam	6400	5.43	3	2	**0.30**
7S	Loam	6400	4.78	3	2	**0.33**
8S	Loamy sand	10000	6.03	3	1	**0.42**
9S	Sandy Clay Loam	6400	4.76	2	3	**0.34**
10S	Sandy Loam	8100	3.74	3	3	**0.53**

Table 6: Values of the parameters used to obtain K for each farm.

Old value	New value
0-0.06	0
0.06-0.18	2
0.18-0.3	4
0.3-0.41	6
0.41-1	8

Table 7: Reclassification values for TWI.

according to the farmer. However, the sum of available P from slurry and FYM results in the highest value. This may link to some studies that highlighted that farmers normally undervalue the N:P:K content of slurry and FYM [31,32].

Slurry storage and timing of application

The information provided by the farmers shows that aspects associated with slurry storage as well as the timing of manure spreading, are far from meeting the Code of Good Agricultural practices required for farms in NVZ areas. Similarly to the figures for slurry store construction [48], 90 of the study farms had slurry storage systems built before 1991. Therefore, they are not obliged to meet the SSAFO requirements for providing slurry storage capacity for a four month period. This information is closely related to the manure application frequency. In fact, 60 of the farms spread manure, at least, once per month during the winter period.

The factors mainly considered by farmers when spreading are associated with wet soil conditions as well as storage capacity. All of the farmers assured that they do not spread manure when is raining or soil is too wet, as this may cause a loss in traction for the tractor as well as soil erosion. However, it should be stressed that all of them spread manure yet soil is snow covered or frozen.

The reasons given by farmers are the better traction for the tractor as well as the limited slurry storage capacity.

The information above shows that limited capacity to store manure exerts a strong influence on the frequency of its applications. This makes sense if we take into account that 40 of the farms only had slurry storage capacity for a maximum of 6 weeks, and only 20 can provide slurry storage capacity for a four month period. It is not surprising then, that, 70 of the farmers would be willing to change the frequency of spreading manure in case of larger capacity to store it.

On the other hand, the rest of the farmers maintain that spreading large amounts of manure during a short period (after January) could be counterproductive, as livestock start

grazing in spring. However, Hooda [49] demonstrated that limited slurry storage capacity and hence, slurry application during winter months is linked to high P concentrations in drainflow. In spite of this, it is discouraging that 90 of the farmers do not consider increasing the capacity to store manure. Among the reasons given are the milk price drop and the lack of grants for this issue.

Application method

Another aspect is the method used to spread manure. This is one of the P source factors considered in the P Index assessment. Slurry application in all the field blocks is carried out with a splash plate system whilst FYM is spread using a manure spreader. Only Farm 5 used the shallow-injection technique to spread slurry in spring (Figure 4). These data show the lack of effective techniques for application of manure what results in an increase of the risk of P loss according to the P Index assessment. The reason for this is associated with the exposure of manure to precipitation events and subsequent nutrients loss in runoff [50,51].

Livestock density

The LU/Ha values were most of them similar or lower in comparison with the average stocking density over all lowland forage area in England (0.58) [52]. Only the result for Farm 2 almost reached 2 LU/Ha. According to Chesterton [52] 2 LU/Ha is considered to be a problem in a NVZ area. In fact, Edwards [30] pointed out that livestock numbers is an important factor contributing to the variation of P surplus. Nevertheless, Pearson's correlation tests showed no significance relationship ($P < 0.05$) between P Index and LU/Ha.

Mitigation measures

The effectiveness of some agricultural stewardship measures for ameliorating nutrient water pollution have been demonstrated

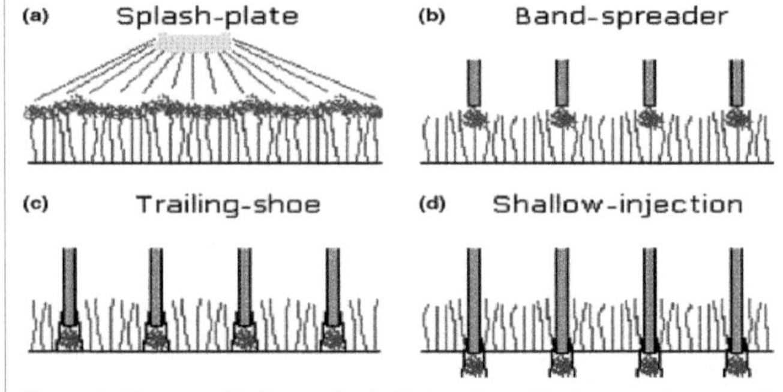

Figure 4: Slurry application methods "(a) on top of herbage and over entire spreading width with splash-plate method; (b) in lines on top of herbage with band-spreader method; (c) in lines below herbage, but above the soil surface, with the trailingshoe method; and (d) below the soil surface (approximately 5 cm) with the shallow-injection method"[51].

Sample Id	P (ppm)	Soil Test P (ppm)	P availability Index (extractable phosphate ppm of soil)	P Index (ADAS)	ADAS Interpretation	N:P:K Mass ratios	Mineral F. application rate (kgN ha^{-1})	Total P (FYM+Slurry) application rate (kg ha^{-1})
1B	73.62	14.7	71-100	5	Very high	34:20:10	40	8.18[a]
1T	33.97	6.8	26-45	3	Medium to high	No chemicals	-	8.18[a]
2S	81.98	16.4	71-100	5	Very high	25:05:05	170	8.4
3S	79.29	15.9	71-100	5	Very high	25:05:05	250	19.2
4S	60.13	12	46-70	4	High	20:10:10	50	23.75
5S	37.21	7.4	26-45	3	Medium to high	25:00:14[b]	125	17.3[a,c]
6S	46.65	9.3	46-70	4	High	20:10:10	123	No data
7S	91.69	18.3	71-100	5	Very high	24:00:15	250	49.3[d]
8S	38.82	7.8	26-45	3	Medium to high	20:10:10	250	23.75
9S	44.76	9.0	26-45	3	Medium to high	20:10:10	125[e]	27.58
10S	25	5	16-25	2	Slightly low to medium	No data	-	7.66[a]

[a]No data for FYM application rate
[b]This mass ratio corresponds to April. 92.5 kgN ha^{-1} of urea (48% N) are applied in March, whilst 92.5 kgN ha^{-1}(24:00:14) are applied in June.
[c]17.3 corresponds to the slurry application rate in spring, whilst 11.5 is the slurry application rate in winter.
[d]49.3 corresponds to the manure (FYM&slurry) application rate in spring, whilst 45.4 is the figure for winter.
[e]125 corresponds to the application rate over all the field blocks, except for moorland where does not apply and meadow where the mineral F. Application rate is 500 kgN ha^{-1} (20:10:10).

Table 8: Fertiliser application rates and P availability indices (ADAS) for the correspondent soil samples.

across numerous studies [31,35,53,54]. Regarding possible measures implemented to reduce the risk of nutrient loss to freshwaters, 40 of the farms had fenced off its waterways. Thus, cattle had access to waterbodies in 60 of the farms what poses a risk for water quality. As stated above, sample 1B was taken from a near river bank trampled by dairy cattle. The impact of cattle poaching around river banks was stressed by Walling [55] as an important sediment source that links to P loss to watercourses.

In addition to this, 60 of the farmers spread fertiliser within 5 to 6 meters of watercourses where there are no riparian buffers. This spreading distance from watercourses considering the no existence of riparian buffers could be clearly insufficient to ameliorate the risk of nutrients reaching the surface waters. In this regard, Heathwaite [31] highlighted the small removal efficiency of Total Phosphorus (TP) using a 10 metre buffer strip, when organic fertilisers are applied.

Mapping of risk mapping

The implementation of the Pennsylvania P Index on the study farm holdings resulted in large areas being classified as "no risk". In particular, 80 of the farms had more than half of the grassland being assigned a value of 0. These field blocks correspond to rough grazing where most of it is moorland. Farmers assured not to spread fertiliser on this land types, from what they were considered to have a Soil Test P <200 mg P mg^{-1}. This information was taken into account when soil sampling was carried out. Therefore, no soil samples were taken from those land types. The consequence of this was that "no spreading areas" scored a value of 0 for Soil Test P rating (Figure 3). The remaining field blocks were mostly classified as having a low risk of losing P to freshwaters (Table 9).

Moreover, the application of mineral fertiliser only in spring as well as different amounts of manure applied by some of the farmers depending on the time of year, were taken into account in order to classify the surface area of each farm in risk categories (Figures 5, 6 and 7a).

Although Table 9 shows the overall figures for each risk class, it is recommended to observe Figure 5. In this regard, P Index for farms 1 and 2 did not differ between the two studied periods (PIV_W & PIV_S). The reason is that the small amount of mineral P fertiliser applied in spring does not result in more areas included in higher risk categories. Furthermore, these two farms maintain the same manure application rates during the two study periods.

Another group is formed by farms 4, 5, 7 and 10. Figure 5 for these farms, shows slight increase of areas included in higher risk categories during the PIV_W period.

These farms have in common the small amounts of mineral P fertiliser applied in spring (Table 8). The most obvious PIV increase occurs in farm 7 due to the high application rate of manure in winter. Although the amount of manure applied to soil, is even slightly higher in spring, the Pennsylvania P Index considers the risk of P loss is 40 less between April and October. This is the reason why adding fertiliser in spring may not pose a higher risk of P loss in this time of year. The opposite effect occurs on farms 3, 6, 8 and 9, where there is a higher risk of P loss during the PIV_S period. In this respect, farms 8 and 9 deserve special attention as the increase of surface area in the very high risk category (>100 according to P Index Rating) is dramatic. The reason seems evident if we look at the figure for mineral P application rate. In this sense, these two farms have the highest inputs of mineral P. This substantial increase of P inputs can clearly be observed if we compare Figures 6 and 7a. The red area indicating "very high risk" is much larger in Figure 6, for both farms. Furthermore, the results shown on the maps represent complex information through a very intuitive visualization. For example, Farm 9 can pose a high risk of P delivery to Semerwater Lake.

Moreover, the P risk maps obtained for each farm are not only the result of fertiliser application rates and soil analysis results, but also transport factors which magnitude varies in each area or pixel. The importance of transport factors in the risk maps, can be observed if we take into account that farms with higher figures for "Soil Test P" are not necessarily the ones with larger areas in high risk categories.

P Index Value	Rating	% of land (PIV_W)	% of land (PIV_S)
0	No risk	67.25	67.25
<60	Low risk	23.14	22.26
60-80	Medium risk	4.39	2.86
80-100	High risk	2.11	2.07
>100	Very high risk	3.12	5.57

Table 9: General interpretation of the Pennsylvania P Index [39] and total % of land in each risk category. Calculating the Pennsylvania P Index on the farm holdings located in Wenslaydale.

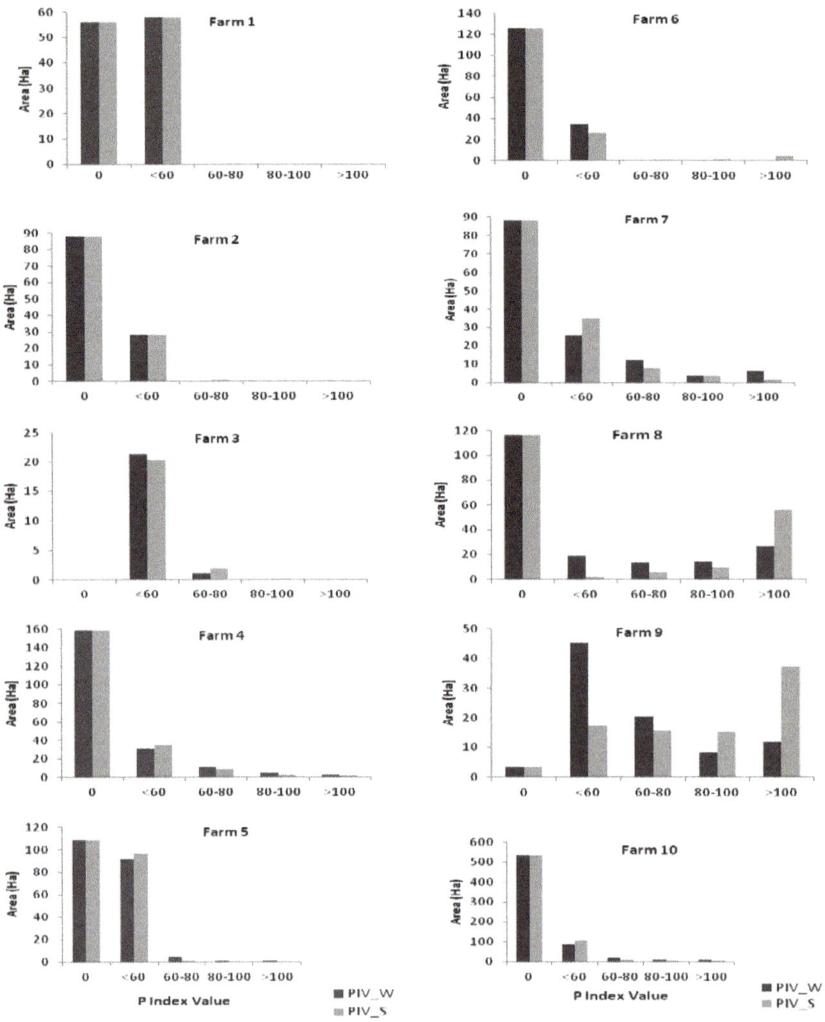

Figure 5: Clustered column charts showing total surface area (hectares) included in each risk class. PIV_W shows the Phosphorus Index Value from November to March. PIV_S shows the Phosphorus Index Value from April to October.

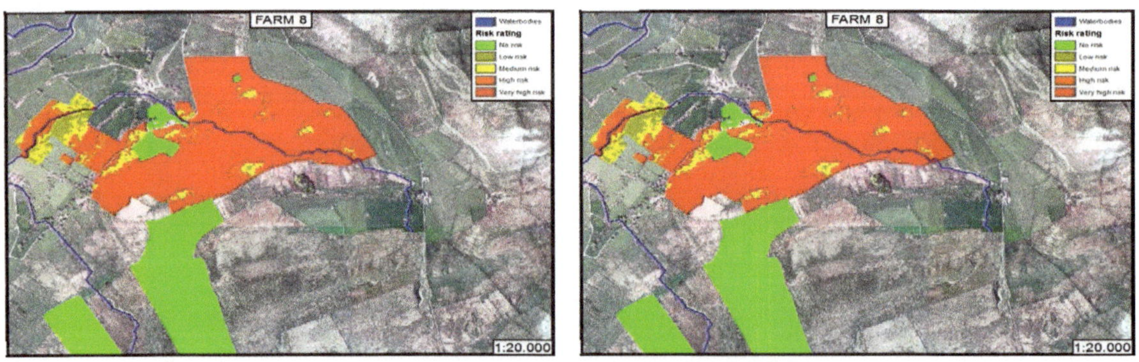

Figure 6: P risk maps obtained for the period between April and October (PIV_S).

In particular, soil loss values or soil erosion greatly influences the risk category of each pixel. The use of a 5 m. resolution to obtain the different factors included in the RUSLE equation, guarantees accurate results.

Nevertheless, the rainfall runoff erosivity factor obtained for the study area could be improved if rainfall intensity data had been available. The use of the empirical formula given by Arnoldus [42] could have increased the systematic error of the model due to the use of statistical techniques to estimate R [56].

However, the use of Fuzzy methods was crucial to represent the continuous properties of soil as well as the geomorphology in order to add as much detail and information as possible [57]. On the other

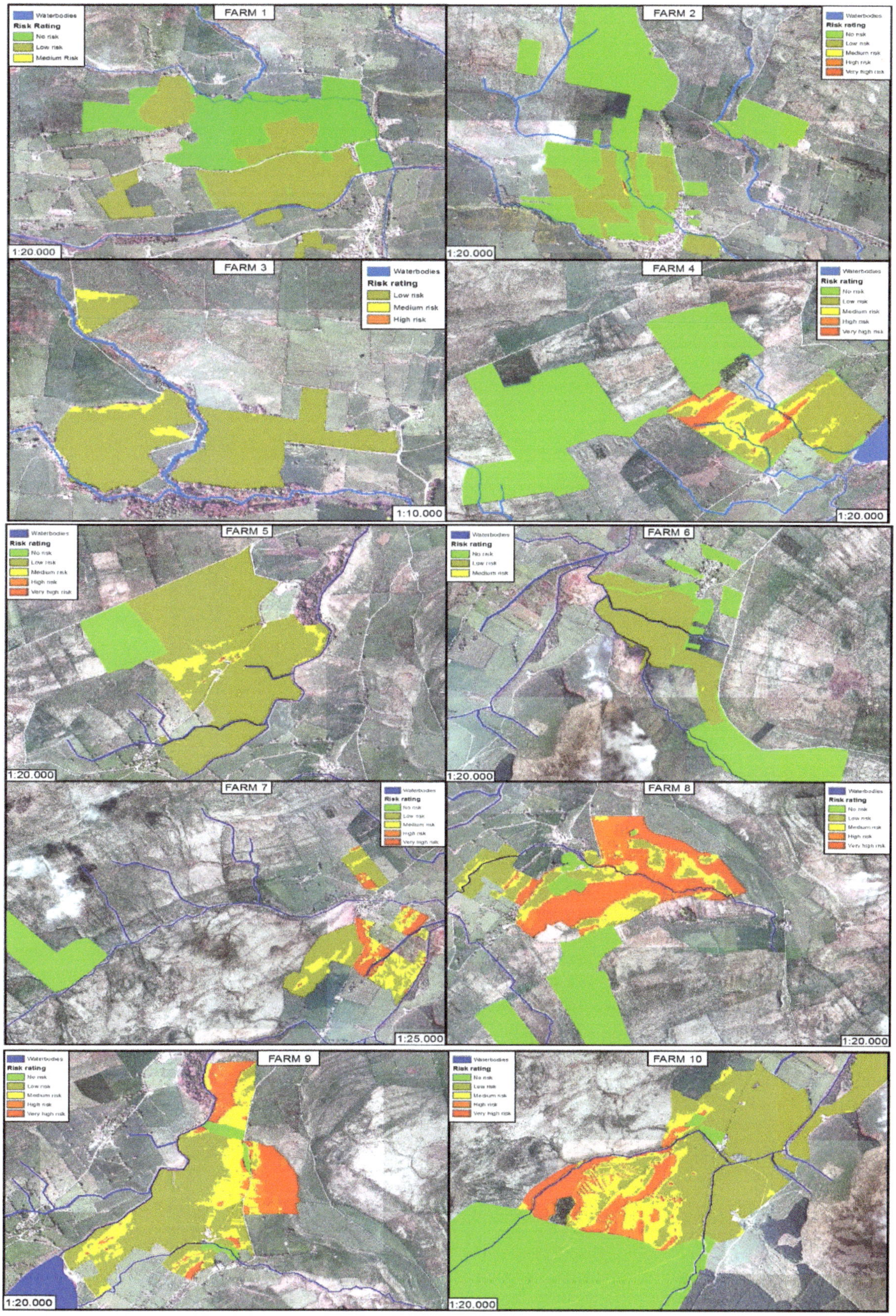

Figure 7a: P risk maps obtained for the period between November and March (PIV_W).

hand, the maps are also the result of using Boolean methods in order to include values associated for example with sub-surface drainage.

Discussion

Landscape connectivity

To meet the objectives of the WFD, at the catchment scale, requires identifying the sources of diffuse pollution on small tributaries [58,59]. Thus, the approach taken in this study was to determine the degree of risk posed by livestock farms in a sub catchment, classified as non NVZ. This classification based on the Nitrate Directive, set clear boundaries to ensure that regulation resources focus on areas where nitrate pollution is known to be a problem (NVZ).

However, some studies have questioned the effectiveness of these policies, specifically the Nitrate Directive [14,22], as they do not consider the upstream-downstream linkages, particularly the nutrient spiralling concept. In this regard, River Ure rises very close to the study area and then, flows eastward through a NVZ. This river is the major pathway for the transport of nutrients from the study area to NVZs. Meybeck [60] in a study carried out in the Seine river basin, demonstrated a continuous rise in P concentrations downstream. Thus, important nutrient loadings may result from the Critical Source Areas (CSA) identified in the study farms and reach these NVZs. This shows that water pollution problems should not exclusively be identified from concentration data such as the threshold value of 11.3 mg N l^{-1} in freshwaters, established by the Nitrate legislation. Water eutrophication in lakes has been demonstrated to occur depending on the N:P mass ratio [30]. The appropriate method should then quantify nutrient loadings or flux rates coming from point and non-point sources of pollution [61], although the bioavailability of P fractions in freshwaters and the time of year (spring/summer) are major factors that influence the risk of water eutrophication in flowing waters [10,62].

Managing source factors

The ECSFDI has been the main mechanism in England that aims at reducing nutrient loadings at the catchment scale. However, meeting the WFD objectives should not only depend on a voluntary scheme targeting farmers. In this sense, the problems to obtain data or records about fertiliser inputs are not supposed to occur in farms included in NVZ, as they are subject to cross compliance under the Nitrate Pollution Prevention Regulations 2008 (Statutory Instrument 2008/2349). The requirement, in NVZ, of making a nutrient plan based on soil analysis could have provided, for example, detailed information about the amount of manure applied to soil. As stated above, fertiliser inputs data is crucial to identify CSA using the P Index assessment.

Moreover, the efficacy of the ECSFDI to reduce the risk of P loss to watercourses is difficult to assess. Continuous long term excess P inputs and hence, P accumulation is not a recent episode in an area that has a long tradition of livestock farming. In fact, all the soil samples taken in this study have moderate (26-45 mg kg^{-1}) to high (>45 mg kg^{-1}) Olsen extractable soil P contents. These figures show large P surpluses (>index 2) according to ADAS. Furthermore, these P surpluses determine the soil retention of P inputs due to a lower sorption capacity [63]. This is associated with the risk of P leaching due to increasing amounts of P into the labile pool [64].

In spite of the lack of some data linked to fertiliser inputs, the risk maps showed a higher risk of P loss for some farms during the PIV_W period. This is strongly associated with the application rate of manure to soil in this time of year [49]. The role played by regulations is determinant in these results and in turn, makes difficult to understand the current policies. The obligation of meeting good status of waters by 2015 becomes a hard task when there are farms that can even be exempt from meeting the SSAFO requirements. As stated above, 90 of the slurry store systems were constructed before 1991. Furthermore, the location of these farms in a non NVZ exempts them from providing storage for organic manures for at least 5 months. This rule aims at preventing the spread of manure in the winter period and agrees with studies that have demonstrated a decreased level of P in winter due to larger capacities to store slurry [65]. In addition to this, the existing limitations of these farms to store slurry pose even a higher risk of nutrients loss to watercourses due to the effects of local conditions. In this sense, the humid weather in the study area, bring about a substantial number of slurry applications under wet conditions according to the farmers. These farms practices are considered in the risk model. The P Index assessment reflects different risk values depending on the timing of spreading manure and mineral fertilisers. Higher values are assigned to manure applications in winter, as this is associated with factors such as reduced soil permeability during freezing due to ice formation [66].

Moreover, despite no mineral P was applied in any of the farm fields in winter, P Index assessment assigns the same weight to manure and fertiliser application in the winter period. However, Wood [67] reported higher concentrations of Total P loss in surface runoff for winter application of manure in comparison with results obtained for soils where only P mineral fertiliser had been applied. This is associated with higher risk of losing Soluble Reactive Phosphorus (SRP) through leaching when manure is applied in winter months [68].

Spreading techniques

The risk model applied in this study takes into account not only the time of year but also the method used to apply fertiliser. Nevertheless, it should have been desirable to find not only one farm using a technique similar to the shallow-injection method in order to obtain different results of the impact that these methods can cause on the risk maps. Despite this limitation, figures for Farm 5 obtained from the risk maps (Figure 5) showed lower levels of risk associated with the use of shallow-injection method in comparison with splash plate methods. This is linked to larger amounts of P absorbed by the upper soil layers due to the direct contact of slurry with the soil [50,69]. Thus, this method contributes to decreasing the SRP loss to freshwaters [69].

Regarding the time of year when different techniques such as splash plate or swallow injection are used, deserves special attention. Although this model represents the risk areas of P loss to watercourses, this does not mean that it shows the risk of ecological impact on the streams network. In this sense, it should be considered that the SRP loss occurred in winter months does not cause the same effect on rivers than in spring/summer months due to different rates of biological activity [10,62]. Therefore, the use of the shallow-injection method in spring/summer substantially reduces the risk of waters eutrophication.

Moreover, there is some strong evidence for suggesting that financial pressures are more important than environmental demands. In this regard, it is hard to persuade farmers to use more effective techniques when this involves an important cost [51] and they are not obliged by regulations. In fact, farmers assured to use splash plate methods as the alternatives are more expensive.

Source characteristics

The P Index assessment considers the most important factors associated with diffuse agricultural P pollution [27] however, existing

P sources not included in the group of diffuse sources may represent a significant problem when P loss from livestock farms is assessed. During the farm visits, it was observed that some of the farms posed a high risk of P loss due to runoff coming from hardstandings that receive fresh faecal material very frequently (Figure 8). This runoff, highly contaminated in nutrients [70] is the result of management operations, in particular, movement of livestock wastes and rainfall events. Also important is the rise of the proportion of P in soluble form transported in this type of run-off. The presence of tracks, and overflows coming from these hardstandings result in shorter interactions between subsoil and runoff and hence, larger amounts of SRP reaching freshwaters [49]. In addition to this, it should be stressed that this type of management operations occur throughout the year and hence, SRP delivered to surface waters in spring/summer periods contributes to increasing the risk of rivers eutrophication [10].

Transport management

All the soil samples taken from the study farms had pH values considered acidic (Figure 7b). Regarding this, P sorption and release processes in these acidic soils are greatly influenced by clay content which in turn, links to P losses by leaching [15]. In this study, soil texture alongside soil organic matter which is another sink of P [15], were taken into account in order to estimate the resistance of soil to be detached and transport by rainfall and runoff [71]. This is helpful to estimate the risk of P loss associated with particulate material (PP). In fact, Walling [72] reported important relative contributions to the total PP flux coming from grassland areas in UK catchments.

However, the modified Danish P Index adds another transport factor in order to assess the leaching potential based on the type of soil [73]. According to Andersen [73] the soils assessed in this study should be assigned the highest risk of P leaching due to its low P binding capacity (organic soils). Nonetheless, the P Index implemented in this study first requires to be examined according to local conditions in order to introduce any possible modification.

Another relevant aspect associated with soil erosion is derived from cattle trampling stream banks. This type of event is likely to occur in all those farms that were reported to have not fenced off its waterways and when the rates of biological activity are at its highest. This means that the cattle housing period coincides with the water eutrophication risk at its lowest (winter months).

The implemented risk model is based on the RUSLE which in turn, was developed to predict sheet and rill erosion. However, this does not include the effects of concentrated runoff or soil erosion events derived from animal trampling [74]. The clearest example of this type of event was reported to occur in Farm 1 (Figure 9).

In this sense, the risk map obtained for Farm 1 (Figure 10) when the ecological risk is at its highest, lacks of including the risk of PP derived from existing stream bank erosion.

Furthermore, it seems that the importance of the effects of PP on rivers ecosystems changes in spatial and temporal scales, as the key factor is the bioavailability of P for algal proliferation. Jarvie [10] highlighted the small relative contribution of PP derived from agriculture to water eutrophication in lowland rivers in comparison with sewage effluent sources. Therefore SRP desorption from sediment particles represents an ecological risk in individual reaches of rivers where SRP concentrations are low at times of eutrophication risk [4,10,75].

Moreover, transport factors linked to sources not included in the group of diffuse sources were found in Farm 4 (Figure 11). In this regard, steep slope and short distance of these intermediate sources to watercourses significantly increase the risk of nutrients loss [33]. Furthermore, it should be pointed out that Farm 4 received in 2009 a report from the ECSFDI where several recommendations were made in order to prevent polluted water from entering the nearby stream.

However, no improvements have been undertaken. This farm can be considered as an example that cast doubt on the current policies that aim at reducing water pollution at the catchment scale.

Conclusion

This study has yielded that there is significantly more room for improving the management in livestock farms located in non NVZs.

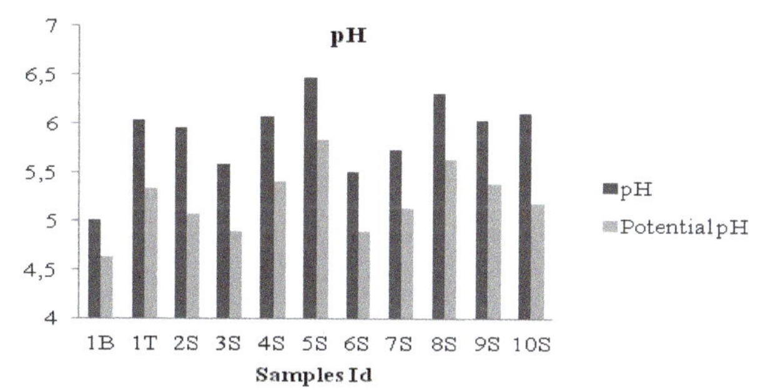

Figure 7b: pH values for the study farms.

Figure 8: Runoff coming from the slurry store area.

Figure 9: The image taken from Farm 1 shows clear signs of stock access and soil erosion.

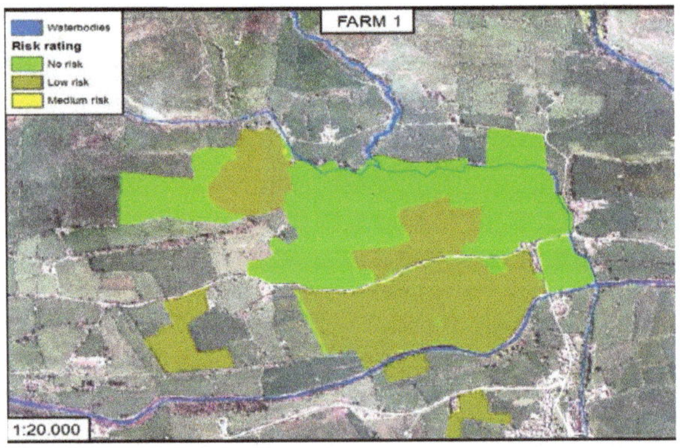

Figure 10: P risk maps obtained for the period between April and October (PIV_S).

Figure 11: Effluent from all structures flowing into the stream.

Most of the farms apply manure without taking into account the nutrient content of manure as well as the soil nutrient pools. The manure application rate would clearly exceed the restrictions in NVZs, as farmers are not allowed to spread organic fertilisers in NVZs during winter months due to the higher risk of nutrients loss to freshwaters.

In the case of the study farms, the limited store capacity determines the high frequency of manure applications, as most the farmers would be willing to reduce these application rates. Furthermore, the humid weather conditions along with the methods used by farmers to spread manure are factors that exert a strong influence on the risk of P loss in the study area. Nevertheless, the risk pose by rainfall events can be improve in the implemented risk model as long as rainfall intensity data are available. It should be stressed that large amounts of nutrients are flushed out from lands during a few heavy rainfall events [76]. This is particularly relevant when manure is applied and left exposed on the surface (splash-plate method).

Although this study has not demonstrated a significant relationship between P application rates and P Index, it should be pointed out that P surpluses found in the soil are not surprising according to the nutrient management undertaken in most of the study farms. Nevertheless the lack of accurate data for manure application rates could have been decisive for the statistical results. Furthermore this links to the lack of conditions such as current Cross-Compliance rules that apply to NVZs. All these factors contribute to P accumulation in soil and hence, increasing the risk of P loss to freshwaters.

Moreover, the risk maps obtained in this study are the first step to identify the areas where resources should focus in order to achieve the WFD objectives. In fact, the two risk maps obtained for each farm (PIV_S&PIV_W) can provide more accurate information about the risk of P loss in winter months as well as the problematic areas when the ecological risk is higher (spring/summer period).

However, the P index assessment used in this study should be tested in order to examine its capability to rank measured annual diffuse total P losses from the Upper Wensleydale catchment. The results obtained should be analysed in order to introduce possible modifications in the P Index, according to local conditions such as major P loss pathways in the catchment area.

Regarding the current policies that aim at reducing nutrients loadings in freshwaters, we could say that financial pressures on farmers are the major obstacle to implement most the measures recommended by the ECSFDI. In this sense, it would be desirable to increase the financial rewards and incentives if we take into account the current drop in milk prices, according to the farmers. This could help to implement mitigation measures such as increase the slurry storage capacity, and reduce the number of intermediate sources of nutrients. Furthermore, these measures should be accompanied by environmental requirements, as voluntary schemes are not resulting in substantial improvements in livestock management. Finally, it could be necessary to include an increase threat of prosecution.

Acknowledgement

The author thanks Yorkshire Dales National Park Authority for access to data from the study area. Jamie Duncan from Environment Agency who helped by providing meteorological data. I am also grateful to David Higgins from Yorkshire Dales Rivers Trust as this research had never been accomplished without his collaboration and cooperation. Lastly, I offer my regards to all lecturers and supervisor for their unlimited support during the study period.

References

1. Allan JD, Castillo MM (2007) Structure and function of running waters. In: Stream Ecology, Springer, US.
2. Edwards AC, Kay D, McDonald AT, Francis C, Watkins J, et al. (2008) Farmyards, an overlooked source for highly contaminated runoff. J Environ Manage 87: 551-559.
3. Curley EM, O´Flynn MG, McDonnell KP (2010) Phosphorus Loss in Subsurface Flow from Agricultural Lands after Manure Application to Miscanthus x giganteus-Impacts on Grounwater Quality. International Journal of Agricultural Research 5: 268-275.
4. Magbanua F, Townsend C, Blackwell C, Phillips G, Matthaei C (2010) Response of stream macroinvertebrates and ecosystem function to conventional, integrated and organic farming. Journal of Applied Ecology 47: 1014–1025.
5. Hatch D, Goulding K, Murphy D (2002) Agriculture, Hydrology and Water Quality. In: Nitrogen, CAB International, UK.
6. Withers PJ, Lord E (2002) Agricultural nutrient inputs to rivers and groundwaters in the UK: policy, environmental management and research needs. Sci Total Environ 282-283: 9-24.
7. Silgram M, Anthony SG, Fawcett L, Stromqvist J (2008) Evaluating catchment-scale models for diffuse pollution policy support: some results from the EUROHARP Project. Environmental science & policy 11: 153–162.
8. Heathwaite AL (2003) Making process-based knowledge useable at the operational level: a framework for modelling diffuse pollution from agricultural land. Environmental Modelling & Software 18: 753–760.
9. Biggs BJF (2000) Eutrophication of streams and rivers: dissolved nutrient-chlorophyll relationships for benthic algae. The North American Benthological Society 19:17-31.
10. Jarvie HP, Neal C, Withers PJ (2006) Sewage-effluent phosphorus: a greater risk to river eutrophication than agricultural phosphorus? Sci Total Environ 360: 246-253.
11. Neal C, Jarvie HP, Neal M, Love AJ, Hill L, et al. (2005a) Water quality of treated sewage effluent in a rural area of the upper Thames Basin, southern England,

and the impacts of such effluents on riverine phosphorus concentrations. Journal of Hydrology 304: 103-117.

12. Gough MW, Marrs RH (1990) A comparison of soil fertility between semi-natural and agricultural plant communities: Implications for the creation of species-rich grassland on abandoned agricultural land. Biological Consemation 51: 83-96.
13. Mainston CP, Parr W (2002) Phosphorus in rivers--ecology and management. Sci Total Environ 282-283: 25-47.
14. Kay P, Grayson R, Phillips M, Stanley K, Dodsworth A et al. (2012) The effectiveness of agricultural stewardship for improving water quality at the catchment scale: experiences from an NVZ and ECSFDI watershed. Journal of Hydrology 422-423: 10-16.
15. Leinweber P, Turner BL, Meissner R (2002) Agriculture, Hydrology and Water Quality. In: Phosphorus, CAB International, UK.
16. European Comission (2010)The EU Nitrates Directive.
17. DEFRA (2009) Protecting our Water, Soil and Air. A Code of Good Agricultural Practice for farmers, growers and land managers, Department for Environment, Food and Rural Affairs, UK.
18. The Water Resources (Control of Pollution) (Silage, Slurry and Agricultural Fuel Oil) (England) Regulations 2010.
19. Uusi-Kämppä J, Braskerud B, Jansson H, Syversen N, Uusitalo R (2000) Buffer Zones and Constructed Wetlands as Filters for Agricultural Phosphorus. Journal of Environmental Quality 29: 151-158.
20. Chave P (2001) An Introduction. In: The EU Water Framework Directive, IWA Publishing, London.
21. Kaika M, Page B (2003) The EU Water Framework Directive: Part 1. European Policy-making and the changing topography of lobbying. European Environment 13: 314-327.
22. Worrall F, Spencer E, Burt TP (2009) The effectiveness of nitrate vulnerable zones for limiting surface water nitrate concentrations. Journal of Hydrology 370: 21-28.
23. Bechmann ME, Berge D, Eggestad HO, Vandsemb SM (2005) Phosphorus transfer from agricultural areas and its impact on the eutrophication of lakes—two long-term integrated studies from Norway. Journal of Hydrology 304: 238–250.
24. McDowell RW, Nash D, George A, Wang QJ, Duncan R (2009) Approaches for quantifying and managing diffuse phosphorus exports at the farm/small catchment scale. J Environ Qual 38: 1968-1980.
25. Owens LB, Bonta JV, Shipitalo MJ, Rogers S (2011) Effects of winter manure application in Ohio on the quality of surface runoff. J Environ Qual 40: 153-165.
26. Collins AL, McGonigle DF (2008) Monitoring and modelling diffuse pollution from agricultura for policy support: UK and European experience. Environmental science & policy 11: 97–101.
27. Cherry KA, Shepherd M, Withers PJ, Mooney SJ (2008) Assessing the effectiveness of actions to mitigate nutrient loss from agriculture: a review of methods. Sci Total Environ 406: 1-23.
28. Haygarth PM, Chapman PJ, Jarvis SC, Smith RV (1998) Phosphorus budgets for two contrasting grassland farming systems in the UK. Soil Use and Management 14: 160-167.
29. Domburg P, Edwards AC, Sinclair AH, Chalmers NA (2000) Assessing nitrogen and phosphorous efficiency at farm and catchment scale using nutrient budgets. Journal of the Science of Food and Agriculture 80: 1946-1952.
30. Edwards AC, Withers PJA (1998) Soil phosphorus management and water quality: a UK perspective. Soil Use and Management 14: 124-130.
31. Heathwaite AL, Griffiths P, Parkinson R.J (1998) Nitrogen and phosphorus in runoff from grasslandwith buffer strips following application of fertilizers and manures. Soil Use and Management 14: 142-148.
32. Chambers BJ, Smith KA. and Pain BF (2000) Strategies to encourage better use of nitrogen in animal manures. Soil Use and Management 16: 157-161.
33. Gburek WJ, Sharpley AN, Heathwaite L, Folmar GJ (2000) Phosphorus Management at the Watershed Scale: A Modification of the Phosphorus Index. J. Environ. Qual 29:130-144.
34. DEFRA (2011) Fertiliser Manual (RB209).
35. Hamzaa MA, Andersonb WK (2005) Soil compaction in cropping systems. A review of the nature, causes and possible solutions. Soil and Tillage Research 82: 121–145.
36. Nix J (2003) Farm Management Pocketbook (34th Edn) Imperial College London, Wye Campus.
37. Rowell DL (1994) Methods and Applications. In: Soil Science, Longman, UK.
38. Ball DF (1964) Loss-on-ignition as an estimate of organic Soils Matter and organic carbon in non-calcareous. Journal of Soil Science 15: 1.
39. Sharpley AN, Weld JL, Beegle DB, Kleinman PJA, Gburek WL, et al. (2003) Development of phosphorus indices for nutrient management planning strategies in the U.S. Journal of Soil and Water Conservation.
40. Shiferaw A (2011) Estimating soil loss rates for soil conservation planning in the borena woreda of south wollo highlands, Ethiopia. J Sustain Dev Afr 13: 3.
41. Morgan RPC, Davison DA (1991) Soil Erosion and Conservation, Longman Group, U.K.
42. Arnouldous HMJ (1980) An approximation of the rainfall factor in the USLE in assessment of Erosion, Wiley Chichester, England
43. Wischmeier WH, Smith DD (1978) Predicting rainfall erosion losses, a guide to conservation planning. U.S. Dept. of Agriculture.
44. Penna D, Tromp-van Meerveld HJ, Gobbi A, Borga M, Dalla Fontana G (2011) The influence of soil moisture on threshold runoff generation processes in an alpine headwater catchment. Hydrol Earth Syst Sci 15: 689-702.
45. Leh MD, Chaubey I, Murdoch J, Brahana JV, Haggard BE (2008) Delineating runoff processes and critical runoff source areas in a pasture hillslope of the Ozark Highlands. Hydrological Processes 22: 4190-4204.
46. Leh MD, Chaubey I (2009) GIS-based predictive models of hillslope runoff generation processes. Journal of the American Water Resources Association 45: 4.
47. Qin CZ, Pei AXZT, Li BL, Scholten T, Behrens T, et al. (2011) An approach to computing topographic wetness index based on maximum downslope gradient. Precision Agric 12: 32-43.
48. DEFRA (2008) Cross Compliance monitoring framework – indicator data sheet Department for Environment, Food and Rural Affairs, UK.
49. Hooda PS, Moynagh M, Svoboda IF, Edwards AC, Anderson HA, et al. (1999) Phosphorus Loss in Drainflow from Intensively Managed Grassland Soils. J Environ Qual 28: 1235-1242.
50. Cuttle S, Macload C, Chadwick D, Scholefield D, Haygarth P, et al. (2006) An Inventory of Methods to Control Diffuse Water Pollution from Agriculture (DWPA) USER MANUAL. Defra report, Project ES0203. Pp: 115.
51. Lalor STJ, Schulte RPO (2008) Low-ammonia-emission application methods can increase the opportunity for application of cattle slurry to grassland in spring in Ireland. Grass and Forage Science 63: 531-544.
52. Chesterton C (2009) Environmental impacts of land management. Natural England.
53. Shuh-Ren J, Yin-Feng L, Der-Yuan L, Tze-Wen W (2001) Nutrient removal from polluted river water by using constructed wetlands. Bioresource Technology 76: 131-135.
54. Fisher J, Acreman MC (2004) Wetland nutrient removal: a review of the evidence. Hydrology and Earth System Sciences 8: 673-685.
55. Walling DE (2005) Tracing suspended sediment sources in catchments and river systems. Sci Total Environ 344: 159-184.
56. Alemseged TH, Rientje THM (2005) Effects of LIDAR DEM resolution in flood modeling: A model sensitivity study for the city of Tegucigalpa, Honduras. ISPRS WG III/3, III/4, V/3 Workshop "Laser scanning 2005", Enschede, Netherlands.
57. Burrough, PA, Macmillan RA, Deursen W (1992) Fuzzy classification methods for determining land suitability from soil profile observations and topography. Journal of Soil Science 43: 193-210.
58. Bowes MJ, Leach DV, House WA (2005) Seasonal nutrient dynamics in a chalk stream: the River Frome, Dorset, UK. Sci Total Environ 336: 225-241.
59. Heathwaite AL, Quinnb PF, Hewett CJM (2005) Modelling and managing critical source areas of diffuse pollution from agricultural land using flow connectivity

simulation. Journal of Hydrology 304: 446-461.

60. Meybeck M (1998) Man and river interface: multiple impacts on water and particulates chemistry illustrated in the Seine river basin. Hydrobiologia 373: 1-20.

61. Anderson DM, Burkholder JM, Cochlan WP, Glibert PM, Gobler CJ, et al. (2008) Harmful algal blooms and eutrophication: Examining linkages from selected coastal regions of the United States. Harmful Algae 8: 39-53.

62. Neal C, Jarvie HP (2005b) Agriculture, community, river eutrophication and the Water Framework Directive. Hydrol. Process 19: 1895-1901.

63. Hooda PS, Edwards AC, Anderson HA, Miller A (2000) A review of water quality concerns in livestock farming areas. Sci Total Environ 250: 143-167.

64. Stewart JWB, Tiessen H (1987) Dynamics of soil organic phosphorus. Biogeochemistry 4: 41-60.

65. Bishop PL, Hively WD, Stedinger JR, Rafferty MR, Lojpersberger JL, et al. (2005) Multivariate analysis of paired watershed data to evaluate agricultural best management practice effects on stream water phosphorus. J Environ Qual 34: 1087-1101.

66. Zuzel JF, Pikul JL (1987) Infitration into a seasonally frozen agricultural soil. J Soil Water Conserv 42: 447-450.

67. Wood BH, Wood CW, Yoo KH, Yoon KS, Delaney DP (1999) Seasonal surface runoff losses of nutrients and metals from soils fertilized with broiler litter and commercial fertilizer. J Environ Qual 28: 1210-1218.

68. Leinweber P, Meissner R, Eckhardt KU, Seeger J (1999) Management effects on forms of phosphorus in soil and leaching losses. European Journal of Soil Science 50: 413-424.

69. Siddique MT, Robinson JS (2003) Phosphorus sorption and availability in soils amended with animal manures and sewage sludge. J Environ Qual 32: 1114-1121.

70. Dils RM, Heathwaite AL (1996) Phosphorus fractionation in hillslope hydrological pathways contributing to agricultural runoff. Advances in hillslope processes 1: 229-251

71. Bryan RB (2000) Soil erodibility and processes of water erosion on hillslope. Geomorphology 32: 385–415

72. Walling DE, Collins AL, Stroud RW (2008) Tracing suspended sediment and particulate phosphorus sources in catchments. Journal of Hydrology 350: 274-289.

73. Andersen HE, Kronvang B (2006) Modifying and Evaluating a P Index for Denmark. Water, Air, and Soil Pollution 174: 341-353.

74. Kinnell PIA (2004) Runoff, sediment concentration and predicting erosion on hillslopes within catchments. 13th International Soil Conservation Organisation Conference, Brisbane.

75. Reaney S, Lane S, Heathwaite L, Dugdale L (2011) Risk-based modelling of diffuse land use impacts from rural lamdscapes upon salmonid fry abundance. Ecological Modelling 222: 1016–1029.

76. Kronvang B, Laubel A, Grant R (1997) Suspended sediment and particulate phosphorus transport and delivery pathways in an arable catchment, Gelbzk stream Denmark. Hydrological Processes 11: 627-642.

Effects of Soil on Degradation of *Robinia pseudoacacia* Forests in the Yellow River Delta in China

Hong W*, Claire D*, Yu Z, Yin S, Yu L and Yi Z

School of Earth Sciences and Engineering, Hohai University, No. 1 Xikang Road, Nanjing 210098, China

Abstract

Soil quality has significant importance for the growth and sustainability of plants. However, due to the variability and diversity of soil characteristics, many trees suffered death according to their confrontation degree. Over the past several decades, *Robinia pseudoacacia* forests in the Yellow River delta of China, lose health and died without an obvious cause. This study focuses on evaluating the role of soil characteristics (moisture content, soil salinity content, soil bulk density, soil texture (the percentages of soil sand, soil silt, and soil clay) and pH value on the deterioration of health level of *Robinia pseudoacacia* forests in the area. To do so, three health levels such as healthy, medium dieback, and severe dieback forest were firstly classified based on the United States Department of Agriculture Forestry Bureau of crown condition classification guide and in situ survey, then soil properties in vertical direction were analyzed by five sampling points for each forest type from surface to the depth of 260 cm with eight layers (0-20 cm, 20-40 cm, 40-60 cm, 60-100 cm, 100-140 cm, 140-180 cm, 180-220 cm and 220-260 cm) for healthy and moderate dieback and 0- 220 cm depth with seven layers for severe dieback because the water occurred after 220 cm depth. The results indicated that there are significant differences in soil moisture content and soil conductivity among three forest health conditions. For a vertical change of soil characteristics only soil particle sizes (sand, silt and clay) had a significant difference in three forest health conditions. For system roots, absorptive roots were observed down to 230 cm depth for healthy *Robinia pseudoacacia* forest but moderate and severe dieback *Robinia pseudoacacia* forests, roots are found in the surface layer.

Keywords: Soil characteristics; Forest health conditions; Tree dieback; *Robinia pseudoacacia* forest; Yellow river delta

Introduction

The plant cannot grow in poor soil. Soil conditions had a strong influence on plant growth [1]. The structure of soil, lays a significant role in plant growth [2] thus well development, bad development or decay of the plant. Soil moisture is one of the most important abiotic factors determining vegetation growth, variability and regeneration [3] but it has a significant impact on growth and productivity of some plants through overcoming of the water stress condition [4] it affects the potential growth of some plants species [5]. Plants growing in well-aerated soils are less stressed by drought or excess water [6]. Lower bulk density is normally optimal soil conditions for many plants and ensures a high crop productivity [7]. Soil salinity declines the growth rates of the plant [8] and its extreme levels can cause plant death [9]. This problem of salinity has strongly affected Yellow River Delta [10,11] and there is no significant difference in soil salt content among different land-use types in the region [11] but high salinity generally appeared in the topsoil [12]. The salinity becomes a threat to many tree species to survive.

R. pseudoacacia is one of the forest species grown in the Yellow River Delta and has been widely planted in the middle of 1980's, but the dieback or dead of *Robinia Pseudoacacia* had been noted in some areas of Yellow River Delta in the early 1990's [13,14]. However, *Robinia pseudoacacia* have a certain resistance to the condition of salt [15], but the plant grows better under conditions of low salinity [15].

Within different studies, the dieback or dead *Robinia Pseudoacacia* plantation were detected through the analysis of different health levels [13,14], three health level classes that are healthy, medium dieback, and severe dieback with 41.46%, 36.09%, and 22.45% respectively observed based on tree crown symptoms within the case study [16].

Robina pseudoacacia is able to grow in a variety of soil conditions [17] and it has the ability to vary its growth patterns, thrive in many regions, and grow at very aggressive rates [18], but in the dry condition, it is not easy to adapt [14,19] because the plant need soil water to improve photosynthetic capacity and biomass accumulation [20]. Moreover, *Robinia pseudoacacia* avoids compact soils and those that are waterlogged for long periods of time [17]. Thus, the plant seems to be limited by water supply. *Robinia pseudoacacia* can grow almost in any soil acid or alkaline except that which is permanently wet [21].

The goal of this paper is to explore the cause of dieback or dead *Robinia pseudoacacia* forest plantation of Gudao forestry area in the Yellow River Delta, China, to provide effective management and protection measurement.

Study area

The study was undertaken in the Yellow River Delta located in Dongying City, Shandong Province of East China with 36°55′N-38°16′N, 117°30′E-119°20′ E with approximately 5400 Km^2 of total area [16].

The Yellow River Delta belongs to the temperate continental monsoon climate and has four distinct seasons such as hot and humid summer and a cold and dry winter.

The area is topographically flat with a gentle slope. It is high in the south and west, low in the north and east. The highest elevation is 28

***Corresponding authors:** Hong W, School of Earth Sciences and Engineering, Hohai University, No. 1 Xikang Road, Nanjing 210098, China
E-mail: hongwang@hhu.edu.cn

Claire D, School of Earth Sciences and Engineering, Hohai University, No. 1 Xikang Road, Nanjing 210098, China, E-mail: sabeclaire@yahoo.fr

m in the southwest and the lowest is 1 m in the northeast. The natural gradient is 1/8000-1/12000 [22].

The main soil types in this area are calcaric fluvisols, salic fluvisols, and gleyic solonchaks, under which *Robinia pseudoacacia* is able to growth [17]. According to Ref. [23], three main factors were found to be the cause of dieback of *Robinia pseudoacacia* trees in this study area: soil texture, soil salt content, and ground water levels.

The area has 393 plant species and varieties in the region; most of them are salt-tolerant [24] such as *Tamarix chinensis*, *Suaeda salsa*, *Salicornia europaea* [25]. Artificial *Robinia pseudoacacia* forest is among them and is planted in Gudao, Junmachang, Abandoned yellow river and Natural reserve (Figure 1). Sampling points locate in Gudao forest area, the majority of trees having more than 25 years old.

Methods

Classification of health conditions of *Robinia pseudoacacia* forests

Health conditions of *Robinia pseudoacacia* forests were classified into three levels: level 3 for healthy or slight dieback, level 2 for moderate dieback and level 1 for severe dieback or death according to the United States Department of Agriculture Forestry Bureau of crown condition classification guide [26] through five trees vigor index (leaf transmittance, crown width, crown density, mortality, live crown ratio) and are reported in Table 1.

Field investigation and sampling procedure

Three types of the health status of *Robinia pseudoacacia* were studied, and for each type, five sample points were randomly collected along the vertical direction of the river because more serious dieback was observed more close to the river and along the river direction [16].

From April 25-29, 2015, soil samples were collected in the foil samples at various depth intervals: 0-20 cm, 20-40 cm, 40-60 cm, 60-100 cm, 100-140 cm, 140-180 cm, 180-220 cm and 220-260 cm for each sampling point but sample for the severe dieback forest soil, were collected up to 180-220 cm because groundwater occurs after the 220 cm depth, and the distribution of absorptive roots (diameter less than 2 mm) were observed and measured during soil sampling periods. We dug a rectangle hole of 80 cm of small side and 1.5 m of large side and 230 cm of deep. In sample collection, soil samples were taken on the side for seven layers and for the last one soil sample was taken at the bottom.

Soil properties analysis

Soil properties analysis included soil moisture content, soil salinity, soil texture, soil bulk density and pH value. The soil salinity content was determined by soil electrical conductivity. Electrical conductivity is the common method in soil salinity measuring [27], by using TOLEDO Seven2GOTM METTLER with measurement of 1:5 in the soil solution. Soil moisture content was determined by the difference of the soil weight before drying and after drying per dry weight [28]. The soil bulk density was calculated by the ratio of dry weight and volume of ring knife.

According to United States Department of Agriculture [29] on soil textural classification, sand, silt and clay particle size was classified through calculation of the percentage of their particles size. The percentage content of each particle size was obtained by using the LS13320 laser particle size analyzer.

Recorded data were analyzed by using Microsoft excel 2013 and IBM SPSS Statistics v22 software. SPSS software has been used to analyze soil characteristics among forest condition levels and different sampling layers with ANOVA method and calculate the correlation coefficient of forest condition and soil characteristics by the method of Pearson correlation analysis. The comparison of soil characteristics and sampling depth made to analyze the change in three levels of forest status. Normalization was used to analyze the vertical variation of soil characteristics in the different health condition of *Robinia pseudoacacia* forest.

Results

Soil characteristics among different forest health conditions

The soil characteristics of the study area (soil bulk density, soil moisture content, soil salinity content, pH value, silt, sand, and clay) were analyzed by three forest health levels. The result (Table 2) showed that there are significant differences in soil moisture content and soil conductivity among three forest health conditions. Other soil characteristics (sol bulk density, sand, silt, clay, and pH value) did not show significant differences between the three forest health statuses.

According to Pearson correlation analysis, positive and negative correlations were observed (Table 3). *Robinia pseudoacacia* health condition showed the significant correlation relationship positively with the soil sand content and negatively correlated with soil moisture content, the soil salinity content, clay, soil bulk density, and silt but *Robinia pseudoacacia* health condition had no significant correlation with pH value. The good health condition of *Robinia pseudoacacia* is

Indicator	Level 3	Level 2	Level 1
Live crown ratio	>90%	70%-85	50%-65%
Crown density	>80%	50%-70%	20%-40%
Crown diameter	>55%	26%-54%	1%-25%
Dieback	0%-5%	10%-25%	>30%
Foliar transparency	0%-20%	30%-50	>60%

Table 1: The CCCG a tree vigor indicators and classification thresholds a CCCG: Crown Condition Classification Guide; b Values of 100% were recorded as 99%, and intermediate values were upgraded to the next full 5% grouping.

	SMC	SBD	Sand	Silt	Clay	SC	pH value
H	6.25^a	1.34	50.16	47.13	2.71	0.06^a	8.06
M	17.75^b	1.35	49.15	47.58	3.27	0.13^b	7.99
S	26.72^b	1.23	29.34	54.28	3.88	0.16^b	7.04

Table 2: Two ways-ANOVA on soil characteristics among three forest condition of *Robinia pseudoacacia* (H=Healthy forest, M=Medium dieback forest, and S=Severe dieback forest). SMC: Soil moisture content (%), SBD: Soil bulk density (g/cm^3), SC: Soil conductivity (us/cm), soil texture: the percentage of soil sand, silt and clay content and pH value a: p<0.05, b: p<0.01.

	Forest condition	SMC	SBD	Sand	Silt	Clay	SC	pH value
Forest condition	1							
SMC	-0.900**	1						
SBD	-0.386**	0.269**	1					
Sand	0.304**	-0.304**	0.066	1				
Silt	-0.289**	0.290**	-0.076	-0.999**	1			
Clay	-0.427**	0.420**	0.071	-0.822**	0.796**	1		
SC	-0.543**	0.523**	-0.047	-0.520**	0.509**	0.568**	1	
pH value	0.002	-0.104	0.145	0.417**	-0.406**	-0.478**	-0.175	1

**Correlation is significant at the 0.01 level (2-tailed)

SMC: Soil moisture content (%), SBD: Soil bulk density (g/cm^3), SC: Soil conductivity (μS/cm), Soil texture: the percentage of soil sand, silt and clay content and pH value

Table 3: *Robinia pseudoacacia* health level and soil characteristics correlation.

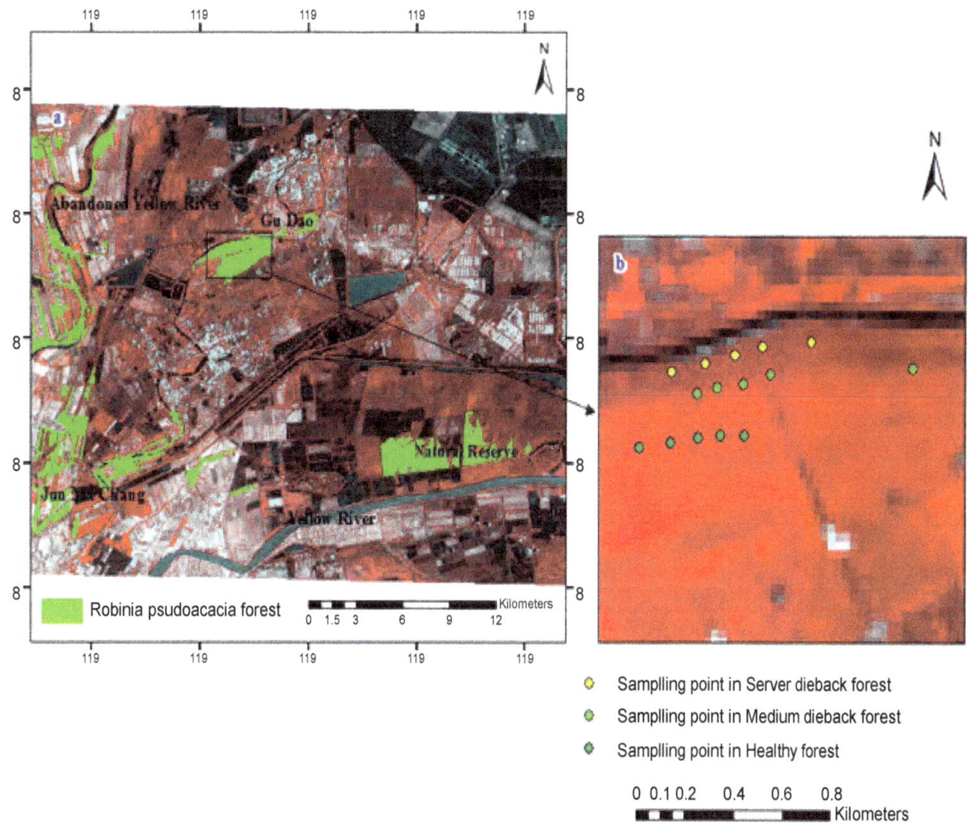

Figure 1: Study area Location. The false color composite image covering the study area was made with Landsat 8 OLI bands 5, 4, 3 vs. RGB. a. Four *Robinia psedoacacia* forest areas in the Yellow River Delta; b. The position of sampling points in GuDao.

seen by means of the increase of soil sand percentage but inversely associated with an increase in soil moisture content, soil salinity, soil bulk density, soil clay and silt percentages.

Vertical change of soil characteristics at different forest health conditions

Vertical change of soil characteristics included soil bulk density, soil moisture content, soil salinity content, pH value, silt, sand, and clay were analyzed among three forest health levels by one-way ANOVA. The result shows only soil particle sizes (sand, silt, and clay) had a significant difference in three forest health conditions. Soil moisture content had a significant difference in healthy and severe dieback forestry. Soil conductivity had a significant difference in medium dieback forest and severe dieback forest. pH value had a significant difference in healthy and medium dieback forest (Table 4).

The results (Figure 2) showed soil characteristics variation at different sampling depth. Percentage of soil sand had uptrend presenting significant increase with increasing in sampling depths while percentage of soil silt and soil clay show downtrend which present significant decrease at a significance level of 0.05. Soil bulk density with fluctuation trend shows no significant difference among three forest health conditions. At healthy forest, soil moisture content is significantly high on surface layer because the soil directly accepts the rainfall supply, coupled with the surface litter and understory grass root barrier, soil moisture decreases continuously but in the middle layer, soil moisture content were lower and seems to reach the valley due to the surface soil gravity water which leads to the formation of dry layer in the middle; at severe dieback forestry, soil moisture content significantly increased with increasing in sampling depths due to groundwater recharge and to its location near and along the river through the continuity of river flows.

Soil properties	Health forest		Medium dieback		Severe dieback	
	F	Sig.	F	Sig.	F	Sig.
SMC	36.698	0.000	3.133	0.012	4.716	0.002
SBD	1.311	0.277	1.533	0.192	3.695	0.008
Sand	15.627	0.000	59.803	0.000	33.451	0.000
Silt	15.397	0.000	61.284	0.000	29.881	0.000
Clay	13.825	0.000	15.011	0.000	34.588	0.000
SC	1.171	0.347	2.312	0.050	2.228	0.000
pH value	7.945	0.000	6.369	0.000	4.575	0.071
Significance level: 0.05						

SMC: Soil moisture content (%), SBD: Soil bulk density (g/cm^3), SC: Soil conductivity (μS/cm), Soil texture: the percentage of soil sand, silt and clay content and pH value

Table 4: One-way ANOVA test on soil properties at different sampling depths on the three health levels of *Robinia pseudoacacia* forest stands.

It is also observed from (Figure 2) that high salinity generally appeared in the topsoil (0-20 cm) at healthy forestry or sub-topsoil (20-40 cm) at severe dieback forestry.

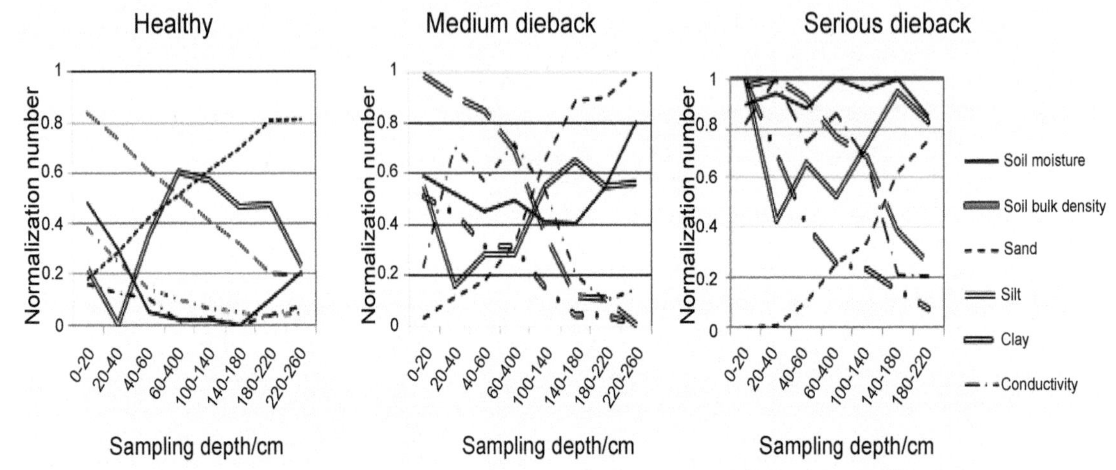

Figure 2: Vertical variations of soil properties at three health levels of *Robinia pseudoacacia* forests.

The distribution of absorptive roots in different forest health conditions

The root system is a major part of the plant. Roots offer the significant advantages of fundamental processes in plant development, they are involved in the acquisition of water and nutrients, synthesis of plant hormones, and storage functions [30].

In this study, the healthy forest of *Robinia pseudoacacia* is characterized by the deep root system, absorptive roots were observed down to 230 cm depth but for moderate and severe dieback *Robinia pseudoacacia* forests, roots are not deepened. Their absorptive roots were found in less than 60 cm deep due to sticky and light gray soil layer that is near 40 cm deep and little roots passing through but little roots tried to deep to 120-180 cm depth and most of them showed edema like, and were fostered by touch. In addition, the moderate and severe dieback *Robinia pseudoacacia* forests present much clay soil in the surface layers that alter the normal deeper growth of *Robinia pseudoacacia* roots.

Discussion

The soil moisture rates in *Robinia pseudoacacia* forest varied periodically with seasons, the soil moisture rate is lower in the dry season as precipitation is rare and high evaporation but higher in the rainy season [29,30]. In this study, samples were taken in dry season and the soil moisture decreased with soil depth in healthy *Robinia pseudoacacia* forest because of soil water loss by transpiration of trees, understory grasses and evaporation from the ground surface [31-33], while increase in severe dieback forest because of lower groundwater table (less than 2 m at our sampling period) the reason for this phenomenon was that all severe dieback *Robinia pseudoacacia* forests located near river (less than 100 m) and also distributed along river direction (Figure 1). Moisture could limit soil respiration when the soil was too wet [34].

Our results contradict the majority of literature on forest dieback which generally considers increased forest mortality related to climatic stress from drought and high temperatures [35-37] but were not different from Ref. [38-40] results which consider forest decline and massive tree mortality to be the resulted from persistent wetness.

This study found that high salinity generally appeared in the topsoil (0-20 cm) at healthy forestry or sub-topsoil (20-40 cm) at severe dieback forestry. And it is found that the higher soil salinity content the higher deterioration degree of forest consequently inhibition of root growth [13]. [28,41] have convincingly demonstrated that soil salinity significantly affects the *Robinia pseudoacacia* forest health. However, our result compared to soil moisture and soil salt content and found that soil moisture had a strong effect on degradation of the forest.

By *in situ* survey, severe dieback forests were frequently seen near the ditch and road and along the river (Figure 3a and 3b) because forest roots soak in water and soil are wet and near the road, there might be an increase in high surface runoff thus increases moisture content. Also with micro-morphology consideration, trees are healthy in higher places than those in lower places (Figure 3c). Topography plays an important role in the spatial organization of soil moisture at different scales.

To better understand the cause of degradation of *Robinia pseudoacacia* forest, the distributions of absorptive roots were conducted because roots of most plants cannot function or even remain alive for drought condition. However, *Robinia pseudoacacia* can develop deep roots for drought resistance [42], which were also found in the healthy forest stands (absorptive roots were observed down to 230 cm depth) in our study area but many roots of severe *Robinia pseudoacacia* are located in the surface layer, the area saturated with water. Soil with high moisture content restricts rooting than those with low moisture content [43]. Periods of water saturation lead to poor aeration [6] and can cause damage to many roots due to the lack of soil aeration [43], inevitably reduce photosynthetic rates and induced carbon starvation [40]. Carbohydrate plays an important role in plant respiration and of plant growth [41,42]. Moreover, the soil layer near 40 cm deep was sticky and light gray that alter may root too deep the soil consequently, crown dieback increased [26].

Changes in bulk density affect available water and air capacity, and strongly influence permeability, drainage rate, and penetration by plant roots [44-46]. In this study, soil bulk density was high in severe dieback forest and seems to be one factor that hinders normal plant rooting and leads to the poor development and growth of the plant [47]. The bulk density can be evaluated using relative compaction. As compactness becomes greater, soil hardness becomes harder leading to

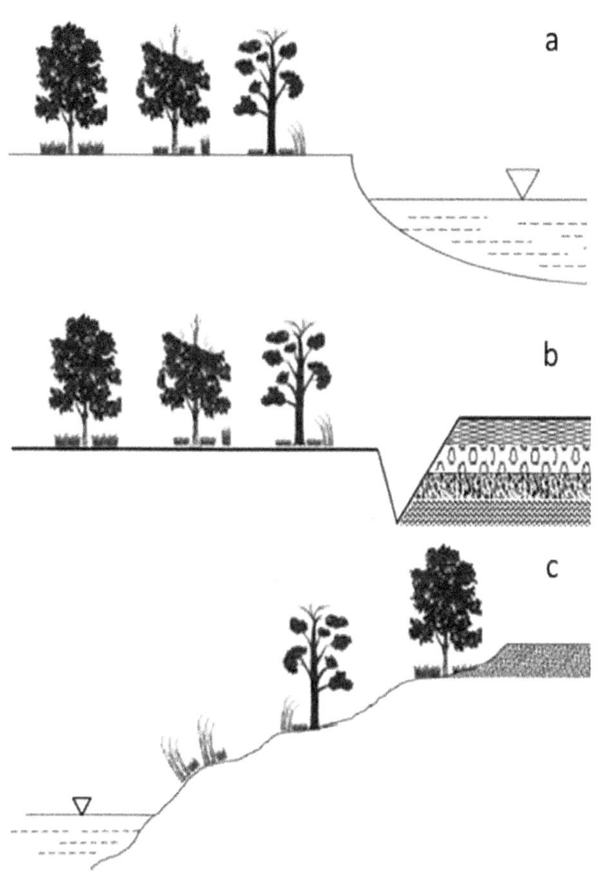

Figure 3: Schematic representation of three health conditions of forest plots: (a) Severe dieback near the river; (b) Severe dieback near the ditch and road; and (c) Healthy at the higher plot and dieback at the lower plot.

a rise in mechanical stress in plant roots interpenetrated in soil [48,49]. Moreover, soil with the high bulk had a significant higher penetration resistance than the low bulk density [50].

Conclusion

The study of the cause of *Robinia pseudoacacia* dieback in Yellow river delta especially in Gudao forest as an area of focus through analysis of soil characteristics in different soil depth, showed that the main reason for the significant decrease in trees health level from the healthy to severe dieback *Robina pseudoacacia* forest were soil moisture content and soil conductivity. However, soil moisture and soil salinity show a higher correlation with soil texture and soil bulk density. Also, it was found that topography had a great influence on health conditions of *Robinia pseudoacacia* forest. The future research should be done on plants well grown in saturated soil and high salinity that can be grown with *Robinia pseudoacacia* because the soil respiration varies due to the vegetation type [51] or should look to the way *Robinia pseudoacacia* be moved in high places, as well as away from the river and road. Ref [53]'s study surveyed soils near and within 3, 9, and 19-year-old plantations of poplar (*Poplus simonii*) in the central part of Naiman Country, eastern Inner Mongolia, China and found that there are changes in the soil properties following afforestation including the accumulation of fine particles (clay and silt), soil moisture level, amounts of available N and available P and soil organic carbon in the surface horizons and increased slowly in the first 9 years period and then rapidly between 10 and 19 years [52] thus the same measure is suggested for our case study.

Acknowledgements

The research was supported by the National Science Foundation of China (41471419 and 40871230) and the State Key Laboratory of Resources and Environmental Information System.

Author Contributions

Hong Wang proposed the idea of the topic and organized the writing. Claire Dusabemariya contributed to laboratory analysis and organization of the writing. Zhao Y, Song Y, Kaiyu L and Zhong Y contributed to laboratory analysis and the field survey data.

References

1. Lu TO (2005) Effects of Some Soil Properties on the Growth of Hybrid Poplar in the Terme-Gölard: Region of Turke. Turk J Agric 29: 221-226.
2. Passioura JB (1991) Soil Structure and Plant Growth. Aust J Soil Res 29: 717-728.
3. Wang S, Fu BJ, Gao GJ, Yao XL, Zhou J (2012) Soil moisture and evapotranspiration of different land cover types in the Loess Plateau, China. Hydrol Earth Syst Sci, pp: 2883-2892.
4. Anju SV, Koppa AG (2013) Influence of in situ soil moisture conservation measures on growth and productivity of Acacia auriculiformis. Karnataka J Agric Sci 26: 170-171.
5. Aslanidou N, Smiris M (2001) Distribution of soil moisture content and its effect on the potential growth of the overstory species in North-East Chalkidiki. Silva Gandav 66: 40-56.
6. Bot A, Benites J (2005) The importance of soil organic matter.
7. Khan B, Ablimit A, Mahmood R, Qasim M (2010) Robinia pseudoacacia leaves improve soil physical and chemical properties. Journal Arid L 2: 266-271.
8. Bennett DL (1995) Using the EM38 to measure the effect of soil salinity on Eucalyptus globulus in south-western Australia. Agric Water Manag 27: 69-86.
9. Imbert D, Dulormne M (2015) Impact patterns of soil salinity variations on the survival rate, growth performances, and physiology of Pterocarpus officinalis seedlings. Trees 29: 119-128.
10. Weng YL, Gong P, Zhu ZL (2010) A Spectral Index for Estimating Soil Salinity in the Yellow River Delta Region of China Using EO-1 Hyperion Data. Pedosphere 20: 378-388.
11. Fang H, Liu G, Kearney M (2005) Georelational Analysis of Soil Type, Soil Salt Content, Landform, and Land Use in the Yellow River Delta, China. Environ Manage 35: 72-83.
12. Yu J, Li Y, Han G, Zhou D, Fu Y, et al. (2014) The spatial distribution characteristics of soil salinity in coastal zone of the Yellow River Delta. Env Earth Sci 72: 589-599.
13. Yao L, Fei L, Gui S, Liu Q, Liu G (2008) Remote sensing for monitoring on the health of artificial Robinia Pseudoacacia Forests in the Yellow River Delta. Natl Nat Sci Found China 176: 1-3.
14. Liu Q, Liu G (2009) Using tasseled cap transformation of CBERS-02 images to detect dieback or dead Robinia Pseudoacacia plantation. Inst Electr Electron Eng, pp: 11-15.
15. Dirr MA (1976) Selection of trees for tolerance to salt injury. J Arboric, pp: 209-216.
16. Wang H, Pu R, Zhu Q, Ren L (2015) Mapping health levels of Robinia pseudoacacia forests in the Yellow River delta, China, using IKONOS and Landsat 8 OLI imagery. Int J Remote Sens 36: 1114-1135.
17. Vítková M, Tonika J, Müllerová J (2015) Black locust-Successful invader of a wide range of soil conditions. Sci Total Environ 505: 315-328.
18. Sabo AE (2000) Robinia pseudoacacia invasions and control in North America and Europe. Student On-Line 6: 1-9.
19. Jin TT, Liu GH, Fu BJ, Ding XH, Yang L (2011) Assessing adaptability of planted trees using leaf traits: A case study with Robinia pseudoacacia L. in the Loess Plateau, China. Chinese Geogr Sci 21: 290-303.

20. Liu X, Fan Y, Long J, Wei R, Kjelgren R, et al. (2013) Effects of soil water and nitrogen availability on photosynthesis and water use efficiency of Robinia pseudoacacia seedlings. J Environ Sci 25: 585-595.

21. Gilman EF, Watson DG (1994) Robinia pseudoacacia Black Locust.

22. Yuan-Xiu G, Gao-Huan Liu, Jin-Feng W (2001) Saline-alkali land in the Yellow River Delta : amelioration zonation based on GIS. J Geogr Sci 11: 313-320.

23. Liang Z, Long Y (2010) Theory and Technology on Robinia Pseudoacacia Cultivation. China For, pp: 1-26.

24. Jian-Feng Z, Qi-Xiang S (2005) Causes of Wetland Degradation and Ecological Restoration in the Yellow River Delta Region. For Stud China 7: 15-18.

25. Xianzhao L, Chunzhi W, Qing S (2013) Screening for Salt Tolerance in Eight Halophyte Species from Yellow River Delta at the Two Initial Growth Stages. Int Sch Res Not, p: 8.

26. Schomaker ME, Zarnoch SJ, Bechtold WA, Latelle DA, Burkman WA, et al. (2007) Crown-Condition Classification: A Guide to Data Collection and Analysis.

27. Maas EV, Hoffman GJ (1977) Crop salt tolerance-current assessment. American Society of Civil Engineers for publication 103: 115-134.

28. Imada S, Iviatsuo N, Acharya K, Yamanaka N (2015) Effects of salinity on fine root distribution and whole plant biomass of Tamarix ramosissima cuttings. J Arid Environ 114: 84-90.

29. United States Department of Agriculture (1987) USDA Textural Soil Classification. Soil Mechanics Level I, pp: 3-11.

30. Schiefelbein JW, Benfey BN (1991) The development of plant roots: new approaches to underground problems. Plant Cell 3: 1147-1154.

31. Huang Y, Wang Y, Zhao Y, Xu X, Zhang J, et al. (2015) Spatiotemporal Distribution of Soil Moisture and Salinity in the Taklimakan Desert Highway Shelterbelt. Water 7: 4343-4361.

32. Gao P, Yang HL, Zhang GC, Zhou ZF (2008) Correlation of eco-hydrographic benefit and height increment of Robinia pseudoacacia stand with climatic environmental factors in Yellow River Delta Wetland of China. J For Res 19: 215-218.

33. Sawada H, Araki M, Chappell NA, LaFrankie JV, Shimizu A (2007) Forest Environments in the Mekong River Basin.

34. Davidson EA, Belk E, Boone RD (1998) Soil water content and temperature as independent or confounded factors controlling soil respiration in a temperate mixed hardwood forest. Glob Chang Biol 4: 217-227.

35. Allen CD (2009) Climate-induced forest dieback: an escalating global phenomenon? Unasylva 60: 43-49.

36. Allen CD, Macalady AK, Chenchouni H, Bachelet D, McDowell N, et al. (2010) A global overview of drought and heat-induced tree mortality reveals emerging climate change risks for forests. For. Ecol. Manage 259: 660-684.

37. Guarin A, Taylor AH (2005) Drought triggered tree mortality in mixed conifer forests in Yosemite National Park, California, USA. For Ecol Manage 218: 229-244.

38. Dale VH, Joyce LA, McNulty S, Neilson RP (2000) The interplay between climate change, forests, and disturbances. Sci Total Environ 262: 201-204.

39. Malhi Y, Aragão LEOC, Galbraith D, Huntingford C, Fisher R, et al. (2009) Exploring the likelihood and mechanism of a climate-change-induced dieback of the Amazon rainforest. Proc Natl Acad Sci USA 106: 20610-20615.

40. Rozas VF, García-González I (2012) Too wet for oaks? Inter-tree competition and recent persistent wetness predispose oaks to rainfall-induced dieback in Atlantic rainy forest. Glob Planet Change 94-95: 62-71.

41. Jianfeng Z, Shangjun X (2009) Research on Soil Degradation of Robinia pseudoacacia Plantation under Environmental Stress. pp: 5-10.

42. Ranney TG (1990) Response of five temperate deciduous tree species to water stress. Tree Physiol 6: 439-448.

43. Fan C, Su C (2008) Effect of soil moisture content on the deformation behaviour of root-reinforced soils subjected to shear. Plant Soil 324: 57-69.

44. Kozlowski TT (1992) Sources and Sinks in Woody Plants Carbohydrate. Bot Rev 58: 107-222.

45. Loescher WH, Mccamant T, Keller JD (1990) Carbohydrate Reserves, Translocation, and Storage in Woody Plant Roots. HortScience 25: 274-281.

46. Archer J, Smith P (1972) The relation between bulk density, available water capacity, and air capacity of soils. J Soil Sci 23: 475-480.

47. Thomas PT (1989) Effect of varying soil moisture and bulk density of Teak, Eucalypt and Albizia root growth.

48. Carter MR (1990) Relative Measures of Soil Bulk Density To Characterize Compaction in Tillage Studies on Fine Sandy Loams. Can J Soil Sci 70: 425-433.

49. Guo Y (2011) The relationship between ancient trees health and soil properties. African J Biotechnol 10: 17997-18004.

50. Goodman M, Ennos R (1999) The Effects of Soil Bulk Density on the Morphology and Anchorage Mechanics of the Root Systems of Sunflower and Maize. Ann Bot 83: 293-302.

51. Mohanty BP, Skaggs TH (2001) Spatio-temporal evolution and time-stable characteristics of soil moisture within remote sensing footprints with varying soil, slope, and vegetation. Adv Water Resour 24: 1051-1067.

52. Han G, Yu J, Li H (2012) Winter Soil Respiration from Different Vegetation Patches in the Yellow River Delta, China. Environ Manage, pp: 39-49.

53. Yasuhito S, Ichiro T, Tong-Hui Z (2004) Changes in Soil Properties after Afforestation in Horqin Sandy Land, North China. Soil Sci Plant Nutr 50: 537-543.

Management of Traditional Agroforestry Practices in Gununo Watershed in Wolaita Zone, Ethiopia

Madalcho AB[1]* and Tefera MT[2]

[1] *Jigjiga University College of Dry Land Agriculture (Natural Resource Management Department), Jigjiga, Ethiopia*
[2]*Mekelle University College of Dry Land Agriculture and Natural Resources, Mekele, Ethiopia*

Abstract

This study was conducted in Gunugo watershed at Wolayitta zone to document indigenous knowledge on agroforestry practices management, and assess the socioeconomic factors affecting tree planting in agroforestry practices. Participatory rural appraisal, and transect walk was used to get qualitative data, whereas, household survey with semi-structured questionnaire was used through systematic sampling method to collect quantitative data. Among different socioeconomic factors, only family size and past participation in agroforestry training were significantly and positively related to tree planting activities. The common management practices on woody species include fertilizer application, pruning, coppicing, prescribed burning, thinning, pollarding, protection from animal damage, mulch application, crop residue application, and watering. After giving due recognition to the indigenous agroforestry management knowledge in the study area. This study has suggested some improvements such as, introduction of new agroforestry practices, and replacement of some of the tree species in order to reduce detrimental interaction of the components. In addition, training is needed to fill the knowledge and skill gap on tree plantation, and management.

Keywords: Homegarden; Indigenous knowledge; Management; Parkland; Woodlot

Introduction

In developing countries especially Africa, rapid population growth, decline in per capita food production and environmental degradation are the main problems. Consequently, the need for intensification of agricultural production coupled with population growth forces poor farmers to expand their cultivation to hilly and marginal areas. This aggravate the degradation of natural resource and unsustainability [1]. The forest resource of Ethiopia is under tremendous pressure due to continuous population growth, rudimentary farming techniques, land use competition, and land tenure [2-4]. In relation to this, agroforestry can help to reduce pressure on remnant natural forests from deforestation and enhances soil fertility [1,5]. The interventions of agroforestry has more to do with agricultural sustainability as it enhances higher yield, production stability, and others such as livelihood support, positive environmental impacts [1].

The concept of agroforestry puts woody perennials, including trees, shrubs, bamboos, etc. as pillars for the system/practice. Accordingly, agroforestry system can be classified mainly as agrisilvicultural, silvopastoral and agrosilvopastoral [6]. Within each system of agroforestry there are different agroforestry practices. Agroforestry system consists of one or more agroforestry practices that are practiced extensively in a given locality or area; usually there is biological, ecological and economic interactions among the components [6,7]. Whereas, an agroforestry practice indicates specific land management actions on a farm and other management units in spatial and temporal scheme [8].

The common traditional agroforestry practices in tropical region are scattered trees on crop fields, homestead tree planting and multi-story home garden [6,9,10] described agroforestry practice of Ethiopia as largely agrisilvicultural with spatial mixed arrangement, which is mainly practiced for soil fertility replenishing function for poor soil. Others [11-14] also described the traditional agroforestry practices in different part of the country as Coffee Shade based scattered trees on the farm land, home gardens, woodlots, farm boundary practices, and trees on grazing lands.

Including Wolaytta area where this study was conducted, agroforestry is a major component of Ethiopian farming systems [3,15] and recently taken as one of the development objectives in PASDEP of national development policy of the country [16,17]. It becoming one of the common features in watershed management especially in the highlands of Ethiopia. This is also true in Gununo watershed, that traditional agroforestry practices have been a main feature in the watershed and serving numerous protective and productive functions in both up streamers and lower catchments since, woody perennials have huge potential for this.

In the agroforestry system, woody perennials are either deliberately retained or planted on the farmland [7]. Different agroforestry systems require different periods of time to develop and manage. Therefore, depending on different benefits obtained from the system, farmers could employ different kinds of component management in the system.

The common managements in tropical agroforestry system are pruning, prescribed burning, thinning, pollarding, grass mulch application, crop residue application, watering, and coppicing [6,18,19]. Pollarding is a cutting of crown of a tree for the purpose of harvesting wood, reduce shade, or protection from browse animals [18]. Prescribed burning is a practice of deliberate burning of plant materials on the surface of ground, while thinning is an intermediate cutting of stunted stands to adjust the stand density and produce better yield [6,18]. Thinning is very common practice for woodlot agroforestry practices

***Corresponding author:** Madalcho AB, Jigjiga University College of Dry Land Agriculture (Natural Resource Management Department), Jigjiga, Ethiopia
E-mail: bajigoaklilu05@gmail.com

[18]. Mulching is a practice used for covering the soil surface to protect it from the direct impact of rainfall [18]. In addition, coppicing is a cutting practice of certain tree species that are close to ground level for the purpose of producing new shoots from the stump while pruning is a practice of cutting tree branches or roots [6,18].

Factors affecting different management decision such as tree planting are mainly related to farming system [20], household characteristics, rules and regulations imposed by state and community in local or national level and different biophysical circumstances, and tree characteristics [20,21]. The management decision in agroforestry is related to different on farm and off farm factors; including issues related with land security and external support services such as credit, input suppliers, and extension [21].

In the management of agroforestry the indigenous knowledge of local people is important, and in order to scale up the different agroforestry practices an appreciation of indigenous knowledge is needed [7]. Indigenous knowledge includes different sets of complex practices. The discovery of knowledge in managing resource is made by local people. Then, the knowledge exhibited and experienced will be transferred to generations with some modifications [22]. Therefore, understanding the historical development of indigenous systems is decisive in the design of ecologically desirable agroforestry production systems [23]. In general indigenous knowledge of local people are not simply producers, they are also engaged in pursuit of knowledge. Most development interventions in the past failed due to lack of giving adequate attention to indigenous knowledge [24].

The objective of this study was to document indigenous knowledge on agroforestry practices management, and assess the socioeconomic factors affecting tree planting in agroforestry practices.

Methodology

Study area description

Wolayitta zone is in southern nations, nationalities and peoples (SNNP) region of Ethiopia with a total land area of 4537.5 square kilometers is located between 6°4´N to 7°1´N latitudes and 37°4´E to 38°2´E longitudes. It is located 22 km from Sodo town and about 330 km from Addis Ababa. The watershed has an area of about 544 hectare with three rural Kebeles namely: *Demba Zamine* (middle elevation), *Doge Hanchucho* (lower elevation) and *Chew kare* (upper elevation).

Soil of the watershed is Eutric Nitosol according to FAO/UN classification system [25]. Soil erosion in watershed is severe due to conversion of natural forests to other land uses. The study area has high population pressure i.e., around 450 person per km^2 [26] and an average land holding is about 0.25 ha per household and drive farmers to cultivate slope lands [27]. Agroforestry is one of the major land uses at the area. Different species (tree crops and woody species) such as *Enset ventricosum Musa accuminata, Moringa oleifera* and *Brassica oleracea* serve as primary food source while *Croton macrostachyus* and different Acacia species are the dominant trees in the degraded natural forest of Wolayitta [26]. From the agroforestry practices in the watershed, homegarden, parkland and woodlot agroforestry practices are the dominant ones. The management of these agroforestry practices are traditional based on indigenous knowledge obtained from their life long practices.

Sampling and data collection methods

During this study time, different qualitative and quantitative data was collected with sequential procedure (began with qualitative and followed by quantitative data collection). The qualitative data assessed the history of the study area regarding the practices of agroforestry, different characteristics of trees, and field observation on the management of agroforestry system as discussed below. These data were collected from primary data source through participatory rural appraisal (PRA) with tools such as observation, transect walk, and group discussion. House hold survey was conducted to collect quantitative data. In addition, secondary data from official documents and available literatures were employed to support different qualitative and quantitative data collected.

Group discussion

The group discussion involved local chairman of the Gununo watershed, official manager of the Gununo woreda, youth representative, woman representative of the area, three natural resource management development agents (DAs), four elders in the area, and three model farmers, a total of fifteen people. The natural resource management DAs, elders and model farmers were involved from the three Kebeles within the watershed i.e., *Demba Zamine, Doge Hanchucho* and *Chew kare/ Gununo*. The discussion took a week and respondents were allowed free to reflect their view concerning issues related to agroforestry system management practices in the area.

Household survey

To collect quantitative data on management practices of agroforestry and factors affecting number of trees planted in agroforestry, the following formula after Watson (2001) [28] was used, and the sample size of respondents were determined by using equation below (Equation 1):

$$n = \frac{\left(\dfrac{P[1-P]}{\dfrac{A^2}{Z^2} + \dfrac{P[1-P]}{N}} \right)}{R} \quad (1)$$

Where n=sample size required, N=number of people in the population, P=estimated variance in population, as a decimal, A=Precision desired, expressed as a decimal, Z=Based on confidence level, R=Estimated response rate, as a decimal. Consequently, the estimated variance in population (P) of 15%, estimated Precision of 5%, Confidence level of 95% and Estimated Response rate of 95% was used to calculate the sample size.

Accordingly, by using systematic sampling, 50 HHs were sampled. After getting the lists of names from each of three Kebeles involved in the study, semi-structured questionnaires were distributed. The name of the first person was chosen randomly so that the next sampling unit (person) was selected at each 10th person. Then, with the help of DAs each sampled person was interviewed by semi-structured questionnaire.

Data analysis

The qualitative data was narrated based on the existing conditions and knowledge on the literature, whereas, quantitative data collected from household questionnaire survey was also analyzed by a multiple regression and descriptive statistics with help of SPSS Inc., Chicago, USA (2007). The collected data was summarized using descriptive statistical tools like tables and figures.

Results

Factors affecting number of trees planted in agroforestry

Characterization of woody species by local people: In relation to agroforestry products and other environmental roles such as, soil and water conservation (SWC), the local people have identified suitable trees. Consequently, trees such as *Eucalyptus globules*, Avocado, *Cordia africana*, Mango (*Mangifera indica*), *Gravillea robusta, Podocarpus falcatus, Cupresus luistanica* and *Ficus* species are considered to be preferred suitable trees in decreasing order. The characteristics of suitable trees are mainly either based on monetary value or SWC functions. Tree characteristics such as, having deep root/shallow, competition for light and nutrient, allelophatic effect, contribution for nutrient improvement and rate of decomposition (from litter) are the main selection criteria for categorizing woody species as suitable. Local people know the suitable tree/shrub species for the specific agroforestry practice in which they establish. In fact there is no "good or bad tree", local people call unsuitable trees as bad, and suitable trees as good trees. Accordingly, *Eucalyptus* species are taken as both good and bad tree in different points of view. In point view of local people, it is good in generating income. Since it is fast growing and coppiced more frequently for firewood selling and construction purpose, when planted as woodlots. Whereas, it is hardily possible to integrate with other crops for soil fertility maintenance. Its interaction to other integrated other component is always negative, especially in the moisture constrained area. Its leaf is not decomposed easily, and the soil fertility reduction and soil moisture completion is very high. Following *Eucalyptus, Cupressus lusitanica* and *Prunus africana* are considered as bad (unsuitable species) for integrating in agroforestry system/practices. In the area, Avocado *(Percia Amercana)* and Mango *(Mangifera indica)* is chosen for both fertility maintenance and household income generation value, and its contribution continues all year round. Therefore, the criteria to select and plant trees in agroforestry practices are highly related to the local people's indigenous knowledge to select those good (suitable) trees.

Factors affecting tree planting	P- value
Elevation level	0.572
Sex	0.991
Education	0.192
Age	0.318
Distance from market	0.442
Family size	0.005
Land holding	0.158
Number of animals per house hold	0.581
Past participation in agroforestry training	0.040

Table 1: Factors affecting number of tree planted in agroforestry in Gununo Watershed, Wolaita.

There are different biophysical (elevation level), and socioeconomic factors such as, sex, education, age, distance from market, family size, land holding, number of animals per house hold and past participation in agroforestry training have been assessed. From among those, only family size and past participation in agroforestry training were significantly and positively related to tree planting activities in the last five years (Table 1).

Family size per house hold in the area include children (age less than 15), productive age (15-64years) and elders (age greater than 64). In the family where large productive age class, the practice of tree plantation is very high compared to the family with small size and unproductive age class dominated.

The knowledge and culture of tree plantations are transferred from the family of their ancestors. The knowledge is not even similar to each households in the watershed, it vary based on the past habit family attitude towards the need of trees in the farming system. In addition to that, those who have got training on tree plantation in the agroforestry system in different sort of occasions have better understanding and positive outlook on tree plantation. The have planted and retained more trees on their farmlands.

Management practices in agroforestry: There were different kinds of management observed in agroforestry practice of Gununo

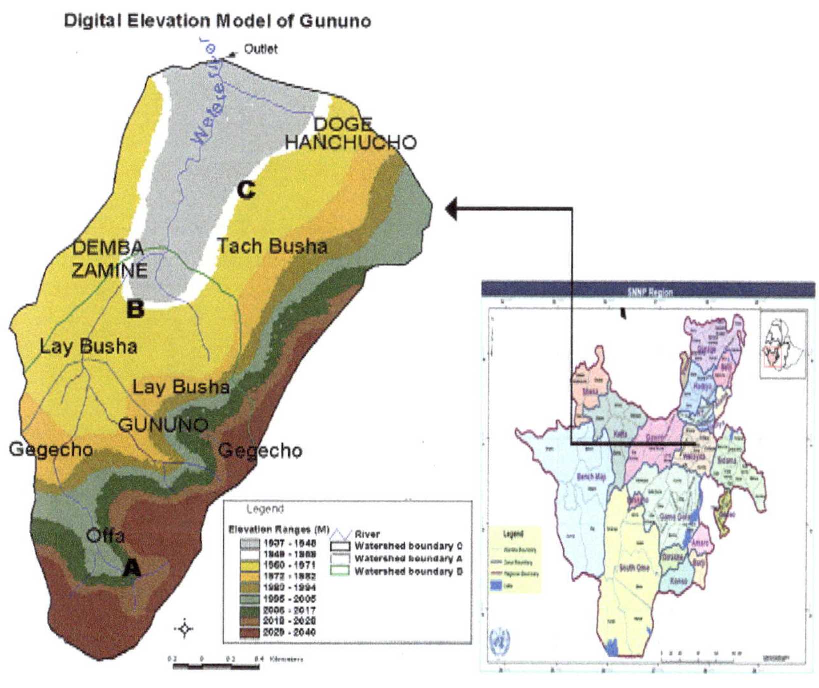

Figure 1: Map of Wolayitta Zone and Gununo watershed, Southern Ethiopia (adopted from Mowo *et al.*, 2011).

Watershed. These managements are shown in Figures 2 and 3. About 95% respondents mentioned an existence of management variations among agroforestry practices. The common management practices include fertilizer application (mainly manure, but also DAP and urea), branch pruning and coppicing, prescribed burning, thinning, pollarding, protection from animal and human damage, grass mulch, crop residue and watering.

The main criterion for such management intensity was mentioned to be related to criterion such as the effect of these managements on agroforestry products (e.g., on fruit and crops i.e., in parkland) and on sustainable land management, to reduce and land degradation reclaim degraded lands. In general, the criterion used to select some management practice in the watershed is mainly to reduce negative interaction between components and maximize the overall function of the system per land management unit. Consequently, management practices in home garden and woodlot are mainly given to increase fruit products and market values, respectively while in parkland it is given to increase survival and yield of agricultural products tree canopy.

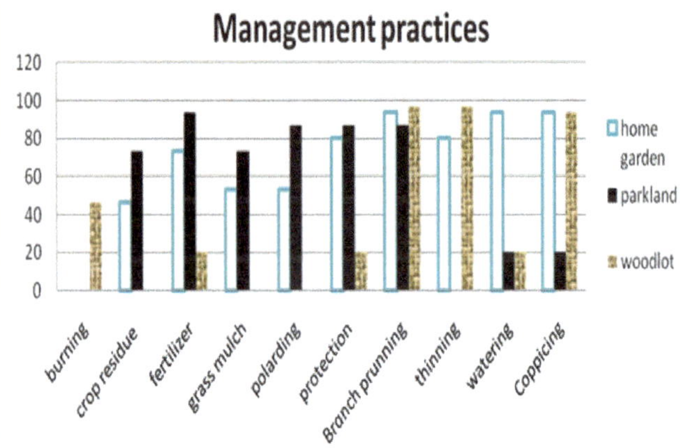

Figure 2: Management of agroforestry practices within agroforestry type in Gununo Watershed.

From the common management practices in the watershed, 93% of coppicing, pruning and watering, 80% of protection from animal intervention and thinning, and 73% of fertilizer usage is applied in homegarden compared to parkland and woodlot agroforestry practices. In addition, 53% of grass mulch and pollarding, and 467% of crop residue application is used in homegarden compared to parkland and woodlot agroforestry practices. The common management practices such as fertilizer application (93%), pruning, pollarding and protection (87%), grass mulch and crop residue application (73%), and watering (20%) are implemented in parklands. Similarly, the common management practices in woodlot include pruning and thinning (97%), coppicing (93%), prescribed burning (44%) and fertilizer application, protection and watering (20%), (Figure 2). The application of these management practices varied among woody species and age class, however the big difference is there between agroforestry practices. Burning was given for only woodlots. It was given on dry seasons to increase the leaching of nutrients on the ground, and to facilitate the dormancy breaking after cutting the tree. Similarly, crop residue, grass mulch application and watering were given mostly in dry seasons to increase the moisture content of the soil. Thinning and coppicing was applied at the base of the bole of woody species depending on their age class i.e. species greater than three years or at their mature stage were mostly chosen for such managements.

Figure 3: Management practices showing Grass mulch application on the top, and protection of agroforestry practices from animal damage (using of stall feeding/ partial captive grazing for animals in the system) in Gununo Watershed.

Discussion

Factors affecting number of trees planted in agroforestry

It is mentioned that only family size and past participation in agroforestry training were affecting the number of trees planted in agroforestry. House hold (HH) with higher family sizes was planting more trees than less family sizes. This could be attributed with labor availability, and most of the age distribution per households in the watershed was found in productive age (age between 15 to 64 years). Higher family size would have better labor to plant trees because most of household members in this family size were in productive age. People found in this productive age were younger, eager to plant trees, able to provide better management for the planted trees, better in economic condition as most of them work an off farm activities. In addition to the agricultural activities, they have different hand works as a means of income generation, which help them to buy seedlings, and participate in different agroforestry trainings that increase their knowledge about the importance of trees. Family size was previously reported to affect tree cultivation in Pakistan i.e. larger family size cultivated more trees [29].

Similarly, people who participated in agroforestry training planted more trees than who did not participate. This is related with level of knowledge i.e., different trainings given for local people would increase an understanding on agroforestry component interaction and management, importance of woody species, and ways of increasing the survival rate of seedlings/saplings. Training/lack of extension was also reported to drastically affect house hold on farm tree growing in Philippines [30].

However, other factors such as sex, land holding, house hold income, proximity to road were affecting tree growing decision in Ethiopian highlands [31] whereas, land size, age, gender, tenure security, education, income, and agro-ecology were among factors that increased the tendency to plant trees in Tigray [32]. Variation in factors affecting tree planting among different areas could be due to difference in socio economic, demographic or ecological conditions. Therefore, it is suggested that agroforestry and natural resource related trainings, incentives for tree plantings and seedling distribution need to be further given for local people in order to increase tree planting per HH.

Management practices

The dominant management practices of the study area such as, thinning, pruning, coppicing, pollarding and others are in line with the commonly used managements in tropical agroforestry [6,18,33]. These management practices were shown to vary in intensity and existence among agroforestry practices. The application of any management scheme in the area is linked with the indigenous knowledge of the people in the watershed. They perceive how well the woody species react to different managements and its effect on different biophysical settings such as soil fertility. For instance, prescribed burning was only given for woodlot. In point view of the local people in the watershed, prescribed burning was given for woodlot to increase release of nutrients to soil and facilitate the vibrant shoots from the stamp. However, it was not given on parkland and home garden due to its danger on damaging some of the components in the system as there is no clear cut of components. The burning could damage agricultural and tree crops and associated high value trees.

On the other hand, crop reside was only given in parkland and woodlot. This is attributed to availability of crop reside in these agroforestry practices: which woodlot lack as all the biomass is harvested for eucalyptus. Fertilizer application is also relatively high in parklands because of less nutrient availability, species diversity and existence of agricultural crops in parklands. However, less attention was given for woodlot in fertilizer application as eucalyptus is deep root and can get nutrients and water from deeper soil horizons while relatively high nutrient availability in home garden makes relatively less application of fertilizers. Grass mulch was relatively highly given for parklands in order to increase the moisture availability as there is less litter accumulation unlike that of home gardens. Additionally, high practice of pollarding in parkland is attributed to the need of light by the associated crops in parklands than home garden where mostly shade tolerant species are integrated. The need for more protection for parklands is because crops associated in parklands are highly susceptible for grazing animals than home gardens and woodlots. Relatively high practice of coppicing in home garden and woodlot is taking place. Most of the tree/shrub species exist in these agroforestry practices are coppicing species. The reason of coppicing is mainly related to the desire of land owners to get regeneration of new shoots from the stamp to optimize the productivity.

Relatively no thinning and less coppicing in parkland is attributed to the low tree/shrub density, and low coppice species composition in the component. The high watering on home garden is linked with the proximity of home garden to homesteads than parkland and woodlot.

In general, variation in the management intensity was observed in the watershed, and this is also true in different areas and eco regions [6,18,33]. The variation can be attributed to difference in socio economic, cultural and biophysical and environmental settings/criteria [21]. In addition, it could be related with some of the factors affecting adoption of agroforestry in tropics such as sex, age, livestock population, education level, growing of trees, species preference, market, family size, farm size, etc. [21,20,29,34].

Conclusion and Recommendation

The common agroforestry management practices in the watershed were fertilizer application, pruning, prescribed burning, thinning, pollarding, protection from animal and human damage, grass mulch, crop residue, watering and coppicing. Additionally, only family size and past participation in agroforestry training were significantly and positively related to tree planting activities.

This study indicated that agroforestry practice could be one option to address the problems of deforestation and related resource degradations in Gununo watershed, Wolaita. The indigenous knowledge on agroforestry system and practice management being applied in the watershed should have to get recognition. However, to cop up with the dynamic situations in resource need, and the continued deterioration of the biophysical condition, it need some improvement. Therefore, the following recommendations are given based on the findings of this study:

Replacement of traditional agroforestry practices by new agroforestry technologies such as hedgerows intercropping, fodder banks, fertilizer trees along terraces and soil bunds is important.

Replacement of Eucalyptus with other suitable trees that best fit to be integrated in different agroforestry practices and valuable in terms of products and services is needed.

Promoting farmer managed natural regeneration (FMNR) by protecting cropland from animal damage to enhance natural regeneration particularly in parkland agroforestry practices is highly needed.

References

1. Kang BT, Akinnifesib FK (2000) Agroforestry as alternative land-use production systems for the tropics. Natural Resources Forum 24, pp. 137-151.
2. Richard CJ (2006) Deforestation and the environmental Kuznets curve: An institutional perspective. School of Agricultural and Veterinary Sciences, Charles Sturt University. Elsevier B.V. Australia.
3. Bishaw B (2001) Deforestation and land degradation in the Ethiopian highlands: A strategy for physical recovery. Northeast African studies. 8.
4. USAID (2008) Ethiopia Biodiversity and Tropical Forests 118/119 Assessment. United States Agency for International Development.
5. Gustavo SG, Fonseca AB da, Harvey CA, Gascon C, Vasconcelos HL, et al. (2004) Agroforestry and Biodiversity Conservation in Tropical Landscapes. Island Press, USA, pp. 537.
6. Nair PKR (1993) An introduction to agroforestry. Kluwer Academic publisher with cooperation ICRAF, Dordrecht, London. p. 489.
7. World Agroforestry Centre (ICRAF) (2006) Agroforestry for improved livelihoods and Natural resources conservation. An Agroforestry Policy Brief. Kenya, Nairobi.
8. Gholz HL (1987) Agroforestry; Reality, possibilities and potentials, martingus, Nijhoff publishers in cooperation with ICRAF pp. 21-22.
9. Nair PKR (1984) Tropical agroforestry systems and practices. International Council for Research in Agroforestry (ICRAF) Chapter No. 14.tropical resource ecology and development. John Wiley, Chichester, England. June 1984–123.
10. Young A (1989) Agroforestry for soil conservation. International Council for Research in Agroforestry. CAB International p. 318.
11. Bekele A (2007) Profitable Agroforestry innovation for Eastern Africa: Experience from 10 agroclimatic zones of Ethiopia, India, Kenya, Tanzania and Uganda, worldagroforestry center, p. 353.
12. Abebe T (2005) Diversity in homegarden agroforestry systems in Southern Ethiopia PhD thesis Wageningen University, Wageningen.
13. Negash M (2002) Socio economic aspects of farmers eucalypt planting practice in the Enset coffee based agroforestry system of Sidama, Ethiopia: The case of Awassa and shebedino district Msc thesis.
14. Ashagre Y (1997) The contribution of tree to the soil chemical properties in the Croton macrostachyus based indigenous agroforestry system in North western Ethiopia, Msc Thesis, skinnskatteberg.
15. Jama B, Zeila A (2005) Agroforestry in the drylands of eastern Africa: a call to action. ICRAF Working Paper – no. 1. Nairobi: World Agroforestry Centre.
16. Teshome A (2006) Agriculture, Growth and Poverty Reduction in Ethiopia: Policy Processes around the New PRSP (PASDEP). A paper for the Future Agricultures Consortium workshop, Institute of Development Studies.
17. FDRE (Government of the federal democratic republic of Ethiopia) (2003) Rural development policy and strategies. Ministry of Finance and Economic Development Economic Policy and Planning Department. Addis Ababa.
18. Tengnas B (1994) Agroforestry extension manual for Kenya. Nairobi: International Centre for Research in Agroforestry. Nairobi, Kenya. P. 188.
19. Kilewe AM, Kealey KM, Kebaara KX (1988) Agroforestry Development in Kenya Proceedings of the Second Kenya National Seminar on Agroforestry held in Nairobi, Kenya.
20. Bongers G (2010) Dynamics in People-Tree Interactions in Farm Fields: Farmers' perspectives in Meskan District, Ethiopia. Master Thesis. Wageningen University.
21. RWEDP (1995) Integrating woodfuel production into agroforestry extention programs in South East Asia. RWEDP Report No.21. FAO. April 1995, West Java, Indonesia.
22. *Oudwater N, Martin A (2003) Methods and issues in exploring local knowledge of soil. Geoderma 111: 387-401.*
23. Miller RP, Nair PKR (2006) Indigenous agroforestry systems in Amazonia from prehistory to today. Agroforestry systems, 66: 151-164.
24. King KFS (1980) Agroforestry and development of tropical forestry. UNEP, meeting of experts on tropical forests.Nairobi.
25. Tegene B (1992) Farmers perceptions of erosion hazards and attitudes towards soil conservation in Gununo, Wolaita, Southern Ethiopia. Etiopian journal of development research, 14.
26. WZFEDD (2008) Zonal Basic Socio-Ecnomic & Demographic Information. Wolayitta Soddo.
27. Amede T, Belackew T, Geta E (2001) Reversing the Degradation of Arable Land in the Ethiopian Highlands. Managing Africa Soils. No. 23. iied.
28. Jeff W (2001) How to Determine a Sample Size: Tipsheet 360, University Park, PA: Penn State Cooperative.
29. Muhammad I, Asadullah K, Mitsuhiro I, Muhammad A, Hassan S (2011) Identifying factors affecting agroforestry system in Swat, Pakistan. African Journal of Agricultural Research, 6: 2586-2593.
30. Denyse JS, Lasco Rodel D (2008) Smallholder Tree Growing for Rural Development and Environmental Services: Lessons from Asia. Springer Science and Business Media BV, Netherlands, p. 493.
31. Ewnetu Z, Bliss JC (2010) Tree growing by small holder farmers in the Northern Ethiopia. Small scale forestry in a changing world: opportunities and challenges and the role of extension and technology transfer IURFO conference. June 2010.
32. Gebreegziabher Z, Mekonnen A, Kassie M, Köhlin G (2010) Household Tree Planting in Tigrai, Northern Ethiopia: Tree Species, Purpores, and Determinants. Environment for development, January, 2010.
33. Bishaw B, Abdelkadir A (2003) Agroforestry and Community Forestry for Rehabilitation of Degraded Watersheds on the Ethiopian Highlands. Wondo Genet College of Forestry Debub University Awassa, Ethiopia International Symposium on Contemporary Development Issues in Ethiopia, July 11-12, 2003, Addis Ababa, Ethiopia.
34. El Tayeb MS (2008) Factors Affecting Adoption of Agroforestry Farming System as a Mean for Sustainable Agricultural Development and Environment Conservation in Arid Areas of Northern Kordofan State, Sudan. Saudi Journal of Biological Sciences, 15: 137-145.
35. Mowo J, Adimassu Z, Masuki K, Lyamchai C, Tanui J, et al. (2011) The Importance of Local Traditional Institutions in the Management of Natural Resources in the Highlands of Eastern Africa. Working Paper No 134. Nairobi: World Agroforestry Centre.

Vegetative Community Development Over 30 Years within Pine Plantations on Reclaimed Mine Land in East Texas

Christian CL, Oswald BP*, Williams HM and Farrish KW

Division of Environmental Science, Arthur Temple College of Forestry and Agriculture, Stephen F Austin State University, Nacogdoches, TX 75962, USA

Abstract

Reclamation of surface mines to plantation forests is a management option that reestablishes both economic and ecological functions after mining. This study investigated vegetative community characteristics (composition, richness, importance) over time in loblolly pine (*Pinus taeda* L.) plantations established over a thirty-year period on reclaimed lignite coal surface mine land in East Texas, United States. The open landscape of newly planted loblolly pine plantations on reclaimed mine land was amenable to shade-intolerant herbaceous and grass species but, when the canopy closed, favored woody species (trees, shrubs, vines) within two decades after stand establishment. Given that these plantations were established on sites generally described as dry to mesic uplands, species composition was generally congruent with East Texas ecology. Community composition, species richness and species importance were discussed, and loblolly pine growth data and vegetative community characteristics of unmined loblolly pine plantation sites in East Texas were compared to reclaimed sites. Surface mined lands reclaimed to plantation forests are ideal locations to conduct further research on how biodiversity and other ecosystems services can be improved while maintaining intended economic and ecological purposes.

Keywords: Loblolly pine; *Pinus taeda*; Reclamation; Surface mining; Ecosystem services; Biodiversity

Introduction

Reclamation of surface mines to plantation forests reestablishes both economic and ecological functions after mining. Increasing consumer demand for wood products necessitates using a greater proportion of land for plantation forests [1]. In 2010 reports, planted forests accounted for 7% of forests globally, up from 3% in 1999 reports and 8.3% of U.S. forests, up from 6% in 1990 [1,2]. These planted forests were included in the 30% of all US forests with designated primary use for production of "wood, fiber, bioenergy and/or non-wood forest products" [2]. Plantation forests consisting of pine species with significant economic demand are a commonly utilized post-mining land use. Restoration of ecosystem services is a desirable outcome of mine reclamation. Although differences exist in comparison to natural forests, plantation forest ecosystems can provide long-term forest-based ecosystem services such as wood production, habitat for native flora and fauna, watershed protection and improved water quality and carbon storage [1,3,4]. Also, biodiversity in plantation forests is of increasing interest [1,5-8].

Many studies have investigated soils, vegetation, water quality and more on un-reclaimed and reclaimed mine lands both pre- and post-SMCRA (Surface Mining Control and Reclamation Act of 1977) [3,4]. Over time, reclamation strategies have changed in order to effectively address environmental issues common on mined land (e.g. acid mine drainage, erosion, lack of natural revegetation) and achieve more successful post-mining land uses [3,4]. Long-term studies of vegetative community development on reclaimed mine land have been invaluable in adjusting reclamation practices, and several (11 to 55 years post-mining) have been conducted within the US [9-13]. Although no long-term studies for East Texas (Piney woods vegetation area) have been performed, a few were conducted in east-central Texas (Post Oak Savannah/Black land Prairie vegetation area) [14-16]. Several short-term vegetation-related studies on reclaimed mine land were conducted in East Texas; these included research on short-term survival and growth of various native pine species in mine soil as well as effects of fertilization rates, cover crops, ectomycorrhizal inoculation, stock type and seed source on seedlings and young trees planted on reclaimed mine land [17-24]. However, none of these studies addressed plant communities as a whole or over time.

In Texas, current lignite coal mine permits cover nearly 132,000 ha of land, including 1,363 ha disturbed by mining activity in 2014 [25]. In East Texas, loblolly pine (*Pinus taeda* L.) plantations are commonly established on reclaimed mine land as a designated post-mining land use ("forestry"). Some literature was available for vegetative community ecology of loblolly pine plantations in East Texas [26-31], but information about long-term vegetative community development of pine plantations on reclaimed mine land is limited to this study [32]. The objectives of this study were to determine vegetative community characteristics (composition, richness, importance) over time in loblolly pine plantations established over a 30-year time frame (1980 to 2009) on reclaimed lignite coal surface mine land in East Texas, USA. This study provided baseline ecological information about these plant communities that may assist land managers and researchers in adjustment of vegetative reclamation techniques as desired to improve reclamation success, economic benefits and provision of ecosystem services on these reclaimed mine lands.

Methods

Study area

This study was conducted on Luminant Mining Company

***Corresponding author:** Oswald BP, Division of Environmental Science, Arthur Temple College of Forestry and Agriculture, Stephen F Austin State University, Nacogdoches, TX 75962, USA, E-mail: boswald@sfasu.edu

property at Beckville and Tatum mines in Panola County, Texas, USA (approximately lat 32°10' N, long 94°20' W). Post-SMCRA reclamation activities occurred within the study area over the past 30 years, including post-mining establishment of loblolly pine plantations. The study area was characterized by irregular, gently rolling to hilly forestland; high precipitation, humidity and temperatures; acidic sand and sandy loam soils; mixed land use (includes timber, pasture, farm land); and major over story species consisting of pines, oaks, hickories and maples [33]. The "subtropical" climate was characterized by annual rainfall of 42 to 46 in (≈107 to 117 cm), mean annual temperature of 66°F (≈19°C), and mean frost free period of 230 to 245 days [34].

Generally, pre-mine soils consisted of soil associations Sacul-Bowie, Fuquay-Troup and Nahatche-Mantachie-Urbo, described as follows: Sacul-Bowie, "gently sloping to moderately steep, slightly acid to medium acid, loamy soils on uplands"; Fuquay-Troup, "gently sloping to moderately steep, slightly acid, sandy soils on uplands"; Nahatche-Mantachie-Urbo, "nearly level, slightly acid to strongly acid, loamy to clayey soils on bottom lands" [35]. A mixed overburden reclamation technique was used at these mines, whereby soil overlying coal resources was removed and set aside, coal was extracted, and the removed soil was used to re-fill excavated areas without any effort to restore pre-mining soil profile (i.e. generally, weathered surface soils became mixed with deeper, un-eathered soils).

Study sites

Fifty-six reclaimed mine sites planted as loblolly pine plantation stands were sampled (30 Beckville, 26 Tatum). For each mine (Beckville and Tatum), study sites were randomly selected to include stands covering a 30-year period (1980-2009) and were grouped into six age categories (i.e. 1 to 5 year, 6 to 10 year, 11 to 15 year, 16 to 20 year, 21 to 25 year, 26 to 30 year). If available, five stands were randomly chosen within each age category (at least one per year as possible). No stands aged 26 to 30 years were available at Tatum. Eleven sites were thinned (11 Beckville, 4 Tatum), ranging in age from 14 to 30 year with only two sites younger than age 22. One Beckville stand (BP8301) contained thinned and unthinned portions.

Each mine's data was analyzed separately for statistical analyses because available study sites were not evenly distributed by age among the two mines, and data skewing toward one mine or the other may have occurred with combined data. Study sites were not chosen based on stand management techniques, and, given the length of time covered by this study and frequent changes in post-SMCRA reclamation techniques, management methods may have varied among study sites (e.g. initial cover crop, seedling type, planting density, thinning). Other than items specifically discussed, management techniques were not accounted for. Study sites in which thinning activity had occurred were not excluded from this study as thinning is an inevitable management event in pine plantations. This study does not cover a true chrono sequence of 30 years, which would require repeated sampling of the same physical locations; a pseudo-chrono sequence was constructed using available reclaimed pine stands of various ages to represent a 30-year timeline. The phrase "over time" throughout this manuscript should be viewed based on the above explanation. The terms "younger", "middle-aged" and "older" were applied to stands within the following age ranges, respectively: 1 to 10 year, 11 to 20 year and 21 to 30 year.

Vegetation

One plot was established in each stand using a modified-Whittaker plot design (20 m × 50 m rectangular plot containing nested rectangular subplots of three sizes) (Figure 1) [36,37]. The 50 m side of the plot was situated parallel to the slope of the site to ensure that vegetative samples captured as much heterogeneity as possible. Understory, mid-story and over story strata were sampled for vegetation parameters and defined by vegetation height: understory (≤ 1 m), mid-story (>1 m to ≤ 6 m) and over story (>6 m) (Figure 1). Vascular vegetation was identified to species, as possible, for all strata, following the USDA PLANTS Database for scientific name and authority [38]. Visual estimates of species cover (%) were recorded for understory and mid-story strata, and over story species cover was represented by basal area, which was calculated from diameter at breast height (dbh) using the formula $\pi^{*}(dbh/2)^2$. Stem counts (density) were recorded for mid-story and over-story strata and converted to stems ha-1. Height of each over story individual was measured by range pole (≤ 7 m) or clinometer (>7 m). It should be noted that diameter, height and stem count data were collected for all loblolly pine individuals, regardless of strata; for individuals, shorter than breast height (1.37 m), root collar diameter was measured and used in place of dbh as needed.

Species composition (i.e. presence), species richness, and species importance values (IV) were obtained from field data. Species IV are unit less numbers that indicate the overall contribution of an individual species to a community relative to all other species in the community [39]. The original definition of IV is the sum of relative cover, relative density, and relative frequency [39], and was used for mid-story and over story. For understory, IV was calculated as the sum of relative cover and relative frequency. For each species, the total species IV was calculated as the sum of understory, mid-story and over story IV.

Data analysis included nonmetric multidimensional scaling (NMS), an ordination statistical technique performed using PC-ORD 6 statistical software [40]. Species codes displayed on ordination graphs are defined in Table A-1 (Supplementary Information, Appendix). City-block distance measures were used for two reasons: 1) the sparsity of the

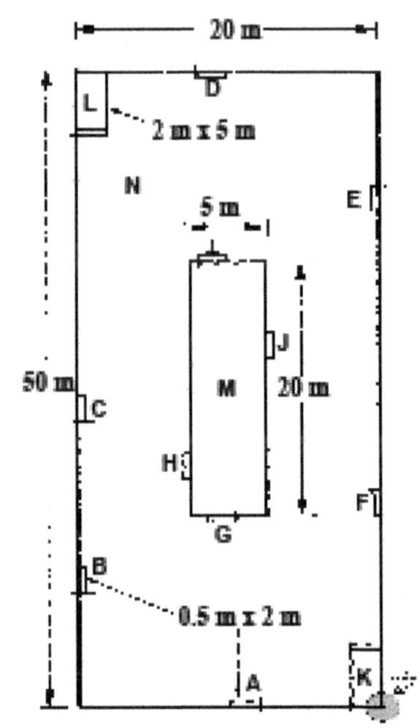

Figure 1: Layout of the modified-Whittaker plot design (Stohlgren et al.). Alphabetical labels and location of GPS coordinates were added. Vegetation was sampled in subplots and strata based on vegetation height as follows: understory (≤ 1 m), "A" through "J"; midstory (>1 m to ≤ 6 m), "K" through "M"; overstory (>6 m), "M" and entire plot "N".

matrices of the datasets and 2) zeros in the dataset did not necessarily mean the numerical zero. Both Sorensen and Jaccard city-block distance measures were used to ensure that solutions obtained were similar for the two measures. In order to determine the appropriate number of axes, Autopilot was run a minimum of three times using random seeds for each of the Sorensen and Jaccard distance measures. If solutions among the Sorensen and Jaccard distance measures were similar, then Sorensen was used in the manual and final analyses. Running several different analyses ensured that a qualitatively inconsistent solution was not chosen as the final solution. Randomization tests were included to assess the strength of the data pattern. Final stress values, randomization test p-values, scree plots, and plotted ordination solutions were examined for overall qualitative consistency among all solutions. Then, a minimum of three manual NMS analyses were run using the number of axes recommended by Autopilot using Sorensen distance measures. In situations where it appeared that a different number of axes from what Autopilot recommended might be more appropriate, three manual analyses were also run using this alternative axis number. For example, in many cases, two axes were recommended, but stress values and other information indicated that three axes might also be appropriate. Mantel tests were run to compare the two axis quantities in order to determine whether or not they conveyed similar information. If the two different axis quantities provided similar information, the least number of axes was used in the final solution in order to simplify interpretation of results.

Results

Predictably, mean loblolly pine height, diameter and basal area showed increasing trends over time (Figures 2-4). However, a decrease in basal area values was observed in the oldest age category sampled at each mine (Figure 4). Density increased over time until a noticeable decrease after two decades of growth (Figure 5).

Species composition

A strong chronological trend in species composition was observed along Axis 1 (represents site age), with older sites generally grouped on the left side of the graph and younger sites on the right (Figures 6 and 7). Over time, species composition became more similar between stands, as evidenced by loosely grouped younger stands and tightly grouped older stands. At Tatum, one 9-year old stand was grouped with the oldest stands and had unique species composition as compared to other younger sites.

Generally, woody vine species (e.g. saw greenbrier (*Smilax bona-nox* L.), Virginia creeper (*Parthenocissus quinquefolia* (L.) Planch.), poison ivy (*Toxicodendron radicans* (L.) Kuntze)) were strongly associated with older sites (species lines point toward older sites), and herbaceous and grass species (e.g. coastal bermudagrass (*Cynodon dactylon* (L.) Pers.), Johnsongrass (*Sorghum halepense* (L.) Pers.), camphorweed (*Heterotheca*

Mine	Understory stratum	Midstory stratum	Overstory stratum	All strata combined
Beckville (30 stands, 1 to 30 yr old)	97	43	12	104*
Tatum (26 stands, 1 to 24 yr old)	86	21	5	86**

*49% of species present in <4 stands; **50% of species present in <4 stands.

Table 1: Combined plant species richness data for reclaimed mine sites planted as loblolly pine (*Pinus taeda* L.) plantation stands on two lignite coal surface mines in Panola County, Texas, USA.

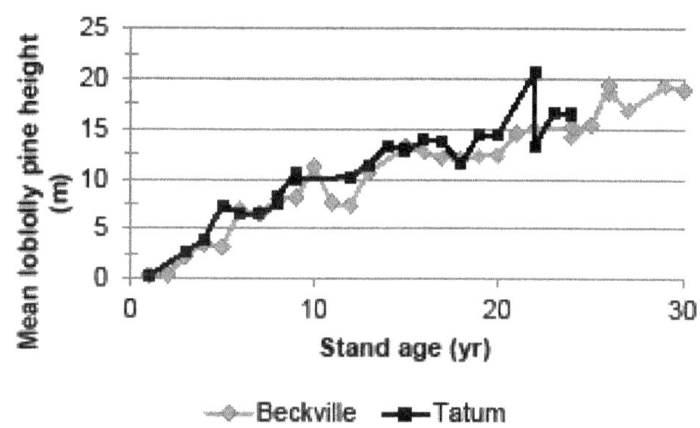

Figure 2: Mean height (m) of loblolly pine (*Pinus taeda* L.) individuals in reclaimed mine sites planted as loblolly pine plantation stands aged 1 to 30 years at Beckville and Tatum lignite coal surface mines (Panola County, Texas, U.S.A.).

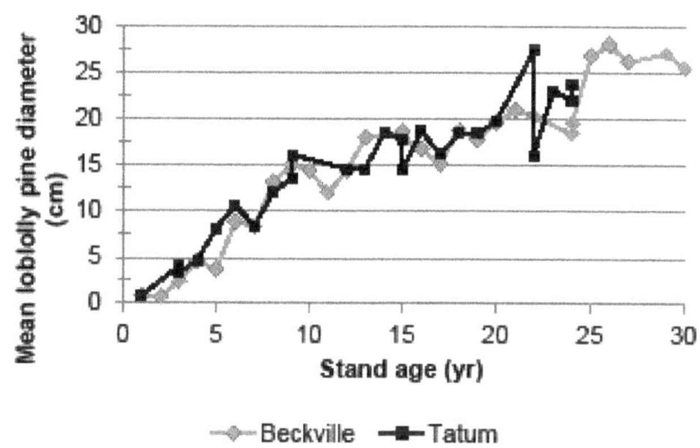

Figure 3: Mean diameter (cm) of loblolly pine (*Pinus taeda* L.) individuals in reclaimed mine sites planted as loblolly pine plantation stands aged 1 to 30 years at Beckville and Tatum lignite coal surface mines (Panola County, Texas, USA).

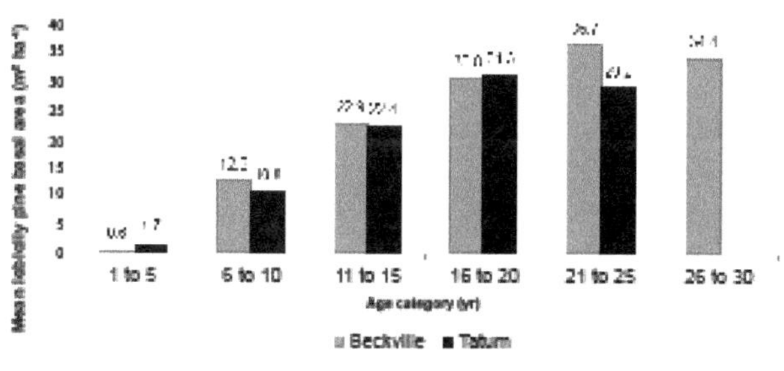

Figure 4: Mean loblolly pine (*Pinus taeda* L.) basal area (m^2 ha^{-1}) by age category in reclaimed mine sites planted as loblolly pine plantation stands aged 1 to 30 years at Beckville and Tatum lignite coal surface mines (Panola County, Texas, USA).

subaxillaris (Lam.) Britton & Rusby)) were strongly associated with younger sites (species lines point toward younger sites) (Figures 6 and 7). Woody shrub and tree species were generally not associated with younger sites (e.g. American beautyberry (*Callicarpa americana* L.), yaupon (Ilex vomitoria Aiton), eastern redcedar (*Juniperus virginiana*

L.), wax myrtle (*Morella cerifera* (L.) Small), water oak (*Quercus nigra* L.), sweetgum (*Liquidambar styraciflua* L.)) (species lines point away from younger sites). The variability in species composition shown on Axis 2 was generally related to herbaceous species found in the understory (e.g. goatweed (*Croton capitatus* Michx.), Canada goldenrod (*Solidago canadensis* L.), Carolina horsenettle (*Solanum carolinense* L.)). Ordination graphs of species presence data were updated to note whether stands were thinned (Figures 8 and 9). Most thinned Beckville stands and three of four thinned Tatum stands were 1) grouped closely together and 2) plotted among unthinned older and middle-aged sites; these data indicated thinned stands were similar in species composition to each other and to unthinned sites of similar age.

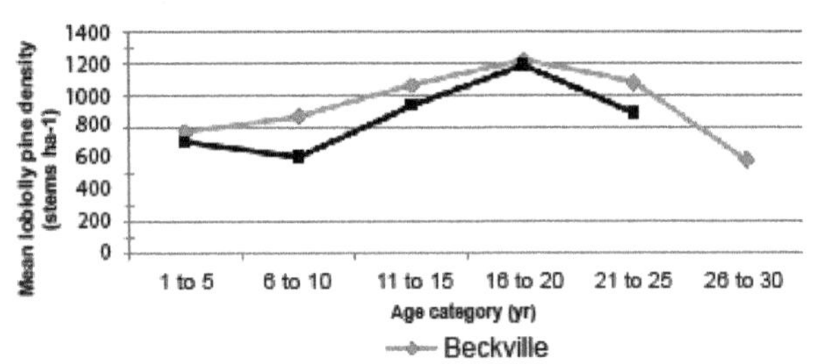

Figure 5: Mean loblolly pine (*Pinus taeda* L.) density (stems ha^{-1}) by age category in reclaimed mine sites planted as loblolly pine plantation stands at Beckville and Tatum lignite coal surface mines (Panola County, Texas, USA).

Species richness

Total species richness for all strata combined was 104 at Beckville and 86 at Tatum (Table 1). Generally, species richness was greatly increased by rarer species that only occurred in 20% or less of all sites at each mine (71 species at Beckville, 60 species at Tatum) (Supplementary Information, Appendix, Table A-2). Species richness increased over time at both mines, especially in the understory, as a greater variety of volunteer species became established (Figure 10 and 11). After 25 years, an increase in mid-story richness was noticeable at Beckville. Over story richness did not increase until approximately two decades after stand establishment due to loblolly pine dominance. Over story species observed with loblolly pine included tree species such as sugarberry (*Celtis laevigata* Willd.), American elm (*Ulmus americana* L.), sweetgum, southern red oak (*Quercus falcata* Michx.) and cherrybark oak (*Quercus pagoda* Raf.). However, though categorized as over story species, it should be noted that their numbers were few and their mean heights were generally at least 10 m less than loblolly pine. Smaller-stature tree species observed in over story included eastern redbud (*Cercis canadensis* L.), eastern redcedar and red maple (*Acer rubrum* L.).

Species importance value

Similarities existed between the two mines in regard to important species with loblolly pine and eastern redcedar being the most important tree species and southern dewberry (*Rubus trivialis* Michx.), Johnsongrass, bahiagrass (*Paspalum notatum* Flueggé) and Canada goldenrod being the most important understory species, especially in

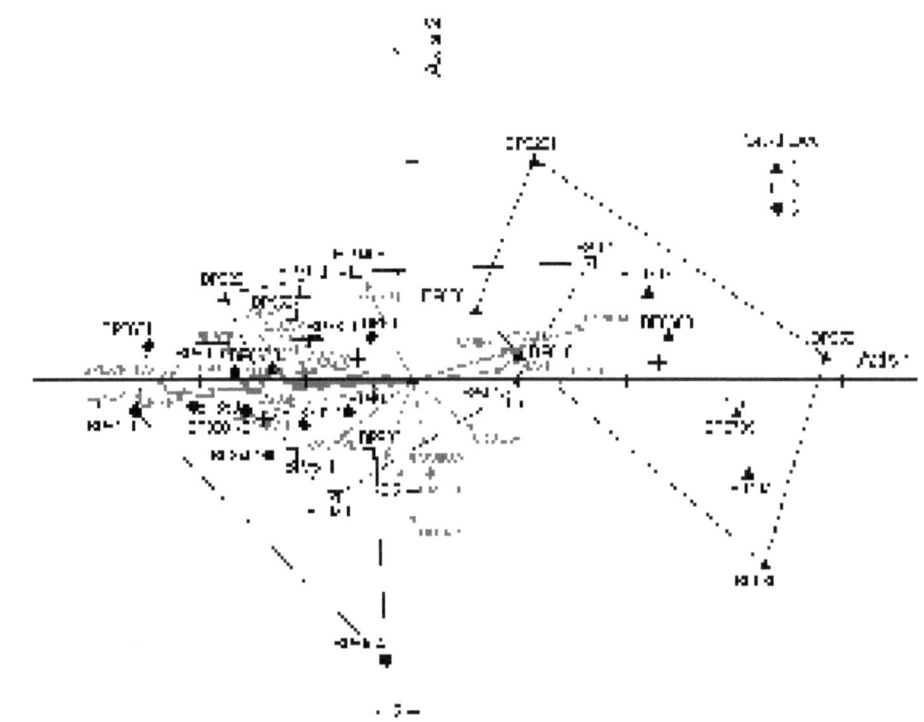

Figure 6: Nonmetric multidimensional scaling (NMS) ordination of reclaimed mine sites planted as loblolly pine (Pinus taeda L.) plantation stands (outlier BP0906 omitted) at Beckville lignite coal surface mine (Panola County, Texas, U.S.A.) based on species presence (i.e., species composition) and including all vegetation strata (understory, midstory, overstory). Stands represented by triangles and stand name (BP=Beckville pine, first two numbers indicate year of stand establishment between 1980 and 2009, last two numbers indicate randomly assigned stand number). Stands with similar species composition plotted relatively close together and dissimilar stands plotted relatively far apart. Species represented by gray lines and species codes (first three letters each of generic name and specific epithet). Species lines represent strength (direction and magnitude) of a species' association with an axis. Site legend: 1=sites aged 1 to 10 years; 2=sites aged 11 to 20 years; 3=sites aged 21 to 30 years.

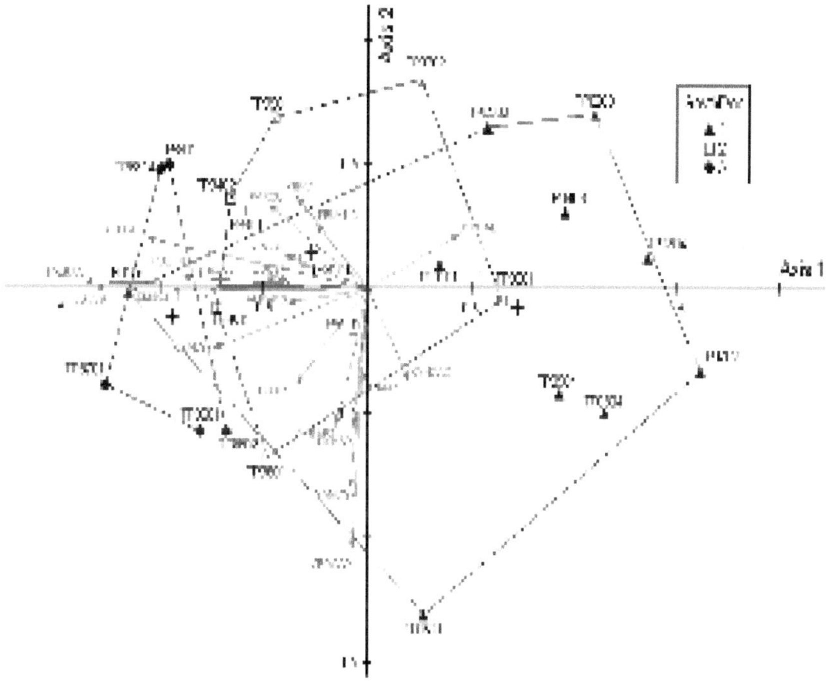

Figure 7: Nonmetric multidimensional scaling (NMS) ordination of reclaimed mine sites planted as loblolly pine (*Pinus taeda* L.) plantation stands (outliers TP9001 and TP0602 omitted) at Tatum lignite coal surface mine (Panola County, Texas, USA) based on species presence (i.e., species composition) and including all vegetation strata (understory, midstory, overstory). Stands represented by triangles and stand names (TP=Tatum pine, first two numbers indicate year of stand establishment between 1986 and 2009, last two numbers indicate randomly assigned stand number). Stands with similar species composition plotted relatively close together and dissimilar stands plotted relatively far apart. Species represented by gray lines and species codes (first three letters each of generic name and specific epithet). Species lines represent strength (direction and magnitude) of a species' association with an axis. Site legend: 1=sites aged 1 to 10 years; 2=sites aged 11 to 20 years; 3=sites aged 21 to 30 years.

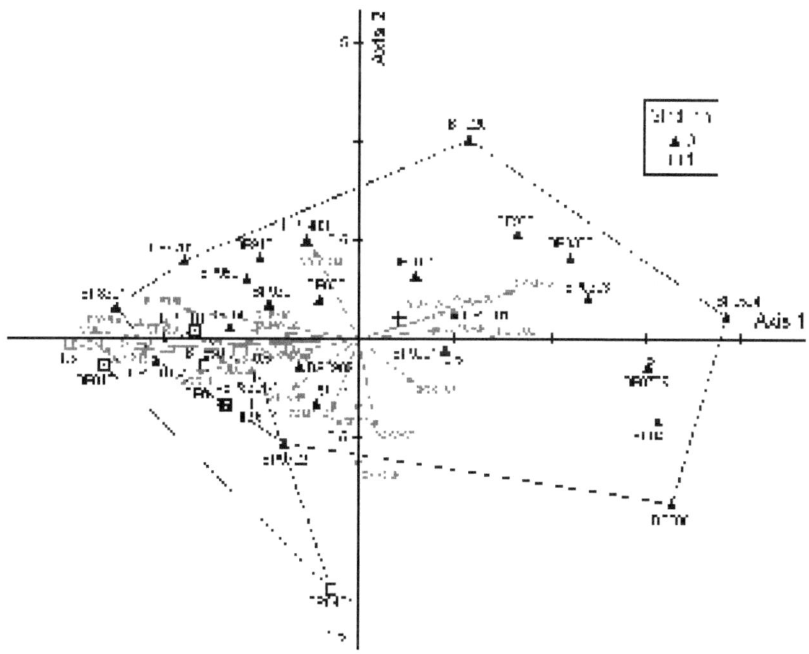

Figure 8: Nonmetric multidimensional scaling (NMS) ordination of reclaimed mine sites planted as loblolly pine (*Pinus taeda* L.) plantation stands (outlier BP0906 omitted) at Beckville lignite coal surface mine (Panola County, Texas, USA) based on thinning activity and species presence (i.e., species composition) and including all vegetation strata (understory, midstory, overstory). Stands represented by triangles and stand name (BP=Beckville pine, first two numbers indicate year of stand establishment between 1980 and 2009, last two numbers indicate randomly assigned stand number). Stands with similar species composition plotted relatively close together and dissimilar stands plotted relatively far apart. Species represented by gray lines and species codes (first three letters each of generic name and specific epithet). Species lines represent strength (direction and magnitude) of a species' association with an axis. Site legend: 0=unthinned; 1=thinned at some point in time after stand establishment.

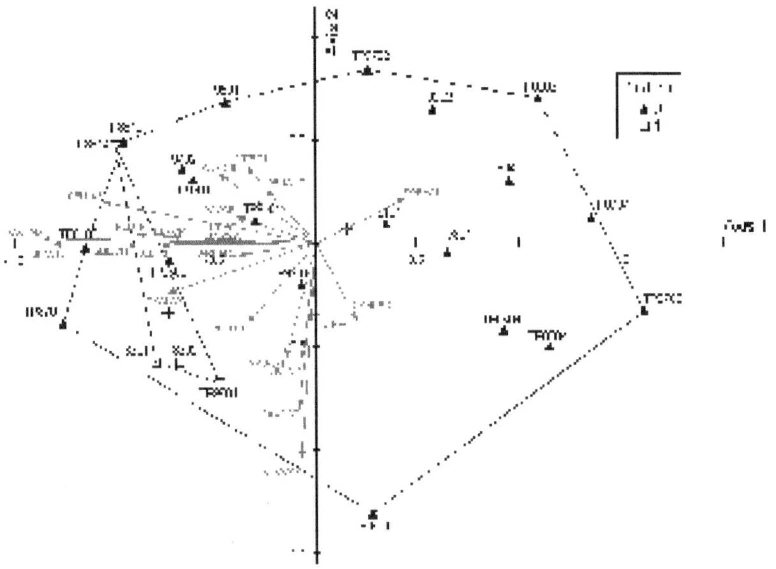

Figure 9: Nonmetric multidimensional scaling (NMS) ordination of reclaimed mine sites planted as loblolly pine (*Pinus taeda* L.) plantation stands (outliers TP9001 and TP0602 omitted) at Tatum lignite coal surface mine (Panola County, Texas, U.S.A) based on thinning activity and species presence (i.e. species composition) and including all vegetation strata (understory, midstory, overstory). Stands represented by triangles and stand name (TP=Tatum pine, first two numbers indicate year of stand establishment between 1986 and 2009, last two numbers indicate randomly assigned stand number). Stands with similar species composition plotted relatively close together and dissimilar stands plotted relatively far apart. Species represented by gray lines and species codes (first three letters each of generic name and specific epithet). Species lines represent strength (direction and magnitude) of a species' association with an axis. Site legend: 0=unthinned; 1=thinned at some point in time after stand establishment.

younger sites (Tables 2 and 3). Several woody shrubs were especially important in the mid-story stratum of middle-aged sites at both mines, including eastern baccharis (*Baccharis halimifolia* L.), yaupon, American beauty berry and wax myrtle. In some cases, woody vine species (e.g. Japanese honeysuckle (*Lonicera japonica* Thunb.), poison ivy) were more important in older stands due to a more developed mid-story and greater understory cover.

Discussion

The open landscape of newly planted loblolly pine plantations on reclaimed mine land was amenable to shade-intolerant herbaceous and grass species but shifted to a closed canopy within two decades after stand establishment, which favored woody species (trees, shrubs, vines). At Beckville, among the 25 species with greatest understory IV (out of 97 species observed in understory) was a mixture of shade-intolerant herbaceous species, woody vines and large tree species, highlighting the shifting species importance over time as a closed canopy forest environment developed. Dominance of woody shrubs in the mid-story at both mines was also indicative of formation of closed canopy forest with multi-layer structure, especially apparent with dominance of shade-tolerant shrubs like American beautyberry and wax myrtle. Important Beckville mid story shrub species varied in cover over time; shade-intolerant eastern baccharis was more abundant in younger stands, and yaupon and wax myrtle were more abundant in shaded older stands.

Over time, species composition became more similar, and species commonly associated with forests in East Texas became successfully established, often within two decades. Given that these plantations were established on sites generally described as dry to mesic uplands, species composition was generally congruent with East Texas ecology, as described by Van Kley [41]. For instance, the assemblage of important species at Beckville (loblolly pine, eastern redcedar, yaupon) and dominance by oaks and blackberry (Rubus sp.) were similar to area ecology. Southern red oak and water oak accounted for 73% of total oak IV for Beckville, which was logical since these species tolerance ranges include drier upland to mesic sites. Tatum species composition did not closely match Beckville because sampled Tatum stands covered a shorter time span (24 years versus 30 years); differences between mine sites over time will likely lessen.

Species richness data indicated that, while loblolly pine was dominant throughout the 30-year period, over time, other over story species made inroads and the mid-story stratum filled in. At Tatum, it is likely that an additional five to ten years of growth in the oldest stands will realize greater over story richness, as was observed in the oldest Beckville stands. Mean mid-story richness increased in the oldest Tatum stands and included species likely to reach over story height in the future (e.g. sugarberry, eastern redcedar, southern red oak). Increased understory richness corresponded with mean loblolly pine height reaching around 10 meters, which occurred between 10 to 15 years after stand establishment; increased shading due to a taller canopy led to occurrence of a greater variety of species in the understory stratum and exclusion of shade-intolerant understory species.

Although ordination graphs based on species presence indicated thinned stands were similar in species composition to each other and to unthinned sites of similar age, further analysis of Beckville site BP8301 revealed vegetative differences between thinned (BP8301-1) and unthinned (BP8301-2) portions of the site. A greater variety of woody vine species was present in the unthinned plot; differences from the thinned plot included understory presence of peppervine (*Ampelopsis arborea* (L.) Koehne), presence of trumpet creeper (*Campsis radicans* (L.) Seem. ex Bureau) in all three strata and over story presence of poison ivy. Eastern baccharis was found in both understory and mid-story of the thinned portion while winged sumac (*Rhus copallinum* L.) was found in understory and mid-story of the unthinned portion;

Common name	Scientific name	Number of stands where observed	Total species importance value
Loblolly pine	*Pinus taeda* L.	30	13,409
Eastern redcedar	*Juniperus virginiana* L.	20	4,299
Yaupon	*Ilex vomitoria* Aiton	20	4,127
Eastern baccharis	*Baccharis halimifolia* L.	19	4,114
Poison ivy	*Toxicodendron radicans* (L.) Kuntze	18	2,879
Southern dewberry	*Rubus trivialis* Michx.	26	2,831
Wax myrtle	*Morella cerifera* (L.) Small	13	2,512
Southern red oak	*Quercus falcata* Michx.	15	2,505
Johnsongrass	*Sorghum halepense* (L.) Pers.	22	2,474
Canada goldenrod	*Solidago canadensis* L.	23	2,471
Sugarberry	*Celtis laevigata* Willd.	15	2,134
Virginia creeper	*Parthenocissus quinquefolia* (L.) Planch.	15	1,921
Bahiagrass	*Paspalum notatum* Flueggé	17	1,885
Woodsorrel	*Oxalis* sp. L.	17	1,858
Coastal bermudagrass	*Cynodon dactylon* (L.) Pers.	16	1,830
Rosette grass	*Dichanthelium* sp. (Hitchc. & Chase) Gould	18	1,829
Blackberry	*Rubus* sp. L.	14	1,767
Japanese honeysuckle	*Lonicera japonica* Thunb.	12	1,732
Saw greenbrier	*Smilax bona-nox* L.	17	1,709
Water oak	*Quercus nigra* L.	14	1,621
Common persimmon	*Diospyros virginiana* L.	13	1,537
American beautyberry	*Callicarpa americana* L.	13	1,511
Little bluestem	*Schizachyrium scoparium* (Michx.) Nash	14	1,439
Goatweed	*Croton capitatus* Michx.	13	1,361
Privet	*Ligustrum* sp. L.	10	1,336
Sweetgum	*Liquidambar styraciflua* L.	7	1,306
Muscadine	*Vitis rotundifolia* Michx.	11	1,107
Peppervine	*Ampelopsis arborea* (L.) Koehne	9	1,106
Sericea lespedeza	*Lespedeza cuneata* (Dum. Cours.) G. Don	10	1,061
Winged elm	*Ulmus alata* Michx.	8	1,050
Camphorweed	*Heterotheca subaxillaris* (Lam.) Britton & Rusby	9	935
Wild carrot	*Daucus carota* L.	8	812
Thistle	*Cirsium* sp. Mill.	8	804

Table 2: Total species importance values (understory, midstory and overstory importance values combined) for plant species observed on more than 20% of reclaimed mine sites planted as loblolly pine (*Pinus taeda* L.) plantation stands aged 1 to 30 years at Beckville lignite coal surface mine (Panola County, Texas, USA).

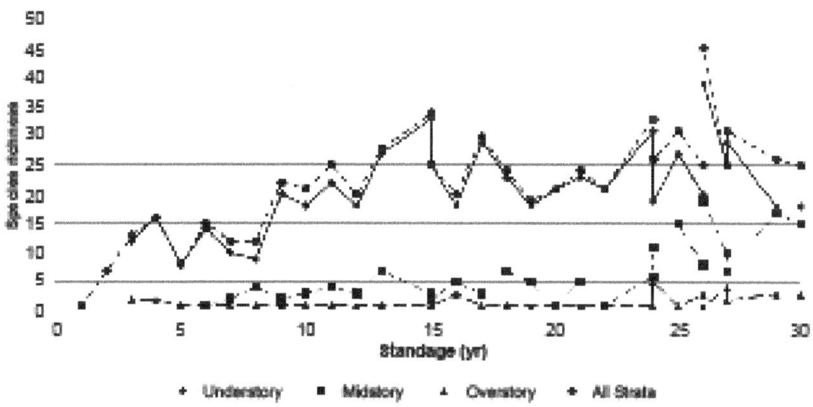

Figure 10: Species richness over time by stratum for reclaimed mine sites planted as loblolly pine (*Pinus taeda* L.) plantation stands aged 1 to 30 years at Beckville lignite coal surface mine (Panola County, Texas, U.S.A).

eastern baccharis is a shrub species more indicative of open and/or disturbed sites while winged sumac tends to be present in more shaded environments. Although several understory species were common to both plots, the thinned plot contained several herbaceous species indicative of open and/or disturbed sites (thistle (Cirsium sp. Mill.), goatweed, everlasting (Gamochaeta sp. Weddell), switchgrass (*Panicum*

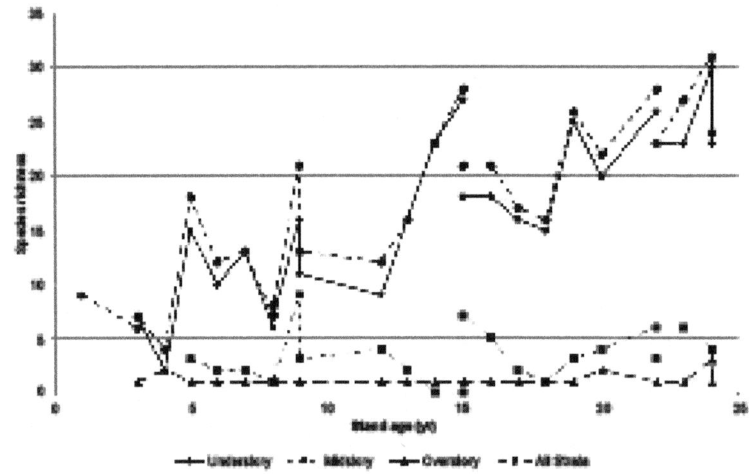

Figure 11: Species richness over time by stratum for reclaimed mine sites planted as loblolly pine (*Pinus taeda* L.) plantation stands aged 1 to 24 years at Tatum lignite coal surface mine (Panola County, Texas, USA).

Common name	Scientific name	Number of stands where observed	Total species importance value
Loblolly pine	*Pinus taeda* L.	26	10,700
Bahiagrass	*Paspalum notatum* Flueggé	23	2,962
Eastern redcedar	*Juniperus virginiana* L.	13	2,461
Johnsongrass	*Sorghum halepense* (L.) Pers.	19	2,442
Southern dewberry	*Rubus trivialis* Michx.	17	2,060
Canada goldenrod	*Solidago canadensis* L.	19	1,989
American beautyberry	*Callicarpa americana* L.	9	1,760
Sugarberry	*Celtis laevigata* Willd.	15	1,703
Wax myrtle	*Morella cerifera* (L.) Small	8	1,611
Yaupon	*Ilex vomitoria* Aiton	10	1,584
Woodsorrel	*Oxalis* sp. L.	15	1,551
Virginia creeper	*Parthenocissus quinquefolia* (L.) Planch.	14	1,551
Poison ivy	*Toxicodendron radicans* (L.) Kuntze	15	1,547
Coastal bermudagrass	*Cynodon dactylon* (L.) Pers.	12	1,462
Saw greenbrier	*Smilax bona-nox* L.	12	1,320
Goatweed	*Croton capitatus* Michx.	9	1,118
Eastern baccharis	*Baccharis halimifolia* L.	8	1,101
Honeylocust	*Gleditsia triacanthos* L.	8	1,100
Sweetgum	*Liquidambar styraciflua* L.	8	1,000
Little bluestem	*Schizachyrium scoparium* (Michx.) Nash	9	999
Peppervine	*Ampelopsis arborea* (L.) Koehne	6	812
Southern red oak	*Quercus falcata* Michx.	6	721
Blackjack oak	*Quercus marilandica* Münchh.	7	704
Blackberry	*Rubus* sp. L.	6	701
Privet	*Ligustrum* sp. L.	6	603
Muscadine	*Vitis rotundifolia* Michx.	6	601

Table 3: Total species importance values (understory, midstory and overstory importance values combined) for plant species observed on more than 20% of reclaimed mine sites planted as loblolly pine (*Pinus taeda* L.) plantation stands aged 1 to 24 years at Tatum lignite coal surface mine (Panola County, Texas, USA).

virgatum L.), bahiagrass, crowngrass (Paspalum sp. L.), little bluestem (*Schizachyrium scoparium* (Michx.) Nash), Canada goldenrod) while the unthinned plot contained several tree and shrub species that could be expected in the understory of a closed canopy forest (sugarberry, common persimmon (*Diospyros virginiana* L.), yaupon, eastern redcedar, water oak).

Data source	Age range (yr)	Number of trees measured (n)	Diameter (cm)				Height (m)			
			Mean	s	Min.	Max.	Mean	s	Min.	Max.
Beckville	6 to 30	2290	19.1	5.3	10.2	39.9	12.9	3.4	4.6	21.2
Tatum	6 to 24	1685	18.3	4.6	10.2	36.1	12.7	3.3	4	23.9
Unmined*	4 to 30	236	21.1	6.1	10.2	38.9	16.6	4	5.7	25.2

Table 4: Comparison of loblolly pine (*Pinus taeda* L.) growth data for reclaimed mine sites planted as loblolly pine plantations (Beckville and Tatum Mines in Panola County, Texas, USA) and unmined, semi-intensive plantation locations in the East Texas region, USA. Mined and unmined plantations were of similar age range and varied management. Diameter measured as dbh (diameter at breast height), and trees smaller than 10.2 cm dbh not included.

Site name		Species richness
Cherokee	CR mean	10.7
	CR-1*	6
(1-year post-thinning)	CR-2*	12.6
	CR-3*	6
	CR-4*	13.4
	CR-5*	15.4
	SU mean	11.8
	SU-1*	11
	SU-2*	12.5
	SU-3*	9.9
	SU-4*	10.3
	SU-5*	15.3
Beckville	Beckville mean	27.4
	BP-9301	30
	BP-9403	20
	BP-9507	25
	BP-9511*	34
	BP-9702	28
Tatum	**Tatum mean**	**21**
	TP-9301	17
	TP-9402	21
	TP-9501	21
	TP-9502	28
	TP-9601*	23
	TP-9702	16

Table 5: Species richness for unmined, thinned, 13 to 17-year-old loblolly pine (*Pinus taeda* L.) plantation stands in Cherokee County, Texas, USA (Ott; Oswald et al.) and reclaimed mine sites planted as loblolly pine plantation stands (thinned and unthinned, 13 to 17 years old) on lignite coal surface mine land (Beckville/Tatum, Panola County, Texas, USA).

At Beckville, loblolly pine was often the only over story species present during the first 25 years after establishment. In most stands older than 25 years, loblolly stems had been thinned, and this allowed richness to increase in both mid-story and over story strata, especially as species like sugarberry, cherrybark oak and eastern redcedar reached over story height. Mid-story density also increased after thinning, given that loblolly pine dominance also suppressed the growth of mid-story species until about 25 years post-establishment, which was when mid-story richness increased. At both mines, decreased over story density in older stands was generally attributed to thinning activity that occurred in 60% of Tatum stands older than 20 years and 100% of Beckville stands 25 years and older. Three of four thinned Tatum stands had similar species composition to each other and were fairly dissimilar to other unthinned stands in their understory composition. These three stands contained unique combinations of herbaceous species not found in unthinned stands (goatweed, everlasting and sedge (Carex sp.) in all three sites and annual ragweed (*Ambrosia artemisiifolia* L.) and rosette grass (Dichanthelium sp.) in two sites); goatweed, everlasting and annual ragweed are commonly found in open and/or disturbed sites.

Loblolly pine growth data were compared to diameter and height of loblolly pine grown on various unmined semi-intensive plantations in East Texas (Table 4) [42]. Although mean diameter and mean height for mined Beckville/Tatum sites were less than unmined sites (approximately 2 to 3 cm less in diameter and 4 m less in height), growth data were encouragingly comparable overall, suggesting that these mined sites produced merchantable trees near par to unmined sites.

A study in Cherokee County, Texas, USA (Piney woods vegetation area) examined biodiversity in 13 to 17 year old loblolly pine plantations that had undergone mid-rotation thinning within the past year [30,31]. As feasible, data of Beckville/Tatum sites and Cherokee sites were compared to ascertain whether vegetative similarities existed between mined and unmined loblolly pine plantations. Out of 108 species/genera observed on Cherokee sites, 36 (33%) were also found in Beckville/Tatum sites. A likely reason for the other two-thirds of Cherokee species lacking in Beckville/Tatum stands was placement of Cherokee plots on tertiary stream slopes that included measurements from more moist soil conditions on mid-slope and slope bottom areas while Beckville/Tatum plots were on upland sites.

Half of the important herbaceous and shrub species/genera at Cherokee were also important at Beckville/Tatum (yaupon, wax myrtle, wood sorrel, Virginia creeper, blackberry, poison ivy, greenbrier (*Smilax sp.* L.)), which revealed some substantial similarity in composition between the mined and unmined plantations. Species richness in both thinned and unthinned Beckville/Tatum sites was higher at both 1-year and 3-years post-thinning than at Cherokee (Table 5). Some of this difference in richness was likely due to the short amount of time since disturbance by thinning activity in Cherokee sites as mean site richness improved by approximately 5 species between 1 and 3 years post-thinning; most 13 to 17-year-old Beckville/Tatum stands had never been thinned.

Data from Ott, Oswald et al., and Coble and Hilpp are only some of the data available on pine growth and community development aspects of loblolly pine plantation ecosystems in East Texas [30,31,42]. Further detailed vegetation community data for loblolly pine plantations on reclaimed Beckville/Tatum mine land in East Texas are available in Christian [32], and other data exist for East Texas pine plantations [26-29]. A meta-analysis or other synthesis of these data is suggested to attain further understanding of these plantation communities and further discern how mined and unmined plantations compare to one another.

Despite the popular application of terms such as “impoverished”, “tree farm” and “biological desert” to plantation forests, it is important to realize that the plantation forest’s state of affairs is hardly bleak. Many studies have provided evidence of divers ecological functions occurring within plantation forests, which is preferable to relatively meager ecological functions provided by deforested and other degraded lands [1,5-8]. The biodiversity of plantation forests spans many taxa and, even if not a defined ecosystem type, plantation forests are functioning communities of flora and fauna. Plantation forests can contribute positively to biodiversity by providing habitat where native organisms can more easily recolonize, adding additional large scale features to surrounding landscape and acting as buffer zones and corridors between natural forest and non-forest. Surface mined lands reclaimed to plantation forests are ideal locations to conduct further research on how biodiversity and other ecosystem services can be improved while maintaining intended economic and ecological purposes.

Conclusions

Overall, reclaimed mine land in the study area was found to be amenable to establishment of productive loblolly pine plantations based on vegetative community characteristics and development over time. Major species were congruent with those common in forest communities of East Texas, and loblolly pine was observed to establish and maintain dominance as was intended in these plantations; neither of these points were observed to be grossly affected by past mining activity. Richness in all three vegetation strata increased over time despite the reclaimed mine sites being initially planted as a loblolly pine monoculture. Pine growth data was found to be fairly comparable to growth data in unmined plantations, supporting the validity of reclaimed mine lands as suitable locations for productive pine plantations. As compared to unmined loblolly pine plantations, reclaimed sites exhibited some similarity in overall vegetative composition, greater plant species richness, and congruence of important herbaceous and shrub species. Further study of the East Texas loblolly pine plantation ecosystem in both mined and unmined settings would be valuable in attaining continued improvement of economic benefits, ecosystem services and biodiversity of this increasingly utilized landscape type.

Acknowledgements

Funding was provided by the Luminant Environmental Research Program. The authors gratefully acknowledge Luminant Generation Company LLC and the Division of Environmental Science, Arthur Temple College of Forestry and Agriculture, Stephen F. Austin State University for their support of this research and Melissa Bozarth and Brittany Compton for their assistance in the field.

References

1. Hartley MJ (2002) Rationale and methods for conserving biodiversity in plantation forests. Forest Ecology and Management 155: 81-95.
2. Food and Agricultural Organization of the United Nations (FAO) (2011) State of the World's Forests 2011. Food and Agriculture Organization of the United Nations, Rome.
3. Zipper CE, Burger JA, Skousen JG, Angel PN, Barton CD, et al. (2011) Restoring forests and associated ecosystem services on Appalachian coal surface mines. Environmental Management 47: 751-765.
4. Skousen J, Zipper CE (2014) Post-mining policies and practices in the Eastern USA coal region. Int J Coal Sci Technol 1: 135-151.
5. Hansen AJ, Spies TA, Swanson FJ, Ohmann JL (1991) Conserving biodiversity in managed forests, lessons from natural forests. Bioscience 41: 382-392.
6. Carey AB, Curtis RO (1996) Conservation of biodiversity: a useful paradigm for forest ecosystem management. Wildlife Society Bulletin 24: 610-620.
7. Stephens SS, Wagner MR (2007) Forest plantations and biodiversity: a fresh perspective. Journal of Forestry 105: 307-313.
8. Brockerhoff EG, Jactel H, Parrotta JA, Quine CP, Sayer J (2008) Plantation forests and biodiversity: oxymoron or opportunity?. Biodiversity and Conservation 17: 925-951.
9. Brenner FJ, Werner M, Pike J. 1984. Ecosystem development and natural succession in surface coal mine reclamation. Minerals and the Environment 6: 10-22.
10. Holl KD, Cairns J (1994) Vegetational community development on reclaimed coal surface mines in Virginia. Bulletin of the Torrey Botanical Club 121: 327-337.
11. Burger JA, Torbert JL, Schoenholtz SS, Kreh RE (2000) Performance of three pine species after eleven years on pre- and post-law mined sites in Virginia. Proc of the 17th Annual Meeting of the American Society for Mining and Reclamation. pp: 226-230.
12. Rodrigue JA, Burger JA (2001) Forest and site productivity on pre-SMCRA mined land. Paper presented at the 2001 National Meeting of the American Society for Surface Mining and Reclamation, Albuquerque, New Mexico, June 3-7, 2001. Pub. by ASSMR, 3134 Montavesta Rd., Lexington, KY 40502.
13. Holl KD (2002) Long-term vegetation recovery on reclaimed coal surface mines in the eastern USA. Journal of Applied Ecology 39: 960-970.
14. Skousen JG, Call CA, Knight RW (1990) Natural revegetation of an unclaimed lignite surface mine in east-central Texas. The Southwestern Naturalist 35: 434-440.
15. Gorsira B, Risenhoover KL (1994) An evaluation of woodland reclamation on strip-mined lands in east Texas. Environmental Management 18: 787-793.
16. Westerman CA (1997) Vegetation trends in reclaimed areas at Gibbons Creek Lignite Mine, Grimes County, Texas. Proceedings of 1997 National Meeting of the American Society for Surface Mining and Reclamation, Austin, Texas, May 10-15, 1997: 77-81.
17. Bryson HL (1973) Early survival and total height, and foliar analyses of eleven tree species grown on strip-mine spoils in Freestone County, Texas. MS Thesis. Stephen F. Austin State University, Nacogdoches, Texas.
18. Mask CL (1983) Row seeding of forest tree species on lignite spoils in east Texas. MS thesis. Stephen F. Austin State University, Nacogdoches, Texas.
19. Kee DD (1984) The effect of cover crop and fertilizer rate on the growth and survival of loblolly pine in east Texas mine spoil. MS thesis. Texas A&M University, College Station, Texas.
20. Wood GA (1985) Two-year survival and growth of loblolly pine seedlings from two Texas seed sources on lignite minesoils. MS thesis. Stephen F. Austin State University, Nacogdoches, Texas.
21. Shupe MA (1986) The effects of nitrogen and phosphorus fertilizer on a young loblolly pine plantation on lignite mine spoil in east Texas. MS thesis. Stephen F. Austin State University, Nacogdoches, Texas.
22. Toups BG (1986) Comparison of site quality for loblolly pine on selected mined and non-mined soils in Panola County, Texas. MS thesis. Stephen F. Austin State University, Nacogdoches, Texas.
23. Musgraves JR (1995) Survival of longleaf pine (Pinus palustris Mill.) on mine reclamation sites using different regeneration methods. MS thesis. Stephen F. Austin State University, Nacogdoches, Texas.
24. McGuire MA (1998) Effects of stock type, fall nursery fertilization and ectomycorrhizal inoculation on survival of longleaf pine (Pinus palustris Mill.) seedlings planted on lignite minespoil. MS thesis. Stephen F. Austin State University, Nacogdoches, Texas.
25. Railroad Commission of Texas (2015) Annual coal production (webpage link to Excel file). Available online at last accessed Nov. 6, 2015.
26. Nixon ES, Willett RL, Butts MC, Burandt CL (1981) Early seral development following partial clearcutting in east Texas. Texas Journal of Science 33: 26-32.
27. Burnett FE (1987) The role of non-planted vegetation within non-old-field loblolly and slash pine plantations in east Texas. MS thesis. Stephen F. Austin State University, Nacogdoches, Texas.
28. Nixon ES, Brooks AR (1991) Species diversity following clearcutting in eastern Texas. Texas Journal of Science 43: 399-403.
29. Taylor RM (1996) Successional trends and competition within unmanaged loblolly and slash pine plantations of east Texas. MS thesis. Stephen F. Austin State University, Nacogdoches, Texas.
30. Ott MG (2002) A comparison of vegetative biodiversity in the understory of managed pine plantations treated with fertilizer, fire and herbicides in East Texas. Ph. D. dissertation, Stephen F. Austin State University, Nacogdoches, Texas, United States. 73.

31. Oswald BP, Ott B, Williams HM, Farrish KW, Van Kley JE (2009) Understory vegetative diversity of post-thinned pine plantations treated with fertilizer, fire and herbicide in East Texas. International Journal of Forestry Research Volume 2009, Article ID 572480, 6.

32. Christian CL (2013) Vegetative community development on reclaimed lignite coal mine land in east Texas. MS thesis. Stephen F. Austin State University, Nacogdoches, Texas.

33. Gould FW (1962) Texas plants, a checklist and ecological summary. The Agricultural and Mechanical College of Texas, Texas Agricultural Experiment Station.

34. Diggs GM, Lipscomb BL, Reed MD, O'Kennon RJ (2006) Illustrated Flora of East Texas, Vol. 1: Introduction, Pteridophytes, Gymnosperms, and Monocotyledons. Botanical Research Institute of Texas, Fort Worth, Texas.

35. Dolezel R (1975) Soil Survey of Panola County, Texas. United States Department of Agriculture, Soil Conservation Service, in cooperation with Texas Agricultural Experiment Station.

36. Stohlgren TJ, Falkner MB, Schell LD (1995) A modified-Whittaker nested vegetation sampling method. Vegetatio 117: 113-121.

37. National Institute of Invasive Species Science (2010) Updated May 25, 2010. Modified Whittaker Plot. Accessed June 4, 2010.

38. United States Department of Agriculture, Natural Resources Conservation Service (2013) The PLANTS Database National Plant Data Team, Greensboro, NC 27401-4901 USA.

39. Barbour MG, Burk JH, Pitts WD, Gilliam FS, Schwartz MW (1999) Terrestrial plant ecology. 3rd edn, Benjamin/Cummings, Menlo Park.

40. McCune B, Mefford MJ (2011) PC-ORD. Multivariate analysis of ecological data. Version 6.0. MjM Software, Gleneden Beach, Oregon.

41. Van Kley J (2006) The Pineywoods. Illustrated Flora of East Texas, Vol. 1: Introduction, Pteridophytes, Gymnosperms, and Monocotyledons. Botanical Research Institute of Texas, Fort Worth, Texas.

42. Coble DW, Hilpp K (2006) Compatible cubic-foot stem volume and upper-stem diameter equations for semi-intensive plantation grown loblolly pine trees in East Texas. South. J. Appl. For. 30: 132-141.

Bark Beetle Demography in Sierra Nevada Mixed Conifer: Variability and Influencing Factors

Walker RF*, Swim SL, Fecko RM, Johnson DW and Miller WW

Department of Natural Resources and Environmental Science, University of Nevada, Reno, Nevada, USA

Abstract

Multiple forest health variables were examined in an eastern Sierra Nevada mixed conifer stand, principal among them bark beetle demography. As indicated by pitch tube counts, California white fir *(Abies concolor var. lowiana* [Gord.] Lemm.*)*, the predominant species, was colonized prodigiously compared to Jeffrey pine *(Pinus jeffreyi* Grev. and Balf.*)* and sugar pine *(Pinus lambertiana* Dougl.*)*. Across host species, pitch tube abundance was positively related to white fir prevalence and negatively related to tree species diversity. Bark beetles preferentially attacked trees of small to medium DBH in white fir and sugar pine. Attack intensity in small Jeffrey pine was positively correlated with stand basal area and biomass and that in sugar pine of medium size was positively correlated with basal area also. Minor mistletoe infestations were detected in white fir along with incense-cedar *(Libocedrus decurrens* Torr.*)* and California red fir *(Abies magnifica* A. Murr.*)*, with bark beetle colonization in the smallest and largest white fir positively related to the percentage of this fir thus infested. Mortality in white fir and Jeffrey pine irrespective of size was positively correlated with the pitch tube count in the largest white fir regarding the former and with total tree count regarding the latter.

Keywords: Forest health; Bark beetles; Dwarf mistletoe; White fir; Jeffrey pine; Sugar pine; Fir engraver; Jeffrey pine beetle; Mountain pine beetle

Introduction

Prevalent among vegetative communities at the mid elevations of the Sierra Nevada is a mixed conifer forest cover type that includes among its variants one on the eastern slopes featuring California white fir, Jeffrey pine, incense-cedar, and sugar pine along with occasional California red fir where cold air drainages prevail [1]. The silvics of these species differ markedly, perhaps most prominently in their tolerance of shade with white fir the most tolerant overall although it is presumed to be only marginally more so than red fir, Jeffrey pine the least tolerant, and incense-cedar and sugar pine considered to be intermediate regarding this characteristic [2]. The tolerance classification for white fir is reflected in its ability to regenerate profusely and persist in the shade of an overstory canopy to greater extent than the other constituents in this cover type, resulting in a propensity for it to increasingly dominate the composition of many mixed conifer stands [3,4], although this compositional shift also reflects a legacy that includes selective harvesting of yellow and white pines during the Comstock mining era [5] and fire exclusion during much of the 20th century [6,7]. Among the concerns over this trend is a perception that forest health deteriorates with overabundance of white fir in mixed conifer stands, particularly on sites where moisture availability is a frequent inducer of stress [8]. To a substantial degree, this reflects a propensity for this species to succumb to attack by the Scolytine fir engraver *(Scolytus ventralis* Le Conte*)*, a bark beetle against which the effectiveness of its resin-based defense mechanisms is questionable [9-11]. Additionally, infestation of white fir by the parasitic fir dwarf mistletoe *(Arceuthobium abietinum* [Engelm.] Hawksw. and Wiens*)*, a common occurrence, can imperil the vigor of the host such that its susceptibility to fir engraver attack is exacerbated [8].

Presented here is a study of bark beetle demography encompassing comparisons of the various insect and tree host species associations residing in an eastern Sierran mixed conifer stand. Also examined were interactions between an array of mensurational and forest health factors and bark beetle colonization.

Materials and Methods

Study site

The stand chosen for study is naturally regenerated, second growth, uneven-aged Sierra Nevada mixed conifer located on the USDA Forest Service Lake Tahoe Basin Management Unit (39.22° N, 120.10° W). Consisting of approximately 8.1 ha, the elevation of this eastern Sierran site is 2050 m, the aspect is generally east, the slope averages 7%, and the average annual precipitation is 81 cm, predominantly snowfall [12]. The soils are derived from volcanic parent material, exceedingly rocky, and of the Jorge-Tahoma Association [13]. Based on dominant and codominant crown class site trees averaging 162 years in age [14], the site quality is class IV according to the Dunning [15] site classification system for Sierra Nevada mixed conifer.

Data collection

Stand attributes were ascertained by measuring trees of pole size and larger (stems ≥ 10.2 cm DBH) within 20 permanent 0.04-ha circular plots established in a square pattern extending over the site in its entirety. All trees in these plots were measured for total height, DBH, and live crown length and tallied by species. Included were free standing dead trees, defined as those exhibiting no live crown, tallied accordingly. Subsequently, tree counts were summed by species and

***Corresponding author:** Walker RF, University of Nevada, Department of Natural Resources and Environmental Science, 1664 North Virginia Street, Reno, NV 89557, USA, E-mail: walker@cabnr.unr.edu

overall within each plot and then again with dead stems segregated, the percentages by species and of dead stems by species as well as overall were determined, and the number of species within each plot was ascertained as a diversity indicator. Also, tree heights and crown lengths were used to calculate live crown percentage, DBH values were used to derive quadratic mean diameter by plot according to the Curtis and Marshal [16] formula, total tree counts and quadratic mean diameters were used to calculate plot basal area according to the formula of Davis et al. [17], and the above-ground tree biomass by species and overall within each plot was determined using the formulas of Gholz et al. [18] and Ter-Mikaelian and Korzukhin [19]. Ultimately, all tree counts along with basal area and biomass values were expanded to reflect an area of 1.0 ha.

Bark beetle prevalence was quantified by counting the pitch tubes on all stand constituents included in the mensurational measurements of each plot. This was accomplished by visually dividing the bole surface into vertically oriented quadrants, counting the pitch tubes in each quadrant, and then totaling the counts for each tree. With ocular aids used as needed, the entire bole length was included in these counts. Principle bark beetle species were identified by the observation of adult and larval forms as well as pitch tube and gallery characteristics as described by Furniss and Carolin [20]. For the purpose of expressing pitch tube counts as the quantity per unit of bole surface area, tree height and DBH were used as approximations of the lateral length and base diameter, respectively, in the geometric formula for the lateral surface area of a right cone, and the counts of individual trees were then divided by their bole surface area thus approximated. Abundance expressed on the basis of the count per unit surface area compensates for the count per tree measure to sometimes overstate attack severity simply because larger trees have more bole surface available to colonize. Nevertheless, for both measures of abundance, values were averaged across and within tree species by plot. Additionally, to better elucidate the influence of tree size on beetle colonization, all stems were segregated by species into six DBH classes, each representing a range inclusive of the low and high values as follows: Class 1, 10.2-20.2 cm; Class 2, 20.3-30.3 cm; Class 3, 30.4-40.4 cm; Class 4, 40.5-50.5 cm; Class 5, 50.6-60.6 cm; and Class 6, ≥ 60.7 cm. Also quantified on all trees included in the mensurational and bark beetle inventories was mistletoe infestation using the Hawksworth [21] rating system with identification of the specific parasitic species based on the descriptions of Scharpf and Hawksworth [22]. Once quantified, the ratings were averaged by species within individual plots and the percentage of stems infested with mistletoe by host species was determined.

Statistical analysis

For purposes of assessing how closely sample means involving compositional and mensurational measures reflected stand attributes, the standard error of the mean (SE) and 95% confidence limits (CL) were calculated from the pertinent plot means and sums, and these statistics were also calculated for the bark beetle demography variables that spanned across the principal tree species represented in the subject stand. Data pertaining to tree mortality, mistletoe infestation, and bark beetle colonization both across and within the DBH classes were subjected to one-way analysis of variance (ANOVA) to test for the effect of host tree species, while within species, one-way ANOVA was also employed to test for the effect of DBH class on beetle colonization. For every ANOVA, effects were considered significant only when $p \leq 0.05$ according to the F test. Subsequently, differences among means were evaluated using the least significant difference (LSD) test with $\alpha=0.05$, and the SE was calculated to provide an additional indication of the variation among the values from which each mean was derived.

Additional statistical analysis consisted of an extensive array of simple linear regression models used to investigate relationships between variables selected as particularly pertinent to the study. These were divided into five subsets, hereafter denoted the mensuration, species representation, mortality, mistletoe, and biomass subsets. The mensuration subset consisted of models incorporating all possible combinations of tree height, DBH, live crown length and percentage, basal area, and total tree count as independent variables with pitch tube counts per tree and per unit bole surface area across and within species, and within each species by DBH class, along with the mistletoe rating by species, the percentage of trees infested with mistletoe by host species, and the dead tree counts and percentages by species and overall serving as dependent variables. For the species representation subset, the independent variables consisted of the percentages of each species plus species diversity among stand constituents while dependent variables were again those identified above regarding the mensuration subset. The mortality subset featured the dead tree counts and percentages by species as the independent components with the overall dead tree count serving as its dependent counterpart and, matched by species, models with pitch tube counts per tree and per unit bole surface area, both combined and by DBH class, as the former with the dead tree counts and percentages constituting the latter. In the mistletoe subset, where independent and dependent variables were also matched by species, the Hawksworth rating and the percentage of trees infested served as the former and were paired with pitch tube counts per tree and per unit bole surface area, both across and within DBH classes, along with the dead tree count and percentage. As for the biomass subset, independent variables consisted of total biomass by species and overall while the dependent variables were the bark beetle and mistletoe measures included as such in the mensuration subset as noted above. With these models, however, the independent and dependent components were not confined to matches within species. For all regression subsets, models were considered significant only when $p \leq 0.05$ according to the F test, and all statistical analyses were performed using SAS Version 9.3 (SAS Institute, Inc., Cary, NC).

Results

Stand characteristics

Over three-quarters of the trees tallied in the inventory of the subject stand were white fir with neither Jeffrey pine, incense-cedar, sugar pine, nor red fir, the remaining species, constituting as much as one-tenth of the its composition (Table 1). Nevertheless, Jeffrey pine, the second most common species, had more than twice the representation of red fir, the least common stand constituent. Mensurationally, the stand was young sawtimber with approximately one-half of tree height supporting live crown on average and it was moderately stocked for Sierran mixed conifer (Table 2). Regarding above-ground biomass, white fir dry weight (mean=249,538.5, SE=29,071.8, 95% CL=± 60,847.9 kg ha^{-1}) greatly exceeded that of all other stand constituents, most apparently red fir (mean=6,159.3, SE=3,172.8, 95% CL=± 6,640.7 kg ha^{-1}) but also Jeffrey pine (mean=29,775.5, SE=9,860.1, 95% CL=± 20,637.4 kg ha^{-1}), incense-cedar (mean=23,536.8, SE=9,631.9, 95% CL=± 20,159.8 kg ha^{-1}), and sugar pine (mean=44,802.9, SE=15,126.0, 95% CL=± 31,659.0 kg ha^{-1}).

ANOVA revealed the species effect on dead tree count to be

Composition (%)					
Statistic	White fir	Jeffrey pine	Incense-cedar	Sugar pine	Red fir
Mean	76.73	8.63	5.45	5.42	3.77
SE	0.04	0.03	0.02	0.01	0.02
95% *CL* (±)	8.05	5.47	3.63	2.41	3.63

[1]Data were collected in 20 plots (*n*=20) from trees ≥ 10.2 cm DBH.

Table 1: Species composition of a mixed conifer stand in the Lake Tahoe Basin selected for assessment of forest health indicators[1].

Statistic	Height (m)	DBH (cm)	Live crown (m)	Live crown (%)	Basal area (m^2 ha^{-1})	Total trees (stems ha^{-1})
Mean	14.13	38.09	7.13	50.10	57.06	530
SE	0.56	1.90	0.28	1.56	3.84	42
95% *CL* (±)	1.18	3.98	0.58	3.26	8.04	88

[1]Data were collected in 20 plots (*n*=20) from trees ≥ 10.2 cm DBH.

Table 2: Mensurational characteristics of a mixed conifer stand in the Lake Tahoe Basin selected for assessment of forest health indicators[1].

significant (p=0.0003), and the LSD test indicated that a relatively high count for white fir significantly exceeded low ones for Jeffrey pine, incense-cedar, and especially sugar pine for which standing dead stems were entirely absent, while the red fir count was the second highest numerically and also surpassed that of the latter species (Table 3). Per hectare, standing dead stems of white fir were 11.0× more abundant than those of Jeffrey pine and 16.5× more prevalent than those of incense-cedar, while the multipliers for red fir in relation to this pine and incense-cedar were 8.3× and 12.5×, respectively. Although unaccompanied by a significant effect as derived from ANOVA, the standing dead percentage was highest overall in red fir and it surpassed that of either Jeffrey or sugar pine according to the LSD test, while white fir and incense-cedar exhibited intermediate, and nearly equivalent, mortality percentages. Proportionally, nevertheless, the mortality rate apparent at inventory was more than twice in red fir that for any of the other species.

Mistletoe infestation

Although mistletoe infestation was exceedingly light overall, ANOVA revealed a significant species (p=0.0067) effect on the Hawksworth ratings (Table 4). In accordance, the LSD test indicated a rating for the white fir that differed significantly from those in either Jeffrey or sugar pine, a reflection of both the relative abundance of fir dwarf mistletoe in the former and the complete absence of the western dwarf mistletoe *(Arceuthobium campylopodum* Engelm.) and sugar pine dwarf mistletoe *(Arceuthobium californicum* Hawksw. and Wiens) in the two pines, respectively, while the ratings for incense-cedar, which reflect infection by its true mistletoe *(Phoradendron libocedri* [Engelm.] Howell), and red fir, another host of the fir dwarf mistletoe *(A. abietinum),* were intermediate in value. Also differing among host species were the percentages of trees in which the various mistletoes were present, led by white and red fir in which infestations occurred most often as compared to Jeffrey and sugar pine where none occurred, while incense-cedar again assumed an intermediate position.

Bark beetle demography

Based on adult and larval forms as well as pitch tube and gallery characteristics, the fir engraver was confirmed as the bark beetle species colonizing white and red fir, the Jeffrey pine beetle *(Dendroctonus jeffreyi* Hopkins) was the principle bark beetle species infesting its namesake host tree, and the mountain pine beetle *(Dendroctonus ponderosae* Hopkins) was the assailant of sugar pine. Because pitch tubes were entirely absent on incense-cedar and red fir occurred in only a small minority of the inventory plots, bark beetle demography measures specific to these two species were omitted from the analyses undertaken in this study and hereafter in this report. Collectively, and based on the remaining species, namely white fir, Jeffrey pine, and sugar pine, the pitch tube count averaged 54.77 per tree (SE=5.44, 95% CL=± 11.38) and 11.85 per m^2 of bole surface area (SE=1.33, 95% CL=± 2.78).

Irrespective of DBH class, pitch tube counts per tree and per unit bole surface area were significantly influenced by species (both p ≤ 0.0001) according to ANOVA, and the LSD test divulged that the counts on white fir significantly exceeded those on either Jeffrey or sugar pine according to both abundance measures (Table 5). By way of further comparison between this fir and the two pines, the count per tree on the former was 13.6× that on Jeffrey pine and 25.6× that on sugar pine, while on a surface area basis, the count on the fir was 32.9× and 51.7× those on Jeffrey and sugar pine, respectively. When considered by DBH class, ANOVA revealed the species effect to persist regardless of tree size according to the count per tree measure, specifically within Class 1 (p=0.0014), Class 2 (p ≤ 0.0001), Class 3 (p=0.0002), Class 4 (p=0.0011), Class 5 (p=0.0363), and Class 6 (p ≤ 0.0001), while with the exception of Class 5, this influence was again apparent based on the counts per unit surface area, specifically within Class 1 (p=0.0017), Class 2 (p=0.0002), Class 3 (p=0.0005), Class 4 (p=0.0008), and Class 6 (p=0.0036). To a large extent, the LSD test again confirmed here the propensity of white fir to accumulate pitch tubes in comparative abundance, disclosing as significant the disparities between its counts by both measures and those in either pine species in all except Class 5 wherein its count per tree surpassed that of sugar pine nonetheless. Within the individual diameter classes, the greatest numerical discrepancy between white fir and Jeffrey pine, regardless of abundance measure, was found in Class 1 where the count per tree for the fir was 71.0× that for the pine and the count per unit surface area was 64.3× for the former compared to the latter. However, the diameter class within which the largest numerical difference between white fir and sugar pine existed varied according to the measure considered, as it was found in

Species	Dead Trees (stems ha^{-1})	Dead Trees (%)
White fir	33^{a} (8)	8.6^{ab} (2.3)
Jeffrey pine	3^{bc} (3)	1.1^{b} (1.1)
Incense-cedar	2^{bc} (2)	9.1^{ab} (9.1)
Sugar pine	0^{c} (0)	0.0^{b} (0.0)
Red fir	25^{ab} (13)	21.1^{a}(12.1)

[1]Within each measure, means sharing a common letter do not differ significantly at α=0.05 according to the LSD test; each mean is based on values from 20 or fewer plots (n ≤ 20) depending on the presence of trees of the pertinent species within individual plots. [2]The SE of each mean is indicated parenthetically.

Table 3: Tree mortality by species in a mixed conifer stand of the Lake Tahoe Basin selected for assessment of forest health indicators[1,2].

Species	Mistletoe (Hawksworth rating)	Infested Trees (%)
White fir	0.24^{a} (0.07)	14.2^{a} (3.2)
Jeffrey pine	0.00^{b} (0.00)	0.0^{b} (0.0)
Incense-cedar	0.11^{ab} (0.07)	10.6^{ab} (7.2)
Sugar pine	0.00^{b} (0.00)	0.0^{b} (0.0)
Red fir	0.06^{ab} (0.04)	18.5^{a} (16.4)

[1]Within each measure, means sharing a common letter do not differ significantly at α=0.05 according to the LSD test; each mean is based on values from 20 or fewer plots (*n* ≤ 20) depending on the presence of trees of the pertinent species within individual plots. [2]The *SE* of each mean is indicated parenthetically.

Table 4: Mistletoe infestation by species in a mixed conifer stand of the Lake Tahoe Basin selected for assessment of forest health indicators[1,2].

	DBH class						
Species	Combined	Class 1	Class 2	Class 3	Class 4	Class 5	Class 6
			Pitch tube counts per tree				
White fir	67.69[a] (5.90)	35.51[a] (5.15)	76.71[a] (6.95)	97.85[a] (10.86)	90.57[a] (14.45)	57.73[a] (9.31)	43.28[a] (5.66)
Jeffrey pine	4.99[b] (1.59)	0.50[b] (0.32)	3.90[b] (1.42)	6.17[b] (0.50)	9.65[b] (4.85)	10.00[ab] (–)	3.13[b] (2.24)
Sugar pine	2.64[b] (0.73)	0.67[b] (0.33)	4.50[b] (2.50)	4.80[b] (1.93)	0.75[b] (0.48)	1.50[b] (1.50)	2.60[b] (0.51)
			Pitch tube counts per m² of bole surface area				
White fir	14.48[a] (1.34)	21.85[a] (3.22)	17.38[a] (1.74)	10.72[a] (1.30)	6.74[a] (1.06)	9.80[a] (5.48)	2.35[a] (0.48)
Jeffrey pine	0.44[b] (0.14)	0.34[b] (0.24)	0.74[b] (0.27)	0.75[b] (0.06)	0.56[b] (0.27)	0.51[a] (–)	0.10[b] (0.07)
Sugar pine	0.28[b] (0.08)	0.42[b] (0.21)	0.70[b] (0.09)	0.48[b] (0.20)	0.06[b] (0.04)	0.08[a] (0.08)	0.07[b] (0.01)

[1]Within each Table component, means sharing a common letter do not differ significantly at α=0.05 according to the LSD test; each mean is based on values from 20 or fewer plots ($n \leq 20$) depending on the presence of trees of the pertinent species and/or class within individual plots. [2]The *SE* of each mean is indicated parenthetically. [3]Denoted parenthetically, "–" indicates that the *SE* was incalculable due to insufficient sample size.

Table 5: Pitch tube prevalence by prominent tree species across and within DBH classes denoted as counts per tree and per unit of bole surface area in a mixed conifer stand of the Lake Tahoe Basin selected for assessment of forest health indicators[1,2,3].

Class 4 regarding the counts per tree, where that in the fir was 120.8× that in this pine, while on a count per unit area basis, it occurred in Class 5, where a multiplier of 122.5× prevailed.

Providing greater illumination of possible size preferences among potential host trees, ANOVA revealed a significant diameter class effect extending to both the count per tree and per unit surface area (each $p \leq 0.0001$) regarding white fir (Table 6). As distinguished by the LSD test, the former measure was higher in Class 3 and 4 than in Class 1, 5, or 6 and was higher in Class 2 than Class 1 or 6 as well. At the most extreme, the count per tree was 2.8× greater in Class 3 than in Class 1, representing the highest and lowest values numerically for this abundance measure, respectively, and was 2.1× higher in Class 4 than in Class 6, which exhibited the second highest and lowest values, respectively. Lower in magnitude, the disparity between Class 2 and either Class 1 or 6 was nonetheless substantial, with the count in the former ≥ 1.8× that in either of the latter. As for pitch tube counts based on bole surface area in white fir, the LSD test denoted as significant the disparities between higher values in Class 1 and 2 than those in all of the other diameter classes plus that between Class 3 and Class 6, and it is perhaps noteworthy that with only a minor deviation involving Class 4 and 5, counts by this measure declined with increasing class designation. Furthermore, the magnitude of the largest difference between classes here was considerably greater than that derived from the count per tree measure, as the value for Class 1 was 9.3× that for Class 6, while other comparisons largely paralleling those detailed above produced substantial discrepancies nonetheless with a count in Class 2 that was 2.6× that in Class 4 and one in Class 3 that was 4.6× the value in Class 6. Unlike the pronounced tree size influence on beetle colonization in white fir, no diameter class effect on pitch tube abundance in Jeffrey pine was disclosed by ANOVA regardless of the measure assessed and the LSD test did not detect any significant differences among the classes for either measure regarding this species. However, specific to the count per unit bole surface area, diameter class imposed an effect on colonization in sugar pine ($p=0.0465$), and the LSD test divulged that this count was significantly higher in Class 2 than in Class 4, 5, or 6 and was such in Class 3 than in Class 4 or 6. Numerically, the magnitude of these disparities was substantial, with a value in Class 2 that was ≥ 8.8× to as much as 11.7× those in Class 4 through 6 and one in Class 3 that was ≥ 6.9× those in Class 4 and 6. Although lacking in a significant influence specified by ANOVA, the LSD test nevertheless detected two significant differences between diameter classes pertaining to the count per tree measure, namely a higher value in Class 3 than in either Class 1 or 4.

Relationships of stand health variables

The first subset of the regressions computed in this study, which was concerned with the relationships of forest health variables to mensurational measures pertaining to tree dimensions and stand density, produced 12 significant models (Table 7). Of those concerned with tree dimensions, all involved live crown with its length the independent variable to which the bole surface area-based pitch tube count across species and DBH classes was negatively related, and its percentage that to which the overall dead tree count and percentage plus the white fir dead tree count and percentage were each negatively related as well. Overall, these models were of modest strength, with a range of 23% to 44% of the variation in the dependent variables explained by that in the independent variables. For models involving stand density measures, the DBH Class 1 surface area-based pitch tube count in Jeffrey pine, the Class 3 sugar pine count per tree, the Hawksworth rating and percentage of infested trees in incense-cedar, and the red fir standing dead percentage were all positively related to basal area, while the Jeffrey pine standing dead and dead percentage were related to the total tree count, and again positively so. These models featured some of the strongest correlations computed in the study, with 49% to 80% of the variation in the dependent variables accounted for by them.

Focusing on host tree prevalence and the diversity therein, the second subset of regressions yielded six significant models (Table 7). Three of these featured the percentage of white fir among stand constituents as the independent component, to which the overall pitch tube count per tree and per unit surface area irrespective of DBH class along with the white fir standing dead count were all positively related. With negative correlations exclusively, however, the surface area-based pitch tube count across host species and DBH classes plus this same count within Class 2 and 6 white fir were related to host species diversity. Collectively, the models in the second subset explained from 20% to 52% of the variation in the dependent variables.

The third regression subset was concerned with tree mortality exclusively regarding dependent variables although with representation of its measures occurring in different form among its independent variables also, and it produced five significant models (Table 7). Four of these coupled, in all possible combinations and with positive correlations prevailing throughout, overall dead tree count and percentage as the dependent components with the white fir standing dead count and percentage as the independent ones, while the fifth

	Species		
DBH class	White fir	Jeffrey pine	Sugar pine
		Pitch tube counts per tree	
1	35.51^{c} (5.15)	0.50^{a} (0.32)	0.67^{b} (0.33)
2	76.71^{ab} (6.95)	3.90^{a} (1.42)	4.50^{ab} (2.50)
3	97.85^{a} (10.86)	6.17^{a} (0.50)	4.80^{a} (1.93)
4	90.57^{a} (14.45)	9.65^{a} (4.85)	0.75^{b} (0.48)
5	57.73^{bc} (9.31)	10.00^{a} (–)	1.50^{ab} (1.50)
6	43.28^{c} (5.66)	3.13^{a} (2.24)	2.60^{ab} (0.51)
		Pitch tube counts per m^2 of bole surface area	
1	21.85^{a} (3.22)	0.34^{a} (0.24)	0.42^{abc} (0.21)
2	17.38^{a} (1.74)	0.74^{a} (0.27)	0.70^{a} (0.09)
3	10.72^{b} (1.30)	0.75^{a} (0.06)	0.48^{ab} (0.20)
4	6.74^{bc} (1.06)	0.56^{a} (0.27)	0.06^{c} (0.04)
5	9.80^{bc} (5.48)	0.51^{a} (–)	0.08^{bc} (0.08)
6	2.35^{c} (0.48)	0.10^{a} (0.07)	0.07^{c} (0.01)

[1]Within each Table component, means sharing a common letter do not differ significantly at α=0.05 according to the LSD test; each mean is based on values from 20 or fewer plots ($n \leq 20$) depending on the presence of trees of the pertinent species and/or class within individual plots.
[2]The *SE* of each mean is indicated parenthetically.
[3]Denoted parenthetically, "–" indicates that the *SE* was incalculable due to insufficient sample size.

Table 6: Pitch Tube Prevalence in Prominent Tree Species by DBH Class Denoted as Counts per Tree and Per Unit of Bole Surface Area in a Mixed Conifer Stand of the Lake Tahoe Basin Selected for Assessment of Forest Health Indicators[1,2,3].

positively related the white fir dead percentage to the DBH Class 6 pitch tube count per tree for this species. Generally among the strongest models produced by the study, as much as 81% of the variation in the dependent variables was explained by that in their independent counterparts, and none of these models explained less than 42% of such variation.

The penultimate regression subset featured measures of mistletoe infestation as the independent component with either bark beetle colonization or tree mortality measures serving as their dependent counterpart, and six such models were significant with positive correlations prevailing exclusively (Table 7). Of these, the surface area-based pitch tube counts for DBH Class 1 and 6 white fir were paired with the percentage of this species infested with mistletoe in two relatively weak models in which only 23% and 35%, respectively, of the variation in the dependent variables was explained. Additionally, however, the red fir standing dead count and percentage were each related to both the Hawksworth rating and infested percentage for this fir in models accounting for 69% to 85% of such variation.

The stand biomass subset, the final one among the five computed in the study, yielded four significant regressions (Table 7). In two of these, the Hawksworth rating and infestation percentage for white fir were both positively related to the above-ground biomass of this species, while of the remaining two, the surface area-based pitch tube count across all species and DBH classes was negatively related to Jeffrey pine biomass in one while the DBH Class 1 Jeffrey pine pitch tube count by this measure was positively related to stand biomass across all species in the other. For the latter of these models, 84% of the variation in the dependent variable was accounted for, but in the previous three, from 32% to 47% was the extent of such explanation they provided.

Discussion

Within the context of the forest conditions prevailing in the eastern Sierra Nevada at present, the mixed conifer stand in which this study was conducted was largely unremarkable. It exhibited characteristics of a common managerial legacy in the region, namely that of the exclusion of fire for a prolonged period, the reflection of a broad policy that encompassed much of the 20th century. Additionally, it has been subjected to occasional partial harvests targeting its yellow and white pine components, which probably intensified in the years leading up to the transfer of ownership from the private to the public sector approximately three decades past, and with minimal extraction of white and red fir, species with little commercial value until the last half of the 20th century. In combination, these factors have culminated in a stand that exhibits a moderate level of stocking for the cover type [23], rendering it somewhat of an aberration regarding density but one that conforms to the long-term trend of rising white fir representation and diminishing Jeffrey and sugar pine prevalence in Sierran mixed conifer forests generally [24]. Despite the moderate stand density, however, a vestige of self-thinning induced by between-tree competition was apparent in mortality that was concentrated in its fir component, although the sheer preponderance of white fir on the site somewhat masked the magnitude of its demise when expressed as a percentage of its overall representation. Any question about the magnitude of the impact of white fir mortality on that of the stand in total, however, was negated by four simple regressions that positively related, in all possible combinations and with moderate to strong correlations prevailing throughout, overall standing dead count and percentage to each of these measures in white fir alone. Another regression model, although weaker, positively related the standing dead count in this fir to its prevalence specifically. It is probable that a factor in the mortality of this species is an annual precipitation for the site that is below the range considered common for white fir [8]. Undoubtedly, such was the case for red fir as well insomuch as precipitation at this site is near the bottom of the range for those typically occupied by this fir, and its limited residence there is largely a reflection of cold air drainage that settles locally [25]. What is also likely is that the mistletoe infestation here in fir generally, and in the incense-cedar as well, was too anemic to be considered a direct causal agent in their mortality other than in isolated cases. Several strong regression models involving red fir, specifically those positively correlating its standing dead stem count and percentage to both its Hawksworth rating and infestation percentage,

Independent variable	Dependent variable	Correlation	Model F test p-value	Model r^2
	Mensuration subset			
Live crown length	Overall combined pitch tube count per m^2	Negative	0.0312	0.2329
Live crown percentage	Overall dead tree count	Negative	0.0014	0.4399
Live crown percentage	Overall dead tree percentage	Negative	0.0046	0.3682
Live crown percentage	White fir dead tree count	Negative	0.0189	0.2698
Live crown percentage	White fir dead tree percentage	Negative	0.0331	0.2283
Basal area	Jeffrey pine DBH Class 1 pitch tube count per m^2	Positive	0.0390	0.8049
Basal area	Sugar pine DBH Class 3 pitch tube count per tree	Positive	0.0474	0.7790
Basal area	Incense-cedar mistletoe rating	Positive	0.0171	0.4862
Basal area	Incense-cedar percentage infested with mistletoe	Positive	0.0171	0.4862
Basal area	Red fir dead tree percentage	Positive	0.0344	0.7132
Total tree count	Jeffrey pine dead tree count	Positive	0.0083	0.6024
Total tree count	Jeffrey pine dead tree percentage	Positive	0.0083	0.6024
	Species representation subset			
White fir prevalence	Overall combined pitch tube count per tree	Positive	0.0073	0.3366
White fir prevalence	Overall combined pitch tube count per m^2	Positive	0.0014	0.4430
White fir prevalence	White fir dead tree count	Positive	0.0353	0.2234
Species diversity	Overall combined pitch tube count per m^2	Negative	0.0113	0.3064
Species diversity	White fir DBH Class 2 pitch tube count per m^2	Negative	0.0480	0.2000
Species diversity	White fir DBH Class 6 pitch tube count per m^2	Negative	0.0084	0.5169
	Mortality subset			
White fir dead tree count	Overall dead tree count	Positive	<0.0001	0.6102
White fir dead tree count	Overall dead tree percentage	Positive	<0.0001	0.6910
White fir dead tree percentage	Overall dead tree count	Positive	0.0005	0.4997
White fir dead tree percentage	Overall dead tree percentage	Positive	<0.0001	0.8079
White fir DBH Class 6 pitch tube count per tree	White fir dead tree percentage	Positive	0.0223	0.4216
	Mistletoe subset			
White fir percentage infested with mistletoe	White fir DBH Class 1 pitch tube count per m^2	Positive	0.0335	0.2275
White fir percentage infested with mistletoe	White fir DBH Class 6 pitch tube count per m^2	Positive	0.0442	0.3462
Red fir mistletoe rating	Red fir dead tree count	Positive	0.0091	0.8485
Red fir mistletoe rating	Red fir dead tree percentage	Positive	0.0210	0.7728
Red fir percentage infested with mistletoe	Red fir dead tree count	Positive	0.0400	0.6920
Red fir percentage infested with mistletoe	Red fir dead tree percentage	Positive	0.0145	0.8098
	Biomass subset			
White fir total biomass	White fir mistletoe rating	Positive	0.0059	0.3506
White fir total biomass	White fir percentage infested with mistletoe	Positive	0.0008	0.4704
Jeffrey pine total biomass	Overall combined pitch tube count per m^2	Negative	0.0087	0.3251
Overall total biomass	Jeffrey pine DBH Class 1 pitch tube count per m^2	Positive	0.0287	0.8398

[1]Each model is based on values from 20 or fewer plots ($n \leq 20$) depending on the presence of trees of the pertinent species and/or class within individual plots.

Table 7: Significant Simple Linear Regression Models Relating Pitch Tube and Mistletoe Abundance Along with Tree Mortality to an Array of Pertinent Influencing Factors in a Mixed Conifer Stand of the Lake Tahoe Basin Selected for Assessment of Forest Health Indicators[1].

would appear to dispute this assertion, but these probably reflect impacts of secondary contributors to a demise mostly caused by the marginality of the host species for this site. Lending further credence to this interpretation is that such regressions did not materialize for white fir, which had the highest rating overall and an infestation percentage near that of the red fir. In fact, the only contribution of the regression analysis to an understanding of the interactions between either white fir or incense-cedar and their respective mistletoes is that higher stand density favored infestation, as the Hawksworth rating and infestation percentage for the latter was related to one stand density measure, specifically basal area, and those for the former were related to a surrogate density measure specific to this fir, that of above-ground tree biomass, all positive correlations of mostly moderate strength. In light of the paucity of Jeffrey pine on this site, a compositional feature of the stand rendered even more distinctive because its silvics are relatively well matched to the site [26], it is perhaps noteworthy that mortality in this species, denoted as either the standing dead count or its percentage, was positively related to another density measure employed in the study, that of the total tree count, in regressions of some strength. And, lending further evidence to the assertion that the mortality in red fir was, to a large extent, caused by between-tree competition for resources exacerbated by its marginal adaptation to the site, which would leave it at a competitive disadvantage, another regression model revealed that the standing dead percentage for this fir was positively, and strongly, correlated with stand basal area.

With bark beetle demography constituting the main thrust of this study, perhaps the most profound finding regarding this subject was the pronounced disparities in the level of colonization sustained by the various tree species represented in the stand, and more specifically that sustained by white fir in comparison to the other two species considered, namely Jeffrey and sugar pine. In part, it is reasonable to attribute the magnitude of the difference to the abundance and close proximity, and therefore ready availability, of this host species, factors that likely facilitated expansion of the fir engraver population. Prior studies by DeMars, Ferrell, and Otrosina [27], Ferrell, Otrosina, and DeMars [28], Walker et al. [4], and Egan et al. [29] have all provided evidence that an important influence on populations of this bark beetle was the proportional prevalence of its host, with greater availability exacerbating attack severity. Insomuch as the overwhelming majority of the pitch tubes found on the tree boles, which served as the quantitative basis here for assessment of colonization level, was found on white fir, two regression models that positively related the overall pitch tube count, i.e. across host species, denoted as both the count per tree and that per unit of bole surface area, to white fir prevalence provided some credence to this assumption. Somewhat related were two other models that negatively related the overall surface area-based count to tree species diversity and to Jeffrey pine biomass. However, suggesting that host availability was not the sole explanation for the exceptional size of the engraver population was that these regressions were not of notable strength and that significant models matching pitch tube abundance in white fir specifically to the prevalence of this host species failed to materialize. Inevitably, this leads to the second of the probable causes of the large engraver population, that of deficiencies in white fir defense mechanisms which are pronounced to such extent that they provide relatively little protection against attack [4, 9-11] and therefore do not exert much control on population levels of this beetle because its reproductive success is only nominally imperiled by them. To a large extent, the vulnerability of white fir is attributable to a resin duct system that is typically underdeveloped relative to those of pine species [30-32] reflecting that its development, rather than perpetual, is initiated only when an attack commences [33] and is largely confined to bark rather than arrayed across both bark and wood as in the pines [34]. An already precarious defense capability is further jeopardized by moisture stress, particularly that chronically induced by drier sites, and a minimum annual precipitation threshold of 89 cm has been established for this species to thrive in the Sierra Nevada [8,35], which is more than that deposited over the long term on the site of this study. Moisture stress induces stomatal closure, thus disrupting photosynthate production by limiting gas exchange, which in turn diminishes the availability of carbohydrates that can be dedicated to the production of defense compounds, primarily oleoresin [36], and such stress likely exacerbates the conundrum of assimilate partitioning because any dedicated to defense is unavailable for growth and reproduction [37]. In a study conducted on a drier site but near that of the one presented here which illustrated the debilitating interaction of defensive mechanisms of dubious efficacy and limited moisture availability in this species, pitch tubes on scattered white fir in a pure Jeffrey pine stand were far more numerous than on the latter species according to both of the abundance measures used herein [38]. As an aside, its relatively subdued presence here belies the status of the Jeffrey pine beetle as the single greatest threat to the health of its namesake host species [26, 39]. Although direct examination of the functionality of the defense mechanisms embodied in the species constituting this stand was beyond the scope of the present study, a relatively weak regression model negatively related overall pitch tube count on a surface area basis to live crown length, suggesting that reduced energy production due to restricted crown development may have resulted in marginally elevated levels of beetle attack. Indirect evidence of the vital role of energy production in stand health may have also been manifested in four other models, specifically two each that negatively correlated dead tree count and percentage both across species and for white fir alone with live crown percentage. Undoubtedly, these reflect in part the distinction employed to detect dead trees, which was a complete lack of live crown, and are moderate in strength at most, but they may be somewhat indicative of the ultimate effect of diminishing vigor on viability as well.

The pronounced inequality between the overall pitch tube prevalence in the white fir and those in the pines here essentially extended to include disparities of substantial magnitude within every one of the six DBH classes. In fact, only in Class 5 were statistically significant differences between the fir and both Jeffrey and sugar pine lacking, and even therein, such was extant between the former and sugar pine according to the count per tree measure. Numerically, the largest difference between white fir and Jeffrey pine occurred in Class 1 regardless of the abundance measure considered, which in and of itself suggests an inclination for the fir engraver to colonize small trees of its host species more so than the Jeffrey pine beetle. In a comparison of this fir and sugar pine, however, the greatest numerical discrepancies occurred in Class 4 on a count per tree basis and in Class 5 according to the surface area-based count, suggesting that the engraver beetle had somewhat more of a proclivity toward colonizing larger host trees than the mountain pine beetle. Nevertheless, somewhat greater clarity regarding the respective host tree size preferences of the three bark beetles in question here was provided by comparisons within species among the diameter classes. Thus evaluated, the preference of the fir engraver was for trees in Class 3 and 4, or white fir from 30.4 cm to 50.5 cm DBH, followed by those of the next smallest size in Class 2, when assessed on the basis of pitch tubes per tree, while the smallest and largest classes were the least colonized overall. Alternatively, when assessed according to the surface area-based counts, targeted sizes were the small stems of Class 1 and 2 with the large Class 6 trees the least preferred. Interpretively, it may be reasonable to explain the discord in these findings as that the count per tree result possibly demonstrates an actual size preference by the fir engraver while that derived from the surface area-based measure probably reflects the size most likely to succumb under concentrated, and therefore lethal, attack. Regardless, what is abundantly clear in either result is that on this site, the fir engraver did not gravitate toward the largest host trees available, which is a departure from the findings of earlier studies [4,27,28] in which a marginal preference of this beetle for larger fir was revealed. Given that resin ducts develop almost exclusively in the bark of white fir as noted previously, and that the thicker bark of larger diameter trees could conceivably accommodate greater resin duct formation, it is possible that the comparative avoidance of larger fir was a de facto avoidance of those with the best developed resin production capability. Extrapolating management implementations from this finding, it suggests that in the selection of stems for removal during thinnings of Sierran mixed conifer, and perhaps especially so for stands of this cover type in which white fir is predominant, eliminating those of smaller stature, the directive of the commonly employed low thinning procedure and one that best mimics the mortality inherent in natural self-thinning [40], would ultimately prove to be the approach most conducive to the enhancement of stand health. Unlike the fir engraver, there was little evidence here to indicate that the Jeffrey pine beetle was discriminating as to the size of its namesake host tree. Although substantial numerical differences among the DBH classes were apparent, more so in the pitch tube quantity per tree than in that per unit bole surface area, excessive variability in the data from which these means were derived rendered them statistically non-significant. Nevertheless, the mountain pine beetle demonstrated clear proclivities regarding the size of the sugar pine it colonized as well as some level of consistency between the two pitch tube abundance measures. Specifically, the highest count per tree occurred in Class 3 followed by that in Class 2, while the highest surface area-based count was found in Class 2 followed by that in Class 3, which collectively would seem to indicate a preference for trees of small to moderate size. However, these findings must be contemplated with caution, as although reasonably well distributed throughout the stand, sugar pines were so widely dispersed that few subject specimens were available from which to collect data. This may partially explain their low overall colonization level, especially when contrasted against the engraver beetle in white fir, which is perhaps most noteworthy because the mountain pine beetle is a formidable damaging agent in sugar pine capable of highly intensive attacks [20,41, 42].

Given the extreme variability typically existing within forest stands in the degree to which their individual constituents are colonized by bark beetles, discerning causation as to localized conditions and beetle responses has often proven to be elusive, but several significant regression models computed in this study provide insight regarding relationships specific to some of the individual diameter classes. Among them, the count per unit bole surface area in Class 1 Jeffrey pine and that per tree in Class 3 sugar pine were each strongly correlated with basal area, so at least for the narrow diameter ranges embodied therein, these species displayed the propensity for higher stand density to elevate beetle populations that has been documented in several western North American forest types [4,43-46]. Closely collaborative here was another strong model featuring a positive correlation between the surface area-based count in Class 1 Jeffrey pine and stand biomass. Conversely, surface area-based counts in Class 2 and 6 white fir were negatively related to tree species diversity, with the former a weak model but the latter much stronger, which indirectly reiterates the aforementioned contribution of host tree availability to expansion of the engraver population. More directly, however, these models infer that a mixed composition presents a complexity for bark beetles in terms of detecting suitable host trees for colonization [47]. The low occurrence of mistletoes in this stand notwithstanding, two models specific to white fir regarding both the dependent and independent components revealed positive, albeit weak, correlations between the pitch tube counts per unit surface area of Class 1 and 6 and the mistletoe infestation percentage. Because of their capacity to weaken trees and therefore compromise their defense mechanisms against a multitude of damaging agents, including bark beetles, dwarf mistletoe infestations are assumed to have considerable diagnostic value in detecting the threat of rapid escalation in beetle populations within western North American forest stands [8,25,48,49]. Nevertheless, in the only significant regression model providing any evidence that fir engraver attacks on white fir caused any of the mortality therein, a positive correlation of moderate strength was detected between its standing dead percentage across the full spectrum of tree sizes and the pitch tube count per tree in Class 6. Regardless, the lack of additional models revealing similar relationships but based on other diameter classes indirectly suggests that the fir engraver population here had not risen to the level known to induce the heavy mortality for which this species is known [4,11], and perhaps further suggests that although a likely contributor, attacks by this beetle were not the sole cause of mortality in its host species.

In summary, this study involved an examination of forest health variables in an eastern Sierra Nevada mixed conifer stand focused on bark beetle demography and its interrelationships with mensurational features as well as other stand maladies. The fir engraver in white fir, the Jeffrey pine beetle in its namesake host species, and the mountain pine beetle in sugar pine constituted the three insect and host associations of interest. Quantified through pitch tube abundance on the tree boles denoted as both the count per tree and that per unit of bole surface area, beetle colonization in white fir greatly exceeded that in either Jeffrey pine or sugar pine regardless of measure considered, likely a reflection of the degree to which the former dominated stand composition, its limited capacity to defend itself against attack, and the stress-induced loss of vigor it incurred due to its marginal adaptation to the site upon which the study was conducted. Pitch tube abundance across host species was positively related to white fir prevalence according to both measures and was negatively related to tree species diversity according to the surface area-based count. Both the fir engraver and mountain pine beetle displayed discriminating behavior concerning the size of their host trees as discerned through the segregation of the diameter dimension into multiple classes encompassing the full range of tree sizes in residence, with both species revealing a propensity to attack small to medium stems rather than the largest ones available to them. Any size preference of the Jeffrey pine beetle was not sufficiently pronounced to induce statistical distinctions. Nevertheless, Jeffrey pine beetle attack intensity in the smallest diameter class was positively correlated with stand basal area and biomass according to surface area-based pitch tube quantities, and the count per tree measure in sugar pine of medium size was positively related to the former as well. Fir engraver attack intensity in the second smallest and the largest host diameter classes was negatively related to tree species diversity and that in the smallest and largest classes was positively related to the percentage of white fir infested with dwarf mistletoe, all cases in which pitch tube prevalence was enumerated on a surface area basis. Extant in white and red fir, the latter an exceedingly minor stand constituent, along with incense-cedar, mistletoe infestations were very light overall, and although red fir mortality was positively correlated with its Hawksworth rating and infestation percentage, it is unlikely that mistletoe infestations were the principal cause of mortality in any of the resident species. Somewhat similarly, within the white fir mortality was linked to the pitch tube count per tree of the largest diameter class, and although considerable mortality occurred in this species, it is probable that deaths resulted from multiple causes of which bark beetle attack was not preeminent. Exemplifying that the latter assumption was likely even more applicable to the other stand constituents was that mortality in Jeffrey pine was found to be induced by higher total stem counts only among factors examined in this study. These findings advance the understanding of forest health indicators, stand attributes, and pertinent interactions thereof as often encountered in Sierra Nevada mixed conifer and similar cover types.

Acknowledgements

Support for this research was provided by the Nevada Agricultural Experiment Station, McIntire-Stennis Cooperative Forestry Research Program, and the USDA Forest Service. The authors are indebted to G. Fernandez, W. Frederick, J. Ray, C. Sarman, and the USDA Forest Service Lake Tahoe Basin Management Unit for their assistance.

References

1. Tappeiner JC (1980) Sierra Nevada mixed conifer In: Eyre FH (ed), Forest cover types of the United States and Canada Washington. DC: Society of American Foresters pp.118-119.
2. Burns RM, Honkala BH (1990) Summary of tree characteristics. In: Burns RM, Honkala BH (eds) Silvics of North America: Conifers. Agricultural Handbook 654, Washington, DC: USDA Forest Service 1: 646-649.
3. Lanner RM (1999) Conifers of California. Los Olivos, CA: Cachuma Press.
4. Walker RF, Fecko RM, Frederick WB, Johnson DW and Miller WW (2007) Forest health impacts of bark beetles, dwarf mistletoe, and blister rust in a Lake Tahoe Basin mixed conifer stand. Western North American Naturalist 67: 562-571.
5. Beesley D (1996) Reconstructing the landscape: An environmental history, 1820-1960. In Status of the Sierra Nevada: Assessments and scientific basis

for management options (Wildland Resources Center Report No. 37, 2: 3-24). Davis: University of California.

6. Arno SF (2000) Fire in western forest ecosystems. In: Brown JK, Smith JK (eds) Wildland fire in ecosystems: Effects of fire on flora (General Technical Report RMRS-GTR-42) Ogden, UT: USDA Forest Service 2: 97-120.

7. Taylor AH (2004) Identifying forest reference conditions on early cut-over lands, Lake Tahoe Basin, USA. Ecological Applications 14: 1903-1920.

8. Laacke RJ (1990) White fir. In: Burns RM, Honkala BH (eds) Silvics of North America: Conifers: Agricultural Handbook 654, Washington, DC: USDA Forest Service 1: 36-46.

9. Ferrell GT (1983) Host resistance to the fir engraver, *Scolytus ventralis* (*Coleoptera: Scolytidae*): Frequencies of attacks contacting cortical resin blisters and canals of *Abies concolor*. Canadian Entomologist 115: 1421-1428.

10. Christiansen E, Waring RH and Berryman AA (1987) Resistance of conifers to bark beetle attack: Searching for general relationships. Forest Ecology and Management 22: 89-106.

11. Berryman AA, Ferrell GT (1988) The fir engraver beetle in western states. In: Berryman AA (ed.) Dynamics of forest insect populations: Patterns, causes, and implications New York, NY: Plenum Press pp. 555-577.

12. NOAA Western Regional Climate Center (2013) Tahoe City, California: Period of record monthly climate summary.

13. USDA Soil Conservation Service (1974) Soil survey: Tahoe Basin area, California and Nevada. Washington, DC: U.S. Government Printing Office.

14. Walker RF, Fecko RM, Frederick WB, Johnson DW, Miller WW (2011) Fuel bed alterations by thinning, chipping, and prescription fire in a Sierra Nevada mixed conifer stand. Journal of Sustainable Forestry 30: 284-300.

15. Dunning D (1942) A site classification for the mixed conifer selection forests of the Sierra Nevada (California Forest and Range Experiment Station Research Note No. 28). Berkeley, CA: USDA Forest Service.

16. Curtis RO, Marshall DD (2000) Why quadratic mean diameter? Western Journal of Applied Forestry 15: 137-139.

17. Davis LC, Johnson KN, Bettinger PS, Howard TE (2001) Forest management (4th edn.) New York, NY: McGraw-Hill.

18. Gholz HL, Grier CC, Campbell AG, Brown AT (1979) Equations for estimating biomass and leaf area of plants in the Pacific Northwest (Forest Research Laboratory Paper No. 41). Corvallis: Oregon State University.

19. Ter-Mikaelian MT, Korzukhin MD (1997) Biomass equations for sixty-five North American tree species. Forest Ecology and Management 97: 1-24.

20. Furniss RL, Carolin VM (1977) Western forest insects (Miscellaneous Publication 1339). Washington, DC: USDA Forest Service.

21. Hawksworth FG (1977) The 6-class dwarf mistletoe rating system (General Technical Report RM-48). Fort Collins, CO: USDA Forest Service.

22. Scharpf RF, Hawksworth FG (1993) Mistletoes. In: Scharpf RF (ed.) Diseases of Pacific Coast conifers (Agriculture Handbook 521), Washington, DC. USDA Forest Service pp.112-135.

23. Dunning D, Reineke LH (1933). Preliminary yield tables for second-growth stands in the California pine region (USDA Technical Bulletin 354). Washington, DC: USDA Forest Service.

24. Dolanc CR, Safford HD, Dobrowski SZ, Thorne JH (2014) Twentieth century shifts in abundance and composition of vegetation types of the Sierra Nevada, CA, US. Applied Vegetation Science 17: 442-455.

25. Laacke RJ (1990) California red fir. In: Burns RM, Honkala BH (eds), Silvics of North America: Conifers. (Agricultural Handbook 654), Washington DC: USDA Forest Service 1: 71-79.

26. Jenkinson JL (1990) Jeffrey pine. In: Burns RM, Honkala BH (eds), Silvics of North America: Conifers (Agricultural Handbook 654), Washington, DC: USDA Forest Service, 1: 359-369.

27. DeMars CJ, Ferrell GT and Otrosina WJ (1988) Host-insect/disease interactions in drought-stressed white fir stands at Lake Tahoe, California. In: PayneTL, Saarenmaa H (eds), Integrated Control of Scolytid Bark Beetles: Proceedings of the IUFRO Working Party and XVII International Congress of Entomology Symposium Vancouver, BC, Canada: College of Agriculture and Life Sciences, Virginia Polytechnic and State University pp.135-146.

28. Ferrell GT, Otrosina WJ, DeMars CJ (1994) Predicting susceptibility of white fir during a drought-associated outbreak of the fir engraver, *Scolytus ventralis*, in California. Canadian Journal of Forest Research 24: 302-305.

29. Egan JM, Jacobi WR, Negron JF, Smith SL, Cluck DR (2010) Forest thinning and subsequent bark beetle-caused mortality in northeastern California. Forest Ecology and Management 260: 1832-1842.

30. Penhallow DP (1907) A manual of the North American Gymnosperms, exclusive of the Cycadales but together with certain exotic species. Boston MA: Ginn and Company.

31. Lewinsohn E, Gijzen M, Croteau R (1991) Defense mechanisms of conifers: Difference in constitutive and wound-induced monoterpene biosynthesis among species. Plant Physiology 96: 44-49.

32. Trapp S, Croteau R (2001) Defensive resin biosynthesis in conifers. Annual Review of Plant Biology 52: 689-724.

33. Haygreen JG, Bowyer JL (1982) Forest products and wood science. Ames, IA: Iowa State University Press.

34. Shigo AL (1986) A new tree biology. Durham NH: Shigo and Trees, Associates.

35. Edmonds RL, Agee JK, Gara RI (2011) Forest health and protection (2nd edn). Long Grove, IL: Waveland Press.

36. Kozlowski TT, Kramer PJ, Pallardy SG (1991) The physiological ecology of woody plants. New York, NY: Academic Press.

37. Franceschi VR, Krokene P, Christiansen E, Krekling T (2005) Anatomical and chemical defenses of conifer bark against bark beetles and other pests. New Phytologist 167: 353-376.

38. Walker RF, Swim SL, Johnson DW, Miller WW, Fecko RM (2014) Bark beetle demography in a Jeffrey pine stand as influenced by mechanized thinning and prescribed fire. Journal of Sustainable Forestry 33: 627-676.

39. Walker LC (1999) The North American forests: Geography, ecology, and silviculture. Boca Raton, FL: CRC Press.

40. Smith DM, Larson BC, Kelty MJ, Ashton PMS (1997) The practice of silviculture: Applied forest ecology (9th ed.). New York, NY: John Wiley and Sons.

41. Struble GR (1965) Attack pattern of mountain pine beetle in sugar pine stands (Research Note PSW-60). Berkeley CA: USDA Forest Service.

42. Kinloch BB, Scheuner, Jr., WH (1990) Sugar pine In: Burns RM, Honkala BH (eds), Silvics of North America: Conifers. (Agriculture Handbook 654) Washington, DC: USDA Forest Service 1: 370-379.

43. McCambridge WF, Hawksworth FG, Edminster CB, Laut JG (1982) Ponderosa pine mortality resulting from a mountain pine beetle outbreak (Research Paper RM-235). Fort Collins, CO: USDA Forest Service.

44. Mitchell RG, Waring RH, Pitman GB (1983) Thinning lodgepole pine increases tree vigor and resistance to mountain pine beetle. Forest Science 29: 204-211.

45. Fiddler GO, Hart DR, Fiddler TA, McDonald PM (1989) Thinning decreases mortality and increases growth of ponderosa pine in northeastern California (Research Paper PSW-194). Berkeley CA: USDA Forest Service.

46. Negron JF, Popp JB (2004) Probability of ponderosa pine infestation by mountain pine beetle in the Colorado front range. Forest Ecology and Management 191: 17-27.

47. Goyer RA, Wagner MR, Schowalter TD (1998) Current and proposed technologies for bark beetle management. Journal of Forestry 96: 29-33.

48. Waters WE (1985) The pine-bark beetle ecosystem: A pest management challenge. In: Waters WE, Stark RW, Wood DL (eds), Integrated Pest Management in Pine-Bark Beetle Ecosystems. Berkeley CA: John Wiley & Sons.

49. Negron JF, Wilson JL (2003) Attributes associated with probability of infestation by the pinion ips, *Ips confusus* (Coleoptera: Scolytidae), in pinion pine, *Pinus edulis*. Western North American Naturalist 63: 440-451.

Collaborative Conflict Management on Ecosystem Restoration Concession: Lessons Learnt from Harapan Rainforest Jambi-South Sumatra-Indonesia

Mangarah Silalahi[1] and Desri Erwin[2]

[1]*Head of Ecosystem Restoration Resource Centre, Burung Indonesia, Indonesia*
[2]*Community Partnership staff of PT REKI, Jambi South Sumatra, Indonesia*

Abstract

Ecosystem Restoration Concession (ERC) is an innovative solution for managing natural forests in Indonesia that integrates forest utilization, environmental services, biodiversity protection and improving the livelihoods of local people. The main aim of ERC is to restore the degraded forest to reach an ecosystem equilibirum. Harapan Rainforest is the first ecosystem restoration concession in Indonesia and it is a site rich with biodiversity. The field implementation process for ERC, however, faces many challenges, and one of the biggest is encroachment. Based on an analysis of a landsat satellite image 7 ETM and Envisat.asat images, from 2005-2012 the total area cleared due to encroachment was ±18,362 ha with an average cleared area as large as 2,623 ha/year.

If encroachment continues to occur and cannot be reduced or stopped completely, an area totaling 98,555 ha is predicted to be lost within the next 10 years. This study describes what the roots of the conflicts are, the modus operandi, the connection between encroachers and supporters, and an appropriate resolution strategy. The encroachment problems are complex and involve the local elites, migrants and speculators. The encroachers are now getting stronger and acquiring legitimacy by gaining support from various mass organizations and government programmes. The decision to implement an encroachment strategy can not be undertaken in a universally agreed-upon or repressive way. The encroachment conflict resolution strategies should use a collaborative management approach and mechanisms which are accepted by relevant parties, resolve the problems permanently, avoid human rights violations, apply international standards and create harmonious relationships. Collaborative Conflict Management is participatory and long process that should implement all steps to create new social contract on natural resources and needs patience and a legitimate process.

Keywords: Ecosystem restoration; Forest and land conflicts; Encroachment; Collaborative Conflict Management

Introduction

Indonesia is known as the country with the third highest levels of biodiversity in the world. The total forested area in Indonesia was 134 million ha in 2011 with 82 million ha of production forests [1]. The total forested area decreases significantly every year because of changes to land use, population growth, deforestation and encroachment. For example in the production forest, 27.8 million ha has degraded and low economic potency [2]. These problems are predominantly caused by failure of forest management systems and paradigms.

In the Dutch Colonial, the forest management system firstly implemented in Java and Madura with *Boschordonantie voor Java en Madoera* 1865 [3]. In Sumatra and Sulawesi, Dutch Colonial used Agrarische Wet 1870 for law basis in forest management, but this law could not be effectively implemented because it had many weaknesses, overlapped with community' land/forest, and was not suited with local socio-culture [3]. In the Dutch Colonial, community had right to cut forest, and all of land forest that couldnot be showed by legal evidence should be given to government. Poffenberger and McGean [4] noticed that the history of forestry in countries such as the Philippines, Indonesia and Thailand involves a shift from traditional (indigenous or local) tenure to state control, and the possibility to back the control over forest by community is using collaborative management. Thus Dutch Colonial has inherited in forest and land conflicts in Indonesia.

After Indonesian Independence, Government of Indonesia intensified forest exploitation management to increase national income in 1967 through ACT No. 5/1967 about the principal provisions of Forestry [4,5]. It was as a basis for forest management with a natural resource exploitation-based management system using an administrative, top-down approach. This type of forest management system resulted that almost all state forest (99%) was managed by the state and corporate organizations. Less than 1% (131.209.34 ha) of the total state forests were managed by the community in 2010 [6]. Forests were seen as little more than places for tress and biodiversity without the presence of people [7]. The government has predominantly thought that the primary product obtained from forest production is wood; however wood makes up less than 5% of total forest productions [8]. In 1999, the paradigm towards forest management changed, and became focused on using a participatory and bottom-up approach, administered through a centralistic government. The changing are (1) local and indigenous people through cooperation could have right for forest utilization, managing forest through village forest, community forest and customary forest, (2) community has right to involve in the forest planning and monitoring, (3) community has right for compensation of their forest and land licensed to company, (4) a part of national forest authority has been given to local government. These changes are based on Forestry law No.41/1999. However the changed forest paradigm could not bring about significant changes for effective

***Corresponding authors:** CMangarah Silalahi, Head of Ecosystem Restoration Resource Centre, Burung Indonesia, Bogor, Indonesia, E-mail: m.silalahi@burung.org

forest management and increased community welfare because the old paradigm had become a "culture" that is difficult to change. For example in the process of determining forested areas the government has failed to implement a participation process that respects the presence of local and indigenous people.

The failure of the forest management system and paradigm is linked with rising levels of conflict over forested areas in Indonesia, which have tended to increase every year. The Land Registration Agency/ BPN recorded that at least 7,491 natural resource conflicts were handled by BPN and the Indonesian police [9]. The Center for International Forestry Research (CIFOR) identified 359 forest conflicts in Indonesia from January 1997- June 2003. The highest frequency occurred in 2000, when there were 153 cases or 43% of the total number. Land conflicts in industrial plantation areas were the highest, reaching 39%, followed by conservation areas (34%) and in Logging Concessions (HPH) with 27% [10]. About 50% of all Indonesian villages (33,000) overlap with state forest areas (Martua Sirait Pers. conference). Almost all state forests handed out as concessions have not gone through the participation process. The Land Reform Agrarian Consortium (KPA) noted that throughout Susilo Bambang Yudhoyono's time as President that commenced in 2004, 618 agrarian conflicts have occurred, in which 44 people died, 731,342 families have been impacted and a total of 2,399,314 ha have been disputed [9]. These conflicts have strong correlations with new licenses for oil palm, forestry, and coal mining which will take place on community land.

Conflicts arising on the ground have been handled mostly by most concession holders (logging and plantation companies) in a rigorous manner, especially under the Suharto regime. This has created a general mistrust among local communities towards companies [10]. CIFOR noted that 48.8 million people live in state forests and approximately 10.2 million of these people are categorized as living below the poverty line. These are communities that settled in the forest during the Dutch Colonial era. Around 20 million people live close to the forest and around 6 million of them rely on forests for their livelihoods [11].

To reduce natural resource conflicts, the Indonesian government uses both litigation (law enforcement) along with non-litigation techniques (negotiation, mediation, reconciliation and arbitration). This law enforcement approach cannot reduce and resolve land conflict, as it is only a temporary solution [9]. Many law enforcement activities can actually create new conflicts and the process is time-consuming. For instance, the law enforcement (community relocation) in Kerinci Seblat National Park in Jambi-West Sumatra created a new encroachment problem in PT. Restorasi Ekosistem area and PT. Agronusa Alam Sejahtera (Timber plantation company) in Jambi. With so many conflicts over natural resources currently in the courts, the Indonesian government has implemented an Alternative Dispute Resolution (ADR) approach to reduce natural resource conflicts with Act No. 30 year 1999, focusing on arbitration and alternative dispute resolution.

To address forestry issues and conflicts the Ministry of Forestry established the National Forestry Council in 2006. Set up primarily to handle forest conflicts, the National Forestry Council has developed a conflict resolution working group with an emphasis on mediation. Besides government, several NGOs and mediation institutions have promoted and facilitated the mediation process including Scale Up, The Forest Trust, Impartial Mediation Network, and LATIN. This marked the beginning of mediation development in Indonesia.

Deforestation and conflicts combined with the new era of autonomy in Indonesia have contributed significantly to deforestation and degradation of production forests. Because of deforestation and degradation, the economic benefits of production forests are low and have been impacted upon by the collapse of logging companies. This was proven by the decreased amount of natural production forest that was managed by logging concessions, from 61.7 million ha in 1993 to 27.7 million ha in 2005 [1]. In 2011, production forests that did not have concession holders made up 49,136,665 ha and they had an open access status [12]. If there aren't any innovative changes to the forest policy in production forests, it can be predicted that by 2020 natural forests in production forests will disappear and in the future only industrial acacia plantations will be found in these forests. Meanwhile, the natural forests especially in dry, low land forests still have a rich level of biodiversity and are a habitat for critical endangered species of mammals.

Concerned about the loss of natural forests in production forests and the loss of flora and fauna, Burung Indonesia together with the Ministry of Forestry initiated the ecosystem restoration concession (ERC) program in 2002. This initiative has changed the forest paradigm from exploiting to restoring, from commodities based to ecosystem based and from single product based to multiple product based management. The new policy on ERC in production forests was established when the Ministry established the Ministry of Forestry Regulation SK 150/ Menhut-II/2004. Through this policy the production forest can be managed with the aim of restoration ecosystem, called "Izin Usaha Pemanfaatan Hasil Hutan Kayu Restorasi Ekosistem (IUPHHK-RE)"/ forest utilization permits on ERC". IUPHHK-RE has four activity pillars which are stabilization of the concession area, rehabilitation of the economic values of production forest, restoration of flora and fauna and developing economic and social aspects [13]. It is hoped that using these 4 pillars, the preservation and maximization of forest products can be achieved and can mitigate the degradation, deforestation and can improve economic and social development.

Regarding to reducing forest/land conflict in production forest, Indonesia government issued Ministry of Forestry Decree No.P.39/ Menhut-II/2013 about empowerment of local community through forestry partnership. The forestry partnership means that cooperation between local community and concession holder in developing capacity and access with equality and mutual benefits (Article 1 point 3 P.39/ Menhut-II/2013). On this decree clearly stated that the forest partnership aim is to reduce forest conflict and empower community within and around production forest including on IUPHHK-RE. The forestry partnership can be achieved by using negotiation and mediation.

Ecosystem Restoration Concession or IUPHHK-RE is an endeavor to restore biotic (flora and fauna) and abiotic (soil, nutrient and hydrological cycle and other natural processes) elements to production forests in order to restore biological equilibrium [14]. Then ERC umbrella regulation was established by the Government of Indonesia through Governmental regulation No. 6/2007 on Forest Arrangement and Formulation of Forest management and Utilization Plans; to Governmental Regulation No.3 /2008 on Amendment to Government Management and Utilization Plans. The aim of ERC is to restore the logged over areas of production forests or open access spaces [14]. Based on Dirjen BUK data these logged over area is 49,136,665 ha in 2011 [12,15]. The ERC permit's aim is to maintain the forest functions, optimize environmental services, non-timber forest products and forest area services.

According to new legislation on ERC, Burung Indonesia and its consortium (The Royal Society for Protection of the Birds/RSPB and Birdlife International) encouraged the ERC initiative in Harapan Rainforest (HRF). IUPHHK-RE Harapan Rainforest is the first ER concession and the second largest ERC in Indonesia [16,17]. Harapan Rainforest has 98,555 ha with two licenses awarded to PT. REKI in 2007 and 2010 [17].

As a new piece of legislation that utilizes a new approach, ERC has received attention from other applicants who are also interested in obtaining a permit. There are 47 applicants for ER concessions from 2007 covering 4,125,115 ha whereas there are 12 ERC holders that have concessions with a total area of approximately 500,000 ha [12]. Because the ERC policy is new and different stakeholders are not yet aware of the many technical regulations needed, so there are many challenges in implementing this ERC policy. Using the case study of the HRF, there are omissions in the policy concerning the regulation of ERC and is still reliant on the wood production paradigm, a lack of technical guidance, problems with encroachment and illegal logging, and many more. One of the biggest problems in the ERC is encroachment. The unit management cannot restore the forest and undertake research and development if encroachment occurs.

Based on mentioned above, many of these conflicts are poorly managed, resulting in negative impacts to both the community and the company, including anxiety, fear, disruption to the company's operations and instability [11,18]. Additionally, conflicts over land and forests are viewed as contributing to unsuccessful logging concessions, timber and oil and one of the key constraints of investment in forestry in Indonesia [18,19]. This underlines the need to develop effective and collaborative conflict management approaches, thereby minimizing this potential damage.

This article describes the affects of encroachment on ERC, conflict and stakeholder analysis, an appropriate resolution strategy, and how to resolve these using the collaborative management. The structure of this article includes the research methods and collaborative management framework, the background of the conflict, results and discussion and the conclusion.

Research Methods and Collaborative Management Framework

One site of ERC- Harapan Rainforest was selected as a case study based on the levels of conflict at this site and efforts to apply collaborative conflict management to resolve these issues. The other criteria is that the Harapan Rainforest was the first ERC and the biggest concession in Indonesia in 2013, and now to be second largest [17], which is now 7 years old and has experienced many different types of conflicts, with many solutions applied to resolve these conflicts, with differing results. In order to understand how stakeholders engage in conflict and conflict resolution, various data acquisition methods were used, namely a literature review, FGD with PT REKI and Community, facilitation of conflict resolution with the conflicting parties and interviews with relevant stakeholders. Field data collection was conducted in May-October 2012. Several FGDs were performed with PT REKI, migrant communities, Malay/ Batin Sembilan Community, Non Governmental Organizations (NGOs) and academicians. Semi-structured interviews were conducted with PT REKI, migrants, Malays/Batin Sembilan community, academicians, NGOs, and government. Following this, conflict resolution meetings were held with the relevant parties several times.

To facilitate the data collection and analysis, researcher uses the logical thinking from empirical facts which is comparing the 3 conflict resolution strategies implemented by PT REKI such as: litigation/ law enforcement, non litigation (negotiation and mediation) and do nothing. The evaluation effectiveness of these three strategies are based on ecological benefits, mutual benefit and access for community, relationships both of conflicting parties, meeting with national and international law/regulation.

In the context of different tenure situations, and to manage complex social and institution as well as silvicultural, and issues like in forest management in Indonesia, collaborative management is the good approach to reach sustainable forest management [20]. To reach the high quality of collaborative management result needs high ladder participation. According to Carter and Gronow [20] Collaborative management is "loosely defined as a working partnership between the key stakeholders in the management of a given forest–the key stakeholders being local forest users and state forest departments, as well as parties such as local governments, civil groups and non-governmental organizations, and the private sector" .

According to Scott [21] collaborative forest management should inclusive all of parties, must flexible and open of inputs. He offered several principles of collaborative forest management which are implemented by forest authorities, sustainable forest management is the major objective, high ladder participation, process orientation and implementation output, fair deal all of parties, fair distribution and benefits, address real issues, responsible for agreement, permanently result, all interest group must be involved, and agreement should be arrive at through consensus.

Considering the forest conflicts and how to resolve it with permanently and bring benefit to conflicting parties, as well as manage the complex social and economic issue, I adopted the Collaborative Conflict Management (CCM process) of Castro and Engel [22,23] to analyze the conflict resolution process in Harapan Rainforest. CCM is an iterative process, able to speed-up or slow-down the process quickly and efficiently. The process is divided in to 4 milestones with ten steps and specific activities. The steps in this approach create a "process pathway". Process Pathways help mediators/facilitators progress towards CCM and achieving beneficial results. The four milestones and ten steps can be seen in the Figure 1.

Background on Conflicts and Sites

Harapan Rainforest is located in the border of Jambi and South Sumatra provinces with 98,555 ha (Figure 2). This site was proposed to become the ER concession as this area is the last remaining dry low land rainforest in Sumatra, with rich biodiversity and is the habitat of large mammals such as the Sumatran tiger and Sumatran elephant and this area is also categorized as an Endangered Bird Area [24]. Based on Holmes [25] Sumatran low land forest will be disappeared in 2015, and its ecosystem becomes the most critical endangered ecosystem in South East Asia. With protecting this forest landscape, means that protects 20% of remaining Sumatran lowland forest.

Originally, the location of HRF that was proposed by Burung Indonesia and its Consortium to the Ministry Forestry was smaller than the total area of these two logging concessions which is only 101,355 ha of from 214,387 ha. The smaller proposed area was aimed at avoiding conflict with community. Following this, the Ministry of Forestry released the first PT REKI 'permit totaling 52,170 ha in the South Sumatra in Musi Banyuasin District through SK.293/Menhut-II/2007 and PT REKI' and a second permit totaling 46,385 ha in Jambi Province, Sarolangun and Batanghari District with SK. 327/Menhut-II/2010.

In Harapan Rainforest there are approximately 228 families of Batin Sembilan Indigenous people [26] who live in this area and depend on the forest and its natural resources for survival. Historically the Batin Sembilan Indigenous people have practiced swidden cultivation (slash and burn), collected non-timber forest products and hunted and gathered animals to fulfill their need for protein and also to sell as a source of income. Some of this group has settled in villages and others still live a semi-nomadic life. They have a specific economic and social

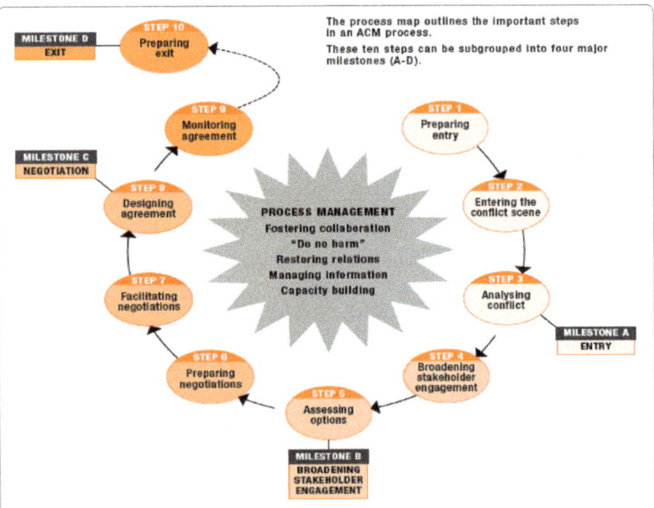

Figure 1: The important steps of collaborative conflict management [22,23].

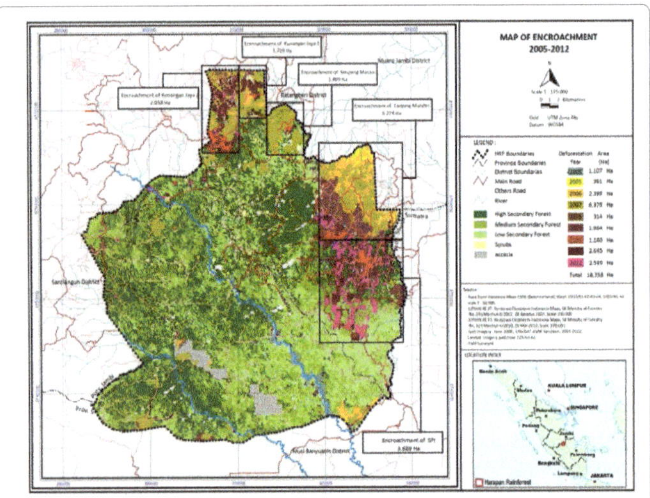

Figure 2: Location map and the encroachment problem in Harapan Rainforest.

system based on their traditions and can be differentiated from other people living in these villages [26]. PT REKI accommodated the Batin Sembilan right through giving 5% of concession to be livelihood zone on PT. REKI' land use zone. There are 3,201 families or 14,400 people including migrants living in and around the HRF in the six villages, including: Sako Suban and Pagar Desa in South Sumatra, Bungku, Tanjung Lebar, Sepintun, and Lamban Sigatal in Jambi Province [27]. Only one of these villages is on the boundary of the Harapan Rainforest which is Sako Suban, the others are out of the Harapan Rainforest concession. The Malay and migrant people tend to plant of oil palm, rubber, rice paddies, and other agricultural plants. In Sako Suban and Pagar Desa villages the community is involved in illegal logging activities.

Many migrants have come to the Harapan Rainforest from Jambi and other provinces since 2007. The largest number of migrants has come from 2007-2010 [17]. PT REKI has not had the right to undertake monitoring and forest protection work during that time as the license for the Jambi area had not yet been received. The presence of new migrants has impacted on the levels of encroachment in this area and on illegal logging activities. When the Government was given the license for the Jambi area in 2010, at least 13,563 ha of PT REKI's ER concession area was opened up by encroachers. This meant that the local government could not undertake monitoring activities in this concession. This period can be described as having open access without any concession holders. Approximately 28% of the ER concession area in Jambi has suffered from encroachment; PT REKI should pay tax to the government for the total concession.

Two of the biggest challenges for ERC in Harapan Rainforest are encroachment and illegal logging. Based on an analysis of a landsat satellite image 7 ETM and Envisat.asat image, from 2005-2012 (Figure 2) the total cleared area due to encroachment and illegal logging was ±18,758 ha with an average cleared area being as large as 3,126 ha/year. If the encroachment cannot be stopped stop, this concession area in Jambi will be lost within the next 10 years. In Jambi area, the main cause of land clearing is encroachment. Meanwhile, in South Sumatra up to the Jambi border, illegal logging predominates. Conflict resolution, however, is directed towards dealing with conflicts arising from encroachment in the Jambi area.

From the total cleared land area, encroachment has sporadically taken place in the northern area, where the bottle caps are located in Kunangan Jaya, which is as large as 3,777 ha, in Simpang Macan, which is 1,409 ha and in eastern area which includes Tanjung Mandiri and Alam Sakti amounting to 6,274 ha and in the Serikat Petani Indonesia' (SPI's) claim area with the total 3,689 ha (Figure 2). In this article resolving conflict mainly discusses the conflict between migrants and PT REKI in two sites which is in Kunangan Jaya and Tanjung Mandiri. The majority of people in these two sites is migrant, only small portion (less then 10%) is Batin Sembilan Indigenous People and Malay-local people. Nevertheless, the intermarriage of migrant and indigenous people and Malay local people has happened in this area, and majority of migrants admitted to acquire land from Indigenous and Malay people [28,29].

Result and Discussion

Rate of deforestation in Harapan Rainforest

There has been a steady increase in the annual amount of land cleared. As Figure 3 below indicates prior to 2005 there was only 871 ha cleared. The amount of cleared land peaked at 6,379 ha in 2007 and dropped down to 696 ha in 2012. But the overall tread has been upward.

Many efforts have been undertaken to decrease the levels of deforestation in Harapan Rainforest, which include non litigation (negotiation, mediation, social development to reach Forestry partnership), and litigation (law enforcement, forest fire monitoring, and forest patrols).

The roots of conflict

To comprehensively understand the land/forest conflict in Harapan Rainforest it is necessary to understand the policy on forests (national and Jambi) and its implications, the role of the government as a regulator, intervention, formal and informal regulations, the situation of land use and inequality of land tenure and the motivations of encroachers.

The roots of this conflict in Harapan Rainforest have a strong relationship with central and local government policy. These policies still prioritize the companies as managers of the production forest creating land tenure inequalities between corporations and the community. This can be demonstrated by the fact that 99% of production forests are under company management and less than 1% are under community management until 2011[6]. Central government

Figure 3: Land clearing rates in Harapan Rainforest from 2005-2012 (Harapan Rainforest Management Unit 2013).

policy on transmigration has failed to anticipate the environmental impacts and the land needs of the second generation. Transmigration programs are linked to conflict and encroachment, for example: transmigration program face conflict within the Bukit Tigapuluh National Park, Tesonilo National Park, and conflict in Alam Sakti-Tanjung Mandiri in the ER concession of Harapan Rainforest.

In Jambi and the area surrounding the Harapan Rainforest, most of the forest and the land have been leased to oil palm and industrial plantation companies. Besides, the forest partnership that is a mandatory of company license under P.39/Menhut-II/2013 do not succeed implemented. So far the forest management system has lacked participation, is driven by a top-down decision making process and involvement of the police and the military to push community [10]. The top-down licensing process does not always take individual conditions and circumstances into consideration. It implies many of the licensing overlaps between local and customary rights are the root of forest and land conflicts.

When there are many overlapping issues between the rights of a company and a community, the government has no responsibility to resolve these issues. Instead, they allow the company to resolve them independently, even though they have already spent money on the licensing process and taxes. In short, it can be said that the government policy typically victimizes both the company and the community. In the case of the Harapan Rainforest, when the government issued a license for PT REKI in Jambi 2010, the total encroacher's area covering around 13,000 ha (Figure 2). The government has let PT REKI to resolve the encroached area independently without support from central and local governments even company has already paid license fee and other responsibilities.

Another root of the conflict is differences between the Agrarian Law No. 5/1960 and other sectoral laws (eg: Forestry law No 41/1999, plantation law No.18/2004). In agrarian and forestry law, customary ownership is recognized, however in forestry law the operational/ technical regulation for implementation customary law are not available. Thus no customary law has been issued by Forestry Minister. In the Village law No. 5/1975 has also recognized the indigenous people right, but the Village Law weakens community land control and the rules concerning customs attached to them. This situation is compounded by the loss of traditional cultural treasures (such as erasing traditional leadership roles and changing this to a uniformed approach) and loss of their economic resources because the control of the forest and land has moved to a private party.

In the democracy transition era, the power of the state for agrarian resources (forest, land and mining) is weakened. All natural resources that have been categorized as common-pool resources (e.g forest) become open access [30]. He explained further that categorized ownership of forest in Harapan was before as a "common-pool customary property right" by Batin Sembilan indigenous people. When this forest was given to logging concession company the ownership shifted to "common-pool private property". Then when PT REKI got the forest concession in 2010, it has been changed into "common-pool quasi conservation private property". Wibowo [30] noted that the effective way to manage common resources/property are collaboratively approach.

The encroachers began to come to HRF when the ex PT Asialog forest has no concession holder which was between 2005-and 2010 [31]. Community viewed that forest ex PT Asialog is open access and local government could not able to protect the forest, local government did not implement their function as monitoring. The government has a chance to open up the forest thereby improving their livelihoods. Local elites, speculators and businessmen have illegally logged, opening up the forest and claiming it as their own and then selling forest products to others. When the law enforcement done, encroachers invite their family and connections to strengthen their power and legitimacy. Newcomers get identity cards and buy the land from the local elite to prove that they are legal.

In 2010, even though PT REKI had obtained ERC licenses, new migrants logging the forest could not be stopped completely. Almost all new migrants after 2010 came to Harapan Rainforest with economic motivations. New migrants can buy forest and land very cheaply from local elit, and then plant oil palms that can add up to 1000% to their land value. They said they are "gambling", and they believe that the more people that relocate here the more power they will have.

Different perceptions, interests and needs of stakeholders in the Harapan Rainforest are some of the root causes of the conflicts. The Batin Sembilan Indigenous people perceive that the Harapan Rainforest is their own. The Local Malay people perceive that they have access/right to its natural resources and as micro climate controllers, PT REKI and board of patrons (Board of patrons are the representatives of Harapan Rainforest initiators (Burung Indonesia, Royal Society for Protection of the Birds, and Birdlife International) or shareholders to be the highest position in strategic and direction in decision making. There number of Board of Patrons are 5 from these tree organization) perceive that Harapan Rainforest is also under their own private ownership. Migrants perceive that this is state forest and should function for the welfare of the community; thus they have right to open forest for their livelihoods. The government perceives the Harapan Rainforest is owned by the state managed by PT REKI. Different interests of these groups include forest restoration, protection, and exploitation for economic development, for business and traditional livelihoods are other roots of the conflicts.

Actors, power relations and chain migration

Total encroachers in HRF are predicted to be approximately 8,000 families (Budi Aulia pers.com). The massive amount of encroachers happen in Sungai Jerat called Claim SPI. Initially, the encroachment problems are very complex and involve the local elites, migrants and speculators. Staff form local government, individual law enforcement apparatus involve and have provided support to the encroachers. The encroachers are now becoming stronger by gaining support from various mass organizations and NGOs /CSOs (STN, CAPPA, PH, AGRA, SETARA, SPI, PRD) that use poverty, land reform and reject the REDD+ campaign (they support encroachers to grab HRF). Meanwhile others bring up issues concerning poverty (PPJ), Indigenous people (AMAN), the environment (WARSI, Gita Buana, SSS) to try and empower the community to reach natural resource agreements. Various government programs, village heads and elite government visitors and local elections have legitimized the existence of encroachers. An analysis of actors and power relations in land/forest conflict in Harapan Rainforest can be seen in Figure 4.

Based on participatory data of land claims coordinated by Forestry District of Batanghari in 2012 most of migrants/encroachers in Kunangan Jaya I are coming from Jambi provinces which are second generation of transmigration locations (Singkut, Pemenang, and Kuamang Kuning) and the rest are coming from other provinces. Meanwhile in Kunangan Jaya II, most migrants are coming from North Numatra (25.74%), Jambi (22.77%), Jawa (12.21%), and the rest are coming from Aceh, Bengkulu, Lampung, Palembang, Kepulauan Riau, Pekan Baru and Kalimantan. It is indicated that there is a good relationship between the actors in Harapan Rainforest and the leaders in each province who invite more people to come to these areas. The second generation of trans-migrants are coming to the ERC in Harapan Rainforest because their father's land listed under the trans-migrant scheme (2 ha) and this is not enough for their children. The successful trans-migrants also want to expand their oil palm land by buying very cheap land in Harapan Rainforest. One hectare of forest in HRF can be bought from Rp.500.000 - 2.000.000. SPI local leaders in Sei Jerat have offered 2 ha for Rp1million to new migrants. Several encroachers have sold their land in North Sumatra and Java and a half a hectare sold there could buy more than 10 ha of forest and land in Harapan Rainforest. Many of the new migrants are speculators who buy and open the forest and sell it again to others at a higher price.

From these actors in Harapan Rainforest, the primary stakeholders are PT. REKI, community and Ministry of forestry/local government. NGOs, local elites, and law enforcers (army, police, judges) are secondary stakeholder. Meanwhile the rest (donors, journalist, investors) are the terrier stakeholders. In the process of conflict resolution trough negotiation and mediation in Harapan Rainforest, there are only primary stakeholders to sit in the negotiation table. However, secondary stakeholders involve in engagement process.

Lands claims and commodities

Based on participatory mapping data coordinated by the Batanghari Forest Agency Team in 2012, those carrying out encroachment are usually not locals to the area in which the encroachment takes place, and may also be owners of capital. Only about 20% of encroachers possess 2 ha of land and fall into the poverty category, 37% are in the middle category which posses 2.25-5.5 ha, whilst 53% of encroachers possess more than 5.5 ha with some even possessing hundreds of ha such as JT (200 ha), NBH (31 ha), GLG (8 ha). It means that majority of encroachers are not really poor but are using their land acquisitions to become more wealthy.

To strengthen their land claims, encroachers get the "SKTT" (Surat Keterangan Tanam Tumbuh/Plant Ownership Certificate) from the village head. This letter explains the ownership of plants above the ground. This letter is not a legal certificate which makes it difficult to use as evidence in court.

Commodities planted in encroachment areas based on the Participatory registration in Kunangan Jaya are oil palm (23%), rubber (7%), rice paddies and agricultural plants (3%), and the remaining encroachment area (67%) is scrub. Encroachers tend to plant oil palm because of the good/stable price, is suitable based on the agricultural conditions, and it generates a high income.

Innovative approaches and negative precedents on the HTR decision

There are many systems that have been put in place by Harapan Rainforest Management Unit (HRFMU) in order to resolve the conflict in a persuasive way. This has included: socialization, registration land participatory, establishment of a government integrated team, negotiation, and mediation. Socialization has been conducted by government integrated team called "KP3RE" (Kegiatan Penanganan dan Penyelesaian Perambahan di Kawasan Restorasi Ekosistem) established by the Jambi Governor in 2012. KP3RE has three divisions such as socialization, partnership and law enforcement. Socialization has been conducted in Kunangan Jaya I and II, Alam Sakti and Tanjung Mandiri. The socialization team was rejected by a member of SPI in Sei Jerat. Registration claim participatory is as an innovative way to know and control the encroachers and obtain land ownership with a higher level of legitimacy. The land registration land participatory has succeeded in Kunangan Jaya I and II and Tanjung Mandiri.

Negotiation process for one group (Narwanto) in Kunangan Jaya I has been succeeded to reach an agreement and mediation process is in ongoing process (Tri Magnuh group). Negotiation process is facilitated by Batanghari Forestry Head. The mediation process with Tri Magnuh group was stuck more than a year from 2012 and has been started again in the November 2013. When the conflicting parties want to publish of its next mediation process in September 2012, on the same time the executive head of PT REKI was changed. The new management does not support mediation process. The Board of Patron PT REKI doesn't have a good understanding of the mediation process, which resulted in the mediation failing to achieve the desired outcomes. At the same time, community in Kunangan Jaya II held a demonstration against the Ministry of Forestry at the local government office. The conflict in Kunangan Jaya involves the acacia and rubber plantations (PT ALN and WN) and has become a national issue. Thus, the Ministry of Forestry office facilitated a meeting with the conflicting parties to resolve the conflicts. Unfortunately, when the community held the demonstration after three months, on 1st February 2013 the Directorate General of *Bina Usaha Kayu* released the letter No S.92/VI-BUHT/2013 deciding Community Forestry Plantation/Hutan Tanaman Rakyat. This decision set a bad precedent in resolving the conflict because it was against the law and did not accommodate the company's interests and Kunangan Jaya I community as well as influenced the mediation process. This decision opposes the Government Regulation No 26/2008, whereas limited production forest and production forest that have concession holders can not be given a Community Forestry Plantation permit. It means that there is an overlap between the two permits. The decision also disturbed the mediation process in Kunangan Jaya I.

The law enforcement processes are done by law apparatus on criminal (burning the assets, burning and cutting the forest). Several time law enforcements are done for criminal cases and for Claim SPI areal due to their presence from in 2011-2013. It means that the

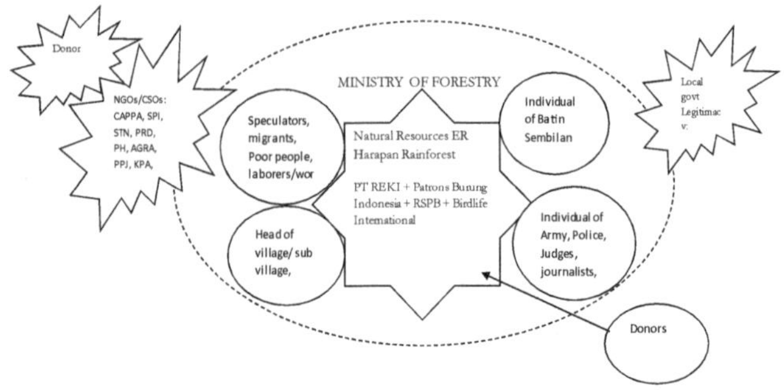

Figure 4: Actors and power relation.

encroachers come after license was hold. Law enforcement processes always increase the escalation of conflict. There were two incidents when the forest patrol camps, cars and motor bikes were burnt by encroachers and they kidnapped the patrol staffs. The law enforcers can temporarily pressure the encroachers, but this action invites more new migrants into Tanjung Mandiri and Claim SPI area. The community invites more people to increase their power. On another place-Sei Jerat with do nothing activities, encroached area become bigger and bigger.

Comparing the conflict resolution strategy in Harapan Rainforest

To analyze the appropriate strategy in conflict resolution based on point 2 the second paragraph, three approaches that has been used in HRFMU are litigation law enforcement, doing nothing, and non litigation using a collaborative method (negotiation and mediation). Litigation (law enforcement) methods to resolve conflicts arising from encroachment, as mentioned above, cannot permanently resolve land clearing disputes as it just stops them temporarily. Law enforcement can create hostility, increase conflict escalation, impact upon victims and cause damage, and the result of court decisions may not be applicable in the field [19]. For example the incident of burning 2 cars, dozens of motor bikes and post guard on 31st July 2013. There are often feelings of injustice on the part of the aggrieved party that are only exacerbated. The litigation process increases the level of hatred, widespread conflict and improves the skills of the next generation of militants. The litigation process always addresses the human rights issues and is based on international principles. This process is used appropriately in the criminal cases with the speculators are targeted and avoids issues human rights. In the case of Tanjung Mandiri and Claim SPI, law enforcement activities impacted upon the staff of REKI who could not access the forest and control new migrants. Law enforcement activities could not develop a partnership scheme that benefited PT REKI, the communities and the forest.

By doing nothing, there is no clear resolution and conflict escalation is stable but the number of encroachers will increase. PT REKI could not manage the forest and land independently. Besides, the company's reputation will worsen and no initiative, not even the ERC regulation will help the company to exercising their policy and strategy for conflict resolution because the status is a production forest, not adaptive and no innovation, and this will cause the loss of the ERC program in Harapan Rainforest and other new ERCs in Indonesia.

Non litigation using negotiation and mediation or another term "collaborative conflict management" will result in an acceptable mutual and workable agreement. The agreement is freely upheld without coercion and covers the rights and responsibilities of the encroachers and PT REKI. CCM creates harmony between encroachers and PT REKI like in Kunangan Jaya and Tanjung Mandiri. Such harmonious relations will also decrease patrolling because local communities will automatically act as "social security". CCM provides a sense of justice and results in-line with international principles and laws such as FPIC and the upholding of human rights adheres with "PHPL" (Indonesian sustainable forest management scheme) on social criteria indicators. Implementation of CCM in resolution conflict will reduce the conflict escalation, can control and give benefit both people, company and forest. A new social contract or agreement can be negotiated based on a collaboration zone in the company's yearly plan that is agreed on by the Ministry of Forestry. Any new forest logging can be stopped, new migrants can be controlled and commodities can be negotiated. The social contract and agreement can bring social and economic benefit for community, and creates a positive image for the company. It can be seen from the case of Kunangan Jaya I whereas the harmonies relationship developed and agreement will bring benefit to both of conflicting parties. In short, collaborative conflict management is a more appropriate way to resolve the forest and land conflicts.

Implementing and adjusting the collaborative conflict management

Conflict can be difficult to resolve but it can be managed to create more equitable solutions. Managing the conflict over common pool resources needs collaborative action [30]. Based on the conflict and several methods that have been initiated by PT REKI with some problems and issues, the encroachment strategy in Harapan Rainforest could not be implemented in a universally applied way. The main approach is to create benefits for both conflicting parties in order to create a new agreement for natural resource management in Harapan Rainforest by using collaborative approach and a legitimate process.

Based on conflict collaborative management [22,23] we analyze the implementing of the Collaborative Conflict Management model in Harapan Rainforest with using the example in Kunangan Jaya I and Tanjung Mandiri as below:

Milestone A: Entry to the conflict: PT REKI has the initiative to contract the third independent facilitators or conflict resolution experts in Kunangan Jaya I and Tanjung Mandiri to do conflict analysis and capture recommendation options. These two independent consultants are expert from Wahana Bumi Hijau and Impartial Mediation Network. The independent consultants came in to Kunangan Jaya I and II and Tanjung Mandiri to understand the root of the conflict, stakeholder analysis, and resolution options. This step is a part of the initial conflict research and is based on a detailed understanding on conflict and consultation with stakeholders in Kunangan Jaya I and Tanjung Mandiri. There are three steps to reach Milestone A that has been implemented in Harapan Rainforest:

i. **Preparing entry:** Independent consultants clarify their roles with the conflicting parties. Research the background of the conflicts, and develop a good strategy to approach the different parties in Kuangan Jaya I and Tanjung Mandiri.

ii. **Entering the conflict scene:** independent consultants meet firstly with the conflicting parties separately (communities, PT REKI, and local government), then with other stakeholders to understand how they see and view the conflict. Independent consultants clarified their role in participating in the process, and ensuring their commitment to work together to resolve the conflicts.

iii. **Analyzing conflict:** Independent consultants analyzed the root of conflicts, conflict escalation, modus operandi, and stakeholder's analysis in Kunangan Jaya I and Tanjung Mandiri. The consultants should get the picture whether both conflicting parties agreed or not for willingness to resolve the conflict. The most important of this step is to know that conflicting parties in both locations are willingness to resolve the conflict/sit together.

In the case of Kunangan Jaya I, and Tanjung Mandiri, independent consultants succeeded to implement these tree steps and get the information needed. Getting based interest negotiation is a part of the work and their intervention will not create harm to either party. The frank communication and a precautionary principle have been taken for move forward the process in to Milestone B.

Milestone B. Broadening stakeholder engagement: There are three steps that have already done in Kunangan Jaya I and Tanjung Mandiri to reach Milestone B:

i. Broadening stakeholder engagement: Independent consultants help conflicting parties to share their positions, their different needs and interests, and choose conflict resolution mechanism. Especially for interests, facilitator should look for the slices of their interests.

ii. Assessing the options: Independent consultants help the conflicting parties to produce an alternative in resolving and managing the conflict. The benefits from each alternative are analyzed and prioritized. The task of these two consultants stopped in this step, then PT REKI and community agreed to choose forestry district Batanghari as facilitator to follow up the object and subject of conflict before entering negotiation and mediation.

iii. Clarify object and subject the conflict: participatory claim land registration has been done in Kunangan Jaya I and Tanjung Mandiri. Forestry district is a leading facilitator for it supported by communities, PT REKI, and NGOs to have legitimacy process. Claim land registration is very important to know clearly the object and subject the conflict for move forward to negotiation/mediation process.

In the case of Kunangan Jaya and Tanjung Mandiri, to reach of Milestone B has been done 3 steps as mentioned above. From the conflict analysis reported by consultants and claim land registration done participatorily, conflicting parties In Kunangan Jaya I and Tanjung Mandiri agreed to choose facilitator and mediator. Group Narwanto in Kunangan Jaya I and PT REKI choose negotiation process facilitated by head of Batanghari Forestry District, Tri Magnuh- group in Kunangan Jaya I and PT REKI agreed to use joined mediators (Independent and government mediator), and Tanjung Mandiri and PT REKI choosed integrated team of Batanghari District government to be a facilitators. After every conflicting parties agreed to sit in one table then the negotiation process can be done. In the process of mediation, Milestone A and B is also called pre mediation.

Milestone C. negotiation: From mentioned above, there are three kinds of conflict resolution process to be done which are negotiation for Narwanto Group, mediation for Tri Magnuh and Tanjung Mandiri. For Narwanto group negotiation process has been reached an agreement (step 9), while for Tri Magnuh Group is in the second roundtable meeting (step 8), and for Tanjung Mandiri is in the initial step (step 7). To reach Milestone B steps to be done are:

i. Preparing for the negotiation: Facilitators and mediators used the conflict analysis reports as a basis for preparation negotiation. Facilitators and mediators engaged conflicting parties in Kunangan Jaya I and Tanjung Mandiri to make a pre requirements e.g no black campaign, no new cutting forest, no law enforcement done, etc and have a willingness to sit together. The facilitators and mediators use precautionary principles and enough preparation. This covers the preparation of people involved, legitimate mandatory, explores strategies, and a plan of type of meeting to be held.

ii. Facilitation negotiation: This part is the most challenging for conflict resolution as it involves the conflicting parties attempting to convince each other to reach an agreement. In the Narwanto group vs PT REKI, there were 4 roundtable meetings that have done to negotiate the conflicting parties' interest. The difficultiest negotiation is when they resolved the existing commodity (oil palm) that is contradict with forestry law. Facilitator showed the skill to change the position in to interest /basic needs and reduce the gap between the parties. The same happen in group Tri Magnuh vs PT REKI, the second meeting tried to deal with existing commodities (oil palm and rubber), but it could not reach an agreement yet and has to be negotiated on the third roundtable meeting. When the roundtable meeting done, several times facilitators/mediators initiated a separate meeting to have hidden interests, influence both of parties, and reducing gap.

iii. Designing the agreement: when Narwanto Group vs PT REKI has been discussed prioritized problems and agreed the resolutions, the facilitators/mediators drafted and got an approval from conflicting parties. The facilitators should encourage the implementation of the agreement. There many agreements that cannot be implemented after the mediation is finished and implementation can be more difficult than reaching an agreement.

This step is also called negotiation and mediation process. The next step is to implement and monitor the agreement.

Milestone D: Exit strategy: The three cases mentioned above have not reached the milestone D yet. Especially for Narwanto Group that has signed agreement with PT REKI, agreed to implement and monitor the agreement for getting mutual benefits and building harmonies relationship. Commonly the function of facilitator and mediator has only to reach an agreement. However, for the natural resources conflicts that is complex, facilitators/mediator should involve in the implementation and monitoring the agreement. This may be due to a setback when the conflicting parties could not implement the agreement or the relationship cannot make a full recovery. Even though the facilitators could not resolve all conflicts at the one time, they should ensure all of parties obey their agreement. Steps 10 and 11 should be done in the case of Narwanto vs PT REKI:

- Monitoring agreement: the facilitator (Head of Forestry District in Batanghari) should use their role to encourage the implementation and monitoring of the agreement.

- Preparing for the exit: The Batanghari Forestry District head should be able to develop a system/mechanism to transition the responsibility to the conflicting parties in the implementation and monitoring of the agreement. It can be transferred to a local facilitator to continue the monitoring process. Facilitators can develop the capacity of conflicting parties to resolve their conflict further.

The management process should cover minimal 9 steps as Castro and Engel [22,23] noted, foster collaboration, do no harm, build trust, restore relations, manage information, and build capacity. In the collaborative conflict management, should be patient and legitimated process. Nevertheless, the steps of CCM should be adjusted based on context and adapt with changing and complex issue.

Conclusion

The root causes of conflicts in Harapan Rainforest are central and local government polices impacting on the inequality of land tenure and not pro poor, the increasing population while the land is stable, developmental approaches, unification of village regulations, conflicting intra-formal laws and conflicts between formal and informal regulations, open access to resources and no concession management. Harapan Rainforest conflict reflects the forest and land conflict in Indonesia with complex and involving many stakeholders. Harapan Rainforest conflict also inherited of Dutch Colonial when implemented the Agrarische Wet 1870 in Sumatra. Implication of Agrarische Wet

1870 implementation was forest and land of Indigenous people in Sumatra including Batin Sembilan were back to colonial government because they could not show the evidence ownership [3]. Then it was followed by Indonesian government after independence.

The biggest problem for forest management in Indonesia is encroachment as impact of conflict over forest and land including in ER concessions. In the case of ERC Harapan Rainforest conflicts, many different parties are involved, each with a different perception, position, needs, interests and motivations. There are various motivation of encroachers which are for fulfill their basic need (poor people), increase their oil palm/rubber plantation (middle class), getting rich/wealthy (speculators, elites). Conflict in Harapan Rainforest has become more complex when the local/traditional elites involved, gaining the support from various NGOs, and local government supports. It needs a high level of conflict and stakeholder analysis to understand the real conflict and innovative solutions.

Based on the PT REKI experience and mentioned above, a conflict resolution strategy should prioritize non-litigation (negotiation and mediation) process with collaborative approach, and be followed by the litigation process (law enforcement) for criminal cases and speculators. This strategy can reduce the deforestation rate in Harapan Rainforest, control new migrant and bring benefits for people and forest including wild life. With using the litigation approach in the beginning of the project, PT REKI could not control new migrants; did harm community, created un-harmonies relationship, and got bad image.

Common pool resource management like Harapan Rainforest involves the interests of many stakeholders at different levels of hierarchy (international, national, provincial, district) and the conflict is very complex, high social and economic pressure, and fast changing situation. To address the conflict in Harapan Rainforest should use the Collaborative Forest Management. The collaborative conflict Management will bring benefit for the forest and community as a requirement of 4 pillars of ERC and can lead to sustainable forest management. Regarding to Conflict Collaborative Management process to reach 4 milestones should be implemented to guarantee the result and mutual agreement that can be implemented. However, it needs to be adapted/adjusted based on contextual problem and a flexible, and the process should be improve continuously.

References

1. Ministry of Forestry (2011) Forestry Statistics of Indonesia, Jakarta.
2. http://www.dephut.go.id/index.php/news/otresults/824
3. Nurjaya I N (2005) Historical Law of forest Management in Indonesia. Jurisprudence Maret 2: 35-55.
4. Poffenberger M McGean B (1996) Village Voices, Forest Choices: Joint forest management in India. New Delhi: Oxford University Press.
5. Peluso Lee N (1990) A History of State Forest Management in Java. In: Keepers of The Forest, Land Management Alternatives in Southeast Asia, Poffenberger M (Ed.), Ateneo de Manila University Press, Philippines. 27-55.
6. Santosa A, Silalahi M (2011) Kajian Kebijakan Kehutanan Masyarakat dan Kesiapannya dalam REDD+(Policy Analysis on Community Forest and its Readiness to REDD+). Forum Komunikasi Kehutanan Masyarakat, Bogor
7. Bursche A (2012) Report on Progress Project Review "Piloting Ecosistem Restoration in Harapan Rainforest Sumatra-Indonesia, KFW, Germany.
8. Darusman D, Siregar IZ (2013) Analysis of ecosystem restoration bussines toward sustainable production forest management: in proceeding 1st scientific discussion on concept, policy and implementation of ecosystem restoration policy: lessons learned, prospect and challenges. Cooperation between IPB and Burung Indonesia, Bogor.
9. Maring PM, Afrizal, Suhendri JS, Rosyani, Rafiq R, et al. (2011) Studi Pemahaman dan Praktik Alternatif Penyelesaian Sengketa oleh Kelembagaan Mediasi Konflik Sumberdaya Alam di Provinsi Riau, jambi, Sumatera Barat, dan Sumatera Selatan.
10. Wulan YC (2004) Analisis Konflik Sektor Kehutanan di Indonesia 1997-2003,Bogor: CIFOR dan FWI.
11. Yasmi Y, Kelley L, Enters T (2010) Conflict over forests and land in Asia: Impacts, causes, and management, Bangkok, Thailand.
12. Dirjen BUK (2011) Pengusahaan Hutan Alam Produksi Indonesia, Jakarta.
13. Hidayanto Y, Walsh TA Asmui, (2012) Ecosystem Licensing Process, Info Brief Ecosystem Restoration, Burung Indonesia, Bogor.
14. Governmental regulation (pp) no. 6/2007 on forest arrangement and formulation of forest management and utilization plans.
15. Utomo AB (2013) The opportunities of ecosystem resotration bussines development, in proceeding 1st scientific discussion on concept, policy and implementation of ecosystem restoration policy: lessons learned, prospect and challenges. Cooperation between IPB and Burung Indonesia, Bogor.
16. http://www.nabu.de/en/aktionenundprojekte/indonesia.
17. http://www.harapanrainforest.org.
18. Enters T, Kelley L, Pescott M, Durst PB (2010) Removing constraints to private sector investment in forestry in Asia and the Pacific. In: Growing green assets Pescott M, Durst PB, Leslie RN (Eds.), Asia-Pacific Forestry Commission, Thailand.
19. Yuliani EL (2006) Step to Change. In: Multi Stakeholder Forestry Bogor, Indonesia, Center for International Forestry Research, Indonesia.
20. Carter J, Gronow J (2005) Recent Experiences in Collaborative Management: A Review Paper, Bogor.
21. Scott P (2000) collaborative forest management-the process, paper presented at the national workshop on community forestry management, Kampala, Uganda.
22. Castro AP, Engel A (2007) Negotiation and mediation techniques for natural resource management: Case studies and lesson learnt, FAO, Rome.
23. Wiyono A (2013) Pendekatan dan Metode Resolusi Konflik, TFT Indonesia-Semarang.
24. Beckert (2014) Contested land: an analysis of multi-layered conflicts in Jambi province, Sumatra, Indonesia.
25. Derek AH (2002) Indonesia: Where have all the forests gone? environment and social development, east Asia and pacific region discussion paper, World Bank, Indonesia.
26. Silalahi M (2013) Fact sheet: The Batin Sembilan and Harapan Rainforest: Ecosystem restoration for whom? ERRC Burung Indonesia, Bogor.
27. Ayat A (2013) Fact sheet villages profile in and around Harapan Rainforest. ERRC-Burung Indonesia, Bogor.
28. Hein, Jonas (2013) Reducing emissions from deforestation and forest degradation (REDD+), transnational conservation and access to land in Jambi, Indonesia.
29. Steinebach (2013) Today we occupy the plantation, Tomorrow Jakarta: Indigeneity, Land and Oil Palm Plantations in Jambi. In: Adat and Indigeneity in Indonesia, Schublin H (Ed.), Indonesia.
30. Wibowo A (2013) Adakah jalan keluar untuk konflik tenurial IUPHHK - PT Restorasi Ekosistem Indonesia,Material Presentation Scientific Second Discussion of Ecosystem Restoration materials, IPB and Burung Indonesia, Bogor.
31. Mangarah S (2013) Lessons learnt: Management of land conflicts in Harapan Rainforest. ERRC-Burung Indonesia – Bogor West Java.

Regeneration and Recruitment Correlate with Stand Density and Composition in Long-Unburned Aspen Stands Undergoing Succession to Conifer in the Sierra Nevada, USA

John-Pascal Berrill* and Christa M Dagley

Department of Forestry and Wildland Resources, Humboldt State University, USA

Abstract

Succession of aspen stands to conifer in the Sierra Nevada Mountains of California and Nevada, USA, is being interrupted by forest managers thinning conifers to sustain aspen stands *in situ*. However, patterns of stand density, species composition, and regeneration prior to management intervention have scarcely been described. We established a grid of sample plots throughout nine aspen stands encircling Lake Tahoe in the central Sierra Nevada. The degree of succession to conifer throughout each aspen stand was spatially heterogeneous. Patches of pure aspen were rare. Stand density index (SDI) reached an upper limit of 1700 in pure aspen. As composition shifted in favor of conifer, SDI approached or attained a maximum of 2500 in some plots. Stand density and species composition data were tested as predictors of conifer and aspen regeneration densities in each plot. Conifer seedlings had an average density of 3261 ha^{-1}; they were most abundant in aspen-dominated areas of any density, and in conifer-dominated areas of higher density. Aspen regeneration had an average density of 3211 ha^{-1} and was one order of magnitude less frequent in areas of pure conifer versus areas of pure aspen, but remained relatively abundant in most areas. Aspen saplings were rare (average density of 42 ha^{-1} for saplings 10-15 cm DBH), especially at high stand densities or in areas where conifers dominated. Our findings suggest that forest managers interested in sustaining aspen stands *in situ* will need to control stand density to promote recruitment of younger aspen to the overstory.

Keywords: Conifer encroachment; Establishment; Mixed-effects models; Negative binomial distribution; *Populus tremuloides*; SDI; Stand density index

Introduction

Succession from quaking aspen (*Populus tremuloides*) to coniferous species is occurring throughout many forests of North America [1-7]. Aspen is a pioneer species adapted to disturbance [8,9]. Some aspen form pure stands and appear to remain stable in that state, but in many cases succession to conifer is expected in the absence of disturbances such as fire that kills young conifers establishing within aspen stands [8]. Longer fire return intervals under active wildfire suppression throughout much of the 20th Century have given conifers time to establish and develop thicker fire-resistant bark with advancing size and age [10]. Conifers may eventually overtop aspen and suppress aspen regeneration and the herbaceous vegetation layer where much of the biodiversity is concentrated.

At the western edge of aspen's natural range, in the Sierra Nevada Mountains of California and Nevada, aspen stands ranked higher in biodiversity than meadows and conifer stands [11]. Pure aspen stands support a diverse herbaceous understory and a greater abundance of bird species than mixed stands where aspen is intermingled with conifers [12]. In the vicinity of Lake Tahoe, in the central Sierra Nevada, aspen are mostly restricted to riparian areas where their vegetation community stabilizes the banks of streams flowing into Lake Tahoe [13].

Forest managers want to maintain an aspen component within the conifer-dominated forests surrounding Lake Tahoe, but have limited options for intervention. Options not widely publicly supported include clearcutting or allowing fire to kill patches of mature trees and create openings for pioneering aspen to invade and occupy. Managers are constrained and must be cautious in their use of prescribed fire as a tool for restoration and management. Concerns over smoke, high fuel loads, and risk of escape and damage to dwellings throughout the wildland-urban interface have prompted forest managers to favor mechanical or manual conifer removal in aspen stands undergoing succession to conifer. Removal of smaller conifers from aspen stands is being undertaken throughout the Lake Tahoe Basin, with the objective of regenerating aspen *in situ* or thinning to relieve crowding.

Basal area (BA) and stand density index (SDI) are measures of stand density that consider tree size and number of trees per unit area. They are useful predictors of aspen tree growth and vigor [14]. Forest managers who are aware of an upper limit to stand density can evaluate stands and prescribe treatments that control stand density (i.e., occupied growing space) to achieve objectives such as enhancing tree growth or promoting regeneration; these 'stand density management regimes' are based on growing space occupancy relative to the maximum for any combination of age classes of trees [15]. Unlike the upper limit of BA which we expect to be higher in stands with older/larger trees, the upper limit of SDI offers the convenience of remaining fairly constant for a range of tree sizes/ages and site qualities for any single species. Less is known about how the upper limit of SDI may vary according to species composition in mixed stands [16].

***Corresponding author:** Dr. John-Pascal Berrill, Department of Forestry and Wildland Resources, Humboldt State University, USA
E-mail: pberrill@humboldt.edu

We sought to describe patterns of stand density, species composition, and regeneration in aspen stands undergoing succession to conifer throughout the Lake Tahoe Basin. This was a unique opportunity to sample an array of long-unburned stands prior to management intervention in the succession process (i.e., conifer removal). Specifically, we aimed to define an upper limit to SDI according to the degree of succession to conifer, and describe patterns of conifer and aspen regeneration and test for correlations with stand density and species composition. We hypothesized that: (i) the upper limit of SDI was significantly higher in areas of aspen stands that had succeeded to conifer; (ii) conifer regeneration was more frequent, and aspen regeneration less frequent, in areas of higher stand density and conifer dominance, and; (iii) aspen saplings whose presence we assumed indicated a higher probability of successful recruitment of aspen regeneration to the overstory were less common in areas of the aspen stands with higher stand densities dominated by conifer.

Methods

Study region

The Lake Tahoe Basin is a 1,310 km^2 watershed located in the central Sierra Nevada Mountains of California and Nevada, USA. Climate in the Lake Tahoe Basin (39°05'N latitude, 120°02'W longitude) is classified as Mediterranean continental with cold winters and summers with cool nights and warm days. Near lake level, at the Lake Tahoe Airport (elevation 1,906 m), August is usually the warmest month with an average maximum of 25.9°C and average minimum of 4.3°C. January is the coolest month with an average maximum of 5.0°C and average minimum of -9.4°C. Minimum temperatures of 0°C or lower occur on an average of 231.8 days annually. Mean annual precipitation varies between years, on average ranging from 660 mm near lake level on the east side of the basin to over 1440 mm on the west side of the basin. Most precipitation falls as snow between November and April (http://www.wrcc.dri.edu).

Field data collection

We selected nine aspen stands undergoing succession to conifer for sampling. These stands encircled Lake Tahoe, ranging in elevation from 1,900-2,260 m above sea level, with three located on the eastern shore of the Lake Tahoe Basin, three in the south, and three in the west [15]. There was no visible evidence of past wildfire in any stand. We established a systematic grid of sample points throughout a one-hectare study area in each stand, with sample points (i.e., plot centers) spaced 10 m apart along transect lines spaced 25 m apart, for a total of 28-30 sample points per study area. The one-hectare study areas were made square (100 × 100 m) or rectangular to fit inside each aspen stand. A square area had three long transects and rectangular study areas had 5-7 shorter transects.

Aspen and conifer regeneration <10 cm diameter at breast height 1.37 m (DBH) were counted in circular 0.004 ha plots (3.57 m radius) at each sample point. Aspen root suckers were tallied if their stem was not joined with adjacent suckers at/above ground level. Conifer tallies excluded seedlings in their first growing season since we expected that many of these seedlings would not survive and become established [17]. We recorded the proportion of area within each regeneration sample plot covered by large rocks (preventing regeneration), allowing for calculation of regeneration counts per unit of unobstructed ground area. Within each one-hectare study area, we mapped tree locations using a tripod-mounted flux gate compass to obtain azimuth and sonar hypsometer for distance from a series of survey points to each tree. Species and DBH were recorded for each aspen >10 cm DBH and conifer >20 cm DBH. Conifer saplings 10-20 cm DBH were so numerous that we counted them within the 0.004 ha regeneration plots instead of mapping and measuring them throughout each study area.

Analysis

Species composition and the upper limit of stand density

Tree data were summarized to describe the relative abundance of each species throughout the nine study areas. Each study area was spatially heterogeneous in terms of species composition and stand density. Therefore, we sought to quantify stand density at localized areas throughout each study area, and examine its relationship with species composition at each of these localized areas. We created a stem location map for each one-hectare study area by converting tree location data from distance and azimuth to an easting (x) and northing (y) for each stem. We used ArcGIS to query each stem location map and derive stand density and species composition in the vicinity of each systematic sample grid point. This was achieved by creating a buffer of 7.98 m radius around each sample point and clipping tree data in each buffer, giving tree data for a series of 0.02 ha circular plots throughout each study area. Tree data from each 0.02 ha plot were summarized, giving BA per hectare and SDI for the aspen and conifer stand components, and other hardwoods when present. SDI was calculated by summing individual tree SDI because the DBH data were not normally distributed: $SDI=\Sigma(0.04DBH_i)^a$ where DBH_i=DBH in cm of the ith tree in the plot, and a=1.605 [17,18]. Species composition in each plot was calculated as the proportion of aspen and proportion of conifer in terms of BA.

We sought to define an upper limit to SDI [19] according to species composition. A scatterplot of total SDI in each 0.02 ha plot versus species composition (ranging from pure conifer to pure aspen) revealed a paucity of data for aspen-dominated areas of higher SDI. Therefore, we established a second series of 0.02 ha sample plots at each study area, systematically sampling in between the first series of sample points, along parallel transects offset 12.5 m from the original transect lines. We also selected six pure aspen stands for sampling at different locations around Lake Tahoe, and established a 0.02 ha circular plot in what appeared to be a fully stocked area within each stand. These six stands covered a wide range of elevations, from 1920m (near lake level) at Burke Creek, Burton Creek, Spring Creek, and Tallac Creek and 1950 m at Fallen Leaf Lake, up to 2415 m at Marlette Lake. The supplemental SDI and composition data were added to the dataset and scatterplot. To define an approximate upper limit of SDI - exclusive of extreme 'outlier' data the SDI data were binned according to species composition, and the 99th percentile of SDI calculated for each bin. We regressed the 99th percentile of SDI against the mean species composition for each bin using PROC REG in SAS [20], defining the approximate upper limit of SDI for any composition. Cook's distance revealed low SDI in the 60-80% aspen bin, supporting collection of additional data. This prompted installation of plots in dense, aspen-dominated mixtures at South Lake Tahoe (1920 m elevation), Spooner Lake (2150 m), Upper North Canyon (2300 m), and 2 km south of Incline Lake at 2440 m elevation.

Regeneration patterns

Count data for regeneration in each 0.004 ha plot were regressed against estimates of stand density and species composition from the 0.02 ha plots centered on the same sample points. The regeneration frequency data had numerous 'zero' counts and a few very high counts (i.e., 'reverse-J' distribution), prompting fitting and testing of four candidate regeneration models: Poisson, negative binomial, zero-inflated Poisson (ZIP), and zero-inflated negative binomial (ZINB) using PROC COUNTREG in SAS [20]. Models were fitted to frequency data for conifer seedlings (DBH <10 cm) and two size classes of aspen: regeneration (DBH <10 cm) and saplings (DBH 10-15 cm). Aspen sapling counts were extracted from the stem map data for trees in each 0.02 ha plot used to estimate stand density and species composition around each systematic sample point.

Finally, we constructed generalized linear mixed-effects models and specified a negative binomial distribution to describe relationships between frequency of regeneration, the dependent variable, and candidate explanatory variables: stand density (SDI or BA) and species composition (aspen BA % or conifer BA %). Models were fitted using PROC GLIMMIX in SAS [21]. The nesting of sample plots within different stands was accounted for by specifying 'stand' as a random effect in the mixed-effects model. Selection of variables for inclusion in the final model was based on likelihood ratio tests comparing the full model against reduced models in terms of model chi-square. The full model included quadratic terms for stand density and species composition (e.g., SDI + SDI^2) and an interaction between stand density and composition. To depict the influence of stand density and species composition and their interactions on conifer and aspen regeneration, and recruitment of aspen to the sapling stage, we used the best generalized linear mixed-effects models to generate average (expected) values of regeneration density (stems ha^{-1}) for a range of stand densities in pure aspen stands, and aspen-conifer mixtures with either 75%, 50%, 25% or 1% aspen BA.

Results and Discussion

Species composition and the upper limit of stand density

Different combinations of hardwood and conifer species were present in each sample stand (Table 1). Pines were much less common than the true firs: red fir (*Abies magnifica*) and white fir (*Abies concolor*). Conifers accounted for between 52% and 95% of stand BA in the nine stands. Six of these stands comprised >75% conifer BA. Hardwoods other than aspen were only found adjacent to stream channels, and rarely exceeded 20 cm DBH.

The scatterplot of SDI and species composition (in terms of percent aspen BA) revealed that within the aspen-conifer stands we selected for sampling, conifers were frequently found in pure patches (at the scale of ~200 m^2 patch size; i.e., 0.02 ha plot size) (Figure 1). Within the nine one-hectare study areas, pure aspen patches were a minority with only 6% of the plots being pure aspen. The majority (70%) of the plots contained a mixture of conifer and aspen. Among mixed plots, almost 3/4 (i.e., 72% of mixed plots) were dominated by conifer in terms of BA.

The upper limit of SDI was higher in conifer-dominated areas. The 99th percentile of SDI in 0.02 ha plots was 2499 in areas of pure conifer and 1716 in pure aspen. The linear regression fitted to binned SDI-composition data defined an approximate upper limit of SDI for any composition. The t-test for the regression slope coefficient ($P<0.0001$) indicated that the upper limit of SDI correlated with species composition in terms of percent aspen. Woodall et al. [22] reported approximately 30% higher maximum SDI for pure aspen and aspen-dominated stands (composition 80-100% aspen) from a nationwide study. Being based on the 99th percentile of data, their estimate could be a reflection of high SDI in a few special areas where aspen were well adapted and had formed pure stands. Alternatively, our upper limit of SDI for central Sierra Nevada stands may be underestimated and should be validated with independent data. Conversely, our maximum SDI values for Sierra Nevada aspen-conifer mixtures (composition 20-80% aspen) were 15-25% higher than those presented by Woodall et al. [22]. This suggests that Sierra Nevada aspen-conifer mixtures are atypical, possibly because they are often restricted to wetter, more productive areas with deeper soils, such as lower slopes and alluvial flats [13]. In our nine sample stands, the most common conifer associates were red fir and white fir that are shade tolerant and support very high stand densities [19]. In other regions, aspen are widespread and may be more negatively affected by a different suite of competitors or found in forests that do not attain such high SDI [23].

Beneath the maximum SDI line, we also present a dashed line at 60% of maximum SDI, considered to represent the lower limit of a zone – or 'threshold' – of imminent competition-mortality above approximately 55-60% SDI [24-26] (Figure 1). Given that a relationship between SDI and competition-induced aspen mortality has not been defined for the Sierra Nevada, it would be prudent for forest managers interested in averting excessive crowding, mortality of aspen, and succession to conifer to maintain stand density below 1000 SDI (i.e., <60% of maximum SDI for pure aspen). Our findings also suggest that in conifer-dominated areas with SDI >1500 (i.e., >60% of maximum SDI for pure conifer), competition-induced mortality would only be "imminent" for conifer whereas aspen in a stand at 1500 SDI would be experiencing intense competition and may soon be extirpated while conifer tree growth proceeded and SDI attained and exceeded the upper limit of approximately 1700 SDI for aspen.

Aspen regeneration and recruitment to the overstory

Conifer seedling densities rivaled aspen regeneration densities, on average, throughout the nine sample stands surrounding Lake

Stand	POTR		PIJE		PICO		ABCO		ABMA	
BC20	44	(4.8)	24	(8.5)	6	(0.6)	326	(86.1)	-	-
BP2	185	(43.2)	22	(9.8)	28	(7.0)	218	(40.0)	-	-
CV05	119	(48.2)	3	(1.0)	52	(17.6)	103	(26.5)	38	(6.7)
CV06	54	(13.6)	-	-	68	(22.0)	69	(30.6)	80	(32.3)
NC03	73	(21.4)	48	(24.7)	-	-	54	(15.0)	110	(38.9)
SHC01	165	(22.2)	17	(24.1)	6	(4.4)	107	(48.4)	-	-
SSP24	122	(22.2)	8	(3.8)	28	(7.1)	371	(62.8)	3	(4.2)
TC01	192	(29.4)	7	(3.6)	-	-	134	(37.7)	58	(25.6)
WA38	58	(12.8)	-	-	63	(28.7)	117	(33.5)	93	(24.9)

Species included: *Populus tremuloides* (POTR), *Pinus jeffreyi* (PIJE), *P. contorta* (PICO), *Abies concolor* (ABCO), and *A. magnifica* (ABMA). Species not included: *Alnus incana* ssp. *tenuifolia* (2 trees ha^{-1} at SHC01), *P. monticola* (1 tree ha^{-1} at CV06) and *Salix spp.* (3 trees ha^{-1} at SHC01 and 26 trees ha^{-1} at TC01).

Table 1: Density (trees ha^{-1}) and species composition (percent of stand basal area; in parentheses) of aspen and conifer trees >20 cm DBH within nine aspen-conifer stands surrounding Lake Tahoe, California.

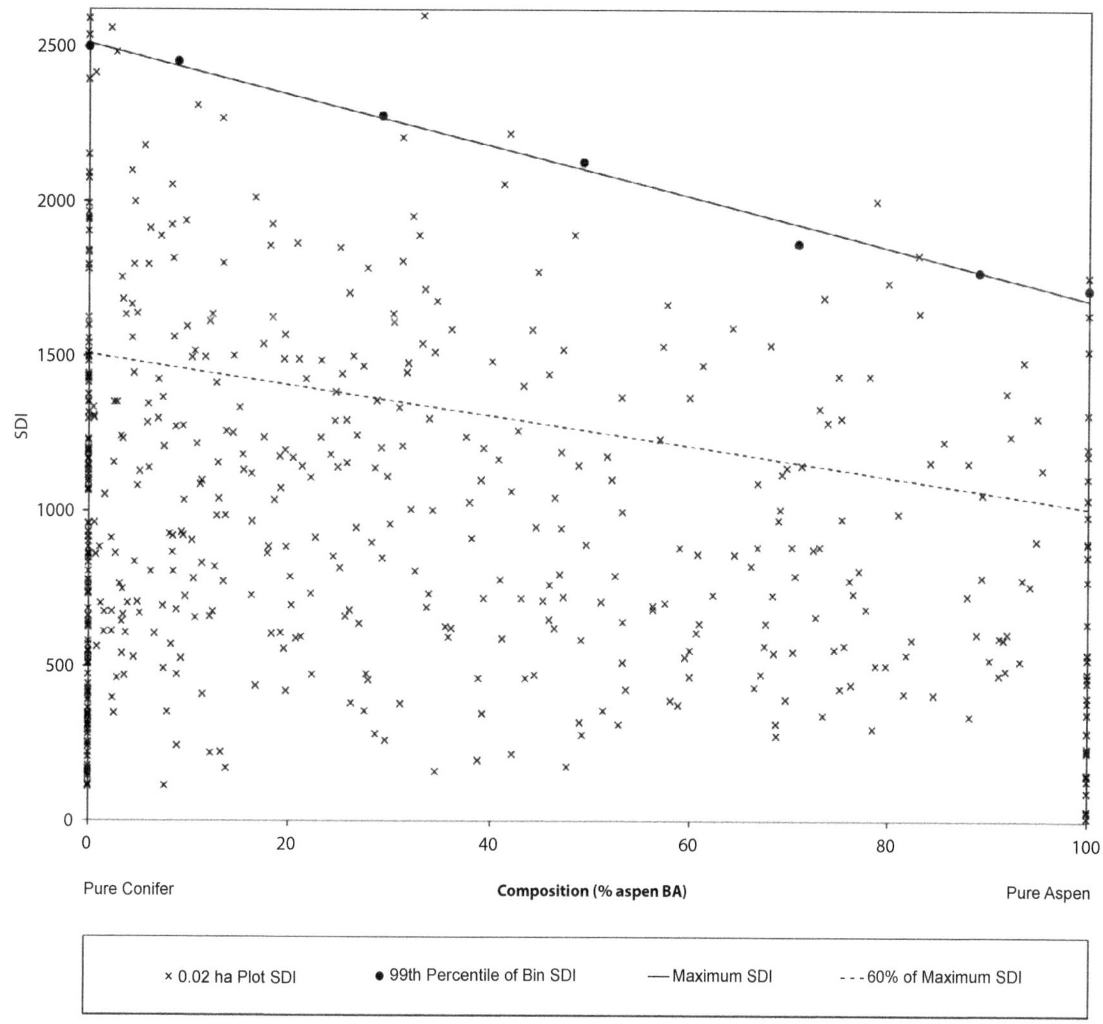

Figure 1: Stand density index (SDI) and species composition in terms of percent aspen basal area (% aspen BA) in 0.02 ha circular fixed-radius plots at 19 locations throughout the Lake Tahoe Basin, California. Maximum SDI for any composition defined by relationship between 99th percentile of SDI and the mean composition (% aspen BA) in each bin (0%, 0.01-20%, 20-40%, 40-60%, 60-80%, 80-99.99%, 100% aspen BA) where SDI_{max} = 2511.2 – 8.264 *%aspenBA* (R^2=0.99).

Tahoe. In the 10-20 cm DBH size class, aspen were outnumbered by conifers of which 5% were pines and 95% true firs. High variability in counts of conifer seedlings (DBH <10 cm) and conifer saplings/poles (10-20 cm DBH) within 0.004-ha plots located throughout the sample stands indicated that conifer regeneration patterns were spatially heterogeneous. Counts of aspen regeneration (DBH <10 cm), saplings (DBH 10-15 cm), and pole-sized aspen (DBH 15-20 cm) also varied between plots. The sharp decrease in densities of aspen saplings vis-à-vis aspen regeneration suggested that few regenerating aspen were advancing to the sapling and pole size classes (Table 2).

On average, aspen saplings numbered only approximately 1.3% of aspen regeneration. Aspen poles numbered <1% of aspen regeneration. Regression analysis of factors correlated with density of regeneration and small trees revealed that models specifying a negative binomial distribution fit the reverse-J frequency distributions better than Poisson-distributed models, ZIP, or ZINB models. Significant over dispersion was detected in all cases. Generalized linear mixed-effects models specifying a negative binomial distribution accounted for the over dispersion (many zero counts, few high counts), and revealed that frequency of aspen and conifer regeneration, and frequency of aspen saplings, correlated with stand density and species composition (Table 3).

Stand BA consistently outperformed SDI as a predictor of conifer regeneration. This could mean that conifer seedlings were more frequent near larger conifers, but we did not explicitly test for the effect of tree size or reproductive maturity of trees in 0.02 ha plots centered on 0.004 ha plots where conifer seedlings were counted. Conifer seedling counts were highest in conifer-dominated areas with higher stand densities (in terms of stand BA) (Figure 2a). At higher stand densities, the expected frequency of conifer seedlings was similar in areas with composition ranging from 0-60% conifer BA (40-100% aspen); however, conifer seedling frequencies were exponentially higher in areas where conifer composition exceeded 60%. Conifer seedlings were least frequent in open areas (low stand BA), especially open areas with proportionally more conifer than aspen (in terms of percent BA). Areas of low stand density in association with widely-spaced aspen often had a layer of organic matter accumulated beneath a diverse herbaceous vegetation layer. These factors may deter or provide a physical barrier to conifer regeneration. Exposed soils or woody shrubs were characteristic of the more xeric areas where conifers were found at low stand densities. In

	Plot size	No. of	Density (stems ha^{-1})			
Species and size class	(ha)	plots	Mean	s.d.	min.	max.
Conifer seedlings <10 cm DBH	0.004	249	3261	6589	0	79250
Conifers 10-20 cm DBH	0.004	249	248	387	0	2750
Aspen regeneration <10 cm DBH	0.004	249	3211	4353	0	26250
Aspen saplings 10-15 cm DBH	0.02	449	42	76	0	550
Aspen poles 15-20 cm DBH	0.02	449	26	64	0	450

Note: aspen sapling and pole data from only 8 stands (BC20 stand with zero saplings/poles excluded).

Table 2: Summary data for density (trees ha^{-1}) of young conifer and aspen in plots systematically located throughout nine aspen-conifer stands surrounding Lake Tahoe, California.

Species and size class	Effect	Estimate	Std. Error	Pr > |t|
Conifer seedling <10 cm	Intercept	-0.9832	0.83	0.2695
n = 249 plots in 9 stands	AScomp$^{0.5}$	2.2885	0.82	0.0057
-2LL = 826.5	(BA+1)$^{0.5}$	0.8157	0.21	0.0001
	BA+1	-0.0384	0.01	0.0023
	Ln(BA+1) x AScomp$^{0.33}$	-0.7254	0.21	0.0006
	Random intercept (by stand)	0.3139	0.18	--
	Scale (dispersion)	1.3461	0.13	--
Aspen regen. <10 cm	Intercept	1.9401	0.41	0.0014
n = 249 plots in 9 stands	AScomp$^{0.5}$	1.7246	0.26	<.0001
-2LL = 817.5	*Ln*(MSDI+1)	-0.8170	0.31	0.0084
	Random intercept (by stand)	0.8955	0.48	--
	Scale (dispersion)	1.2097	0.13	--
Aspen sapling 10-15 cm	Intercept	-3.4500	1.02	0.0118
n = 449 plots in 8 stands	AScomp$^{0.5}$	5.6906	1.33	<.0001
-2LL = 1783.1	(MSDI+1)$^{0.5}$	-2.3267	0.98	0.0181
	(MSDI+1)$^{0.5}$ x ((1-AScomp)+1)^{2}	0.8532	0.29	0.0039
	Random intercept (by stand)	0.2066	0.15	--
	Scale (dispersion)	1.3639	0.18	--

Note: AScomp = aspen BA as proportion of total BA in plot (range: 0-1); MSDI = SDI/1000; -2LL = Model goodness of fit in terms of -2 Residual Log Pseudo-Likelihood. Aspen sapling data from only 8 stands (BC20 stand with zero saplings excluded from analysis).

Table 3: Coefficients for generalized linear mixed-effects negative binomial log-link models of conifer and aspen regeneration (DBH <10 cm) count density in 0.004 ha plots, and aspen sapling (DBH 10-15 cm) count density in 0.02 ha plots systematically located throughout nine aspen-conifer stands surrounding Lake Tahoe, California. Log-link models give expected values for natural logarithm of mean count per plot assuming negative binomial distribution.

these drier areas, desiccation of ephemeral conifer seedlings likely kept seedling densities low [27].

Aspen regeneration (DBH <10 cm) was most frequent in areas of lower stand density with a higher proportion of aspen BA (Figure 2b). Contrary to results from northern California where no correlations were detected [28], we specified a negative binomial distribution for aspen regeneration frequency and found that fewer aspen <10 cm DBH were counted in areas of higher stand density and areas where conifer represented a greater proportion of stand BA. However, our analysis indicated that aspen regeneration was still abundant in dense, conifer-dominated areas (>1000 stems ha^{-1}), but that these densities were 1/10th of those in more open areas of pure aspen.

In aspen-dominated parts of the study areas, aspen saplings (DBH 10-15 cm) were also more frequent in low-density areas and less frequent in denser areas (Figure 2c). As the proportion of conifer increased, however, fewer aspen saplings were found at any density. These findings are consistent with reported correlation of aspen sapling density and canopy openness, and the detection of repulsion between aspen saplings and neighboring white fir in northern California [28]. Smith and Smith [6] found that aspen regeneration was not long-lived and rarely advanced to the sapling stage in the presence of conifer at higher stand densities. In parts of our study areas where composition was less than a third aspen (in terms of BA), stand density no longer appeared to correlate with the occurrence of aspen saplings. In these heavily conifer-dominated areas (i.e., BA >67% conifer), the frequency of aspen saplings appeared insensitive to stand density while declining as the proportions of conifer increased. This finding was unexpected, since conifer-dominated areas of lower density had more aspen regeneration <10 cm DBH than areas of higher density. Additionally, we expected young aspen in areas of lower density to receive more light, and grow into larger size classes, given that growth of aspen >10 cm DBH is more rapid in high light [28] at lower stand densities [14]. The discrepancy between aspen regeneration and recruitment to larger sizes, especially in areas of lower stand density where conifers dominated, merits further investigation. In the interim, our findings suggest that managers of aspen stands in the Sierra Nevada might focus their interventions in places where conifers have come to dominate, removing conifers to maintain stand density below 1000 SDI. Greater reductions in SDI are advised when the objective is to promote recruitment of young aspen to the overstory, to replace senescing aspen or enhance resilience by promoting age-class diversity within stands.

Conclusion

The degree of succession to conifer in aspen stands throughout the Lake Tahoe Basin was spatially heterogeneous. Within the nine one-hectare study areas, patches of pure aspen were rare; most common were mixed patches dominated by conifer. The regression fitted to the 99th percentile of plot SDI revealed that as composition shifted in favor of conifer, SDI increased beyond the upper limit of around 1700 for

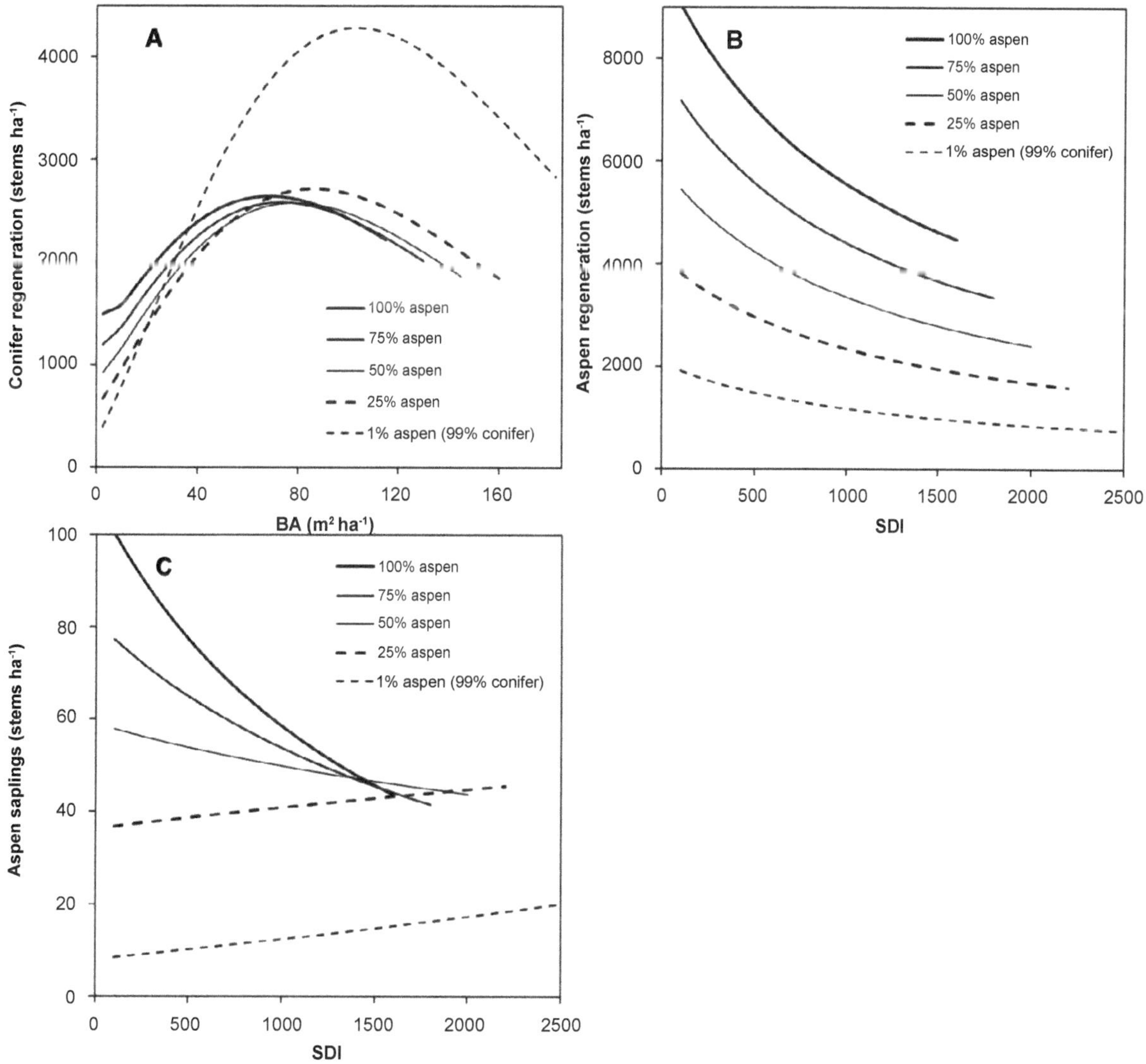

Figure 2: Expected frequency of (A) conifer seedlings (DBH <10 cm), (B) aspen regeneration (DBH <10 cm) and (C) aspen saplings (DBH 10-15 cm) as a function of stand density in terms of basal area (BA) or stand density index (SDI) and tree species composition. Expected mean frequencies obtained (for each level of aspen composition: 100%, 75%, 50%, 25%, 1% aspen BA) using generalized linear mixed-effects models of aspen frequency in multiple sample plots within 8-9 aspen-conifer stands surrounding Lake Tahoe, California.

pure aspen. We infer from this finding that aspen were being extirpated as succession to conifer led to levels of crowding and competition in excess of what aspen could withstand. Our findings suggest that forest managers must control stand density to pre-empt competition-induced mortality among aspen growing in association with conifers that tolerate higher densities.

Counts of regeneration throughout the nine study areas correlated with stand density and species composition. Numerous conifer seedlings were becoming established in aspen-dominated areas of any density, and in conifer-dominated areas of higher density. Our analysis revealed that, on average, aspen regeneration was one order of magnitude less frequent in areas of pure conifer versus areas of pure aspen. However, aspen regeneration was still abundant in most areas. Most surprising was that aspen saplings – whose presence we assumed was associated with stand conditions favoring recruitment of aspen regeneration to the overstory – were rare, especially at high stand densities or in areas where conifers dominated.

Forest managers interested in sustaining aspen stands *in situ* over the long term will need to replace older aspen as they mature and die. If the goal of management is to foster recruitment of younger aspen to the overstory, then heavier removals of conifer (site-wide, or in patches) should provide conditions (i.e., low stand densities dominated by aspen) most favorable for regeneration and growth of young aspen beneath a partial canopy within an existing stand.

Acknowledgments

We are grateful for advice and assistance provided by our collaborators Stephanie Coppeto and Victor Lyon (USDA Forest Service), David Catalano and Mark Enders (Nevada Division of Wildlife), Tamara Sasaki and Silver Hartman (California State Parks), and Judy Clot (California Tahoe Conservancy) and field assistants Chris Harrison, Jesse Jeffress, Nick Knipe, Brandon Namm, Kirk Perttu, Dustin Revell, and Chris Valness. This research was supported using funds provided by the Bureau of Land Management through the sale of public lands as authorized by the Southern Nevada Public Land Management Act. Thoughtful critique by anonymous reviewers motivated us to collect more data which strengthened our analysis, and expand our discussion of management implications.

References

1. Shepperd WD, Bartos DL, Mata SA (2001) Above- and below-ground effects of aspen clonal regeneration and succession to conifers. Canadian Journal of Forest Research 31: 739-745.
2. Gallant AL, Hansen AJ, Councilman JS, Monte DK, Betz DW (2003) Vegetation dynamics under fire exclusion and logging in a Rocky Mountain Watershed, 1856-1996. Ecological Applications 13: 385-403.
3. Di Orio AP, Callas R, Schaefer RJ (2005) Forty-eight year decline and fragmentation of aspen (Populus tremuloides) in the South Warner Mountains of California. Forest Ecology and Management 206: 307-313.
4. Jones BE, Rickman TH, Vazquez A, Sado Y, Tate KW (2005) Removal of Encroaching Conifers to Regenerate Degraded Aspen Stands in the Sierra Nevada. Restoration Ecology 13: 373–379.
5. Kaye MW, Binkley D, Stohlgren TJ (2005) Effects of conifers and elk browsing on quaking aspen forests in the central Rocky Mountains, USA. Ecological Applications 15: 1284-1295.
6. Smith AE, Smith FW (2005) Twenty-year change in aspen dominance in pure aspen and mixed aspen/conifer stands on the Uncompahgre Plateau, Colorado, USA. Forest Ecology and Management 213: 338-348.
7. Calder WJ, Horn KJ, St. Clair SB (2011) Conifer expansion reduces the competitive ability and herbivore defense of aspen by modifying light environment and soil chemistry. Tree Physiology 31: 582-591.
8. Perala DA (1990) Quaking aspen (Populus tremuloides Michx.), in Silvics of North America: II. Deciduous. Burns RM, Honkala BH (Eds.), 555–569, United States Department of Agriculture Handbook no. 654.
9. Barnes BV (1966) The clonal growth habit of American aspens. Ecology 47: 439-447.
10. Sugihara NG (2006) Fire in California's ecosystems. University of California Press, Berkeley, CA.
11. Kuhn TJ, Safford HD, Jones BE, Tate KW (2011) Aspen (Populus tremuloides) stands and their contribution to plant diversity in a semiarid coniferous landscape. Plant Ecology 212: 1451-1463.
12. Richardson TW, Heath SK (2004) Effects of conifers on aspen-breeding bird communities in the Sierra Nevada. Transactions of the Western Section of the Wildlife Society 40: 68-81.
13. Shepperd WD, Rogers PC, Burton D, Bartos D (2006) Ecology, biodiversity, management, and restoration of aspen in the Sierra Nevada. U.S. Department of Agriculture, Forest Service, Rocky Mountain Forest and Range Experiment Station, RMRS-GTR-178, Fort Collins, CO.
14. Berrill J-P, Dagley CM (2012) Geographic patterns and stand variables influencing growth and vigor of Populus tremuloides in the Sierra Nevada (USA). ISRN Forestry 1-9.
15. Long JN, Daniel TW (1990) Assessment of growing stock in uneven-aged stands. Western Journal of Applied Forestry 5: 93–96.
16. Shaw JD, Long JN (2010) Consistent definition and application of Reineke's stand density index in silviculture and stand projection. In: Jain TB, Graham RT, Sandquist J (eds.) Integrated management of carbon sequestration and biomass utilization opportunities in a changing climate, U.S. Department of Agriculture, Forest Service, Rocky Mountain Research Station, RMRS-P-61. Fort Collins, CO.
17. Crotteau JS, Ritchie MW, Varner JM (2013) A mixed-effects heterogeneous negative binomial model for postfire conifer regeneration in northeastern California, USA Forest Science.
18. Shaw JD (2000) Application of stand density index to irregularly structured stands. Western Journal of Applied Forestry 15: 40-42.
19. Reineke LH (1933) Perfecting a stand-density index for even-aged forests. Journal of Agricultural Research 46: 627-638.
20. SAS Institute Inc. (2004) SAS/STAT 9.1 user's guide, Vols. 1-7, Second edition. SAS Institute Inc. Cary, NC.
21. Littell RC, Milliken GA, Stroup WW, Wolfinger RD, Schabenberger O (2006) SAS for Mixed Models, Second edition. SAS Institute Inc. Cary, NC.
22. Woodall CW, Miles PD, Vissage JS (2005) Determining maximum stand density index in mixed species stands for strategic-scale stocking assessments. Forest Ecology and Management 216: 367-377.
23. DeByle NV, Winokur RP (1985) Aspen: ecology and management in the western United States, U.S. Department of Agriculture, Forest Service, Rocky Mountain Forest and Range Experiment Station, GTR-RM-119, Fort Collins, CO.
24. Drew TJ, Flewelling JW (1977) Some recent Japanese theories of yield density relationships and their application to Monterey pine plantations. Forest Science 23: 517-534.
25. Drew TJ, Flewelling JW (1979) Stand density management: an alternative approach and its application to Douglas-fir plantations. Forest Science 25: 518-532.
26. Long JN (1985) A practical approach to density management. Forestry Chronicle 61: 23-27.
27. Gray AN, Zald HSJ, Kern RA, North M (2005) Stand conditions associated with tree regeneration in Sierran mixed-conifer forests. Forest Science 51: 198-210.
28. Pierce AD, Taylor AH (2010) Competition and regeneration in quaking aspen–white fir (Populus tremuloides–Abies concolor) forests in the Northern Sierra Nevada, USA. Journal of Vegetation Science 21: 507-519.

PERMISSIONS

All chapters in this book were first published in FR, by OMICS International; hereby published with permission under the Creative Commons Attribution License or equivalent. Every chapter published in this book has been scrutinized by our experts. Their significance has been extensively debated. The topics covered herein carry significant findings which will fuel the growth of the discipline. They may even be implemented as practical applications or may be referred to as a beginning point for another development.

The contributors of this book come from diverse backgrounds, making this book a truly international effort. This book will bring forth new frontiers with its revolutionizing research information and detailed analysis of the nascent developments around the world.

We would like to thank all the contributing authors for lending their expertise to make the book truly unique. They have played a crucial role in the development of this book. Without their invaluable contributions this book wouldn't have been possible. They have made vital efforts to compile up to date information on the varied aspects of this subject to make this book a valuable addition to the collection of many professionals and students.

This book was conceptualized with the vision of imparting up-to-date information and advanced data in this field. To ensure the same, a matchless editorial board was set up. Every individual on the board went through rigorous rounds of assessment to prove their worth. After which they invested a large part of their time researching and compiling the most relevant data for our readers.

The editorial board has been involved in producing this book since its inception. They have spent rigorous hours researching and exploring the diverse topics which have resulted in the successful publishing of this book. They have passed on their knowledge of decades through this book. To expedite this challenging task, the publisher supported the team at every step. A small team of assistant editors was also appointed to further simplify the editing procedure and attain best results for the readers.

Apart from the editorial board, the designing team has also invested a significant amount of their time in understanding the subject and creating the most relevant covers. They scrutinized every image to scout for the most suitable representation of the subject and create an appropriate cover for the book.

The publishing team has been an ardent support to the editorial, designing and production team. Their endless efforts to recruit the best for this project, has resulted in the accomplishment of this book. They are a veteran in the field of academics and their pool of knowledge is as vast as their experience in printing. Their expertise and guidance has proved useful at every step. Their uncompromising quality standards have made this book an exceptional effort. Their encouragement from time to time has been an inspiration for everyone.

The publisher and the editorial board hope that this book will prove to be a valuable piece of knowledge for researchers, students, practitioners and scholars across the globe.

LIST OF CONTRIBUTORS

Ana Milena Plata Fajardo and Romano Timofeiczyk Junior
Federal University of Paraná, Brazil

Adedayo A. G
Department of Forestry and Wood Technology, Federal University of Technology Akure Ondo state, Nigeria

Sobola Oluronke
Department of Forestry and Wildlife Management, Federal University Wukari, Taraba State, Nigeria

Yaling Tian, Fuliang Cao, Guibin Wang, Wangxiang Zhang and Wanwen Yu
College of Forest Resources and Environment, Nanjing Forestry University, 210037, Nanjing, Jiangsu Privence, China

Lissette Cortés, Jaime Hernández, Diego Valencia and Patricio Corvalán
Faculty of Forestry and Nature Conservation, University of Chile, Chile

Manuel Esteban Lucas-Borja
Higher Technical School of Agricultural and Forestry Engineering. Castilla La Mancha University, Albacete, Spain

LS Kandari
School of Natural Resources and Environmental Sciences, College of Agriculture and Environmental Sciences, Haramaya University, P. Box # 70, Dire-Dawa, Ethiopia

Ashish K. Gharai
Centre for People's Forestry, Tarnaka, Hyderabad, India

Tripti Negi
Wildlife Institute of India, P.O. Box No. # 18, Chandrabani, Dehradun 248 001, Uttarakhand, India

PC Phondani
G.B. Pant Institute of Himalayan Environment & Development, Almora, Uttarakhand, India

Tagel Gebrehiwot
University of Twente, P. O. Box 217, 7514 AE Enschede, The Netherlands

Anne van der Veen
Professor of Governance and Spatial Integrated Assessment, University of Twente, P. O. Box 217, 7514 AE Enschede, The Netherlands

Bo Song and Thomas M Williams
Belle W. Baruch Institute of Coastal Ecology and Forest Science, Clemson University, Georgetown, USA

Jiquan Chen
Department of Earth, Ecological & Environmental Science, University of Toledo, Toledo, USA

Francesca Rinaldi and Ragnar Jonsson
European Commission, Joint Research Centre (JRC), Institute for Environment Sustainability (IES), Forest Resources and Climate Unit, Via E. Fermi, 2749 I-21027 Ispra, Italy

Yonas Yohannes
Wolkite University, Wolkite, P. O. Box 07, Wolkite, Ethiopia

Georg Guggenberger
Institute of Soil Science, Leibniz Universität Hannover, Herrenhäuser Str. 2, 30419 Hannover, Germany

Olga Shibistova
Institute of Soil Science, Leibniz Universität Hannover, Herrenhäuser Str. 2, 30419 Hannover, Germany
VN Sukachev Institute of Forest, SB-RAS, Akademgorodok, 660036 Krasnoyarsk, Russian Federation

Zeleke Asaye
Adami Tulu Agricultural Research Center, Oromia Agricultural Research Institute, P.O.Box:35, Zeway, Ethiopia

M Ramdani, T Lograda and A Zeraib
Laboratory of Natural Resource Valorization, Sciences Faculty, Ferhat Abbas University, 19000 Setif, Algeria

P Chalard and JC Chalchat
Ecole nationale Supérieure de Chimie de Clermont Ferrand, Laboratoire de Chimie des Hétérocycles et des glucides, EA987, 63174 Aubière Cedex 01

G Figueredo
LEXVA Analytique, 460 rue du Montant, 63110 Beaumont, France

Mishra R.K
Department of Wildlife and Biodiversity Conservation, North Orissa University, Takatpur, Baripada, India

Pattanaik S and Mohanty R.C
Department of Botany, Utkal University, VaniVihar, Bhubaneswar, India

Randall W Myster
Biology Department, Oklahoma State University, Oklahoma City, USA

Vicente de P. R. da Silva, Glayson F. B. das Chagas and Vanessa de A. Dantas
Federal University of Campina Grande/Center of Technology and Natural Resources/Academic Unity of Atmospheric Sciences, Av. Aprígio Veloso, 882, Bodocongó, 58109 970, Campina Grande, PB, Brazil

Rafaela S. R. Almeida
State University of Paraíba, R. Domitila Cabral de Castro, 148-572, Bodocongó, 58429-570 Campina Grande, PB, Brazil

Vijay P. Singh
Department of Biological and Agricultural Engineering & Department of Civil & Environmental Engineering, Texas A&M University, College Station, TX 77843-2117, USA

Anne Jurkiewicz Melo and Ivanilza Moreira de Andrade
Graduate Program Biotechnology (PPGBiotec), Campus of Parnaíba, Federal University of Piauí, Brazil

Jefferson Almeida Rocha
Graduate Program Biotechnology (PPGBiotec), Campus of Parnaíba, Federal University of Piauí, Brazil
Center for Natural Sciences, Federal University of Maranhão, Brazil

Maria Francilene Souza Silva and João Antônio Leal de Miranda
Graduate Program Biomedicine. Federal University of Piauí, Brazil

Fabrícia Meireles Meneses da Silva
Anhydrous Brazil Extractions SA, Povoado Rosápolis s/n, Brazil

Santelmo Vasconcelos and Ana Maria Benko-Iseppon
Laboratory of Plant Genetics and Biotechnology, Department of Genetics, Center of Biological Sciences, Federal University of Pernambuco, Brazil

Vicente de P. R. da Silva, Vanessa de A. Dantas and Glayson F. B. das Chagas
Federal University of Campina Grande/Center of Technology and Natural Resources/Academic Unity of Atmospheric Sciences, Av. Aprígio Veloso, 882, Bodocongó, 58109 970, Campina Grande, PB, Brazil

Rafaela S. R. Almeida
State University of Paraíba, R. Domitila Cabral de Castro, 148-572, Bodocongó, 58429-570 Campina Grande, PB, Brazil

Antonio C. L. da Costa
Federal University of Pará, Av. Augusto Corrêa 1, Guamá, CEP 66075-110, Belém, PA, Brazil

Vijay P. Singh
Department of Biological and Agricultural Engineering & Department of Civil & Environmental Engineering, Texas A&M University, College Station, TX 77843-2117, USA

Chandrakant Tiwari and Meena Bakshi
Plant Physiology Discipline, Botany Division, Forest Research Institute, Dehradun, India

Pramanik MK
Research Scholar, Centre for International Politics, Organization and Disarmament (CIPOD), Jawaharlal Nehru University, NewDelhi-110067, India

Suraj Kumar and Munesh Kumar
Department of Forestry and Natural Resources, HNB Garhwal University, Srinagar Garhwal Uttarakhand, India

Bajigo A
Jigjiga University College of Dryland Agriculture (Natural Resource Management Department), Jigjiga, Ethiopia

Tadesse M
Mekelle University College of Dryland Agriculture and Natural Resources, Mekele, Ethiopia

Tsabang N
Center for Research on Medicinal Plants and Traditional Medicine, Institute of Medical Research and Medicinal Plants Studies, Ministry of Scientific Research and Innovation, Yaounde, Cameroon

Fongnzossie E
University of Douala, Cameroon

Donfack D
Faculty of Education, Department of Specialized Education Sciences, Yaounde, Cameroon

Yedjou CG and Tchounwou PB
Cellomics and Toxicogenomics Research Laboratory, NIH-RCMI Center for Environmental Health, Jackson State University, Jackson, Mississippi, USA

Minkande JZ and Nouedou C
General Hospital of Yaounde, Cameroon and Faculty of Medicine and Biomedical Sciences, Cameroon

Van PD
Faculty of Science Engineering, Universitteit Ghent, Coupure Links 653B-9000 Ghent, Belgium

Sonwa
Center for International Forestry Research (CIFOR), Yaounde, Cameroon

Romero Tejero G
Geography Department, University of Leeds, UK

Hong , Claire D, Yu Z, Yin S, Yu L and Yi Z
School of Earth Sciences and Engineering, Hohai University, No. 1 Xikang Road, Nanjing 210098, China

Madalcho AB
Jigjiga University College of Dry Land Agriculture (Natural Resource Management Department), Jigjiga, Ethiopia

Tefera MT
Mekelle University College of Dry Land Agriculture and Natural Resources, Mekele, Ethiopia

Christian CL, Oswald BP, Williams HM and Farrish KW
Division of Environmental Science, Arthur Temple College of Forestry and Agriculture, Stephen F Austin State University, Nacogdoches, TX 75962, USA

Walker RF, Swim SL, Fecko RM, Johnson DW and Miller WW
Department of Natural Resources and Environmental Science, University of Nevada, Reno, Nevada, USA

Mangarah Silalahi
Head of Ecosystem Restoration Resource Centre, Burung Indonesia, Indonesia

Desri Erwin
Community Partnership staff of PT REKI, Jambi South Sumatra, Indonesia

John-Pascal Berrill and Christa M Dagley
Department of Forestry and Wildland Resources, Humboldt State University, USA

Index